Introduction to molecular biology

Carol Bertrand

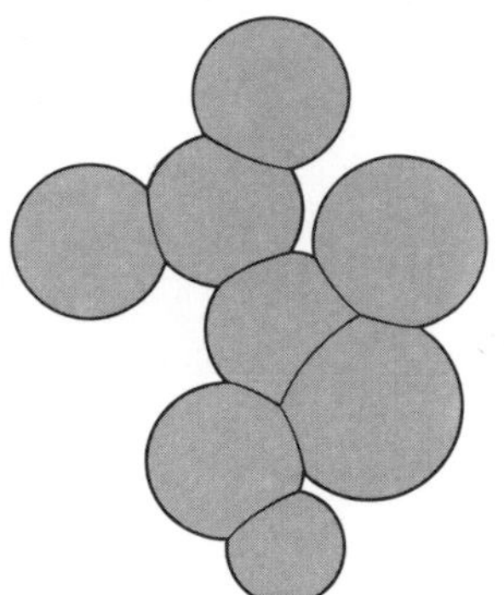

Introduction to molecular biology

Second edition

G. H. Haggis
Electron Microscope Centre
Canada Department of Agriculture
Ottawa

and

Biology Department
Carleton University
Ottawa

including material contributed to the first edition by

D. Mitchie
University of Edinburgh

A. R. Muir
University of Edinburgh

K. B. Roberts
Memorial University, Newfoundland

P. M. B. Walker
University of Edinburgh

a new appendix on X-ray diffraction by

B. M. Blow
Laboratory of Molecular Biology, Cambridge, England

and a section on tumour viruses by

Rose Sheinin
Ontario Cancer Institute, Toronto

Longman

Longman
1724-1974

LONGMAN GROUP LIMITED
London

Associated companies, branches and representatives throughout the world

First published 1964

Second Edition 1974

ISBN 0 582 44745 3

Made and printed in Great Britain by
William Clowes & Sons, Limited
London, Beccles and Colchester

Contents

Acknowledgements

I am much indebted to a number of colleagues who, in addition to the co-authors, read through the various chapters in whole or in part, correcting errors and suggesting improvements: C. H. Amberg, J. Dainty, G. Fulcher, S. Haskill, V. N. Iyer, G. Kaplan, J. Kilmartin, G. Setterfield, L. Siminovitz and H. Yamazaki.

For each electron micrograph reproduced from published work I am grateful to the author, or authors, listed in the Sources of Figures section who were kind enough to send original prints for reproduction, and to the publishers of the relevant journals for permission to use these illustrations. I am also indebted to the Cold Spring Harbor Laboratory Press for permission to include the abstract from Avery's letter, and to A. F. Howatson and A. Massalski for original electron micrographs.

Preface to the first edition

W. T. Astbury, in the Harvey Lectures for 1950, defined his own use of the term Molecular Biology as follows:

> Molecular Biology implies . . . searching below the large-scale manifestations of classical biology for the corresponding molecular plan. It is concerned particularly with the forms of biological molecules and with the evolution, exploitation and ramification of these forms in the ascent to higher and higher levels of organization. Molecular biology is predominantly three-dimensional and structural – which does not mean, however, that it is merely a refinement of morphology. It must of necessity enquire at the same time into genesis and function.

Accepting Astbury's definition as still valid today, we may say that molecular biology is primarily concerned with the structure of proteins, nucleic acids and other large biological molecules, and with the detailed structure of myofilaments, chromosomes, ribosomes, membranes and other cell components. But the study of structure cannot be divorced from the study of function. Our ultimate aim is to understand how the structure of myofilaments determines their interaction in muscular contraction, how the structure of chromosomes and ribosomes leads to specificity of protein synthesis, how the structure of membranes results in active transport of metabolites. This book is not therefore confined to a discussion of bimolecular structure; relevant material from the fields of biochemistry, cytology, genetics, microbiology and physiology is included also, to give some account of the function and activity of large molecules in cells.

The book is intended primarily for students in the biological and medical sciences in their second and third undergraduate years or in their early research years. It may also prove useful to physics and chemistry students interested in recent advances in biology at a molecular level. The reader is assumed to have some background knowledge of physics and chemistry (at about G.C.E. 'A' level) but a prior knowledge of biology is not essential.

The plan of the book is as follows. The first chapter, on the observation of living cells under the phase-contrast microscope, is introductory and not in any sense 'molecular'. Its aim is to set the discussion of later chapters in biological perspective, and to emphasize dynamic aspects of cell structure which can easily be lost sight of in more detailed, but static, studies made with the electron microscope and the X-ray diffraction technique. It is intended primarily for readers unfamiliar with the microscopical appearance of cells, but it may be useful also to those who have studied histology without much opportunity to look at the movement of living cells in culture. After this introductory chapter, our aim is first to show how important proteins are in cellular activity and to discuss their structure and function, then to show how important membranes are in cellular ultrastructure, and discuss their structure and function. The factors which determine protein structure: hydrogen bonds, hydrophobic effects and electrostatic forces, are discussed fairly fully since these same factors determine the structure of lipoprotein membranes and nucleoproteins, discussed in later chapters.

In the second half of the book the same approach is carried over to a discussion of the structure of nucleic acids, and refinements in genetic analysis which have made it possible to relate mutational events to modification of nucleic acid structure. The effects of mutations on the structure of a protein are illustrated by a discussion of the haemoglobins and a chapter is concerned with protein synthesis. More difficult concepts and fuller discussion of experimental techniques are relegated to short Appendices.

G. H. HAGGIS

Preface to the second edition

The main ideas and principles of molecular biology established during the 1950s have not altered significantly since the first edition of this book was prepared ten years ago and still form the core of the new edition. However, at many points we now have more detailed understanding of molecular processes, and these have often proved more intricate and baffling than was expected. Ribosomes, for example, are made up of more than fifty different types of protein molecule, but why so many are needed is not known. DNA duplication appears to require backward synthesis of one of the new DNA strands, in short lengths, which are later joined, which seems a cumbersome way to achieve a simple end.

The material of the first edition has been arranged more logically and trimmed where possible to make room for new material. An attempt has been made to treat of the essence of currently confused problems, not the details. Many new sections have been added as, for example, one on the results of X-ray analysis of the structure and mode of action of lysozyme. The first edition held expectation of 'new contributions to systematics and a new study of evolution based not on gross structure, but on descriptions of protein differences carried down to the finest level'. As an example of developments of this kind, the evolutionary changes of cytochrome c are now treated in some detail. The first edition ended with a chapter on protein synthesis; a new chapter has now been added on *control* of protein synthesis, including a section on the transformation of cells by tumour viruses. There is a new Appendix on the fundamentals of thermodynamics.

In selecting topics to include, and aspects to cover, the main aim has been to keep the book an introduction to *molecular* biology and not attempt to survey the whole of modern biology. Emphasis is on areas where we have now, or are near to, detailed molecular understanding of biological events. It becomes increasingly difficult to draw any borderline between biochemistry and molecular biology – 'the practice of biochemistry without a licence' (Chargaff). The present book aims perhaps to be more free-ranging than most textbooks of biochemistry, on the one hand probing quite far out into cell structure, physiology, genetics and modern cancer research, exploring areas and problems ripe for attack at the molecular level, and on the other hand probing down, in a search for the underlying molecular forces and thermodynamic principles that cause molecular chains to fold and interact as they do. A book of this kind can provide a useful survey, but remains an introduction; it has value mainly if it leads to further reading in areas of individual interest.

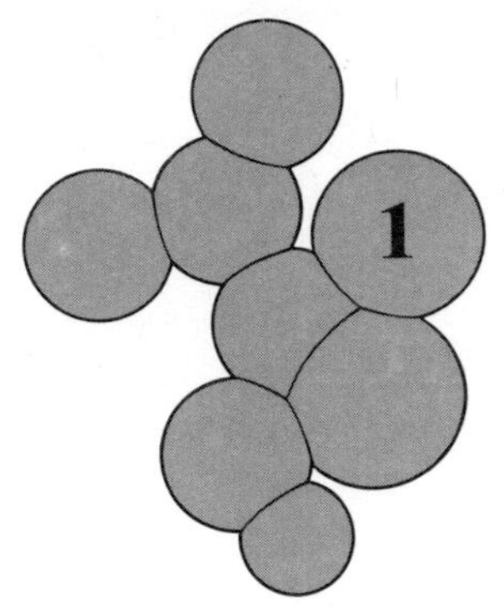

1 Historical background: cells under the light microscope and isolation of cellular components

Biologists had at one time only forceps and scalpels to dissect the structure of animals and plants; they could observe only with their unaided eyes. Using such methods, anatomists in Renaissance Italy made dissections and produced drawings of human and animal anatomy which were of great accuracy and elegance. Later with the perfection, first of lenses and then, in the early nineteenth century, of compound microscopes, the structure of living tissues was described as it was seen when magnified some tens or hundreds of times. It was soon found that the structures dissected by the early anatomists could be resolved under the microscope into their component cells and intercellular material. Consideration then narrowed down further to the structure of the cells themselves, and there is now behind us a hundred years of such cytological investigations.

The early microscopists often studied living preparations, but later, with the chemical industry producing a great variety of dye-stuffs, many turned to observing fixed, stained slices of tissues. Much more detail could be seen in these preparations, for under the lens fresh material is usually transparent and relatively featureless. However, work on living cells complemented these fundamental, classical studies on fixed material. Since 1942 the development of a new kind of microscope has led to an extension of work on living cells. The phase-contrast microscope, first used in biology in that year, works in this way: if two cell components transmit the same amount of light they will be difficult to distinguish in the ordinary light microscope; however, if there is a difference in refractive index between these two components, the phase-contrast microscope will turn this into a difference in brightness, so what appears uniformly transparent in the ordinary microscope may become visible in the phase-contrast microscope.

During more recent years information on even finer structure has been obtained, for the electron microscope has increased the resolution of microscopy several hundred times. As yet electron microscopy cannot be used satisfactorily to examine living cells; the cells or tissues must be fixed, and then very thin sections cut. Using electron microscopy, structures previously unsuspected have been discovered in the cell. There are occasions when it is possible to identify some large molecules within tissues. At this limit of resolution, how-

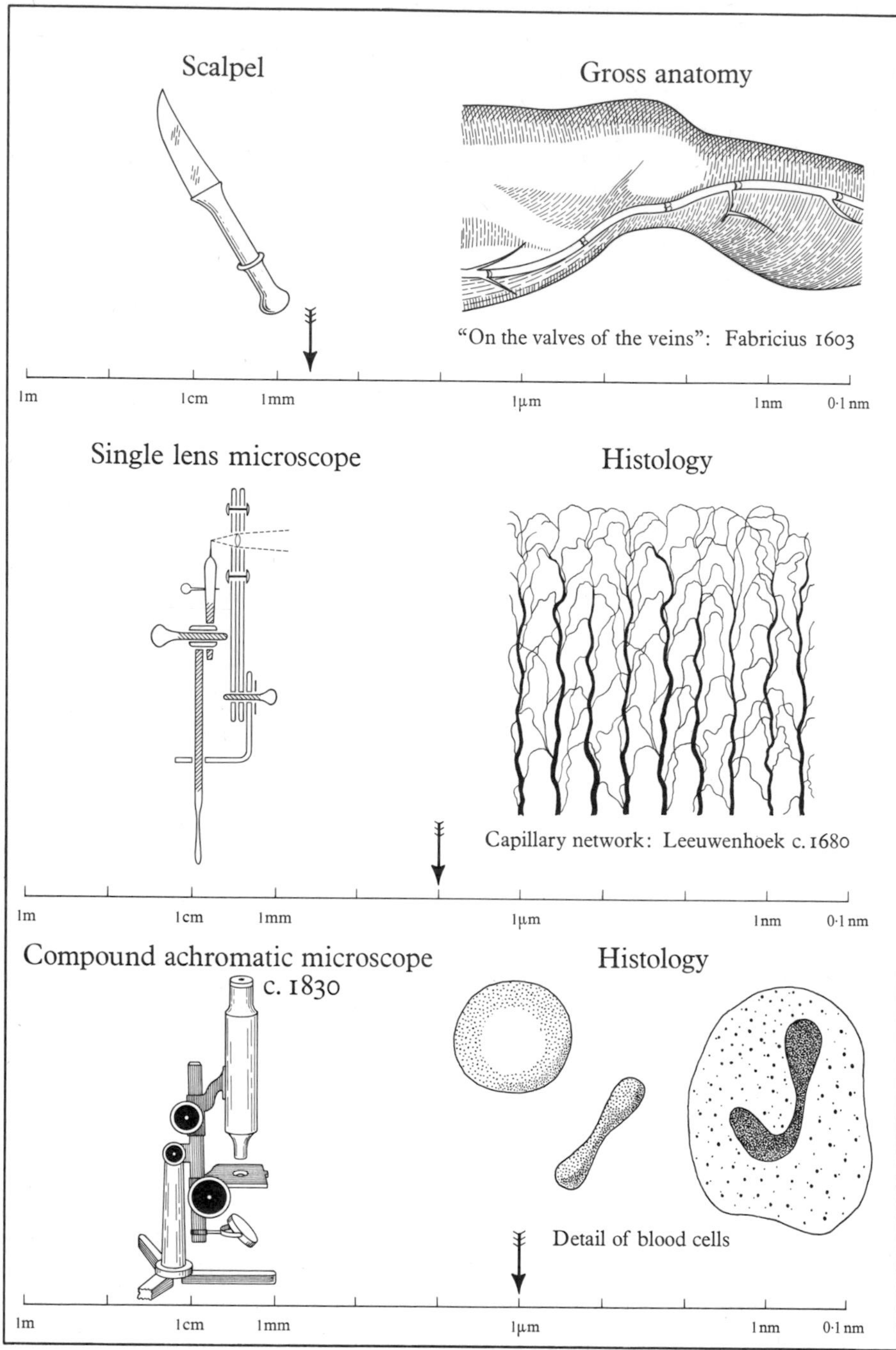

Fig. 1.1 Progress in biology since 1600 illustrated by the study of the fine structure of blood vessels and blood cells. The useful limit of resolution with the naked eye is about $\frac{1}{10}$ mm; lenses and compound microscopes extend this to about 0·2 μm (1 μm = $\frac{1}{1000}$ mm). The electron microscope can resolve useful detail in biological material down to about

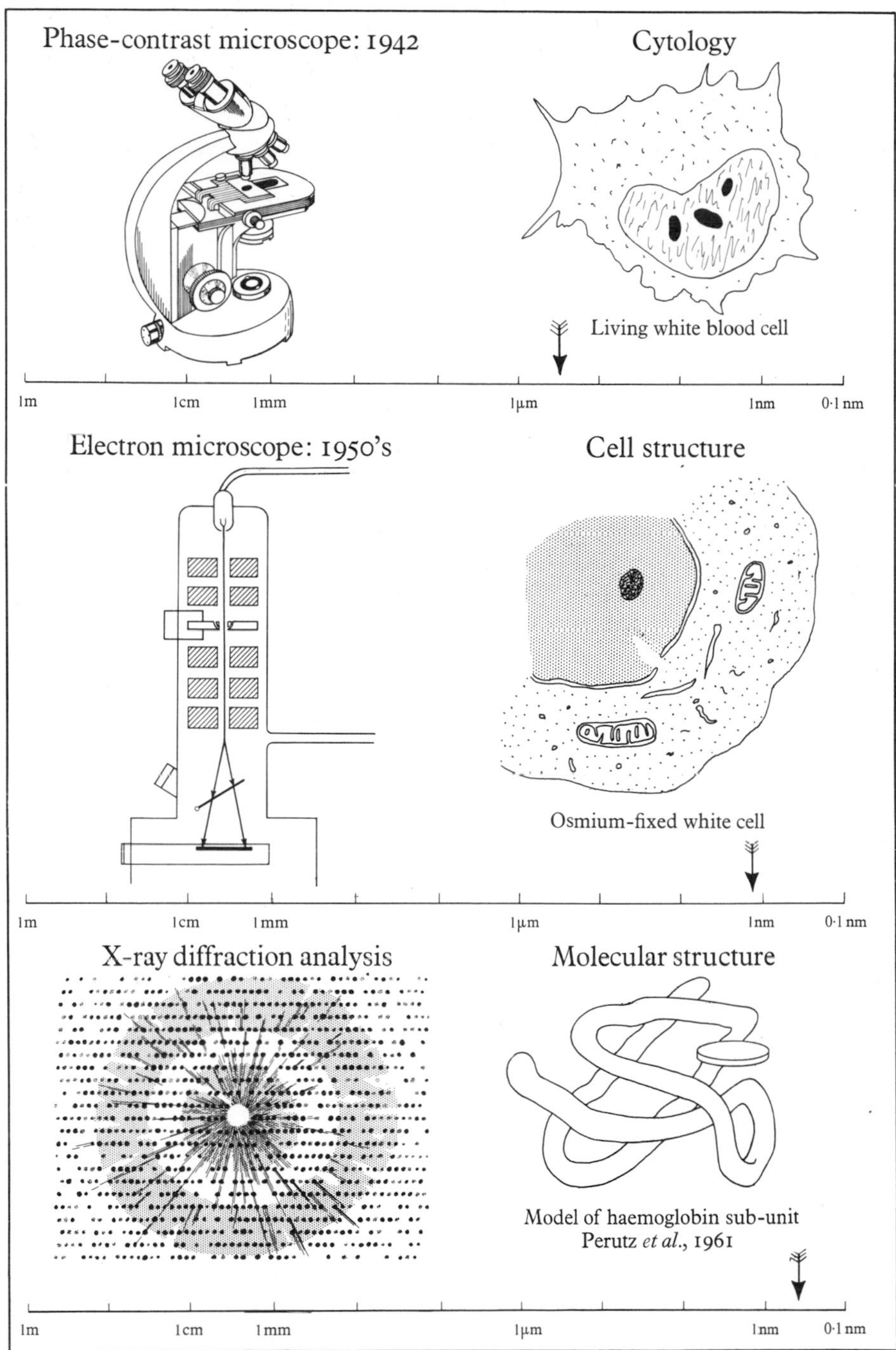

2 nm (1 nm $= \frac{1}{1\,000}$ μm) and shows structural detail of, for example, mitochondria. X-ray diffraction studies on the haemoglobin molecules from red blood cells show structure at a resolution of 0·2 nm, when the positions of the individual atoms can be determined.

ever, a different technique can yield more detail (though only when the molecules can be extracted, purified and crystallized). By the use of X-ray diffraction analysis, the haemoglobin extracted from red cells, for example, has been shown to have a characteristic 4-coiled shape (discussed in Chapter 2). The position of the individual atoms in the haemoglobin molecule has been determined.

The concentration of anatomists, microanatomists and molecular biologists on finer and finer detail may be illustrated by four centuries of work on the structure of the vascular system and the blood within it (Fig. 1.1).

And so biology, which was at one time only associated with the naked-eye observations of organisms, has extended, particularly in recent years, to include observation at finer and finer levels until large molecules can now be seen in the electron microscope, and their detail reconstructed by X-ray analysis, at a magnification of a million times or more. Other chapters in this book will describe the contributions of these methods to biology. But with all the wealth of information now available we must remember that in biology our reference point is always the living organism and its components. It is therefore appropriate to set the scene for observations at a molecular level by describing in this chapter what can be seen of the behaviour and structure of living cells in the light microscope.

When Robert Hooke looked at the cut surface of cork under his microscope he saw that the cell spaces were separated from each other by walls. Their appearance resembled, he thought, a collection of small box-like rooms; that was why he called them cells (1665). It was not until the 1830s that the general existence of cells was recognized in animal tissues. Animal cells were found to be bounded by membranes very much thinner than the cell walls of plants; later it was found that plant cells also have thin membranes inside the comparatively thick cellulose walls.

The membrane that limits the extent of cell cytoplasm and forms a permeability barrier between the cell contents and the medium in which the cell lives is now usually called the *plasma membrane*. It is reinforced in many plant cells by an outer cellulose wall; another reinforcement, found in crustacea, is chitin. The membrane itself cannot be seen in any type of light microscope for it is too thin, but the refractive index of cytoplasm usually differs from that of the fluid outside, and, because of this, the shape of the cell can be seen with phase-contrast illumination.

Cell movement

If a fragment of heart muscle is embedded in a plasma clot and cultured at 37°C, some of the cells of the fibrous tissue (fibroblasts) move out from the muscle fragment into the plasma. If such a culture of living fibroblasts at 37°C is observed patiently, the cells may be seen to change their shape slowly. This movement can be studied by making drawings at, say, three-minute intervals, but the process is most dramatically illustrated when time-lapse cinematography is used: frames are exposed every 10 seconds or so, and the film then run as a normal cine film; the cell movements are thus speeded up many times.

Granular white cells of the blood (granulocytes) move very much faster than fibroblasts and their movement can easily be seen directly without filming. Indeed, under the oil-immersion lens with a magnification of some 1 000 times, it is necessary to adjust the mechanical stage constantly in order to keep a moving granulocyte near the centre of the field. Granulocytes move somewhat like amoebae: at an advancing edge extensions or pseudopodia are formed, while at the same time, at another part of the cell surface, other pseudopodia are withdrawn into the substance of the cytoplasm.

The cytoplasm of cells may show a streaming movement even though the cell is not transporting itself from place to place, and the membranes of most isolated cells in culture are in constant movement. This membrane movement may be seen if monocytes (monocytic white blood cells) are observed in slide cultures. When a preparation is first taken into a hot room (37°C) for observation, the cells are seen to be more or less spherical in shape, if they have previously been at normal room temperature. After some minutes at 37°C they begin to spread out, flatten and adhere to the glass; characteristic pseudopodia are extended. If the thin edge of the flattened cell is observed carefully under phase contrast, it can be seen to be undulating continuously; the cell border is irregular and changing the whole time. The surface of many cells, at least in tissue culture, should not be represented pictorially by a smooth, regularly drawn line, but by a contour as irregular as that of a map of Greece. This would moreover represent the cell surface for one moment of time only; the outline is continually changing.

Movements of the particles and granules found in cells may be seen particularly well in time-lapse films of tissues isolated in culture. Cytoplasmic streaming can also be observed in plants, for instance in the thin layer of cytoplasm stretched around the vacuole in the giant cell of the alga, *Nitella*. The active movement of cell granules must be distinguished from Brownian movement; the latter type of movement is seen in particles suspended freely in fluid cytoplasmic vacuoles and also when there is liquefaction of the cytoplasm, in which case the cell is probably dead. The viscosity of normal cytoplasm prevents obvious Brownian movements of cell contents.

Pinocytosis and phagocytosis

It has been known since 1931 that some cells engulf droplets of the medium in which they live by a process called pinocytosis or cell drinking. Pinocytosis can be studied most easily in amoebae stimulated by the addition of protein or certain ions to the medium, but it is also seen in mammalian cells in tissue culture. Conical elevations of the cell surface appear, and then cup-shaped depressions or tubular invaginations are formed, leading down from these humps into the cytoplasm. At the depths of the pits, droplets of the fluid they contain are cut off and pass into the cytoplasm as vesicles (Fig. 1.2). The size of the drop engulfed in this way may be quite small, down to at least the resolution of the light microscope; from the evidence of electron microscope pictures, pinocytosis goes on at even finer levels (see Chapter 4). Types of mammalian cell as different as monocytes, capillary endothelial cells and the Schwann cells of nerve fibres are capable of pinocytosis. Pinocytosis can also take place in a somewhat different manner. We have already noted that the edge of a flattened monocyte is continuously undulating and changing shape. Under the microscope, the waving edge can be seen sometimes to enclose droplets of fluid which then pass as vesicles more deeply into the cytoplasm.

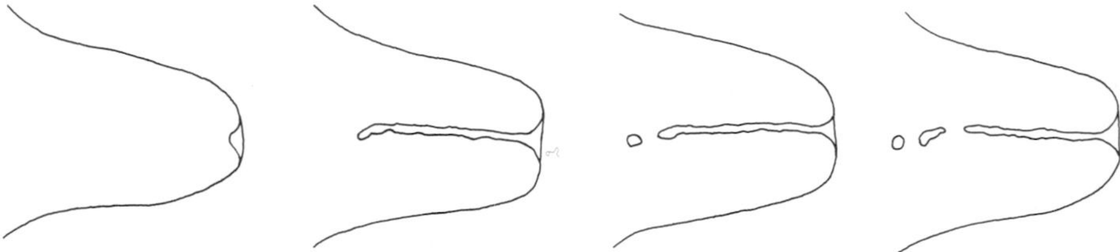

Fig. 1.2 A series of drawings illustrating pinocytosis in amoebae. A short pseudopodium is formed, projecting outwards from the surface of the cell. A re-entrant tubular channel then forms in this pseudopodium. Fragmentation of the bottom end of this channel produces vesicles. (*From* Rustad, 1961.)

The uptake of proteins of high molecular weight (which cannot pass across the plasma membrane) from a fluid environment by some cells can perhaps be explained by pinocytosis. We have seen that intense pinocytotic activity may be induced by proteins. Pinocytotic vesicles are however not filled with droplets of unchanged external medium but contain certain constituents in high concentration. An active adsorption of the inducing protein to the cell membrane is a preliminary step in the formation of vesicles, and this adsorption leads to a selective concentration of the inducing protein in the vesicles. This is shown in the following experiment.

Antibodies are proteins found in the blood of an animal after injection of a foreign protein, and the antibody formed to a given foreign protein will combine specifically with this protein. An antibody is prepared to a protein that can induce pinocytosis. The antibody is coupled chemically with a fluorescent material. The fluorescent antibody will thus combine specifically with the pinocytosis-inducing protein. Amoebae are placed in a medium containing the pinocytosis-inducing protein and are frozen rapidly after varying intervals, so that the cell constituents retain to a large extent the positions they occupied in the active cell. The water in the cells may be removed while they are still frozen, by putting them in a vacuum. The specific fluorescent antibody is now poured over the frozen-dried cells and the preparation looked at in the microscope using ultraviolet light. For cells frozen shortly after exposure to the pinocytosis-inducing protein, fluorescence may be seen on the surface of the amoebae, indicating an adsorption of pinocytosis-inducing protein to the membrane. Later appearances show that the membrane that has adsorbed this protein becomes invaginated, so that multitudes of pinocytosis channels form. The vesicles breaking off these channels then move deeper into the cell. Fluorescent antibodies have been similarly used in other experiments to locate specific proteins in cells and tissues. This technique extends the range of microscopy, providing an exceptionally specific and sensitive stain which may be used in unfixed frozen material.

It has long been known that solid particles are taken into the cytoplasm of such organisms as amoebae. This is also true of certain cells of multicellular species. The process is called *phagocytosis*. Metchnikoff, a Russian zoologist who worked with Pasteur in Paris, described the role of phagocytosis in the defence of the organism against invading bacteria. Much of the phagocytosis of this type, in mammals, is carried out by a group of macrophages scattered throughout the body; these are large cells capable of eating largely. The macrophages of mammals not only ingest foreign particles and bacteria but also engulf the remnants of worn-out red cells and, after trauma, tissue debris. The contents of pinocytosis and phagocytosis vesicles are subsequently broken down chemically.

Cell adhesion

Another property of cell membranes can be seen when cultures of fibroblasts are photographed with time-lapse cinematography. At the edge of a culture, the constant exploratory movements of the cell membranes are most active in those parts of the cell surface that are not in contact with other cells. Contact between cell surfaces seems to inhibit movement. The outgrowth of normal fibroblasts in culture produces a coherent sheet of cells adhering to each other. It is of interest to note that some strains of cells derived from malignant tumours of connective tissue do not show contact inhibition of movement, or adhesion. These sarcoma cells may be seen in culture to move freely over each other.

The specificity of cell contact can be seen in a dramatic way in some experiments carried out by Moscona.* He produced tissue cultures of mammalian fibroblasts and kidney cells. Using trypsin, he separated the cells of both cultures into suspensions of single cells. He then mixed the two types of cell and put them into a shaken culture medium. After some hours the cells sorted themselves out and, in place of randomly mixed kidney and connective tissue cells, there were clumps of the same cell-type. Cells of the same type adhered together on contact. Moreover, in mixed cultures containing chick kidney cells and fibroblasts as well as mammalian cells of the same kinds, tissue specificity was found to predominate over species specificity: chick fibroblasts came together with mammalian fibroblasts while chick kidney cells associated with mammalian kidney cells.

* References to papers, reviews, and books for further reading are listed chapter by chapter, and section by section, in the Bibliography at the end of the book. The papers listed in the Sources of Figures section, which follows the Bibliography, also provide further reading.

The organization of cells into tissues

The differentiation of cells to perform special functions in higher organisms, their anatomical arrangement and the development of these cell patterns in embryogenesis, are of course large topics which cannot be treated in this introductory chapter. It may be useful however to briefly mention some of the main features of tissue organization which form a necessary background to molecular topics discussed in later chapters.

First a certain number of cells, such as blood cells, play a free-lance role and remain without direct association with other cells. Apart from these free-lance cells, there is a general tendency for cells to come together to form close-packed layers. These are termed *epithelia* if they form a boundary between the interior and exterior of the organism and *endothelia* if they separate two internal compartments. In this terminology the cells which line the ducts of pancreas and liver, for example, are epithelia since these ducts run into the lumen of the intestine, which in turn makes direct connection with the outside world. The cells lining the kidney tubule are epithelial cells, while fine capillaries are formed from a thin layer of endothelial cells. Epithelia and endothelia are strengthened in various ways; there are adhesion points called *desmosomes* cementing the cells together (see Chapter 4). Underlying the epithelial layers of cells there is often a *basal lamina*, formerly called a basement membrane, a further strengthening feltwork of protein and polysaccharide fibrils. Basal laminae also surround capillary endothelia. Epithelial cells can be specialized for secretion, like the pancreatic acinar cells which secrete the digestive enzymes into the pancreatic duct, or the chief cells and parietal cells of the stomach epithelium which secrete pepsinogen and acid. Alternatively epithelial cells may be specialized for adsorption, notably the intestinal cells actively taking up the amino acids and sugars formed by digestion of proteins and carbohydrates, or the cells lining the kidney tubule which actively take up Na^+ and H^+ ions to control the salt and pH levels of the blood.

A quite different way in which cells may come together, typified by the muscle of higher organisms, is by fusion of the membranes to form a giant cell. In the case of a muscle cell this finally becomes a long tubular cell sometimes several centimetres long and up to 100μm in diameter containing contractile protein fibrils, and still containing the many nuclei of its constituent cells. Muscle cells are surrounded by strengthening sheaths of protein fibrils, similar to basal laminae and these fibrils extend beyond the muscle to form tendons. The main protein of tendons is collagen and it is in matrices of collagen that hydroxyapatite, the calcium mineral of bone, is deposited in vertebrates. Thus the collagen of a tendon is enmeshed with bone at one end and with the muscle sheath at the other, and transmits to the bone the tension of muscle contraction. Finally we may mention a further type of cell association, the apposition of the long processes of nerve cells to one another, and from nerve to muscle cells, to form functional contacts at which electrical impulses are transmitted from cell to cell.

Mitochondria and other cell inclusions

Most types of cell possess mitochondria; these are small cytoplasmic organelles present in plant as well as animal cells and in unicellular organisms as well as in the cells of metazoa. They vary, in shape, from spheres to filamentous rods and, in size, from 0·5μm or less to a maximum length of 6μm. The number and size of mitochondria in the cytoplasm varies with the type and also, in general, with the activity of the cell in question, more appearing for instance when a secretory cell becomes active. An early indication of damage to a cell is often given by alterations in the shape and size of mitochondria. The detailed structure of individual mitochondria is not visible in the light microscope; they do however present a characteristic appearance in the electron microscope (see Chapter 5).

Mitochondria in a living cell are constantly in movement, those that are rod-shaped bending and twisting, and all moving with the streaming of the cytoplasm. Over a period of time, they may be seen to change shape, lengthen and divide. The incessant movements and structural changes of these cytoplasmic particles in life should be born in mind during the discussion, in later chapters, of their electron microscopic appearance and enzymic activity.

Other cell inclusions besides mitochondria may be seen under phase contrast in the living cell. Among these may be mentioned the secretory granules of glandular cells, vacuoles such as those observed after pinocytosis or phagocytosis, and deposits of stored fat and stored carbohydrates such as glycogen or starch. Moreover differentiated regions of the cytoplasm can be recognised in the microscope such as the Golgi region. Information on this structure was first obtained from the study of fixed cells, treated appropriately with silver and other stains and it was questioned whether it was not merely a preparative artefact. However it can be seen in electron micrographs to have a quite distinct and complex structure (Chapter 5). The Golgi region contains a collection of membranous sacs. It is well marked in glandular cells and seems to be associated with secretory activity.

Another cell inclusion can be seen in the chlorophyll-containing cells of green plants. These chloroplasts, as they are called, have a complex fine structure and molecular organization, a topic that is discussed later in relation to their photosynthetic activity (Chapter 5).

This short survey of the appearance of cells in living organisms and tissues gives but little indication of the diversity of structures found. No mention, for instance, has been made of cilia – flexible, beating extrusions from the cell surface found in bacteria, in unicellular organisms and in a great variety of animal cells. The fine structure of cilia will be discussed in Chapter 5. In muscle cells contractile elements are present which, in skeletal and cardiac muscle, show a characteristic banded structure (see Chapter 3). It is however remarkable that in spite of this diversity a description of the main structures of a cell can be given which, with suitable modifications, applies to most nucleated cells.

In particular, the behaviour of the nucleus during somatic cell division (mitosis) shows remarkable similarities in cells of various types. Descriptions of mitosis in a mammalian tissue cell, or a protozoa, or a plant cell are very similar, although differences in detail do occur. Since the behaviour of fibroblasts has already been described, and as these cells are relatively easy to culture and observe, we give now a description of fibroblasts in division.

The nucleus and mitosis

At the beginning of mitosis a fibroblast will round up to an almost spherical shape and at the same time lose its adhesive properties, the pseudopodia being withdrawn into the cell. A remarkable bubbling of the cytoplasm occurs, and this becomes furiously active in the later stages of cell division. After division the movement dies away and the daughter cells spread out, separate and adopt the spindle shape characteristic of fibroblasts in culture. If spherical fibroblasts are patiently observed in healthy cultures most of them will eventually be seen to divide.

While these cytoplasmic changes are occurring, the nucleus progresses through a series of alterations of great complexity and precision. Before these are described, a comment must be made on the appearance of the intermitotic nucleus in living cells. In fibroblasts, a spherical or oval nucleus can be seen, and within the nucleus, one or more *nucleoli*; otherwise the interior of the nucleus appears featureless. However in certain cells, threads coiled up within the nucleus can be seen by phase-contrast but not by ordinary light microscopy; these threads are attached to the nuclear membrane or to the nucleoli. The nucleoli are usually featureless structures in living cells although occasionally filaments have been observed within them. The nucleoli disappear at an early stage of cell division and do not reappear until the later stages. The intermitotic, or interphase, nucleus has been called the resting nucleus; this is however a misnomer, for it is at this stage that it is in its most actively synthetic state. Indeed it has been shown by chemical estimations on single cells, and by a variety of other techniques, that the nuclear material, including the genetic material, is doubled during this period.

Microscopically, the period of greatest activity of the nucleus is during division. In the first phase of cell division, the interphase threads thicken and shorten, taking on the characteristic appearance of fully formed *chromosomes*. The nuclear membrane now disappears (electron microscope observations show that it breaks up into vesicles too small to be seen in the light microscope). The chromosomes arrange themselves in the central equatorial region of the cell and each chromosome is seen to be split longitudinally. The two component strands of the chromosomes, called *chromatids*, then separate and move in opposite directions away from the equator. After they have separated they are called daughter chromosomes.

A furrow has by now appeared in the cytoplasm and this extends to divide the cell; the chromosomes of what will be the daughter cells lengthen and become thinner. A nuclear membrane is formed in the new cells and nucleoli reappear. The phases of cell division are therefore:

1. The shortening and thickening of the interphase threads to form clearly visible chromosomes within the nuclear membrane – prophase.
2. The arrangements of the chromosomes in the equatorial region – metaphase.
3. The separation of the daughter chromosomes – anaphase.
4. The formation of two daughter nuclei – telophase.

The stages of mitosis are illustrated in Fig. 1.3.

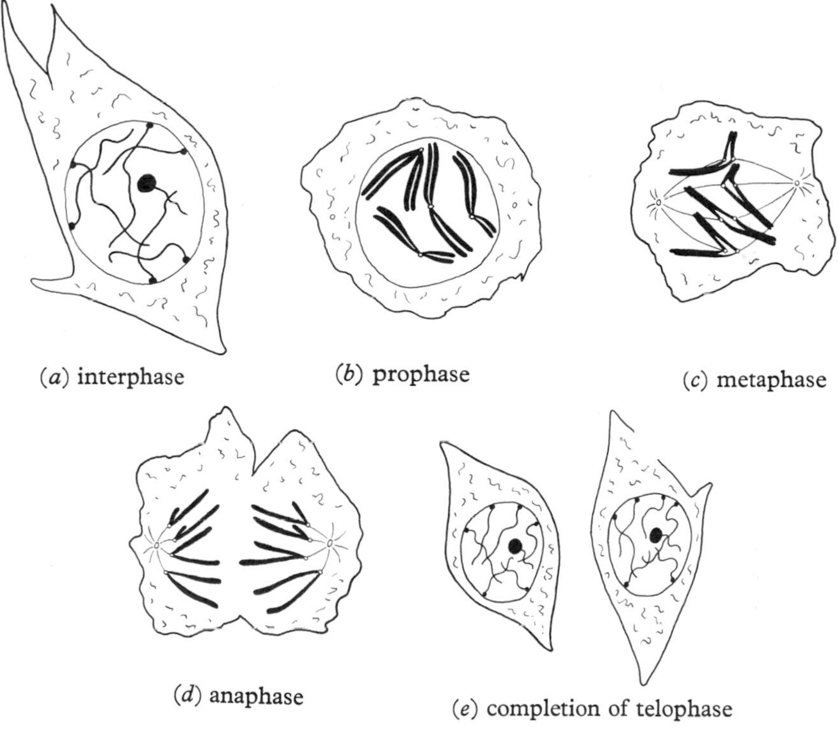

Fig. 1.3 The stages of mitosis.

As the chromosomes appear in prophase, the chromatids of each one can be seen to lie close together, but they are not joined except at one point (Fig. 1.3*b*). At this point, which is in a constant position for each chromosome, the two chromatids are constricted into a region called the *centromere*. The centromeres are arranged at the equator of the cell during early metaphase (Fig. 1.3*c*). This alignment of the chromosomes forms the so-called equatorial plate.

While the chromosomes are moving to the equator, the rest of the cell has not been inactive but is in the process of contributing other parts of the mitotic apparatus including the *spindle*, which must now be described.

In resting cells, and in early prophase, a small body in the cytoplasm in close proximity to the nuclear membrane can be recognized in living fibroblasts under the phase-contrast microscope; this is the *diplosome* (each diplosome is made up of two *centrioles*, as described in Chapter 5). In prophase, the diplosome divides and the two halves pass to the opposite poles of the cell. The diplosomes become the foci of radiating fibres that spread out through the neighbouring cytoplasm, so forming two *asters*. When metaphase is reached and the nuclear membrane has disappeared, the aster strands that pass to the equatorial region form the mitotic spindle. Threads of the mitotic spindle become attached to the centromeres of the chromosomes as they lie orientated forming the equatorial plate (Fig. 1.3*c*).

In metaphase, the centromeres divide and when they have done so, anaphase begins with the separation of the chromosomes. The spindle strands appear to shorten, pulling the daughter chromosomes towards opposite poles (Fig. 1.3*d*). Each daughter chromosome during this rapid movement is bent, and often seems to be trailing behind its centromere as it is pulled towards one of the poles. (In plant cells, this description needs to be modified slightly; there are for instance no asters formed in most plants, although a spindle stretches out between the two poles of the cell.)

At the end of anaphase, the asters become less clearly visible and, marking the beginning of telophase, the chromosomes lengthen again and become thinner. A membrane is formed around the daughter nuclei, and a cleavage furrow deepens to separate the daughter cells. (In plants, the two cells are separated not by a constriction but by the formation of a new cell plate, Chapter 5.)

During metaphase the mitochondria collect around the spindle. As the daughter chromosomes are pulled apart in anaphase all the mitochondria of the cell are grouped around the equatorial region and, with the separation of the two cells in telophase, the bundle of mitochondria is divided between the two cells.

At some period during interphase the lengthened chromosomes must split, for when the prophase chromosomes appear for the next division, they are already divided longitudinally into two chromatids. As we have noted, cytochemical studies show a doubling of the nuclear material during this phase.

The time taken for each stage of the mitotic cycle varies from species to species and cell-type to cell-type; examples are given in Table 1.1.

The chromosome number in typical somatic cells is constant for each species (Table 1.2); there are an even number of chromosomes in a normal cell. In human beings, for instance, there are forty-six chromosomes. These are grouped into 22 pairs, the members of each having a characteristic shape, and a further exceptional pair. The members of this pair are similar in the female (XX) and dissimilar in the male (XY) (Fig. 1.4). These are the sex chromosomes. In cells from women, an extra button of chromatic material is seen to be attached to the

TABLE 1.1 Duration of various stages of mitosis in mouse spleen cells and in neuroblasts of the grasshopper (in minutes).

Phase of the Mitotic Cycle	Mouse spleen cells at 37°C	Grasshopper neuroblasts at 38°C
Interphase	480–1 080	27
Prophase	20– 35	102
Metaphase	6– 15	13
Anaphase	8– 14	9
Telophase	9– 26	57
TOTAL	523–1 170	208

(Data from HUGHES, A. F. W. (1952) *The Mitotic Cycle*, New York, Academic Press; and CARLSON, J. G. and HOLLAENDER, A. (1948) *J. cellular comp. Physiol.* **31**, 149.)

nucleus, no such button being present in cells from the male. This is one of the X chromosomes and is functionally inactive.

Human chromosomes have been intensively studied in recent years, for it has been found that certain clinical disorders are associated with chromosomal abnormalities. In Down's syndrome (mongolism), for example, an extra chromosome is present. Chromosome preparations can be made from the dividing cells present in cultures of fragments of the patient's skin or bone marrow. Such studies have been facilitated by the finding that human lymphocytes, derived from a small sample of blood, will produce satisfactory dividing cells when cultured at 37°C for three days in the presence of a bean extract (a phytohaemagglutinin).

The daughter cells at each division receive a full complement of chromosomes for, as has been seen, each chromosome is split longitudinally prior to cell division. They also receive approximately half the mitochondria and other cytoplasmic contents. Later in this book it will be shown that the chromosomes

TABLE 1.2 The chromosome number of various species of animals.

Species	Chromosome number ($2n$)
Hydra circumcincta	30
Helix aspersa (snail)	54
Daphnia pulex (water-flea)	20
Aphis oenothera (greenfly)	10
Apis mellifica (honey bee)	16
Drosophila melanogaster (fruit fly)	8
Rana temporaria (common frog)	26
Gallus gallus domesticus (domestic fowl)	78
Mus musculus (mouse)	40
Felis catus (cat)	38
Homo sapiens (man)	46

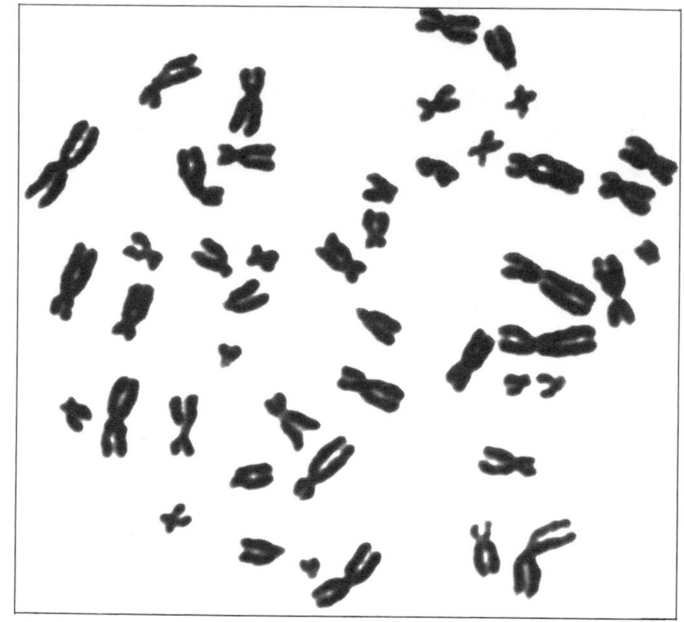

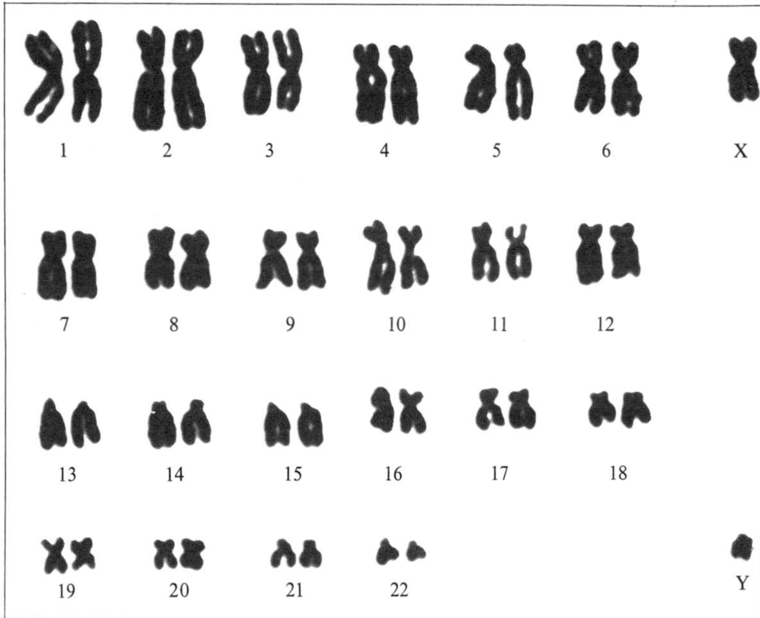

Fig. 1.4 *a*. Chromosomes of a normal male subject. A dividing cell has been squashed between two glass slides; the metaphase chromosomes have spread out and have been stained.

b. A print of the photomicrograph shown in (*a*) has been cut up and the chromosomes arranged in homologous pairs. Note the dissimilar pair of chromosomes – labelled X and Y – characteristic of males.

Both pictures at magnification approximately ×2000 (photographs by courtesy of Dr D. G. Harnden).

form the main pathway for transmitting inherited information contained in the fertilized egg to all the cells of the body. The splitting and sharing of the chromosomes at each division provide the physical basis for this. During the formation of the sex cells, the gametes, special types of cell division occur which result in the spermatozoa or the ova having only half the number of chromosomes found in somatic cells. This process is called meiosis and will now be described.

Meiosis

A fertilized egg (zygote) is a single cell formed by the fusion of an ovum (female gamete) and a spermatozoon (male gamete). It gives rise by successive division to many millions of somatic cells. Each of these, with some special exceptions, is equipped with a full complement of chromosomes – half derived from the mother and half from the father. This mitotic type of cell-division ensures that each daughter cell receives an exact replica of the chromosomal apparatus of the original zygote, with nothing added and nothing subtracted. We shall now consider the special form of cell division by which the gametes are generated. At first sight it may appear unnecessary to have a separate mechanism for this purpose. Why cannot the gametes also be derived by mitotic division from their cellular precursors? A convincing answer can be provided by simple arithmetic. The number of chromosomes in any mitotic descendant of a zygote can be represented by the symbol $2n$, where n chromosomes were initially received from the mother and n from the father. If the gametes were formed by mitosis, an ovum would contain $2n$ chromosomes and be fertilized by a sperm containing $2n$ chromosomes. The cells descended by mitosis from this zygote would therefore each have $4n$ chromosomes, and in the next and subsequent generations the number would become $8n$, $16n$, $32n$, etc. In order to avoid this absurdity there must at some point be interposed a *reduction division* whereby the number of chromosomes is halved. The halving actually occurs during the formation of gametes, so that a sperm or ovum only carries the *haploid* (Greek root: haplo = single) number, n, of chromosomes instead of the diploid (Greek: diplo = double) number $2n$. The manner in which this is achieved is known as meiotic division, and is found in essentially the same form throughout the animal, plant and fungal kingdoms. Meiosis has been defined as the occurrence of two divisions of a nucleus accompanied by one division of its chromosomes.

The onset of prophase in meiosis was traditionally held to differ from the early prophase of mitosis in one respect. It was claimed that in early prophase of meiosis the chromosomes were not yet divided into chromatids on the ground that no double structure of the kind observed in mitotic prophase is microscopically visible. This view is, however, incorrect. Chromosome breaks induced during early meiotic prophase by X-irradiation should, if the chromosomes were at this stage still undivided, show up later as injuries to both daughter chromatids. If, on the other hand, splitting had already occurred, an injury would be represented in only one of a pair of chromatids. The latter pattern of injury was found by S. Mitra in *Lilium longiflorum*, and by varying the stage in the division cycle at which irradiation was applied he was able to pin-point the moment at which chromosome duplication occurred. This turned out to be before the onset of meiotic prophase, so that the chromosomes at prophase must each consist of two chromatids so closely apposed as to give the appearance of a single structure. The difference between this and the situation in early mitotic prophase is thus one of appearance rather than reality.

The way in which the chromosomes of a diploid cell can be grouped in homologous pairs (except for the XY sex chromosomes) is illustrated in Fig. 1.4. During mitotic division with rare exceptions there is no interaction between homologous chromosomes, that is between a given chromosome among the n derived from the mother and the corresponding chromosome among the n derived from the father. In meiosis, by contrast, the chromosomes of a homologous pair are seen to be in close contact with each other, at one or more points along their length, at early prophase. During later stages of the meiotic division cycle the zones of apposition spread along their length from the initial contact points in each direction like zip-fasteners. An extremely intimate approximation of homologous elements is finally achieved along the entire chromosomal length, and there is some shortening and thickening of the paired strands. Each paired complex is known as a *bivalent*, being of dual origin, paternal and maternal.

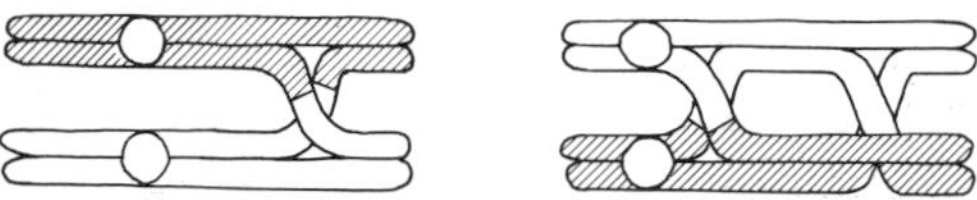

Fig. 1.5 Purely diagrammatic representation of crossing-over. The bivalent on the right exemplifies a 'three-stranded cross-over', since three of the four chromatids are altogether involved. Chromatid material of paternal origin is shown shaded, and of maternal origin unshaded. Centromeres have been left unshaded.

The next event consists of a separation of each bivalent into constituent halves, beginning with a moving apart of the two centromeres. (These are the centromeres of the two homologous chromosomes; the centromeres do not themselves divide in this first meiotic cell division.) As the centromeres diverge, each drags behind it its attached chromatid pair, which can for the first time be seen microscopically as a double structure. Complete separation of chromosomal material of paternal and maternal origin is, however, restricted by the occurrence of what are known as *chiasmata* (the plural form of chiasma, a word of Greek origin used to describe crossing-over, as in the Greek letter chi: χ).

It is convenient to refer to the genetic exchanges associated with chiasma-formation as 'crossing over' but inspection of the diagram of Fig. 1.5 makes it clear that the chromatids themselves do not actually cross over. The arrangement of maternal and paternal material suggests, rather, a process of breakage at corresponding points of homologous chromatids followed by reunion with the 'wrong' strand. Chiasma formation almost invariably accompanies the pairing of homologous chromosomes, although the frequency with which chiasmata are formed varies from one bivalent to another. Different parts of a chromosome are also affected with differing frequency.

The nuclear membrane now disappears and the spindle takes shape. The separation of the centromeres to opposite poles follows a similar pattern to that of mitotic anaphase, except that each carries two chromatids with it. The telophase which brings the first meiotic division to an end is similar to mitotic telophase.

The whole procedure of first meiotic division has resulted in a halving of the number of centromeres per cell (since these still remain undivided) whereas the number of chromatids after the division is equal to the diploid number of chromosomes in somatic cells (Fig. 1.6). The detailed composition of the chromatids, however, is very different after the first meiotic division. In the region of each centromere both attached chromatids consist exclusively of material of the same maternal or paternal origin as the centromere itself. But the remainder of each chromatid is a patchwork of maternal and paternal material as a result of the exchanges associated with the formation of chiasmata. By contrast the chromosomes of a somatic cell can be grouped into homologous pairs, each of which consists of one whole chromosome exclusively derived from the mother and another exclusively from the father.

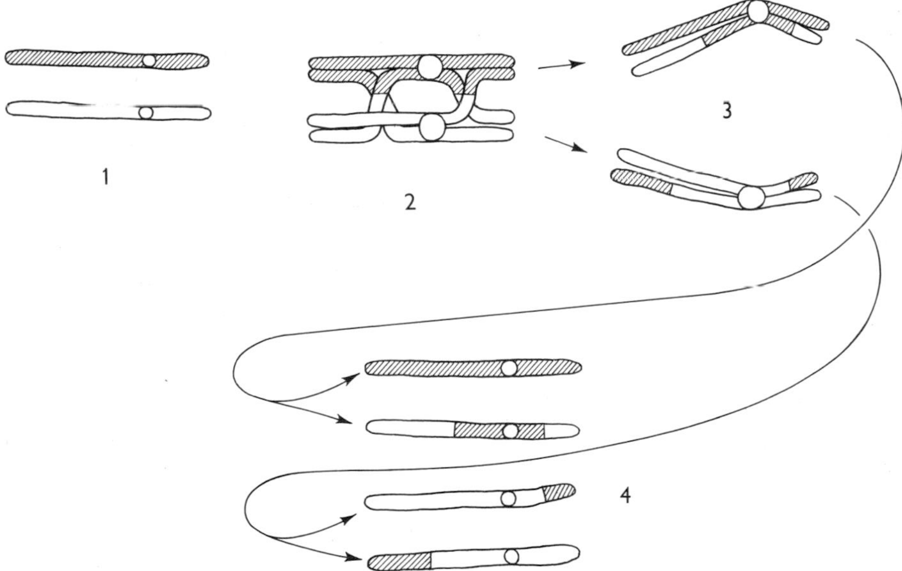

Fig. 1.6 Redistribution of the material of a pair of homologous chromosomes during the first and second meiotic divisions. In the first meiotic division the members of each pair of homologous chromosomes (1) divide with crossing-over (2) and separate to opposite poles (3). In the second meiotic division, the division of the centromeres gives rise to four chromosomes (4), each of which will form part of the haploid chromosome complement of a gamete.

In the next meiotic division, which follows after an exceedingly brief interphase, the observed events are similar to those of mitosis. The difference is that in the second meiotic division the division of each chromosome into chromatids pre-existed throughout the previous cell cycle, and was directly inherited from it, whereas in mitosis it took place during the preceding interphase and hence was of relatively recent occurrence. In the second meiotic division, as in mitosis, each centromere splits into two. The resultant halves migrate to opposite poles, each carrying one chromatid, which can now be called a chromosome. The haploid number of chromosomes is thus finally attained (Fig. 1.6).

We shall see in Chapter 6 that chromosomes carry genetic (inherited) information. Mitosis ensures continuity of inherited information, with one chromosome of each homologous pair in a diploid somatic cell identical with the corresponding chromosome of the original ovum from which the whole organism is derived, and the other chromosome of each homologous pair identical with the corresponding chromosome of the sperm which fertilized this ovum. What is achieved in meiosis, with genetic consequences of profound importance, is a reshuffling of maternal and paternal genetic material, as a consequence of crossing-over, so that no single chromosome of the final gamete is genetically identical with any one of the chromosomes possessed by the original zygote or by the somatic cells mitotically derived from it. Further, since the number and precise localization of chiasmata in the first meiotic division differs from one cell to the next, and since the decision as to which centromere of a pair shall migrate to which pole is made at random, in both the first and second meiotic divisions, it follows that no single gamete will ever constitute an exact genetic match of any other gamete produced by one and the same individual. Thus if we could inseminate a million genetically identical females with sperm taken from a single ejaculate, we would obtain a million *genetically distinct offspring*. In spite of the fact that each would have received half its genetic endowment from the same father, each individual endowment would be found to have been made up of a different assortment of the father's mother's and the father's father's genes. This fact is worth reflecting on, for it forms the material basis of the 'segregation' and 'recombination' of inherited traits upon which a hundred years of genetics have been built and which will be considered in detail in Chapter 6.

Isolation of cellular components

We have seen how the cells of higher organisms can be isolated, cultured and studied as individual living entities. Simpler organisms such as bacteria or amoebae exist as free-living single cells and can be readily grown in culture. For study of the chemistry and the molecules of living organisms the cells themselves must be broken down further, and we consider in the remaining sections of this chapter some of the modern techniques that are used to isolate subcellular organelles and the large molecules of these organelles, and the methods used to purify the individual components in a mixture of large molecules.

It must be appreciated that when we isolate the single cells from a multicellular organism, or a single type of molecule from a cell, we are destroying and grossly disrupting the living material, and isolating for study a small part of the total integrated system. But this is a necessary first step to detailed understanding. By first studying the parts we can return with deeper insight to the complexity of the cell or the whole organism.

As long ago as 1897 the Buchner brothers prepared an extract from yeast that was free of cells, but would nevertheless carry out the fermentation of sugar to alcohol. By this experiment they showed that some ‘ferments’, or enzymes, could be separated from cells, and still catalyse specific reactions which had earlier been thought to be inseparable from living matter. The crystallization of the enzyme urease by Sumner in 1926 and of various proteolytic enzymes by Northrop during the 1930s were major advances. Dixon and Webb (1964), in their monograph, print photographs of 134 crystalline enzymes and list over 600 others which, although not prepared as crystals, have been well characterized.

Each of these enzymes catalyses a chemical reaction or group of reactions. To do this, the enzyme first combines with the reactants, or substrates as they are called; a chemical change then takes place in the substrates and the products of the reaction separate from the enzyme, leaving it free to combine with, and catalyse reaction between, further substrate molecules. Some enzymes combine only with substrates of precise chemical pattern; these are said to have a high specificity, and only catalyse one reaction or a limited number of reactions. Other enzymes have a low specificity and catalyse a larger series of reactions. For instance, the enzyme urease will act only on urea:

$$(NH_2)_2C{=}O + H_2O \xrightleftharpoons{\text{urease}} 2NH_3 + CO_2$$

It will not even hydrolyse the methyl derivative $CH_3NH.CO.NH_2$. Lipases, on the other hand, will catalyse the hydrolysis of a large number of carboxylic acid esters:

$$R.CO.OR' + H_2O \xrightleftharpoons{\text{lipase}} R.COOH + HO.R' \quad \text{(where R and R}' \text{ can vary).}$$

Most enzymes have, like urease, a high degree of specificity, and since most of the complex reactions of the cell are dependent on enzymes, it follows that a large number of enzymes must be present in the cell. It is apparent then that the synthesis of specific enzymes is a central problem in cell biology. Since enzymes are classified chemically as proteins we may say more generally that the synthesis of proteins is a central problem; in this way we include not only enzymes, but other cell proteins. Included among the non-enzyme proteins will be, for example, structural proteins such as collagen, and respiratory proteins such as haemoglobin. A further group of proteins are the antibodies formed in response to infection or the injection of antigens (foreign proteins or polysaccharides) into the tissues. Part of the surface of the antibody molecule is complementary in structure to part of the surface of its antigen, just as an enzyme surface is complementary to its substrate. Specific antibodies can be produced to a wide variety of antigens, and different antibodies differ in their detailed structure.

Enzymes, structural proteins, antibodies and other proteins are chemically composed of long chains of amino acids. The structure of a single amino acid may be written

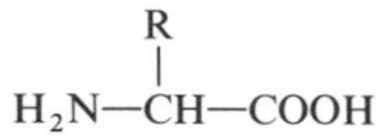

$$\mathrm{H_2N{-}\overset{\displaystyle R}{\overset{|}{C}}H{-}COOH}$$

where R varies from one amino acid to another. In the protein chains the amino acids are linked through the —COOH and $—NH_2$ groups by removal of water

$$\mathrm{{-}\underset{\displaystyle O}{\underset{\|}{C}}{-}OH \qquad H{-}\underset{\displaystyle H}{\underset{\|}{N}}{-}}$$

to give the chain structure

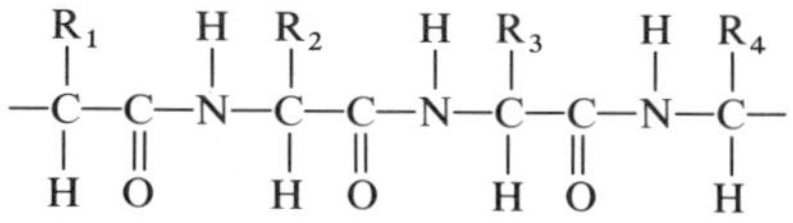

$$\mathrm{{-}\overset{\displaystyle R_1}{\underset{\displaystyle H}{\underset{|}{\overset{|}{C}}}}{-}\underset{\displaystyle O}{\underset{\|}{C}}{-}\overset{\displaystyle H}{\overset{|}{N}}{-}\overset{\displaystyle R_2}{\underset{\displaystyle H}{\underset{|}{\overset{|}{C}}}}{-}\underset{\displaystyle O}{\underset{\|}{C}}{-}\overset{\displaystyle H}{\overset{|}{N}}{-}\overset{\displaystyle R_3}{\underset{\displaystyle H}{\underset{|}{\overset{|}{C}}}}{-}\underset{\displaystyle O}{\underset{\|}{C}}{-}\overset{\displaystyle H}{\overset{|}{N}}{-}\overset{\displaystyle R_4}{\underset{\displaystyle H}{\underset{|}{\overset{|}{C}}}}{-}}$$

In addition to the amino-acid chains, other groups such as the haem group of haemoglobin may be incorporated into protein structure.

Separation of proteins

The first step in the study of proteins is their isolation and purification. Since proteins are delicate compounds this may not always prove easy, but no significant advances can be made until we know some of the properties of the compounds present in cells; this is true for enzymes and other proteins alike. After such an analysis it is possible to investigate what part the proteins thus identified play in the economy of the cell.

Some proteins can be easily prepared almost free of contaminating material. If blood is taken into an anticoagulant and then centrifuged, the red cells are deposited. The supernatant plasma, and the thin layer of white cells which is deposited on top of the red cells, can be pipetted off and replaced by isotonic saline (the terms isotonic, hypertonic and hypotonic are defined in Appendix 4). The red cells may now be re-suspended and again centrifuged. If they are washed in this way a number of times, a deposit of cells is obtained free from contaminating plasma. If distilled water is now added, the red cells will swell osmotically and soon burst, releasing their contents into the fluid. The red cell envelopes, or 'ghosts', and cell debris may be deposited by spinning at high speeds. Since the content of red cells is largely haemoglobin, the supernatant will be predominantly a solution of haemoglobin which by suitable procedures may be crystallized out. In this way considerable amounts of haemoglobin may be prepared with only a small degree of contamination.

Most other proteins are much more difficult to prepare to a similar degree of purity. In the preparation of haemoglobin described above, supernatant plasma was removed from the centrifuged blood. It has been known for a long time that a number of different proteins are present in this plasma, for when ammonium sulphate is added to half saturation a proportion of the protein precipitates; this fraction contains the globulins. When three-quarter saturation with ammonium sulphate is reached, the albumin fraction precipitates. So with ammonium sulphate we can separate protein fractions from plasma, but the fractions still contain a number of different proteins, and further preparative techniques must be adopted to separate pure proteins from these mixtures.

In a buffered solution at a given pH, a protein will be ionized and carry a number of charged groups, notably $—COO^-$ and $—NH_3^+$ groups. At any pH except the unique isoelectric point these do not cancel each other out and the whole molecule therefore carries a net charge. (This point is amplified in Chapter 2 and in Appendix 2. pH, pK, buffered solution and isoelectric point are defined in that appendix.) If a potential difference is applied across a protein solution the protein molecules will move either towards the anode or the cathode. With the buffers commonly used, most proteins will carry a net negative charge and will consequently move towards the anode. The rate at which they move will depend on their size, their net charge and a number of other factors. Thus if the experiment is carried out using a mixture of proteins, under appropriate conditions, the proteins can be separated one from another according to their size, and to the charge they carry. This method is called electrophoresis.

In the Tiselius apparatus a potential is applied across a boundary between the solution containing the proteins and a protein-free buffer solution (moving boundary electrophoresis). Figure 1.7 shows the separation obtained with plasma by this technique. In an alternative method, a drop of the protein solution is placed on a supporting medium such as filter paper soaked in a buffer solution; the current is now switched on and the protein molecules move from the point of application of the protein solution through the buffer in the paper towards the electrodes at a rate dependent largely on their net charge. The method is called zone electrophoresis. When the proteins have separated sufficiently, the paper may be removed and stained with, for example, naphthalene black. Alternatively the method may be used preparatively, by cutting out zones containing particular proteins, which are then soaked out from the paper.

Either moving boundary or zone electrophoresis is capable of distinguishing between the globulins in plasma much more efficiently than precipitation methods using ammonium sulphate. During electrophoresis the proteins are not subjected to any treatment that might cause denaturation or alteration of their structure.

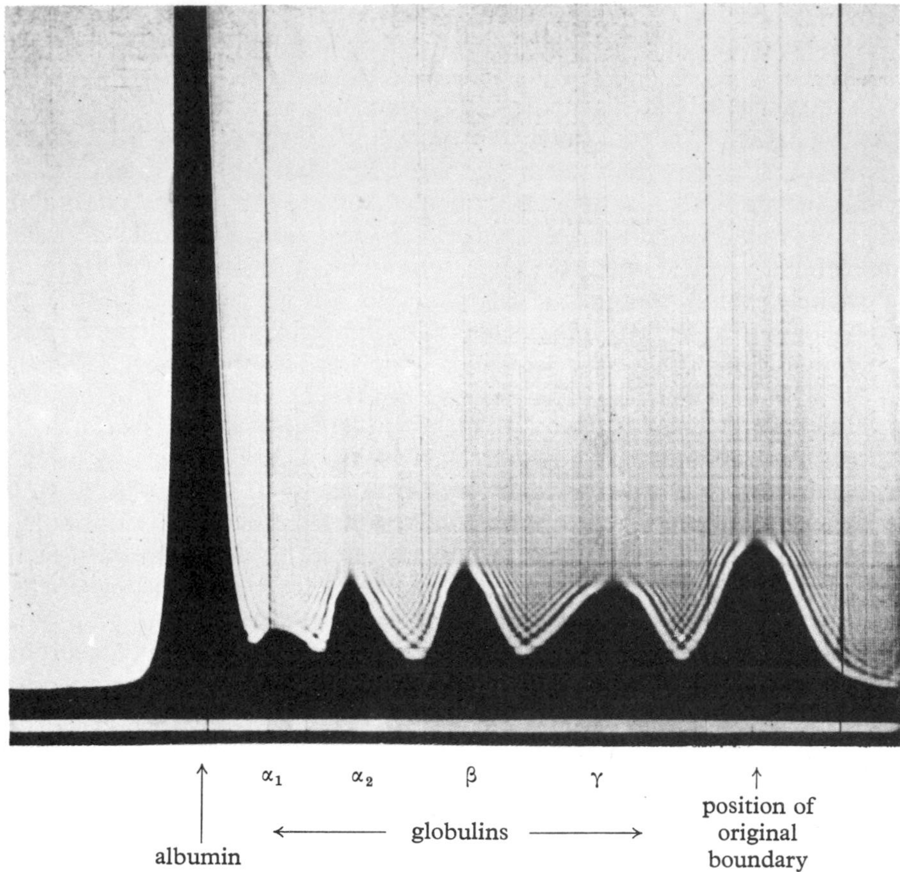

Fig. 1.7 The separation of some of the human plasma proteins by moving boundary electrophoresis. The boundary formed by each component is detected by the corresponding change in refractive index. (*From* Williams, 1960.)

One of the developments of zone electrophoresis has been to use, in place of the filter paper as a supporting medium, a gel prepared from a suitable buffer and acrylamide. The interstices of the gel are small enough to retard the large protein molecules and so, in this technique, the separation of the proteins will depend to a greater degree on their molecular dimensions. Proteins having a similar charge may, in this way, sometimes be separated by gel electrophoresis (Fig. 1.8).

A further technique which relies principally on differential filtering of large molecules is that of gel filtration, also called molecular sieve chromatography. Glass tubes are set upright and packed with a watery, porous gel. The type of gel so far found most generally useful is that made of dextran derivatives. These are polysaccharide chains, cross-linked in various ways; they can be synthesized in a

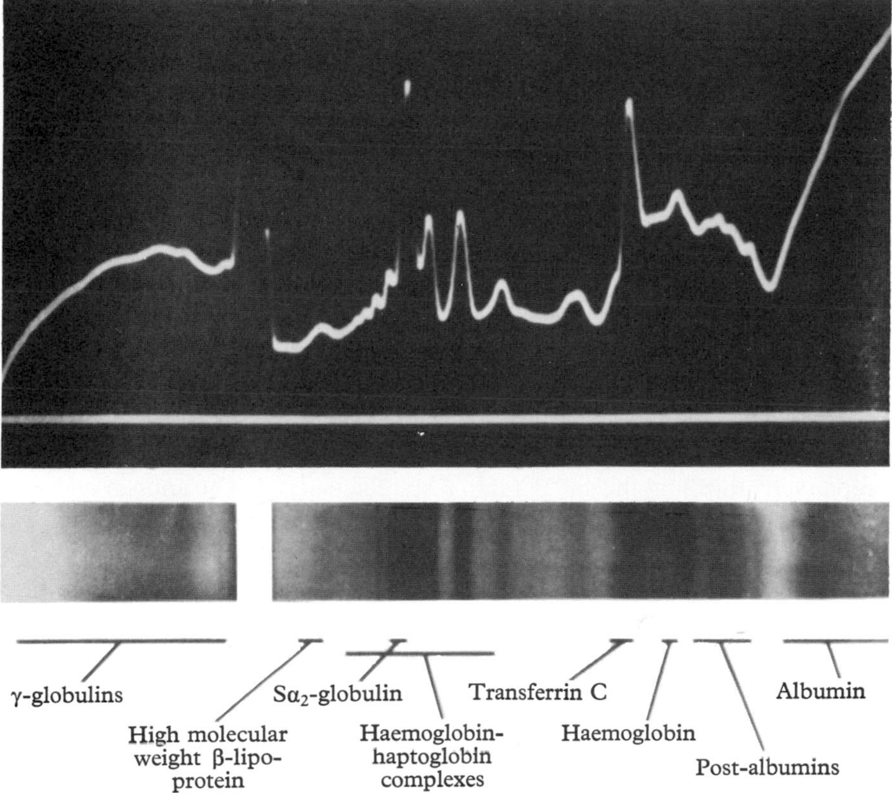

Fig. 1.8 The separation of some of the human plasma proteins by zone electrophoresis in starch gel at pH 8·6. The protein was stained after separation, with amido-black. The lower part of this figure shows a photograph of the gel in transmitted light, and the upper part a photometric scan. (*From* Smithies, 1959.)

wide range of molecular weights, tailored, as it were, to requirements. In this way, different gels with different degrees of porosity can be used. (These are marketed under the trade-name of 'Sephadex'.) The protein mixture, in a buffered solution, is placed on the top of the gel in the column and the different proteins are washed through the pores of the Sephadex gel at rates dependent on three main factors; the composition of the particular gel used, the volume of the sample applied to the gel and the molecular size of the protein. Successive collections of the effluent from the column are made; an automatic fraction collector is normally used for this. Other gels besides Sephadex have been found useful including those made with agar and with polyacrylamide.

A technique used for separating the components of protein mixtures, which is more directly dependent on the molecular weight of the proteins involved, is that of centrifugation. In this technique the proteins in solution are spun at very high speeds and subjected to sedimentation forces many thousands, or even hundreds of thousands, times greater than gravity. Since the proteins will sediment at a rate dependent in part on their molecular weight (and in part on their shape and hydration), the method may be used to obtain an estimate of molecular weight, as well as for the preparation of homogeneous protein.

Having used this battery of techniques on such material as plasma, we must ask what sort of resolution have we obtained; how far, for instance, can we succeed in preparing homogeneous globulin fractions? The answer is that globulin fractions, even those separated by starch gel electrophoresis, are still mixtures, for if one of these fractions is injected into the circulation of an experimental animal, a number of antibodies will be produced, suggesting that a number of different antigenic proteins must be present in the fraction. Gel-diffusion techniques show this up very elegantly. An agar gel is made in a Petri dish, as for bacterial work; wells are cut out of the agar and in one is placed the protein fraction used as the antigen, and in another the antiserum, i.e. the serum prepared by injecting the antigen into a suitable animal. The antigen and antibody will each diffuse through the gel and where they meet will react to produce a precipitate. If the antigen is homogeneous there will normally be one line of precipitation; if there is more than one protein present in the sample then there will be two or more precipitation lines (Fig. 1.9). This method is particularly

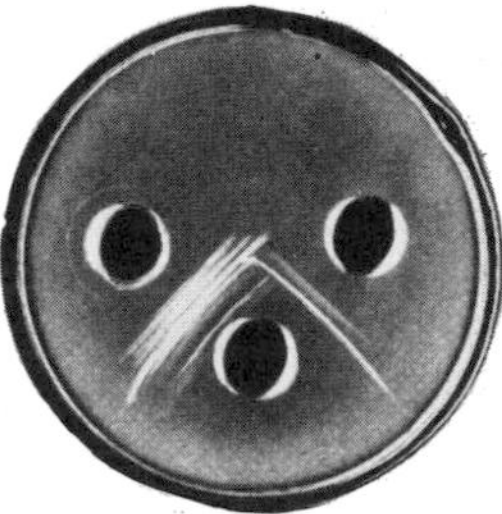

Fig. 1.9 The identification of protein fractions present in human plasma using the agar-gel diffusion technique of Ouchterlony. Human blood serum has been placed in the well cut in the agar-gel on the left, the middle well contains horse antiserum to human serum, while the right-hand well contains purified human plasma γ-globulins. The various proteins have diffused through the gel at different rates and where antigen and complementary antibody meet there is formed a milky line of precipitation. A large number of antigen-antibody precipitation lines have been formed by the reactions between human serum and horse antibodies. The purified γ-globulins form only a few lines. (*From* Williams, 1960. Photograph by David Linton.)

sensitive in detecting the component proteins in a mixture, especially if combined with electrophoresis. The type of result obtained with immuno-electrophoresis is shown in Fig. 1.10.

It is apparent that physico-chemical means of separating proteins are not always successful and that we have, when dealing with such complex mixtures as the globulins, to use in addition a biological method to distinguish some of the component proteins. Some workers at one time doubted that chemical purity had any meaning when applied to the macromolecules of proteins; they suggested that isolated proteins would have a mean molecular composition and structure, but that there would be considerable variation around this mean. We shall see in Chapter 2, however, that proteins *are* made precisely and that chemical homogeneity has a meaning even with such large molecules.

We have been discussing the separation of proteins found in nature in an aqueous suspension (plasma). In the study of intracellular proteins there are further problems of extraction from cell structures and compartments. The possibility of change in the properties of the proteins during extraction must always be kept in mind.

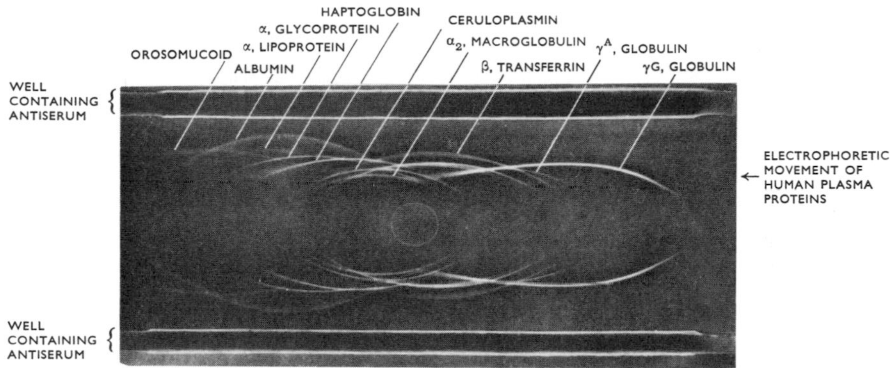

Fig. 1.10 The identification of protein fractions present in human plasma using immuno-electrophoresis. Longitudinal wells have been cut in the gel and may be seen at the top and bottom of the illustration; they contain the horse antiserum. A sample of human plasma proteins has been submitted to electrophoresis along the gel, moving across from the right-hand side. The protein fractions have been separated by electrophoresis and identified immunologically by antigen-antibody precipitation lines. (*From* Williams, 1960.)

Many, perhaps most, of the proteins present in living tissues are enzymes, but others are, for example, used as scaffolding. A list of various proteins with some indication of their roles in the living organism is given in Table 1.3. This list does not pretend to be systematic, or complete, but it does serve to emphasize the great variety of proteins to be found in nature. Their biological activity is dependent on the molecular structure of the particular protein.

The role of enzymes in metabolism

In Table 1.3 all the enzymes have been classified into only four groups, but as we have noted above, these form a most varied and important class of proteins. As Dixon and Webb have put it, 'Life depends on a complex network of chemical reactions brought about by specific enzymes . . .', or as Borek has said, 'We live because we have enzymes. Everything we do – walking, thinking,

TABLE 1.3 Some proteins and biologically important peptides.
Proteins listed as haemoglobin, serum albumin, insulin, etc., show minor variations in structure and properties in different species, and even in different groups of individuals within a species. They thus represent not a single protein but a class of proteins with closely similar properties. Pure proteins have to be specified more precisely, e.g. beef insulin, normal human adult haemoglobin.

Enzymes	Hydrolysing enzymes, e.g. proteolytic and other digestive enzymes. Transferring enzymes, e.g. transaminases. Oxidases and enzymes of dehydrogenase systems. Enzymes with other specific activities.
Structural proteins	Keratins, e.g. proteins of hair, wool and feathers. Elastin and collagen – connective tissue proteins. Silk protein. Muscle proteins, e.g. actin and myosin. Proteins of the mitotic spindle. Proteins of cilia and flagella.
Respiratory proteins	Haemoglobin. Cytochromes. Myoglobin. Haemocyanin and other respiratory proteins found in invertebrates.
Proteins of blood plasma and lymph	Serum albumin. Globulins – a complex containing very many proteins. Fibrinogen – converted to fibrin during clotting.
Antibodies	Proteins formed in response to antigens. These make up an important component of the blood globulin fraction.
Toxic proteins and peptides	Snake venoms. Bacterial toxins. Fungal toxins.
Protein and peptide hormones	Anterior pituitary hormones, e.g. ACTH. Posterior pituitary hormones – vasopressin and oxytocin. Gastro-intestinal hormones, e.g. secretin. Angiotensins. Parathyroid hormone. Thyroglobulin. Insulin. Glucagon.
Milk proteins	Casein. β-Lactoglobulin.
Egg proteins	Vitellin, and other yolk proteins. Egg albumin, and other proteins of egg white.

TABLE 1.3 – *continued*

Proteins peculiar to plants and their seeds	Prolamines. Glutelins.
Nucleoproteins, (proteins found associated with nucleic acids)	Chromosomal proteins – protamines, histones. Ribosomal proteins. Virus proteins.
Other conjugated proteins	Lipoproteins (proteins associated with lipids, e.g. plasma β-lipoprotein). Glycoproteins (proteins associated with carbohydrates, e.g. certain proteins in cell membranes). Phosphoproteins (proteins containing phosphorus, e.g. casein). Many conjugated proteins are included in other sections of this table.
Other proteins	Proteins recognized at present only by immunological techniques.

reading these lines – is done with some enzymic process.' We must now consider briefly the role of enzymes in the metabolic life of the cell.

Enzymes may, as we have seen, be obtained as crystals of protein completely, or very nearly, free of contaminating material. If we wish to study the specificity of an enzyme or its rate of reaction with its substrates, it is highly desirable that we use purified material. After we have studied, in this way, the properties of individual enzymes we can consider how they interact in the cell. Some enzymes, for example, the amylase secreted in saliva, or the proteolytic enzymes secreted by pancreatic cells, act extracellularly. These digestive enzymes break up the

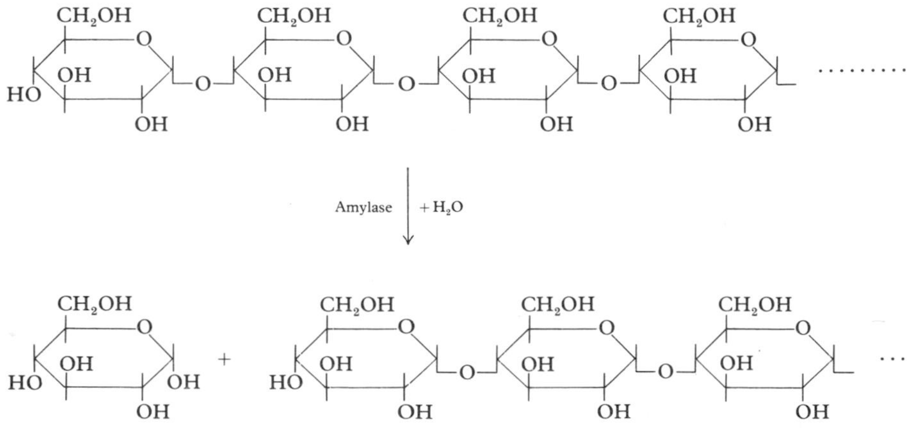

Fig. 1.11 Hydrolysis of starch.

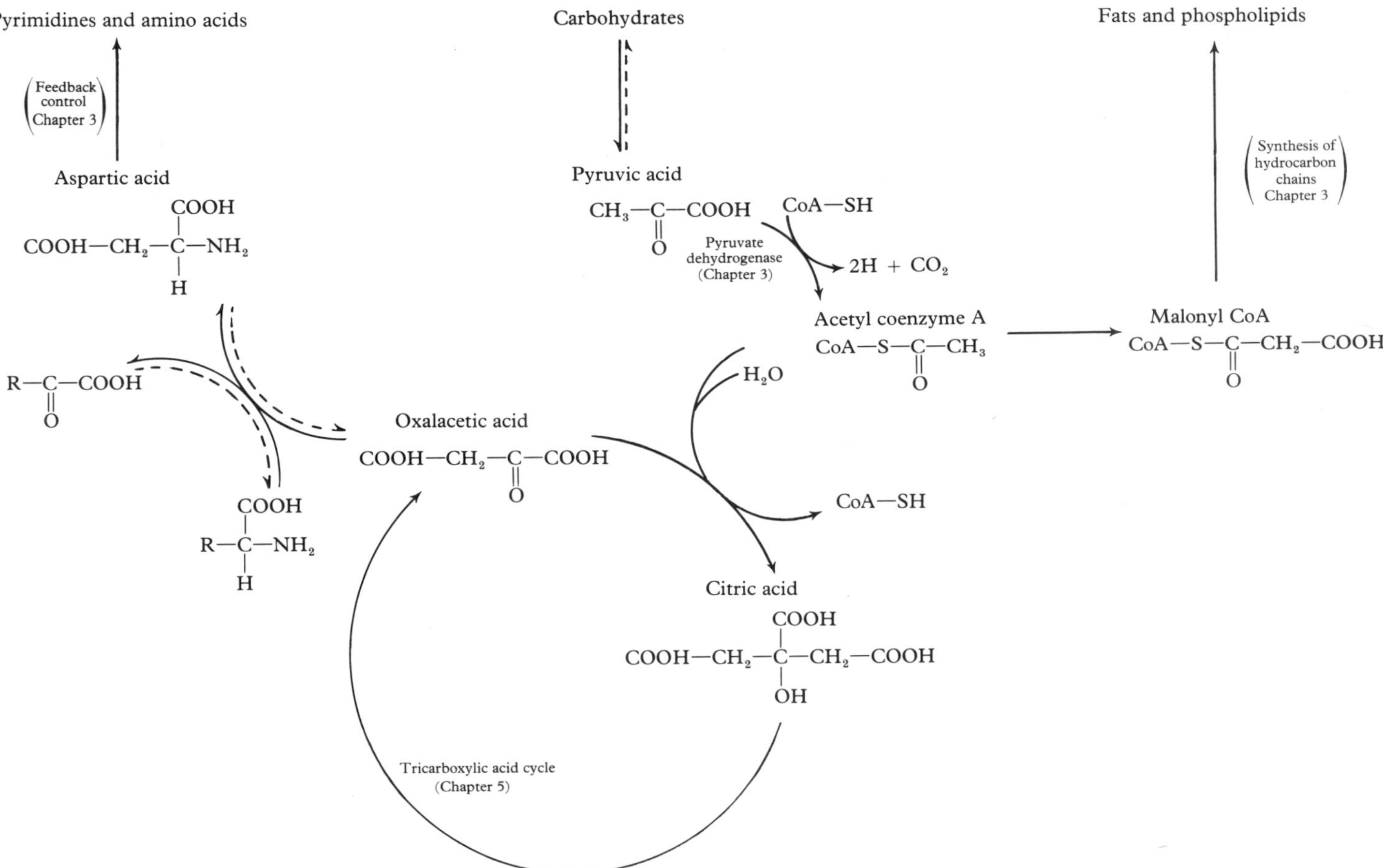

Fig. 1.12 A diagram showing relationships between metabolic pathways discussed in later chapters. At pH ~ 7 most of the acidic groups of these compounds are dissociated, with pyruvic acid present as pyruvate ($CH_2.CO.COO^-$), etc.

polysaccharides and proteins of food to their constituent sugars and amino acids by hydrolysis, in which the bonds between the sugars and amino acids are broken, and —H and —OH groups from a water molecule are joined to the broken bonds (Fig. 1.11). Most enzymes however act within cells, each part of a series of enzymes catalysing an integrated chain of reactions, the sum of which is a recognizable metabolic activity. For instance, the breakdown of glucose to CO_2 and water involves a large number of enzymes, the first in the sequence being hexokinase; this catalyses the transfer of a phosphate group from adenosine triphosphate (ATP) to glucose with formation of adenosine diphosphate (ADP).

For subsequent stages in the overall reaction many other enzymes with different specificities are required, the original carbohydrate being passed through a large number of steps through pyruvate and the tricarboxylic acid cycle (Fig. 1.12) before it is degraded to CO_2 and water. There is more than one pathway that glucose may take in its degradation, the alternative routes each involving a series of more or less well characterized enzymes.

The energy of oxidation is made available to the cell largely in the form of a net gain of ATP. The breakdown of this 'high-energy' phosphate compound to ADP and inorganic phosphate releases energy which may be used in muscular contraction, active transport, or in other metabolic activities. This mechanism of energy transfer through high-energy phosphate compounds is a central feature of animal, plant and bacterial metabolism.

One intermediate compound in the oxidation of glucose is acetyl-coenzyme A (Fig. 1.12). This compound is also an intermediate in the breakdown and synthesis of fatty acids. This is one way in which the systems involved in carbohydrate metabolism are interconnected with those involved in fat metabolism. Other relationships exist between these systems and those of protein metabolism. The cell may, because of such numerous interlocking pathways, be regarded as a single, very complex multi-enzyme system, no metabolic activity going on in isolation.

Intracellular position of enzymes. Homogenization of cells

We have seen that acetyl-coenzyme A is produced during the breakdown of glucose. This compound is oxidized in the cell via a complex series of reactions known as the tricarboxylic acid cycle; each step in the cycle is catalysed by a specific enzyme. The oxidation reactions of this cycle, and the breakdown of fatty acids, are coupled to the phosphorylation of ADP to ATP (see Chapter 5). Under aerobic conditions these are the reactions mainly responsible for the generation of the ATP required for muscular contraction and energy-requiring metabolic reactions. Green suggested in 1947 that the enzymes responsible for these reactions were to be found in the cell in the form of an integrated unit and were not present randomly distributed in the cell cytoplasm. The complex was thought to contain the enzymes of the tricarboxylic acid cycle in relatively constant amounts, the necessary coenzymes, and the enzymes of the oxidative phosphorylation pathway, associated with regeneration of the coenzymes of the cycle and phosphorylation of ADP to ATP.

Meanwhile the method of differential centrifugation of ruptured cells had been developed to a precision which allowed mitochondria to be separated in amounts large enough for the study of their biochemical behaviour. Figure 1.13 illustrates the stages of the method as applied to liver. The structure of the mitochondria isolated in this way can be examined in the electron microscope, and their appearance is similar to that of mitochondria of fixed and sectioned liver cells. When the biochemical activities of the mitochondrial fraction were examined, it was found that many of the enzymes of the tricarboxylic acid cycle and of the oxidative phosphorylation pathway were localized in these particles.

If a particular enzyme is present principally or exclusively in a given fraction after centrifugation, it is likely that the structural components of this fraction contained the enzyme in life. But there is always the possibility of redistribution of the enzyme during homogenization. How can we be sure that the association of these enzyme systems with mitochondria is to be found in the intact living cell? An experiment by Holter provides some evidence for this association. It has long been known that the eggs of sea urchins can be centrifuged intact, without prejudice to their subsequent fate; the cell contents stratify, the yolk granules, for instance, rising to the top of the cell. Holter used a similar technique to separate the cell contents of a large species of amoeba. Such centrifuged cells are still viable. Holter then, by micro-dissection, divided the cell and examined each half for its content of enzymes. The portion that contained the mitochondrial layer showed succinate dehydrogenase activity; this enzyme is one of the tricarboxylic acid cycle enzymes. Where mitochondria were not present there was no dehydrogenase activity.

Yet another quite different approach suggests that succinate dehydrogenase activity is to be found in the mitochondria alone. The technique in this case depends on the breakdown of sodium tellurite at the site of the enzyme, to tellurium. Thin sections are examined in the electron microscope; the tellurium is opaque to the electron beam and so shows up the sites where the enzyme was originally present in the cell. Again the localization of this dehydrogenase is in the mitochondria and not elsewhere.

Further analysis of the biochemical potentialities of the mitochondria has confirmed that they contain all the enzymes of the tricarboxylic acid cycle, and the associated oxidative phosphorylation pathway, and indicates that these intracellular particles are the site of other metabolic reactions such as the breakdown of fatty acids. The respiratory enzymes are associated in constant proportion, even in mitochondrial fragments – a fact which suggests they are similarly associated in the intact mitochondrion.

A variety of other enzymes are found predominantly in the supernatant, and other cell fractions, after homogenization. To take examples from carbohydrate metabolism, hexokinase and most of the enzymes involved in glycolysis are found mainly, if not exclusively, in the supernatant fraction after centrifuging ruptured cells. This might mean either a uniform distribution in the cytoplasm

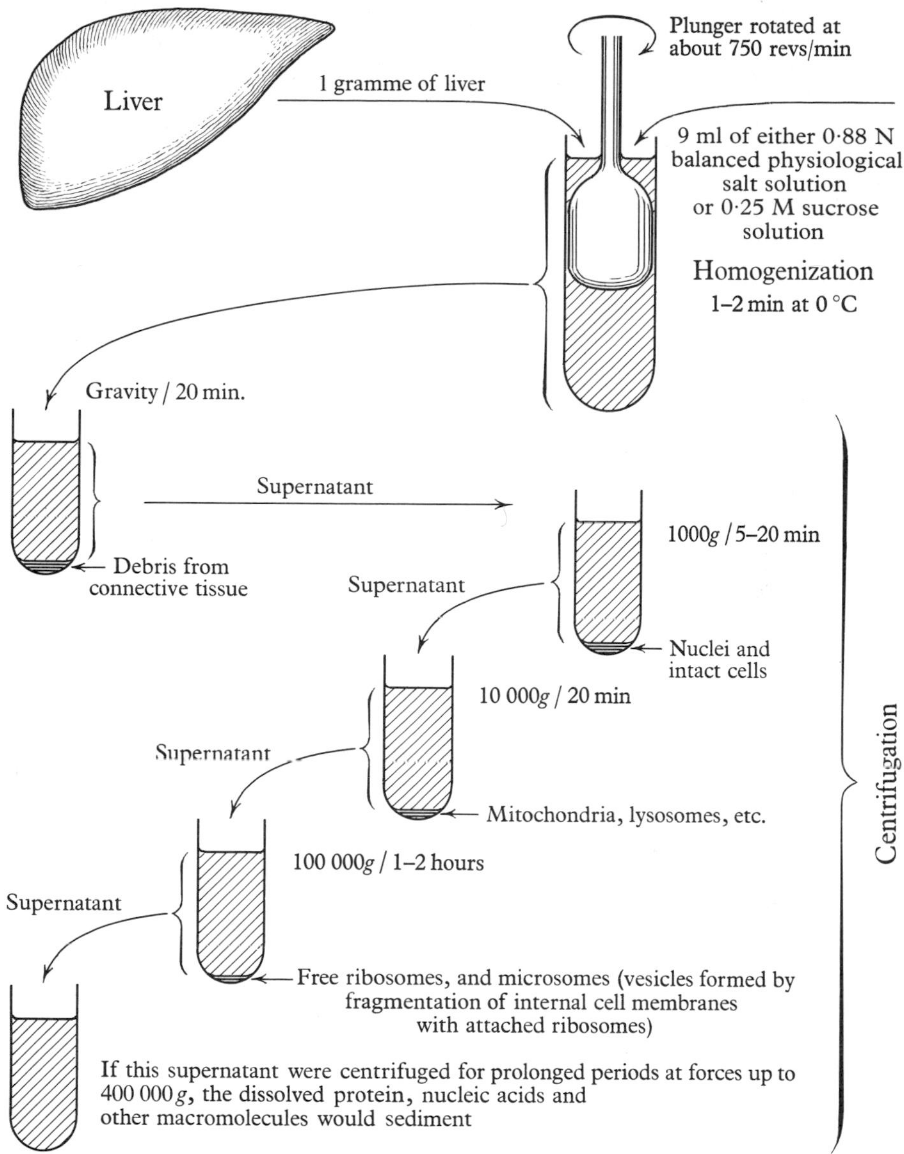

Fig. 1.13 A diagram showing successive stages in the fractionation of cell components. The tissue is first homogenized. While the plunger is rotating the tube is moved slowly up and down, so that the tissue, previously cut into small pieces, is driven through the narrow space between the plunger and the wall of the tube and disrupted by the shearing forces. The connective tissue fibres, and other debris, sediment on standing. The solution above this sediment, i.e. the supernatant, is then centrifuged at a series of increasing rotor speeds. At a centrifugal force of 100 × gravity, the nuclei, and any remaining intact cells, are spun to the bottom of the tube. At 10000 *g* the mitochondria and lysosomes are brought down, and at 100000 *g* the ribosomes and microsomes.

or a *loose* binding to some structure in the living cell. Homogenization studies such as these cannot be expected to resolve this question.

Lysosomes (Fig. 1.13) are membrane-bound sacs containing high concentrations of enzymes hydrolysing the breakdown of proteins and of other biochemically important compounds. These enzymes are released when the sac is broken; this can be done in a test tube using enzymes which hydrolyse lipids of the membrane. It also occurs when the cell dies; hence the self-digestion, or autolysis, of dead cells. In life, phagocytic cells take up particles in phagocytosis vesicles which fuse with cytoplasmic lysosomes to allow intercellular digestion of the particles. Bacteria may be killed by lysosomal enzymes in this way. In the scheme of fractionation outlined, the lysosomes have precipitated with the mitochondria. If the centrifugal force at this stage is slightly less, the lysosomes will come down instead with the microsomal fraction.

The microsomes and ribosomes that sediment at forces of about 100000 g (Fig. 1.13) have an association with protein synthesis discussed in other chapters.

A picture of the metabolic activity of the cell emerges in which some enzymic reactions progress exclusively or mainly in association with certain of the cell organelles. We are here at the meeting-point between biochemistry and the investigation of fine structure, where biochemical processes cannot be understood simply in terms of a series of discrete enzyme reactions, but only in relation to the structural organization of the enzymes of the biochemical sequence. The question of the detailed organization of mitochondrial enzymes is taken up again in Chapter 5.

In the later sections of this chapter we set out to discuss, in general, the isolation of cellular components, and large molecules, but because of their importance we have concentrated so far only on proteins. This emphasis continues through the next two chapters. Later, in discussing cell membranes, we shall consider lipids and carbohydrates and, in discussing chromosomes, have much to say about the isolation and properties of nucleic acids.

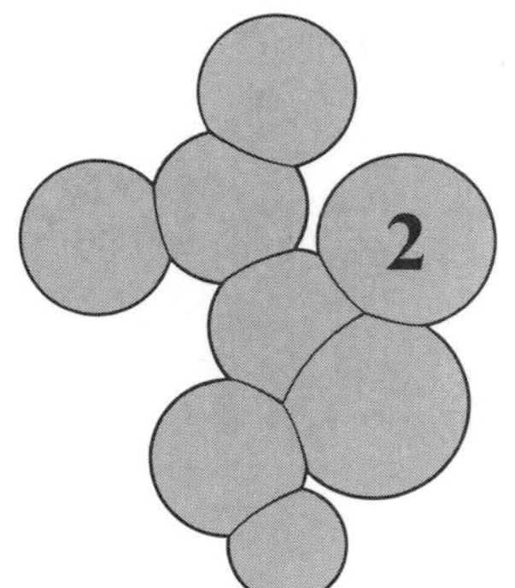

The structure of small protein molecules

Primary structure

We have noted in Chapter 1 that proteins are made up predominantly of amino acids linked in long chains. These are termed *polypeptide* chains. Figure 2.1 shows the chemical structure and a molecular model for the *backbone* of a polypeptide chain, the *side chains* being represented by the letters $R_1, R_2, \ldots, R_n$. Table 2.1 (pp. 34–5) shows the different types of side chain found in proteins. In this figure and table, acidic and basic groups are shown as they exist in aqueous solution at pH ~ 7, that is in the form $-\overset{+}{N}H_3$ and $-COO^-$.

Two amino acids joined together form a dipeptide, three a tripeptide, and so on. Long chains are termed polypeptides; short chains are often referred to simply as peptides. Polypeptide chains of about 50 amino acids (molecular weights ~ 6 000) or aggregates of several chains, with a total molecular weight greater than about 6 000, begin to show distinguishing properties, arising from their greater complexity of structure, which characterize them as proteins. Protein molecules may contain from 50 to 5 000, or more, amino acids. The lengths of the individual chains from which they are made appear to range from 20 to 600, or more, amino acids. The total structure of a protein molecule often includes, in addition to its polypeptide chains a component of different chemical constitution, known as a prosthetic group. The haem component of haemoblobin is an example of a group of this kind.

The relative proportions of the different amino acids in a variety of proteins are shown in Table 2.2 (pp. 38–9). A few generalizations can be made from this table which apply to many fibrous proteins, and most of the non-fibrous, or *globular*, proteins: (a) considering the marked variation in the properties of different proteins (discussed in Chapter 1) the variation in amino-acid composition is surprisingly small; (b) the proportion of hydrocarbon side chains (i.e. side chains containing only carbon and hydrogen atoms) is usually in the range 30 to 50 per cent; (c) the numbers of positively and negatively charged side chains are comparable, at pH ~ 7, so that the net charge on the protein molecule is very much less than the total number of charged groups. These generalizations do not apply to silk, collagen, elastin, keratin, salmine and other protamines, or to histones. The special properties of these atypical proteins depend

TABLE 2.1 The side chains (R-groups) of the twenty primary amino acids.
This table includes chemical formulae, drawings of molecular models, and also 'short-hand' drawings which will be used in later figures.

Glycine (Gly)	$-H$
Alanine (Ala)	$-CH_3$ 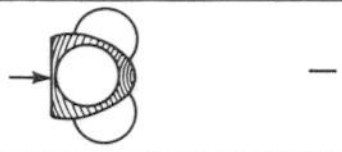
Valine (Val)	$-CH<^{CH_3}_{CH_3}$
Leucine (Leu)	$-CH_2-CH<^{CH_3}_{CH_3}$
Isoleucine (Ileu)	$-CH<^{CH_2-CH_3}_{CH_3}$
Phenylalanine (Phe) 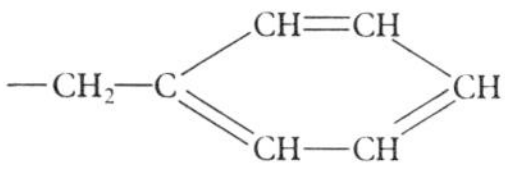 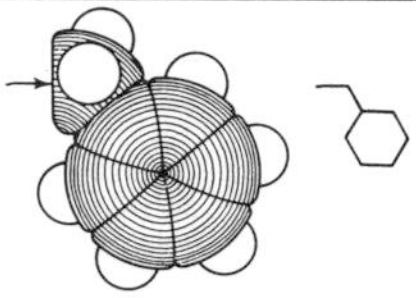	$-CH_2-C<^{CH=CH}_{CH-CH}>CH$
Proline (Pro) 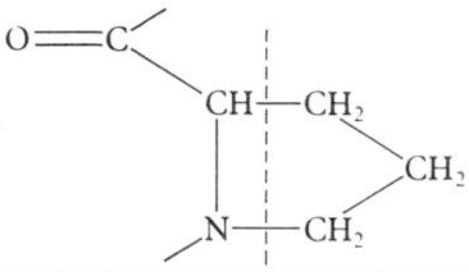 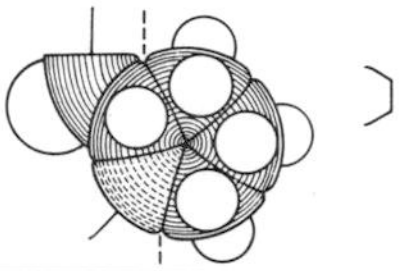	$O=C<$, $CH-CH_2$, CH_2, $N-CH_2$ (ring: $CH-CH_2-CH_2-CH_2-N$)
Tryptophan (Try) 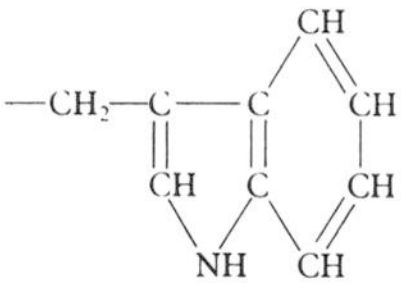 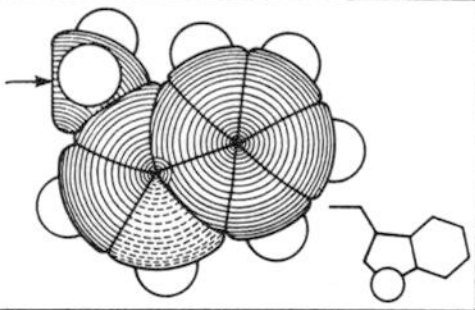	$-CH_2-C-C$, CH, C, CH, CH, NH, CH, CH
Serine (Ser)	$-CH_2-OH$ 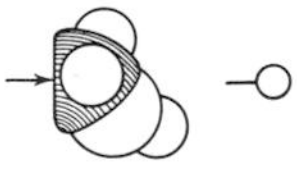
Threonine (Thr)	$-CH<^{OH}_{CH_3}$

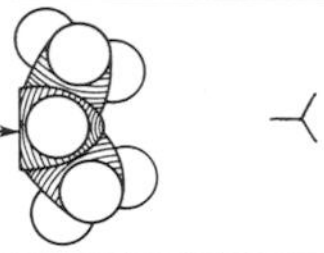

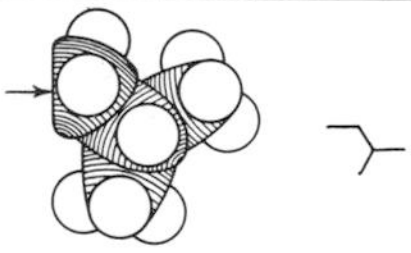

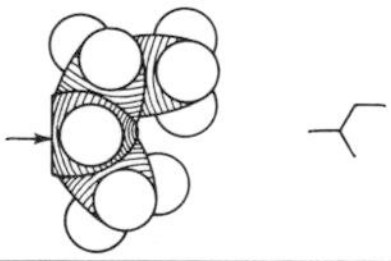

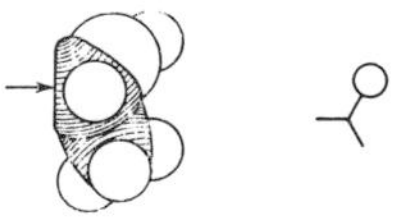

The arrows indicate holes in these models fitting the pegs on the model of Fig. 2.1. The side chain of proline forms a loop, with the end of the chain covalently linked to the backbone nitrogen. In this case therefore the formula and drawing of the whole residue are shown and not just the side chain.

Cysteine (CysH)	$-CH_2-SH$ 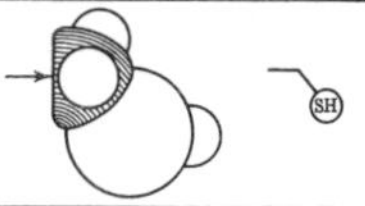
Methionine (Met)	$-CH_2-CH_2-S-CH_3$ 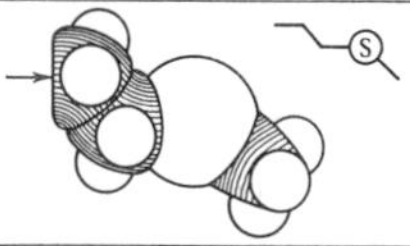
Aspartic acid (Asp)	$-CH_2-C(=O)-O^-$ 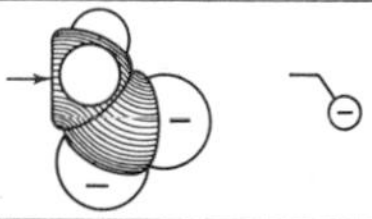
Glutamic acid (Glu)	$-CH_2-CH_2-C(=O)-O^-$ 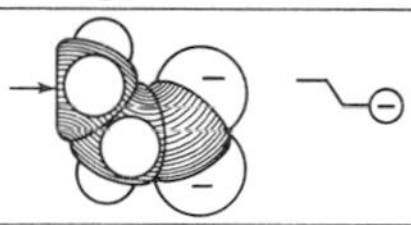
Asparagine (Asp NH_2)	$-CH_2-C(=O)-NH_2$ 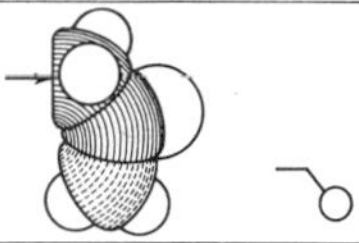
Glutamine (Glu NH_2)	$-CH_2-CH_2-C(=O)-NH_2$ 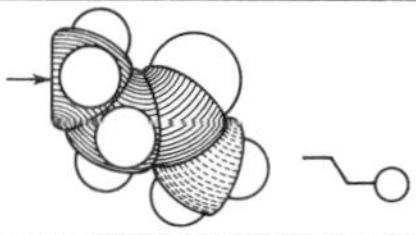
Tyrosine (Tyr)	$-CH_2-C \langle {}^{CH=CH}_{CH-CH} \rangle C-OH$ 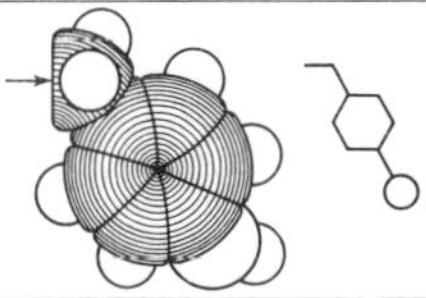
Histidine (His)	$-CH_2-C=CH$, NH, N, CH (imidazole ring)
Lysine (Lys)	$-CH_2-CH_2-CH_2-CH_2-\overset{+}{N}H_3$ 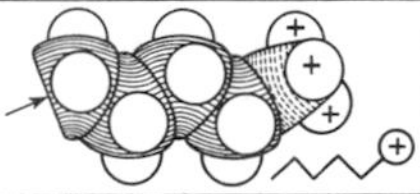
Arginine (Arg)	$-CH_2-CH_2-CH_2-NH-C(NH_2)=\overset{+}{N}H_2$ 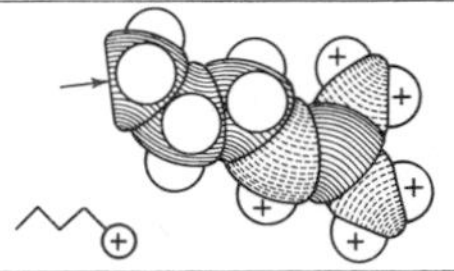

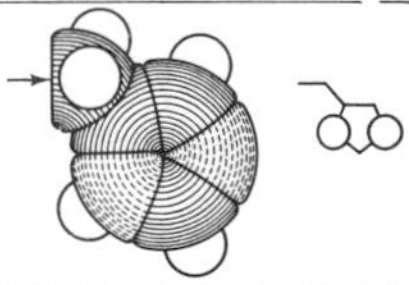

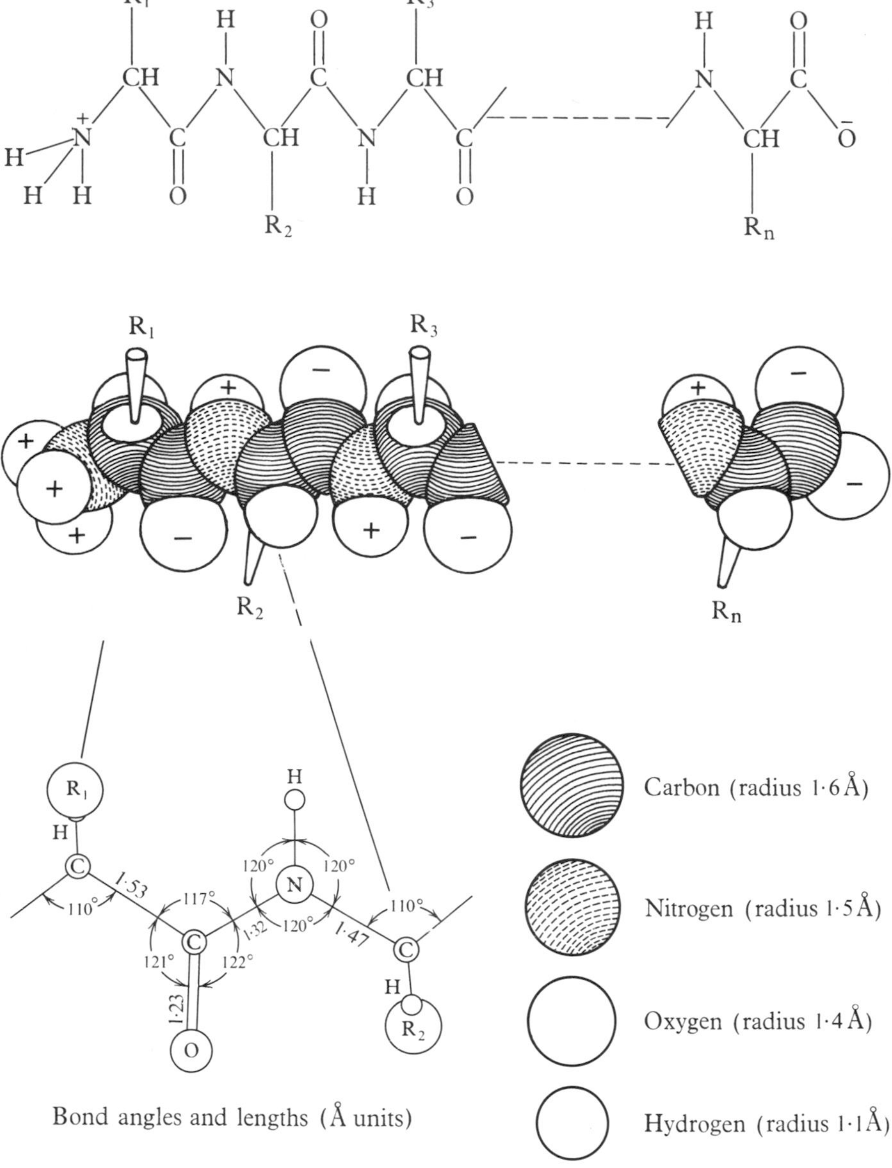

Fig. 2.1 Chemical formula and drawing of a molecular model of the 'backbone' of a polypeptide chain. The points of attachment for side chains (R_1, R_2, . . ., R_n) are illustrated by pegs. To form a molecular model of a complete polypeptide chain one or other of the side-chain models of Table 2.1 must be slipped onto each of the pegs of this backbone model. The shading convention adopted in this drawing will be uniform throughout the book. The + and − signs on the molecular model represent fractional charges of $\frac{1}{3}$ to $\frac{1}{2}$ atomic units (see Appendix 2). The lower drawing shows bond lengths and angles. (1 Angström (Å) $=\frac{1}{10}$ nm)

on the fact that they contain some side chains in very high proportion; their discussion is postponed to the relevant sections of Chapters 3 and 7.

We shall concentrate our attention in this chapter on the proteins of Table 2.2 which show no marked peculiarities of amino-acid composition (i.e. those not mentioned specifically in the previous paragraph). How can molecules of such similar overall composition form such a variety of enzymes of high specificity, and fulfil such a variety of roles in cells? To what extent does specificity reside in the prosthetic group of a protein rather than in its polypeptide chains?

To answer these questions we have to know more about a protein molecule than its overall amino-acid composition. It seems that for a given protein preparation, e.g. pig insulin, adequately characterized and purified, each molecule carries the same *sequence* of amino acids along its component polypeptide chains, and the detail of this sequence is the primary factor determining the structure of the protein molecule and its interactions.

An analogy may be drawn between sequences of amino acids in proteins and sequences of letters in a sentence. This analogy illustrates how from twenty different amino acids it is possible to form an almost endless variety of different sequences, and hence of different proteins, for we know that an almost endless variety of sentences can be formed in English, using twenty-six letters. The analogy illustrates further how proteins of very varied role can be formed, without much variation in the proportion of the different amino acids, for sentences of very varied meaning can be formed with letters in roughly the same proportions arranged in different sequences. In Fig. 2.2 the average frequency of occurrence of different amino acids in *E. coli* proteins is compared with the frequency with which different letters are used in written English. Some like alanine, leucine and serine seem to be either more readily available or, like vowels, specially useful, and are incorporated frequently. Others like tryptophan, or the letter z, find only occasional use.

Some of the amino-acid sequences which have so far been elucidated for naturally occurring peptides and proteins are illustrated in Figs. 2.3–2.5. These include the pituitary hormones oxytocin, vasopressin, and adrenocorticotrophic hormone (ACTH), the protein hormone insulin and the enzyme ribonuclease. The sequences for lysozyme, cytochrome c, myoglobin and haemoglobin are also known and are given in a later section of this chapter and Chapter 9 where these proteins are discussed further. Other proteins for which sequence studies are now complete include the structural protein of tobacco mosaic virus (TMV – see Chapter 8) chymotrypsin, ferridoxin and many others. The formulae of Figs. 2.4 and 2.5 for insulin and ribonuclease represent the *primary* structure of these proteins. The reason for introducing this term will become clear, as further structural aspects are described in later sections. So far no regularities in amino-acid sequences have been discerned in the globular proteins, such as, for example, q being always followed by u, in the analogy with written English. (Silk, collagen and protamines are, again, exceptional in this respect – see Chapters 3 and 7.) A brief account of the way in which the sequences of Figs. 2.3–2.5 are determined is given in Chapter 9.

TABLE 2.2 Amino-acid composition of proteins [1] (expressed as number of amino acid residues/molecule).

The three columns at the right of the table give the percentage of purely hydrocarbon side chains in the protein, the number of groups carrying a positive charge at pH 7, and the number of groups carrying a negative charge at this pH. The number of positive groups is calculated from number of arginine residues + number of lysine residues + half the number of histidine residues (since the histidine group has a pK ~ 7, and approximately half these residues will be in the ionized form at pH ~ 7). The number of negative groups is calculated from number of aspartic acid and asparagine residues + number of glutamic acid and glutamine residues minus the number of amide groups. The cystine content is a measure of the number of S–S bonds in the molecule.

	Molecular weight	Alanine	Glycine	Valine	Leucine	Isoleucine	Proline	Phenylalanine	Tyrosine	Tryptophan	Serine	Threonine	Cystine
Human insulin[3]	6000	1	4	4	6	2	1	3	4	0	3	3	3
Ribonuclease	15000	12	3	9	2	3	4	3	6	0	15	10	4
Lysozyme	15000	10	11	6	8	6	2	3	3	8	9	7	5
β-Lactoglobulin	37000	30	7	18	44	17	17	9	8	3	14	16	3
Egg albumin	45000	35	19	28	32	25	14	21	9	3	36	16	1
Horse myoglobin	17000	15	13	6	22	0	5	5	2	2	6	7	0
Horse haemoglobin	68000	54	48	50	75	0	22	30	11	5	35	24	0
Human serum albumin	69000	—	15	45	58	9	31	33	18	1	22	27	16
Human fibrinogen[6]	160000	66	120	56	86	59	78	45	48	26	107	83	14
Human γ-globulin[7]	160000	—	90	133	114	34	112	45	61	22	175	114	16
Salmine	8000	1	3	2	0	1	4	0	0	0	7	0	0
Calf thymus histone	15000	12	9	4	6	24	5	4	3	0	7	6	0
Pepsin	34000	—	29	21	27	28	15	13	16	4	40	28	2
Chymotrypsinogen	24000	16	20	21	16	9	8	5	3	7	25	22	3
TMV protein[4]	17000	14	6	14	12	8	8	8	4	3	16	16	0
TYMV protein[8]	21300	16	8	15	18	15	19	5	3	2	18	26	—
Human myelin A1 protein[9]	18600	14	26	4	8	4	12	9	4	1	18	8	0
Staphylococcus aureus membrane phosphokinase[10]	17000	14	9	11	23	23	4	10	4	4	8	8	0
Rabbit myosin[5]	— [2]	78	39	42	79	42	22	27	18	4	41	41	4
Rabbit actin[5]	— [2]	71	67	42	63	57	44	29	32	10	56	59	5
Rabbit tropomyosin[5]	— [2]	110	12	38	95	29	0	3	15	0	40	28	3
Silk fibroin	— [2]	334	581	31	7	8	6	20	71	0	154	13	—
Collagen	— [2]	107	363	29	28	15	131	15	5	0	32	19	0
Elastin	— [2]	58	376	118	56	26	136	29	8	—	9	10	2
Wool keratin	— [2]	46	87	40	86	—	83	22	26	9	95	54	49

1. From TRISTRAN, G. R. (1953) in *The Proteins*, ed. H. Neurath and K. Bailey, Vol. 1A, New York, Academic Press, except where specific reference is given.
2. Where there is still uncertainty in the molecular weight of a protein the figures given are amino-acid residues per molecular weight of 100 000, i.e. moles/10^5g protein.
3. SANGER, F. (1960) *Brit. med. Bull.* **16**, 183.
4. TS'O, P. O. P., BONNER, J. and DINTZIS, H. (1958) *Arch. Biochem. Biophys.* **76**, 225.
5. KOMINZ, D. R., HOUGH, A., SYMONDS, P. and LAKI, K. (1954) *Arch. Biochem. Biophys.* **50**, 148.
6. LAKI, K., GLADNER, J. A. and FOLK, J. E. (1960) *Nature, Lond.* **187**, 758.
7. PHELPS, R. A. and PUTNAM, F. W. (1960) in *The Plasma Proteins*, ed. F. W. Putnam, Vol. 1, New York, Academic Press.
8. HARRIS, J. I. and HINDLEY, J. (1961) *J. mol. Biol.* **3**, 117.
9. EYLAR, E. H. (1970) *Proc. Nat. Acad. Sci. Wash.* **67**, 1425.
10. SANDERSON, H. Jr. and STROMINGER, J. L. (1971) *Proc. nat. Acad. Sci. Wash.* **68**, 2441.

Cysteine	Methionine	Arginine	Histidine	Lysine	Aspartic acid + asparagine	Glutamic acid + glutamine	Hydroxyproline	Hydroxylysine	Amide (asparagine + glutamine)	% hydrocarbon	+ at pH 7 (arg. + lys. + $\frac{1}{2}$his.)	− at pH 7 (asp. + glu.)	
0	0	1	2	1	3	7	0	0	6	41	3	4	Human insulin
0	4	4	4	10	15	12	0	0	17	29	16	10	Ribonuclease
—	2	11	1	6	20	4	0	0	18	37	18	6	Lysozyme
3	8	6	4	29	32	48	0	0	28	44	37	52	β-Lactoglobulin
5	16	15	7	20	32	52	0	0	33	45	39	51	Egg albumin
0	2	2	9	18	10	19	0	0	8	44	25	21	Horse myoglobin
4	4	14	36	38	51	38	0	0	36	47	70	53	Horse haemoglobin
4	6	25	16	58	46	80	0	0	44	—	91	82	Human serum albumin
5	27	72	27	101	157	158	0	0	170	58	187	146	Human fibrinogen
10	11	45	26	88	106	128	0	0	127	—	146	107	Human γ-globulin
0	0	40	0	0	0	0	0	0	0	19	40	0	Salmine
0	0	15	2	11	7	5	0	0	10	49	27	2	Calf thymus histone
2	4	2	2	2	41	28	0	0	32	—	5	37	Pepsin
3	2	4	2	13	20	12	0	0	27	44	18	5	Chymotrypsinogen
1	0	11	0	2	18	16	0	0	—	47	13	—	TMV protein
4	4	3	3	8	12	15	0	0	—	48	12	27	TYMV protein
0	2	19	10	12	11	10	0	0	9	33	36	12	Human myelin A1 protein
0	4	5	3	8	8	8	0	0	—	58	14	16	*Staphylococcus aureus* membrane phosphokinase
—	22	41	15	85	85	155	0	0	86	37	134	154	Rabbit myosin
—	30	38	19	52	82	101	0	0	66	41	99	117	Rabbit actin
—	16	42	5	110	89	211	0	0	64	32	155	236	Rabbit tropomyosin
—	—	6	2	5	21	15	0	0	—	77	12	—	Silk fibroin
0	5	49	5	31	47	77	107	7	47	72	83	77	Collagen
—	0	6	0	3	4	22	0	0	—	92	9	—	Elastin
—	5	60	7	19	54	96	0	0	83	—	83	67	Wool keratin

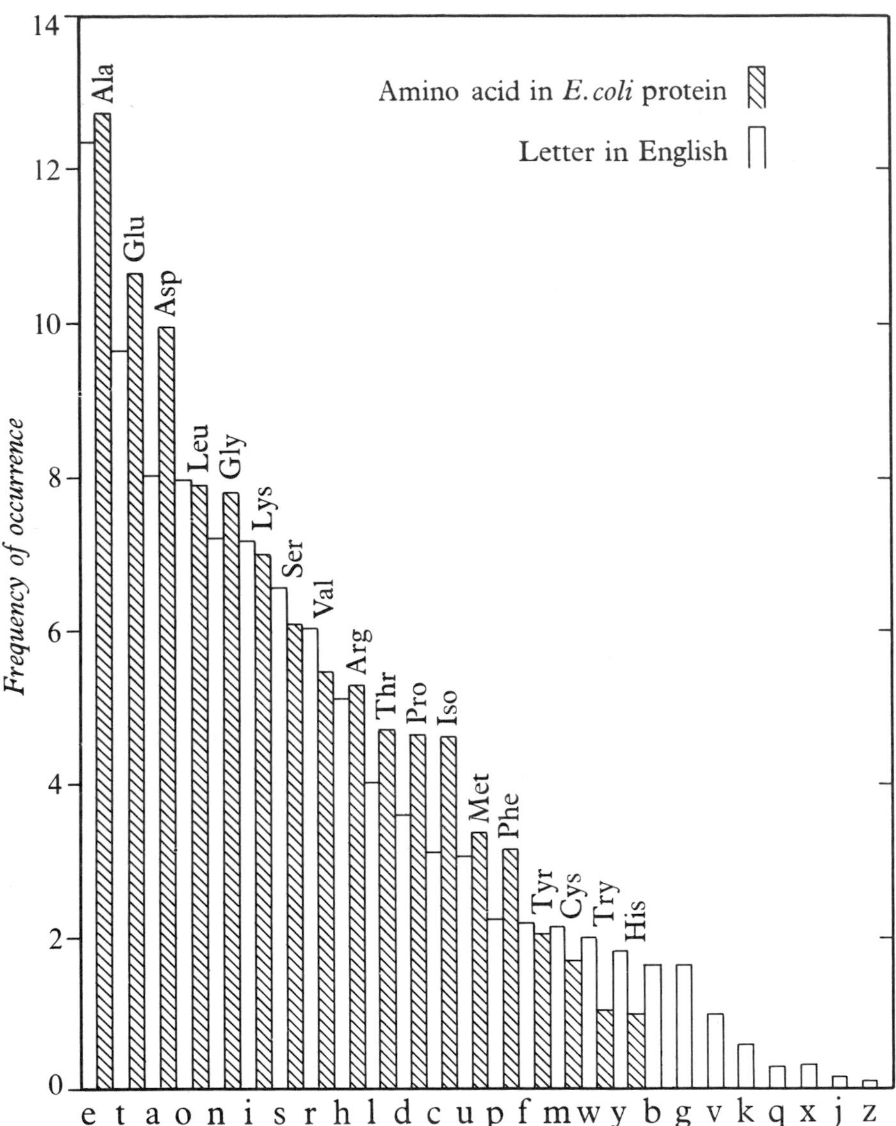

Fig. 2.2 A comparison between the frequency with which different letters are used in written English, and the frequency of occurrence of different amino acids in *E. coli* proteins. (*From* Morowitz, 1958.)

The amino acid cysteine (see Table 2.1) plays a rather special role. Cysteine side chains can be linked by oxidation of the type:

$$\overset{\text{cysteine}}{-CH_2-SH} + \overset{\text{cysteine}}{HS-CH_2-} + O \rightleftharpoons \overset{\text{cystine}}{-CH_2-S-S-CH_2-} + H_2O$$

Such linkages form loops in the peptide chains of oxytocin, vasopressin, insulin and ribonuclease, and in insulin join the two component chains together. The number of S–S links in different proteins is included, in effect, in Table 2.2 under the column cystine (cystine being the amino acid formed by S–S linkage of two cysteines). The two chains of insulin are actually made first as a single chain of 84 residues, a 33-residue peptide being later cut out of the middle by a proteolytic enzyme.

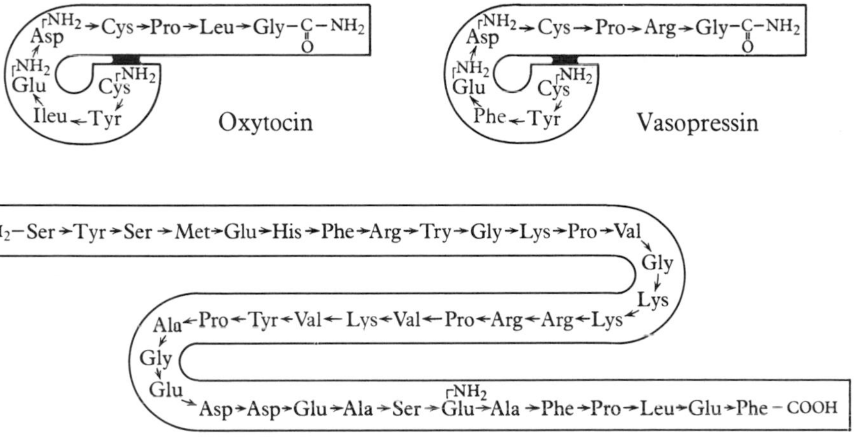

Asp(NH₂)→Cys→Pro→Leu→Gly–C(=O)–NH₂; Cys(NH₂)→Tyr→Ileu→Glu(NH₂)→Asp(NH₂)

Oxytocin

Asp(NH₂)→Cys→Pro→Arg→Gly–C(=O)–NH₂; Cys(NH₂)→Tyr→Phe→Glu(NH₂)→Asp(NH₂)

Vasopressin

NH_2–Ser→Tyr→Ser→Met→Glu→His→Phe→Arg→Try→Gly→Lys→Pro→Val→Gly→Lys→Lys→Arg→Arg→Pro→Val→Lys→Val→Tyr→Pro→Ala→Gly→Glu→Asp→Asp→Glu→Ala→Ser→Glu(NH_2)→Ala→Phe→Pro→Leu→Glu→Phe–COOH

Adrenocorticotrophic hormone (ACTH)
(Sheep)

Fig. 2.3 The primary structure of the pituitary hormones: oxytocin, vasopressin and adrenocorticotrophic hormone (ACTH). (S–S bonds in Figs. 2.3–2.5 are represented by black 'cement' between chains, or between folds of a chain.) In oxytocin and vasopressin the chain is terminated by an amide rather than a carboxyl group.

Now that we have discussed some specific sequences for peptides and proteins in precise terms, we must consider more critically the rather confident assertion with which we introduced this topic. We must question whether it is true that the molecules of a sufficiently purified protein preparation all carry exactly the same amino-acid sequence. Prior to about 1950, before the sequences of Figs. 2.4–2.5 had been determined, the answer to this question was still very much a matter of speculation. As techniques for the purification of proteins were refined, during the first half of the century, preparations previously thought to be homogeneous were fractionated into different components, and preparations apparently homogeneous by physico-chemical criteria were shown to be made up of many fractions of different biological specificity (e.g. serum proteins discussed in Chapter 1). It seemed quite possible that a certain heterogeneity might extend right down to the individual protein molecules, the molecules of a purified preparation appearing identical, so far as their physical, chemical, or even biological properties were concerned, but still with some individual variation in amino-acid sequence or composition.

This viewpoint is no longer tenable. The procedure for sequence determination, developed by Sanger at Cambridge (see Chapter 9), involves breaking the chains into fragments, determining the sequences in these fragments, and reconstructing the original sequence. If there were any significant variation in the

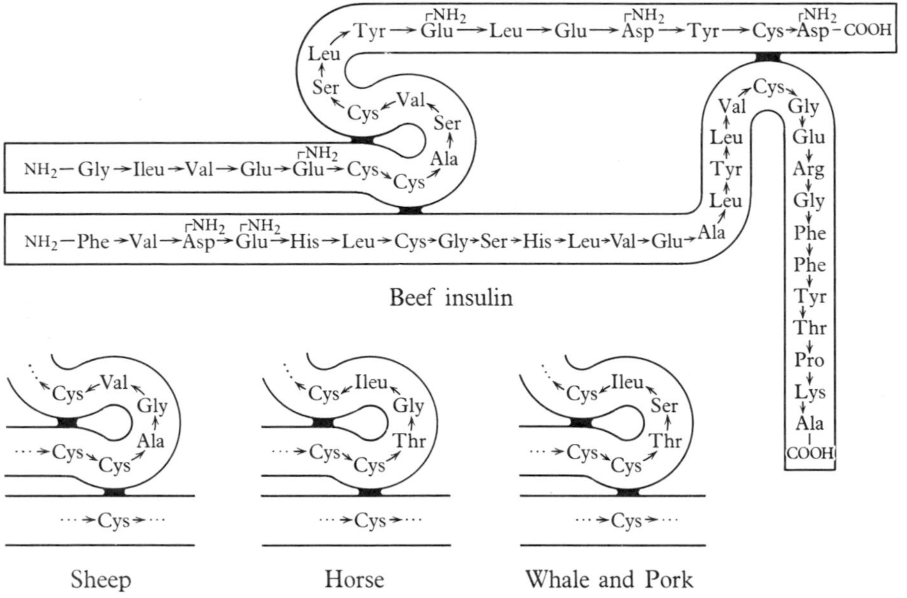

Fig. 2.4 The primary structure of insulin.

sequences of the original molecules of the preparation, it would not be possible to reconstruct a single sequence from the fragments.

Sanger's work on insulin established a new and satisfactory criterion for the purity of a protein preparation, namely that all the molecules in the preparation should have the same primary structure. The fact that it has proved possible to extend his methods to the study of ribonuclease, TMV protein, myoglobin, haemoglobin, and other protein sequences shows that insulin is not a unique protein in this respect. Proteins are always made in cells with specific sequences, rather than by any form of random, or non-specific, joining together of amino acids. That all molecules of a purified preparation carry the same sequence, is nothing more than a definition of what is nowadays considered a pure protein preparation. (For small molecules crystallization is an adequate criterion of purity, i.e. only identical molecules can be incorporated into the crystal structure, but for proteins this criterion fails.)

The insulins of different species are indistinguishable in their hormonal activity, although small differences in amino-acid composition are found, and Sanger showed (Fig. 2.4) that their sequences differ in one or two amino acids. The differences mainly occur in a localized part of the primary structure, and we shall see in Chapter 9 that there are similar small localized differences in the sequences of certain normal and abnormal haemoglobin molecules. These differences between similar proteins raise fascinating questions in genetics and chemical evolution, to which we shall return in later chapters. In Chapter 8, in particular, we discuss protein synthesis, and the way in which amino acids are assembled in specific sequences under genetic control.

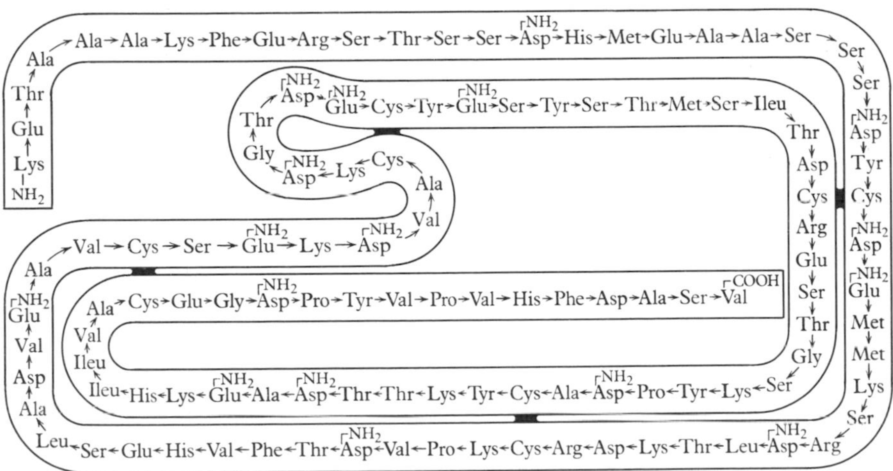

Fig. 2.5 The primary structure of ribonuclease. It should be noted that in the drawings of Figs. 2.3–2.5 contact points between chains, or between different parts of a chain, are indicated correctly, but the chains are otherwise drawn in a quite arbitrary way to fit on to the page. (*From* Spackman, Stein and Moore, 1960.)

We consider now how the details of amino-acid sequence, and the nature of the amino-acid side-chain groups, determine the form of protein molecules and the specificity of their interactions. We shall see that the polypeptide chains coil up, and fold back on themselves, to form a complex three-dimensional molecular structure, and that the determination of sequence, or primary structure, is only the beginning of a description of the protein molecule.

Secondary structure

The backbone structure in Fig. 2.1 is not a rigid one. Although *double* bonds between atoms link them together in a rigid way, *single* bonds allow rotation about the axis of the bond. (Both types allow only small variation in bond lengths and angles.) The single bonds on either side of the peptide $\begin{array}{c} \mathrm{R} \\ | \\ \mathrm{CH} \\ \wedge \end{array}$ group lie at an angle of about 110° (Fig. 2.1) and rotation about bonds set at an angle leads to a fully flexible chain. An attempt is made to illustrate this point in Fig. 2.6, which shows bent rods joined by tubes in which the rods must be supposed free to rotate.

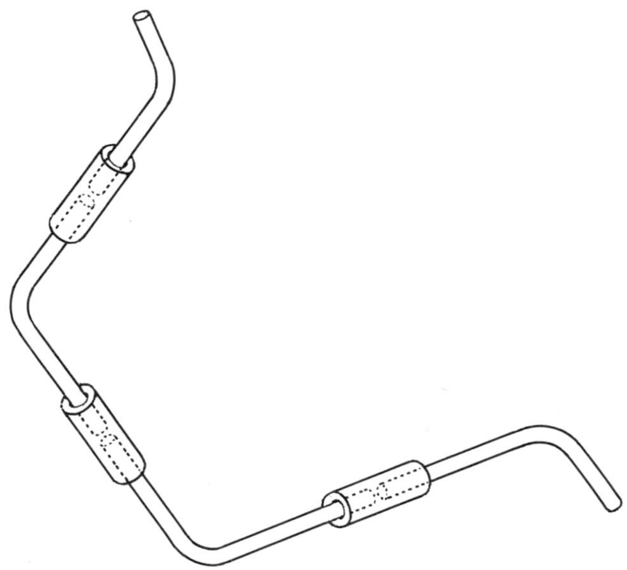

Fig. 2.6 A drawing showing how rotation about bonds set at an angle gives rise to a flexible molecular chain. The bent rods in this drawing must be supposed free to rotate in the short tubes joining them.

At room temperature, or at a body temperature of 37°C, molecules are, of course, in constant rapid movement, and a polypeptide chain in solution might be expected to be in a constantly changing random conformation, like a thread in turbulent waters. For peptides of low molecular weight, and, under certain conditions, for some synthetic polypeptides of relatively high molecular weight, this is in fact what is found. For these molecules, both diffusion rates (which are a measure of the velocity with which the molecules are carried about in the solution by random thermal movement) and sedimentation rates (which are a measure of the velocity with which the molecules settle in the 'high *g*' field of a centrifuge) approximate to those which would be expected theoretically if the molecules were in a randomly folded, constantly changing, form. This is not, however, what is found for the globular proteins. Diffusion rates, sedimentation rates, and a variety of other evidence suggests that in these proteins the polypeptide chains are folded up in some way to form a relatively compact molecule.

Our knowledge of how the chains are folded in protein molecules, and in fact much of our detailed knowledge of the structure of smaller molecules, comes from the technique of X-ray diffraction. Crystals are made up of regularly repeating molecular units, and their component electrons diffract, or scatter, X-rays in directions which depend on the intermolecular spacings in the crystal. The intensities of scattering in various directions depend on the arrangement of electron density in the crystal, and the X-ray diffraction pattern can be used to study the three-dimensional structure of the molecule. (The X-ray diffraction technique is discussed more fully in Appendix 1.)

Protein molecules contain thousands of atoms, and protein crystals give diffraction patterns which contain thousands of diffracted beams (Fig. 2.18). The only way in which the three-dimensional structure of a complete protein molecule has been determined is by a full analysis of such a diffraction pattern, by methods outlined in Appendix 1. These methods of analysis, however, are rather sophisticated and were not developed successfully until the late 1950s. In earlier work on proteins an alternative, and more direct, procedure was used, and that was to build molecular models of the polypeptide chains to try to see how they might fold up, and then to check the suggested structure against the main features of the diffraction patterns.

In many cases, as first noted by Astbury, X-ray patterns show directly that the chains in protein molecules must be folded in a regular way. This applies particularly to the fibrous proteins, such as wool, silk and collagen, which generally cannot be crystallized in a regular three-dimensional lattice, but which can be oriented so that the fibres all run in the same direction. The regular folding of the polypeptide chain then confers order on a smaller scale than that of a complete protein molecule and the regularly repeating motif is only one, or a few, peptide units, so that the model-building approach is perfectly feasible. The X-ray diffraction observed from fibrous specimens consists of lines and arcs which indicate the important spacings between groups of atoms in the regularly folded polypeptide. It is found that many fibrous proteins give prominent diffraction corresponding to inter-atomic spacings of 0·50–0·55 nm, which could not be produced by extended polypeptide chains, stretched out as in Fig. 2.1. Various folds were proposed to account for this 0·50–0·55 nm spacing, based on ways in which molecular models could be folded, but none of those proposed before 1951 were found compatible with all the X-ray diffraction, spectroscopic and other evidence.

As early as 1937, Pauling and Corey had realized that a successful solution to the problem might depend on the accuracy with which the models really reproduced the behaviour of polypeptide chains. Inadequate models might fold in all sorts of ways, quite different from those adopted by polypeptide chains in proteins. A really good model, on the other hand, might be found to fold only in the exact way, or ways, in which a polypeptide could fold, and hence immediately give the true solution. They therefore undertook a detailed study of the

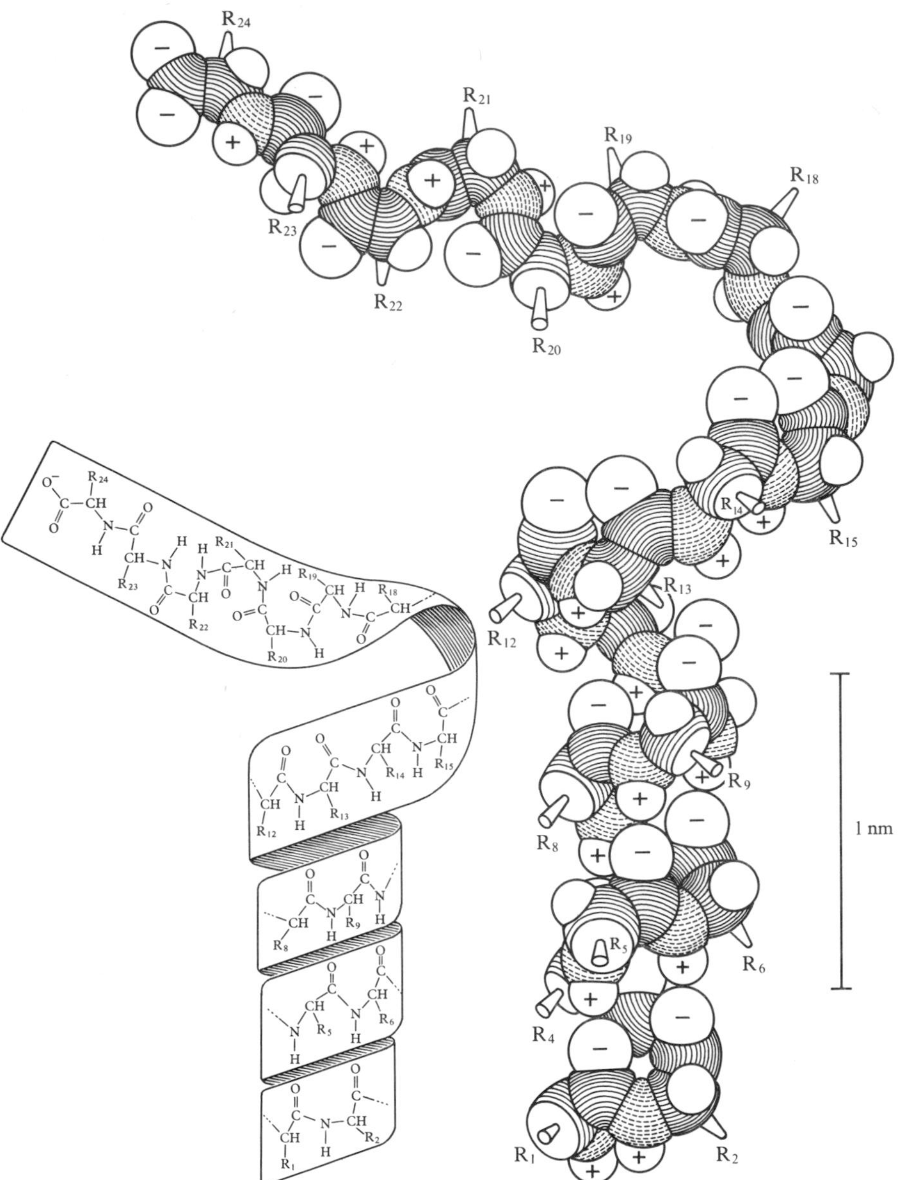

Fig. 2.7 Drawing of a molecular model of the backbone of a polypeptide chain coiling into a right-handed α-helix. As indicated by the ribbon, drawn alongside the molecular model, the lower part of the chain forms two turns of an α-helix. The upper part of the chain is folded in a random way, with the chain fully extended (as in Fig. 2.1) at the extreme upper end.

X-ray diffraction patterns of crystals of amino acids and small peptides, and from these data built accurate models of the type shown in the drawing of Fig. 2.1. Bond lengths, angles, and the radius of each atom in the model were scaled to the bond lengths, angles and volume occupied by similar atoms in the amino-acid and peptide crystals. Their models allowed free rotation about single bonds, and a small amount of flexibility in each bond, as determined from the slight molecular distortions found in the crystals.

In 1951, Pauling and Corey published a series of papers which were the outcome of these far-sighted studies.* They suggested that the most probable way that a polypeptide chain might fold up, or coil, would be into a helix, or long spiral, as indicated in the lower part of Fig. 2.7. This figure shows a model of the backbone of a randomly folded polypeptide chain coiling up into a helix at its lower end. Pauling and Corey discuss in their papers a number of helices and folds which their models suggest as possible in proteins. The structure shown in the lower part of Fig. 2.7 they termed the α-helix. In the α-helix there are approximately $3\frac{2}{3}$ amino acids to each turn, or 11 amino acids in 3 turns of the helix. Figure 2.8 shows a view down the axis of one turn of the α-helix, with side chains now attached to the backbone.

In Fig. 2.7 and in Fig. 2.1, the N—H hydrogen atoms along the polypeptide backbone have been shown as regions of positive charge, and the backbone oxygens as regions of negative charge. (A discussion of how this distribution of charge arises is given in Appendix 2.) In the α-helix (lower part of Fig. 2.7) each positively charged hydrogen along the backbone is attracted to a negatively charged oxygen in the next turn of the helix. N—H and O= groups aligned in this way are said to be forming hydrogen bonds (discussed more fully in Appendix 2).

As indicated in Fig. 2.7 hydrogen bonds are primarily electrostatic in nature. Pauling and Corey had found that these played an important part in determining detail of molecular arrangement in the crystals of amino acids and small peptides which they had studied. They were careful to ensure that the dimensions and angles of these bonds in their models were, like those of the covalent bonds, in close correspondence with the dimensions and angles of comparable bonds in the amino-acid and peptide crystals. Hydrogen bonds can be formed in various ways between O—H and N—H groups (see Table in Appendix 2) and they play an important role in determining the structure of nucleic acids as well as proteins (see Chapter 7).

It may be predicted that in the X-ray diffraction pattern produced by polypeptide chains in α-helical form there should be evidence not only for a molecular repeat pattern in the range 0·50–0·55 nm, corresponding to successive turns of the helix, but also evidence for a repeat pattern at 0·15 nm, which is the distance between successive amino acids, measured along the axis of the helix. After Pauling and Corey had published the results of their model-building

* References to papers, reviews, and books for further reading are listed chapter by chapter, and section by section, in the Bibliography at the end of the book. The papers listed in the Sources of Figures section, which follows the Bibliography, also provide further reading.

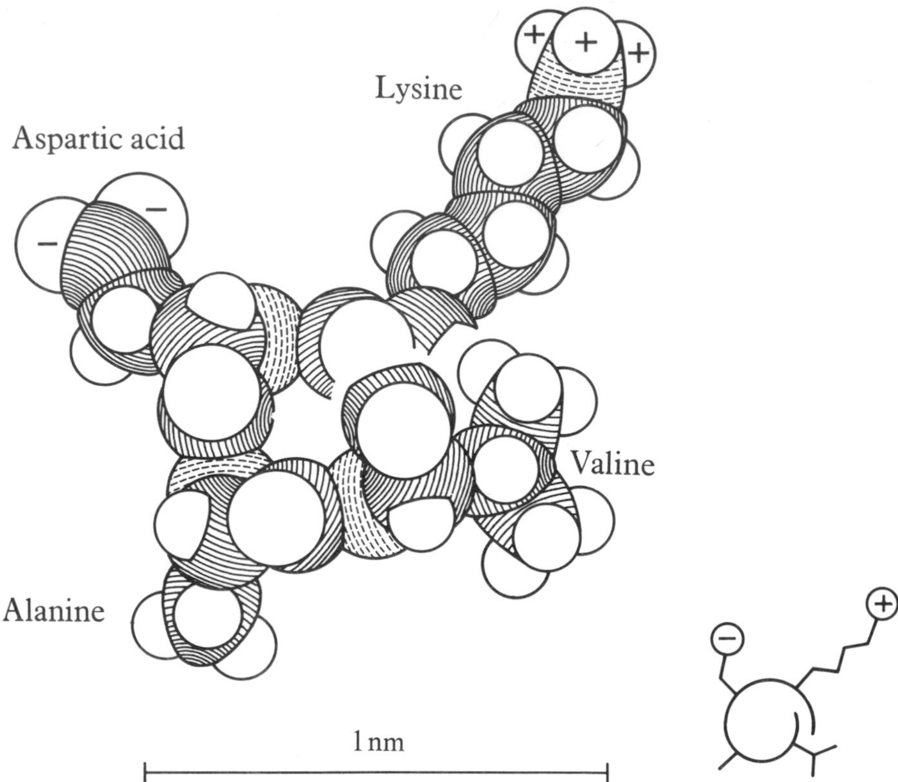

Fig. 2.8 A view along the axis showing one turn of the α-helix. In this model arbitrarily chosen side chains have now been attached to the backbone. In the small diagrammatic picture the side chains are represented by the 'shorthand' drawings of Table 2.1. This small diagrammatic representation of the α-helix will be used in later figures.

experiments, a search was made for evidence of a 0·15 nm repeat distance in X-ray diffraction patterns; reflections corresponding to this spacing were found, not only in new patterns, but also in X-ray diffraction results published earlier. These reflections had been ignored previously because their significance had not been apparent. The α-helix may be tested in other ways also, against X-ray diffraction patterns, and there is X-ray evidence for this coiling in at least a part of the molecular structure of myosin, haemoglobin, myoglobin, hair and quill keratin, and also of synthetic polypeptides in appropriate solvents.

It may be seen from Table 2.1 that all amino acids, except glycine, contain an asymmetric carbon atom and hence can exist in two mirror-image forms (L- and R-). In proteins only L-amino acids are found, and the chain shown in Fig. 2.1 is a chain of L-amino acids. The polypeptide chain can coil up in either a right-handed or a left-handed helix, but it seems that for a chain of L-amino acids the right-handed α-helix forms more readily than the left-handed. The reason for this is to some extent apparent from the molecular models, for in a left-handed helix of L-amino acids the first carbon of the side chains comes rather too close for comfort to the backbone oxygen atoms, and the X-ray diffraction study of myoglobin has now been refined to a point where it can be seen that the α-helical regions of the molecule are coiled up in a right-handed way. The right- and left-handed helices can also be distinguished by their effect on optical rotation. Optical rotation studies, for a variety of proteins, suggest that either only right-handed helices are present, or at least that the right-handed form is much more predominant in proteins than the left.

The α-helical structure is not the only type of coiling or folding which can be adopted by polypeptide chains. For example, the fibrous protein silk is of unusual amino-acid composition (Table 2.2). Almost half its amino acids are glycine, and the remainder mainly the other small amino acids, alanine and serine. Studies on peptides derived from silk by partial digestion show that the predominant sequence pattern is one in which glycines alternate with alanine or serine. This sequence pattern causes the chains to adopt a structure different from the α-helix. They pack together to form pleated sheets in which the chains are almost fully extended, with interchain hydrogen bonding between backbone groups (Fig. 2.9).

We shall see in a later section that a further type of helical folding is also possible, the 3_{10}-helix, which is a slightly tighter coil than the α-helix, and in Chapter 3 we shall consider the chain folding in collagen which is different

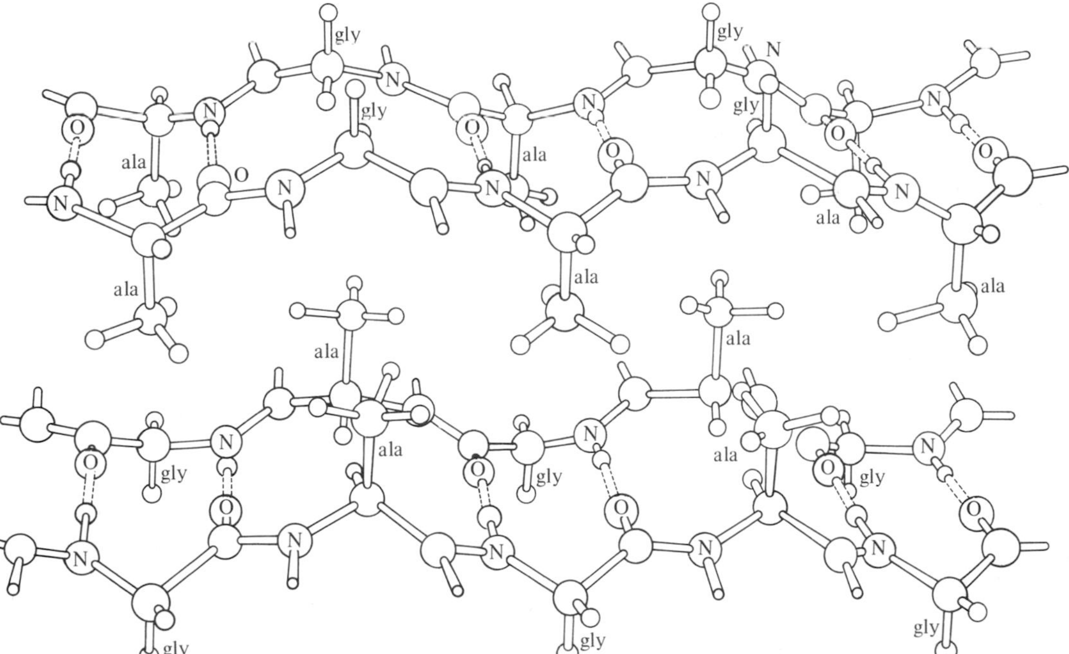

Fig. 2.9 The pleated-sheet structure of silk. (*From* Dickerson and Geis, 1969.)

again from the α- and 3_{10}-helices and the pleated sheet structure of silk. The type of coiling or folding adopted by the polypeptide chains of a protein is referred to as the *secondary* structure of the protein.

The extent of secondary folding

In the various types of secondary structure so far described the polypeptide backbone N—H groups are hydrogen-bonded to backbone C=O groups. In the pleated sheet structure of silk and in collagen these bonds form between different chains. In the α-helix each N—H group is hydrogen-bonded to a C=O group four amino acids back along the same chain and in the tighter 3_{10}-helix to an N—H group three amino acids back. Pauling and Corey made progress in their model-building approach by ignoring other considerations and supposing that inter-backbone bonding was the only way the hydrogen-bonding tendency of these groups could be satisfied. But in fact, if the secondary structure is unfolded or uncoiled in an aqueous environment these hydrogen bonds are broken, but the backbone N—H and C=O groups can still form hydrogen bonds to surrounding water molecules. (The α-helix has only marginal stability, see Appendix 4.) In the globular proteins we shall see further that side-chain polar groups are often positioned in such a way that they can form hydrogen bonds to backbone N—H and C=O groups. These factors allow considerably more flexibility of structure. The structure actually adopted by a given polypeptide chain depends very much on the side-chain groups which we have so far ignored in our discussion of secondary structure.

The effect of side-chain interactions on the form adopted by a polypeptide chain in aqueous solution may be illustrated from studies of the optical rotation of solutions of the synthetic polypeptide polylysine (a polypeptide made up of lysine only) and polyglutamic acid (a polypeptide made up of glutamic acid only). These polypeptides can exist in aqueous solution in either of the forms shown in Fig. 2.7: the randomly-folded, constantly changing, flexible form, or the α-helix. (Optical rotation refers to the rotation of the plane of plane–polarized light passing through the solution. A randomly folded peptide chain produces some optical rotation effect, but this is modified if the chain coils into a helix.) In acid solution when each lysine carries a positive charge (see Table 2.1) polylysine forms a flexible chain, the repulsion between the side chains preventing helix formation. In alkaline solution, when the side chains carry no charge, it coils up into the α-helix. Polyglutamic acid, on the other hand coils into a helix in acid solution (for in this case acid conditions leave the side chains uncharged) and unfolds to the flexible thread in alkaline solutions when carboxyl groups are charged. These effects are illustrated in Fig. 2.10, the presence of the α-helix being deduced from the change in the optical rotation of the solution. (A reader unfamiliar with the physico-chemical ideas underlying this discussion will find in Appendix 2 a definition of pH and a brief description of the way in which protein side-chain groups become charged, or lose their charge, in acid or alkaline conditions.)

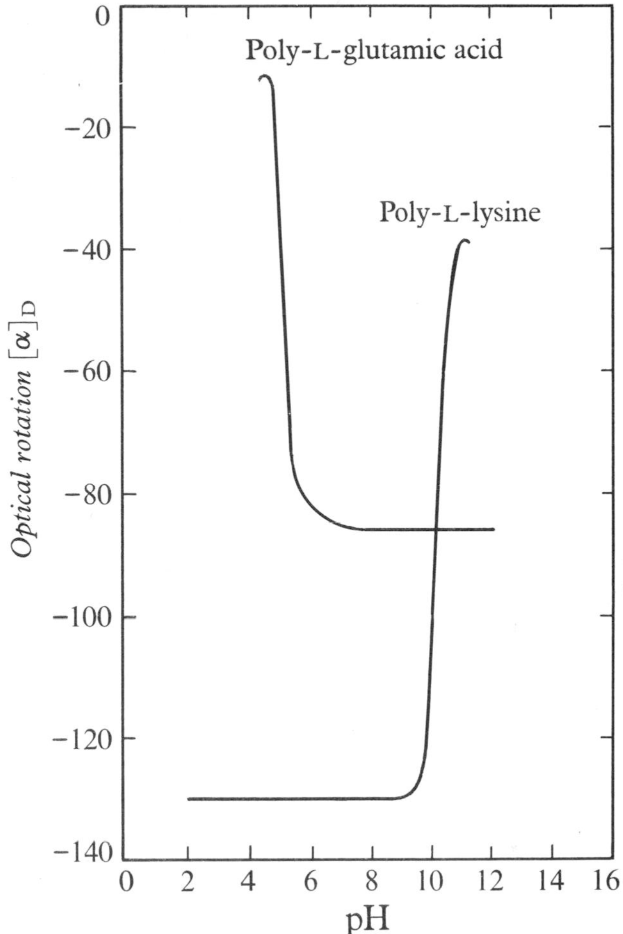

Fig. 2.10 The change in optical rotation with pH, for solutions of polylysine and poly-glutamic acid. A fall in optical rotation to a more negative value of the rotation coefficient $[\alpha]_D$ corresponds to unfolding of the α-helix. (*From* Doty, 1959.)

The change in optical rotation, when a synthetic polypeptide goes from a fully coiled to a completely unfolded form, provides a standard by which to make an estimate of the percentage of the total length of polypeptide chain coiled in helical form in a protein molecule. Results for a variety of proteins are shown in Table 2.3. For all these proteins, except casein, actin and γ-globulin, the optical rotation results suggest the presence of a helical type of secondary structure. However, only one protein, tropomyosin, shows an estimated helical content approaching 100 per cent. For most proteins only parts of the poly-peptide chains are in helical form, presumably where side-chain interactions, or other factors, make this form favourable.

TABLE 2.3 A Table showing the extent of helical structure in proteins, in aqueous solution at pH ~ 7.

The figures may be taken as an estimate of the total helical content only if it is assumed that no left-handed helical regions are present. More precisely, the figures give an estimate of the amount of right-handed helix present in excess of left-handed.

Tropomyosin	[1]	80–100%
Myoglobin Haemoglobin	[3] [3]	60–80%
Myosin Insulin	[2] [1]	45–60%
Egg albumin Lysozyme Fibrinogen	[1] [1] [2]	30–45%
Pepsin Histone	[1] [1]	20–30%
Ribonuclease	[1]	10–20%
Casein Actin γ-Globulin	[2] [4] [5]	0–10%

1. DOTY, P. (1960) in *Biophysical Science*, ed. J. L. Oncley. New York, Wiley.
2. SZENT-GYORGI, A. G. and COHEN, C. (1957) *Science*, **126**, 697.
3. BEYCHOK, S. and BLOUT, E. R. (1961) *J. mol. Biol.* **3**, 769.
4. KAY, C. M. (1960) *Biochim. biophys. Acta*, **43**, 259.
5. WINKLER, M. and DOTY, P. (1961) *Biochim. biophys. Acta*, **54**, 448.

The behaviour of synthetic polypeptides discussed above raises the question of how stable we can expect the three-dimensional structure of a protein molecule to be. There is much evidence to show that the properties of protein molecules change with change in pH and that the molecules undergo a profound modification in structure (irreversible denaturation), if taken beyond certain limits on either the acid or alkaline side, or if heated for a time at temperatures approaching 100°C. We shall consider first further aspects of protein structure, viewing the molecules as static units, and later try to build up a more dynamic picture, particularly of changes in molecular structure which might take place around pH 7, and at 37°C or below, and therefore play some part in normal cell activity. One example of a situation in which structural changes of this kind are thought to be of biological importance is in the 'allosteric' effects which may be involved in the control of enzyme activity (see Chapter 3) and control of gene expression (see Chapter 10).

Tertiary structure

We have discussed one of the two approaches, mentioned earlier, to an understanding of protein structure – model building. Since about 1960 there has been dramatic progress along the other line of approach, in the direct analysis of the complex diffraction patterns of crystalline proteins.

The proteins that were studied first in this way by Kendrew, Perutz and co-workers, were myoglobin and haemoglobin. Later Phillips and co-workers made an interesting study of the enzyme lysozyme, and more recently results have been published for a variety of proteins. Myoglobin is a muscle protein involved in oxygen transport and storage, found in particularly high concentration in the muscles of whales and seals. The myoglobin molecule contains a single polypeptide chain of 153 amino acids and a haem group (Fig. 9.1) similar to that of haemoglobin, which is the site of O_2 binding. The primary structure of myoglobin is given in Table 9.1.

The initial analysis of the X-ray patterns of crystals of whale myoglobin gave the result shown in Fig. 2.11. This represents one myoglobin molecule. The first point to be noted in this figure is the scale marker. The resolution of the X-ray diffraction analysis from which Fig. 2.11 is derived is not sufficient to show up the individual atoms, but only a 'sausage' of high electron density folding back and forth within the molecule. Later refinement of the X-ray diffraction analysis has shown that the straight portions of the sausage of Fig. 2.11 represent α-helical regions (inset drawing of Fig. 2.11).

It may be estimated, from the length of the straight portions of the sausage, that about 70 per cent of the polypeptide chain in myoglobin is in the α-helical form. The α-helix is a relatively inflexible structure, so that there must be a breakdown of α-helical structure in the bent parts of the sausage. X-ray studies at higher resolution show that the secondary structure is irregular in these regions, the backbone taking up a position which allows, as far as possible, the formation of hydrogen bonds between backbone and side-chain groups. It must be noted that Fig. 2.11 shows only the backbone of the myoglobin molecule (and the haem group). In the complete molecule, side chains fill the open spaces of this figure, and it is interaction between side-chain groups which causes the polypeptide chain to take up this particular conformation. These interactions, and the exact position of each atom in the myoglobin molecule, have now been studied by refinement of the X-ray analysis. The relation between positions of amino acids in the primary and tertiary structure of myoglobin is shown in Fig. 2.12.

The successful analysis of the X-ray diffraction pattern of myoglobin was a most important milestone in the study of protein structure. It had been suspected, on the evidence we have noted earlier, that polypeptide chains were folded back on themselves within a globular protein molecule, but it had always been supposed that they must be folded in a more or less regular way, and not in a tangled knot. It seems however that the interaction between the side-chain groups along a polypeptide chain of given sequence will cause the chain to fold

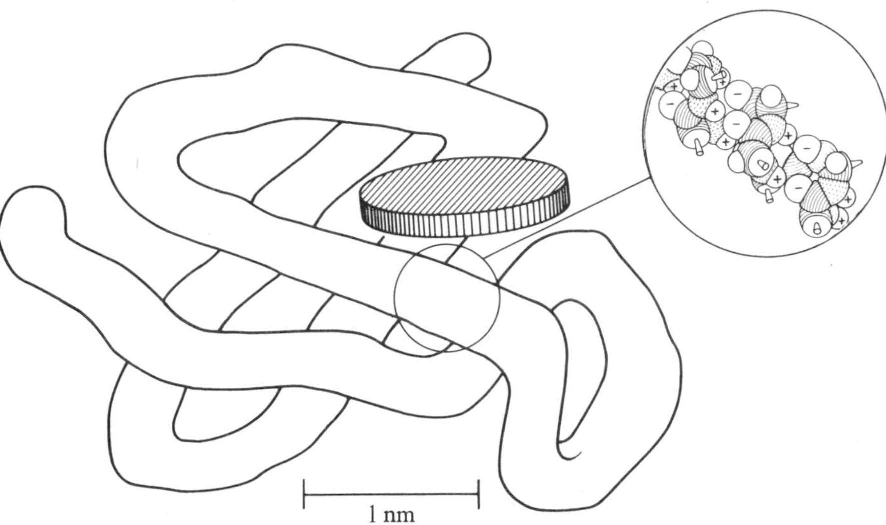

Fig. 2.11 The tertiary structure of myoglobin. The photograph shows an early model of the molecule made from a relatively low resolution (not very detailed) X-ray diffraction study. The folded 'sausage' represents a region of high electron density in the three-dimensional Fourier synthesis (see Appendix 1). The high density region corresponds to the backbone of the single polypeptide chain which makes up the molecule. The sketch below the photograph shows how later, more detailed studies have cleared up some uncertainties which remained as to the exact way in which the chain was folded. The sketch also shows the orientation of the haem group more accurately (the dark disk in the photograph, shaded in the sketch). In the straight portions of the sausage the backbone is folded into an α-helix (compare sketch inset with lower part of Fig. 2.7). (*Adapted from* Kendrew, 1961.)

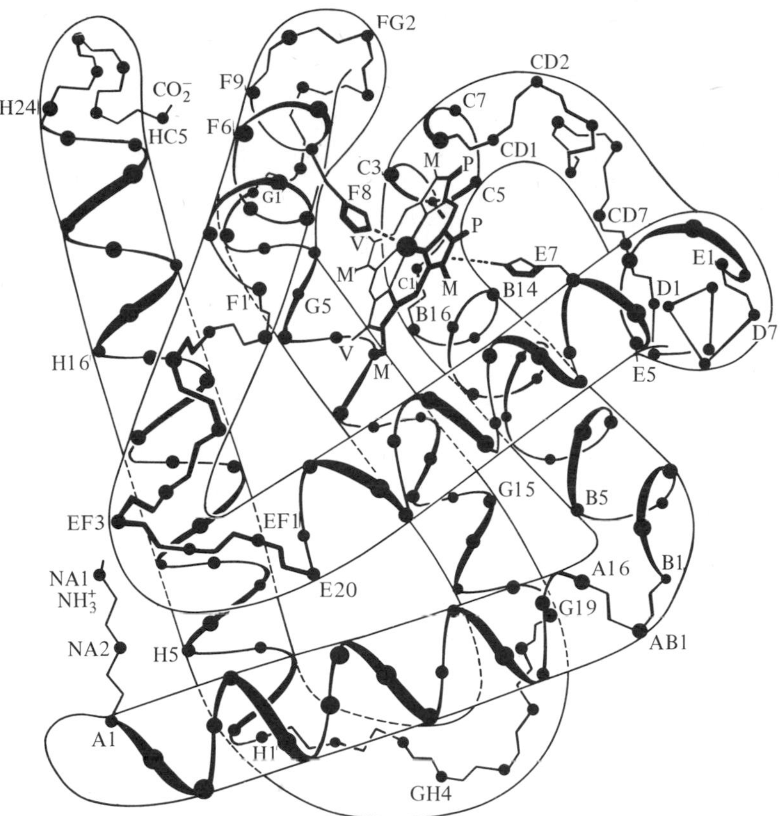

Fig. 2.12 A drawing showing the relation between the positions of myoglobin side chains in the primary structure (Table 9.1) and tertiary structure of the molecule (Fig. 2.11). (*From* Dickerson, 1964.)

up into a *complex irregular three-dimensional form, characteristic of the sequence*. Thus all whale myoglobin molecules (of the primary sequence given in Chapter 9) coil up exactly as in Fig. 2.11. The run of the polypeptide chain through the molecule, as shown in Fig. 2.11 for myoglobin, is known as the *tertiary* structure of a protein. The interactions between side chains that bring about this tertiary folding will be discussed in due course.

The successful X-ray diffraction analysis of lysozyme at high resolution by Phillips and his co-workers has been another important milestone in the study of protein structure. For lysozyme is the first *enzyme* to be analysed in detail and is a rather different protein from myoglobin, containing intrachain S–S bonds and much less α-helical structure, thus considerably broadening our understanding of the factors which cause a polypeptide chain to fold up in a particular way. Figure 2.13 shows the primary structure of lysozyme, Fig. 2.14 its tertiary structure.

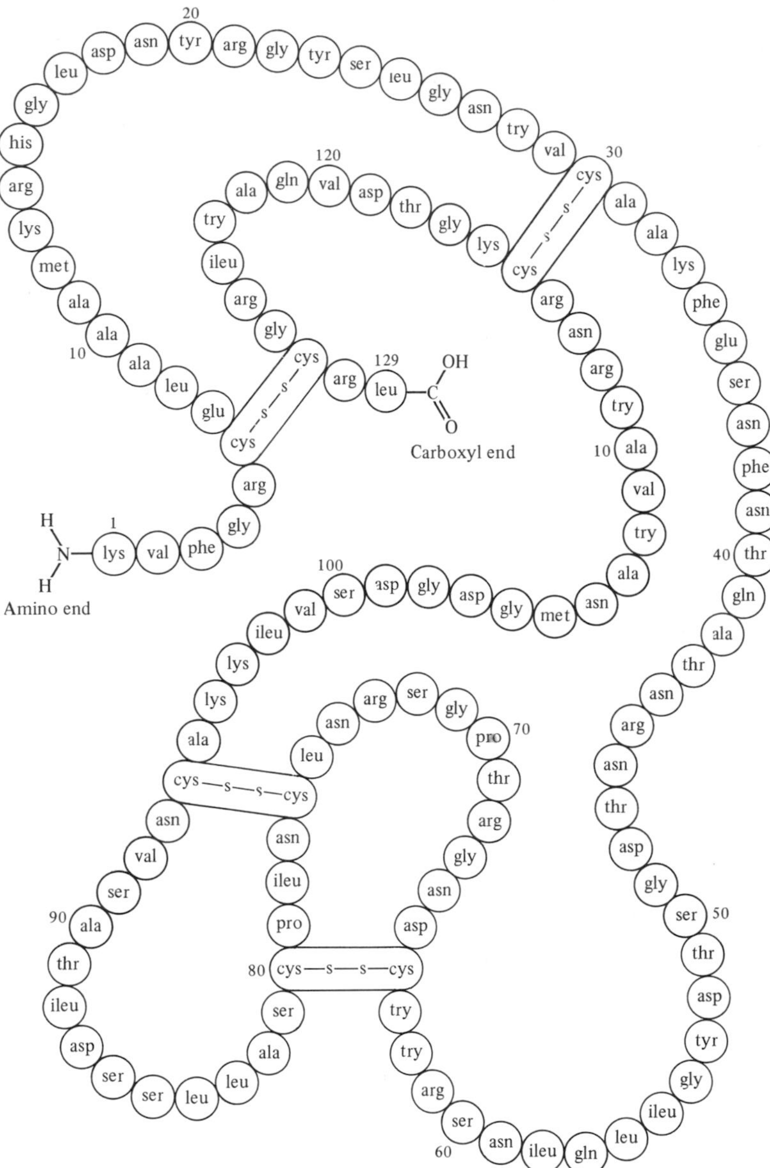

Fig. 2.13 Primary structure of lysozyme (GLN = $GLU{-}NH_2$ and ASN = $ASP{-}NH_2$). (*From* Jolles *et al.* 1965.)

The lysozyme results, in the first place, confirm the optical rotation evidence that stable tertiary structure can be formed with little α-helical content. They show further that helical regions are not always in α-helical conformation. There are short lengths of α-helix along residue sequences 5–15, 24–34 and 88–96, but there is a short sequence 80–85 where a 3_{10}-helix is formed. This is

Fig. 2.14 Tertiary structure of lysozyme. The protein backbone is shown, numbers 1–129 indicating attachment points for the side chains. In following through the numbers sequentially, the eye travels along the backbone from the amino to the carboxyl end. The letters A to F indicate successive sugar rings of the polysaccharide substrate, bound to the enzyme surface. If space-filling atomic models are used the protein molecule becomes a solid object, with mainly hydrocarbon chains in the interior, and with the substrate lying in a cleft of the protein surface. (*From* Dayhoff and Eck, 1967.)

similar to the α-helix but, as noted above, with a slightly tighter turn, so that hydrogen bonds form from N—H groups to C=O groups three residues back along the chain rather than four back as in the α-helix. In both myoglobin and lysozyme α-helical lengths are often terminated by a last turn in this form (at points EF1, G19 and HC1 in myoglobin and at 15 and 122 in lysozyme). In part of the helical regions of lysozyme a variant form is also found intermediate between the α- and 3_{10}-helices, in which the N—H group points midway between the C=O groups three and four amino acids back. It is probable that the α-helix is energetically the more favourable of these forms of coiling (it can be shown with molecular models that in the 3_{10}-helix there are unfavourably close contacts between the nitrogens and surrounding atoms), but some variation is energetically allowable, presumably where it leads to more favourable side chain interactions. These variations in helical folding follow from the electrostatic nature of hydrogen bonds. In all these helices the positive N—H hydrogens abut against the negative C=O oxygens.

In part of the lysozyme molecule (residues 41 to 50) the chain forms a hairpin loop in the pleated sheet secondary structure of silk. This arrangement is again, presumably, favourable to side-chain interactions in this region.

The side-chain interactions in myoglobin and lysozyme can really be studied only with a three-dimensional model of the molecule to hand. However certain generalizations can be made. It seems that one of the main factors which determine the tertiary and secondary folding of both myoglobin, lysozyme and presumably other globular proteins is a tendency for hydrophobic side chains to cluster in the interior of the molecule away from water. (The distinction between hydrophilic and hydrophobic groups is discussed in Appendix 2. The hydrophobic side chains are those which contain only carbon and hydrogen atoms and no groups capable of forming hydrogen bonds with water molecules.) Helical secondary structure, and other favourable side-chain interactions occur where possible, but only in ways that are compatible with this clustering of hydrophobic side chains.

It must be remembered that Figs. 2.11 and 2.14 are *skeletal* pictures of the backbones of the myoglobin and lysozyme peptide chains. When this skeleton is clothed with flesh and space-filling models are used as in Fig. 2.8, with the pegs fitted with space-filling models of the side chains (as drawn in Table 2.1) then these side-chain and backbone groups occupy almost the whole of the interior of the molecule leaving a pocket to accept the haem group in the case of myoglobin and a groove to accept the substrate in the case of lysozyme (binding of substrate and the mechanism of action of lysozyme will be discussed in Chapter 3).

The picture of a globular protein molecule which emerges from these studies is that of a roughly spherical or ellipsoidal molecule with an irregular crinkly surface studded with a variety of chemical groups. These groups at the protein surface may be either polypeptide side-chain groups (i.e. the groups of Table 2.1), or prosthetic groups (such as the haem of myoglobin) built into the protein structure.

Where proteins act as enzymes, substrate molecules are temporarily bound to the protein surface. Differences in primary structure as between one enzyme and another, or between different non-enzymatic proteins, lead to differences in three-dimensional structure, and differences in the groups, and in the exact arrangement of groups at the protein surface. It is in this way, we believe, that differences in primary structure (amino-acid sequence) are translated into differences in function. (This question is taken up in more detail in Chapter 3.) With the variety of groups available, and the possibility of different three-dimensional spatial arrangement of these groups at the protein surface, we can thus understand how hundreds of different enzymes can be formed of varied specificity, and also a wide variety of different non-enzymatic proteins including hundreds of different antibody molecules, from polypeptide chains of somewhat similar amino-acid composition, but varied sequence.

When the terms secondary and tertiary structure were introduced by Linderstrøm–Lang some years ago there seemed a clear distinction between these two aspects of protein structure. Tertiary structure referred to the main run of the polypeptide chain as seen in a low resolution picture such as Fig. 2.11, and secondary structure referred to the detailed conformation of the backbone groups in any given region of the molecule. This distinction is however not now a clear-cut one. Primary structure still has a clear meaning – it is simply the sequence of amino acids along the polypeptide chain. Tertiary structure, if the term is still to be used, for lysozyme for example, must now be redefined to accommodate the results of modern high-resolution studies. Specification of tertiary structure must now mean giving the co-ordinates for the position of each atom in the three-dimensional structure. The term secondary structure remains useful only in a minor way, to refer to the folding or coiling of small parts of the polypeptide chain which happen to be in a regular conformation such as the α-helix or pleated sheet.

The relatively high α-helical content of myglobin and haemoglobin helped in the interpretation of early electron density maps derived from X-ray analysis, so that these two proteins were the first for which analysis was carried to resolution sufficient to reveal atomic detail. We can see now that this historical development put exaggerated emphasis on the α-helix as a component of globular protein folding. In lysozyme, in cytochrome c (see Chapter 9), in insulin and in other proteins now studied in comparable detail there are much more limited regions of α-helix. The lysozyme structure of Fig. 2.14 is more typical of globular proteins in general, than is the myoglobin structure.

Aggregation of protein sub-units (quaternary structure)

The molecular weight of a protein can be determined from the osmotic pressure of the protein solution, or from the rate of diffusion of the molecules, or their rate of sedimentation in the centrifuge. It is a measure of the size of the piece of protein that moves as a unit in solution. But it is frequently found that the molecular weight of a protein depends on conditions of pH or salt concentration. For example at pH$\sim$7 the molecular weight of normal human haemoglobin is approximately four times that of myoglobin, each molecule containing four polypeptide chains and four haem groups. Under mildly acid conditions the molecules split in two parts each containing two polypeptide chains and two haem groups.

The molecular weight of a protein also depends on extraction procedures. Haemoglobin can be extracted very easily from red cells (Chapter 1) and its molecular weight in solution at pH$\sim$7 is probably the same as its molecular weight *in vivo*. But for most proteins we do not know how the size of the molecule in extracted and purified solution is related to the size of the structure into which it is incorporated in the cell. The term 'protein molecule' is normally applied to the form existing at pH$\sim$7, after mild extraction. It must be appreciated however that this choice of conditions is arbitrary. We have to think of protein molecules readily aggregating to form larger molecules and fibres, or readily breaking up into smaller parts. Proteins form complexes with lipids (Chapter 4) and nucleic acids (Chapter 7), and *in vivo* are to a large extent incorporated into membranes and nucleoprotein cell structures.

Although the term *sub-unit* is often used whenever protein molecules break up into smaller parts, we shall restrict our use of this term to the *smallest* units into which a protein can be broken by mild treatments, i.e. without breaking covalent (chemical) bonds. The sub-units of a protein molecule are then normally the individual polypeptide chains, although in certain cases, e.g. insulin (Fig. 2.4) a sub-unit may include two or more chains linked by S–S bonds.

The haemoglobin molecule is made up of four sub-units and X-ray diffraction studies (Perutz and co-workers, 1960) have shown that each sub-unit is very similar to the myoglobin molecule. The four sub-units are arranged as shown in Fig. 2.15. A most interesting detailed comparison can be made, between primary and tertiary structure, for the haemoglobin and myoglobin chains, and discussion of this topic is taken up in Chapter 9.

The complexity of the analysis, and the skill and experience required to achieve this solution (Fig. 2.15) may be judged from the fact that Perutz, with a steadily growing team of co-workers, had been concentrating on this one problem since 1937. Success was achieved by refinement of technique and by the use of fast computing machines. Perutz and Kendrew's insight and persistence in tackling the formidable problem of the analysis of the diffraction patterns of haemoglobin and myoglobin was amply rewarded. As the primary structure of these proteins was unravelled, it became possible to discuss the physiological and pathological behaviour of haemoglobin in a detailed way (see Chapter 9) with precise consideration of the exact position and interaction of every atom in the protein structure. It is this precision of thought which should perhaps be considered the characteristic of the molecular biologist.

Other examples of the aggregation of protein sub-units to form larger struc-

Fig. 2.15 Tertiary and quaternary structure of haemoglobin. The haemoglobin molecule is made up of two α-chains and two β-chains (see Chapter 9), each chain binding a haem prosthetic group. The tertiary structure of these chains, as shown in the sketch, is rather like that of myoglobin, although there are minor differences in the folding, as between the α- and β-chains, and as between these chains and myoglobin, related to differences in primary structure (see Table 9.1). The photograph shows how closely these chains fit together in approximately tetrahedral arrangement in the quaternary structure of the haemoglobin molecule. (The double, and unconnected, use of the α-suffix is confusing, but this is unfortunately now a standard nomenclature: both the α- and β-chains of haemoglobin contain α-helical regions of secondary structure.) The model in the photograph, like the myoglobin model of Fig. 2.11, is showing up the electron dense regions of the molecule, i.e. the backbones of the polypeptide chains. (*From* Perutz *et al.*, 1960.)

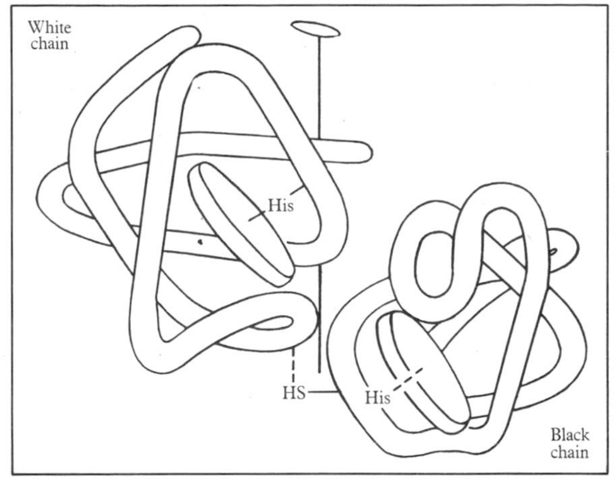
White chain
His
HS
His
Black chain

tures will be given in Chapter 3. We do not know at present what upper limit is set to the length of individual polypeptide chains by the cell's apparatus for protein synthesis. Larger proteins are frequently made up of sub-units of molecular weight 12000–24000 (polypeptide chains of 100–200 amino acids). However, serum albumin, of molecular weight comparable with that of haemoglobin, appears to be made up of a single polypeptide chain, which must thus contain about 600 amino acids.

It has been noted by Bernal that if the folding of the polypeptide chain within a protein sub-unit is described in terms of secondary and tertiary structure, then the arrangement of sub-units in a protein molecule represents a *quaternary* aspect of protein structure, e.g. the approximately tetrahedral arrangement of sub-units in haemoglobin.

Dynamic aspects of protein structure

The X-ray diffraction technique provides a powerful and precise tool for the study of molecules in a crystal, arranged in a regularly repeating array. In diffraction, the more extensive the repeating array, the sharper will be the spots in the diffraction pattern. For crystals of very small size, or for a repeating array broken up by imperfections, the diffraction pattern becomes more blurred and gives less information. Normally, the closer we go to the true biological situation, the less information we can get from X-ray diffraction studies. Muscle filaments are the only intracellular components, in somatic cells, sufficiently well oriented *in vivo* to give X-ray diffraction patterns and, in general, the patterns given by other biological material in which there is some degree of natural orientation (myelin sheath of nerve, fibres of wool or hair) are relatively diffuse and give little detailed information (Fig. 2.16). Collagen, sperm, virus crystals and extracted nucleic acids show patterns with more detail (Figs. 2.16, 2.17 and 2.18). Crystalline proteins give very detailed patterns (Fig. 2.18).

Even with a relatively complete picture of the structure of a protein molecule in a crystal, from the X-ray diffraction pattern, we still have to turn to less precise techniques to find out to what extent the molecule retains this structure on going into solution, or in its special environment in the living cell.

There is a variety of evidence to suggest that the polypeptide chains are coiled and folded into a compact form in native globular proteins, in solution. In particular, the optical rotation estimate of the proportion of the chains of myoglobin and haemoglobin coiled into an α-helix, for the molecules in solution (Table 2.3), is in good agreement with the estimate for the molecules in the crystal, derived from X-ray diffraction studies. It seems reasonable then to suppose that myoglobin and haemoglobin retain, in solution, a structure much like that revealed by the X-ray diffraction studies on crystals, although the molecules may come unravelled to some extent in solution, with some freedom of movement for the ends, and the non-helical parts of the polypeptide backbone.

In Fig. 2.8 it can be seen that the α-helix allows considerable freedom of

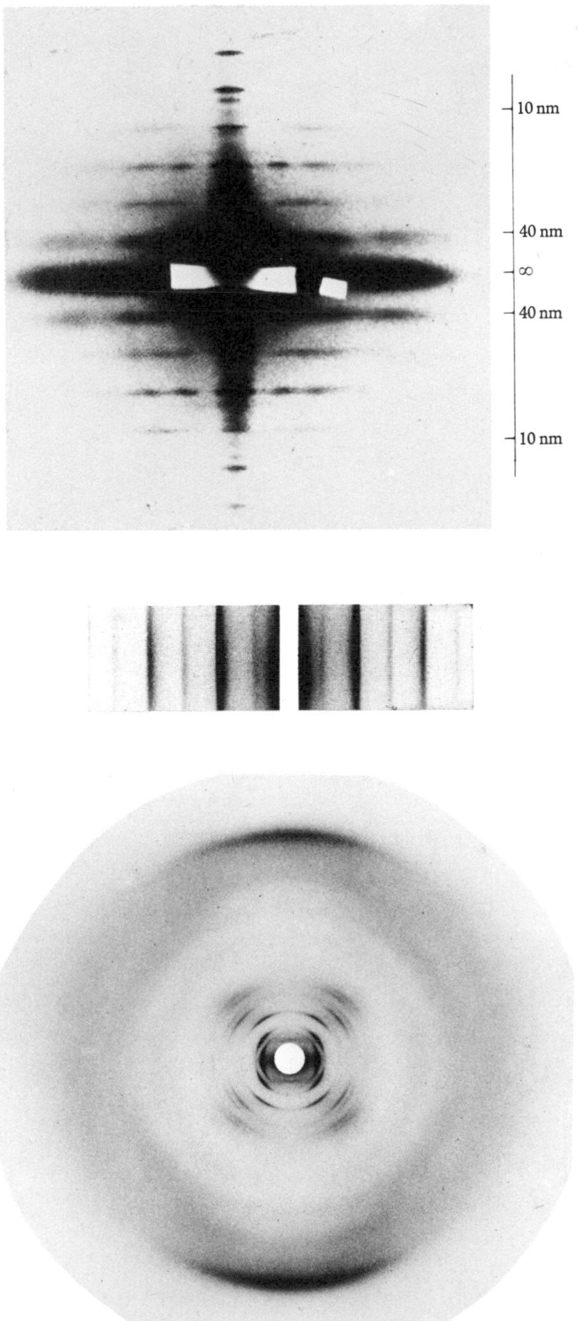

Fig. 2.16 X-ray diffraction patterns produced by fresh untreated samples of muscle and myelinated nerve, and by oriented but otherwise untreated sperm.

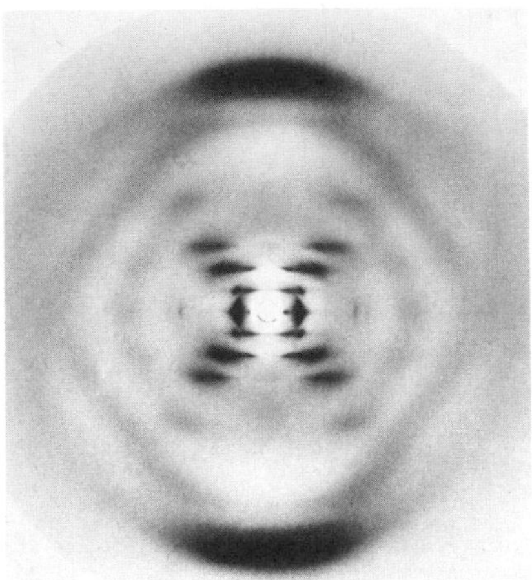

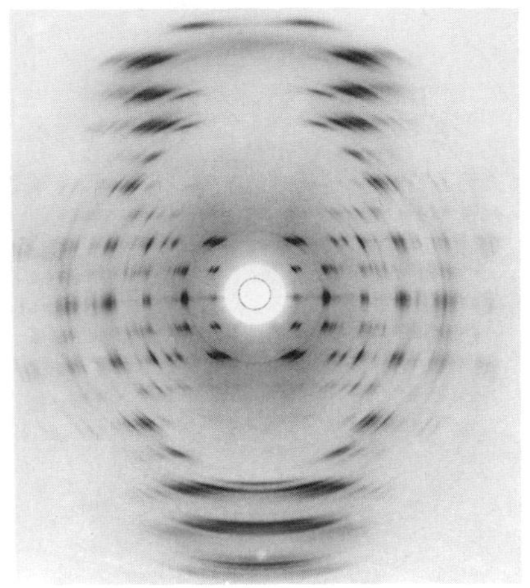

Fig. 2.17 X-ray diffraction patterns of the non-crystalline (B) and crystalline (A) forms of threads drawn out from viscous solutions of extracted deoxyribonucleic acid (DNA) (drawing out the threads tends to orient the molecules parallel to the fibre axis).

movement for the side chains. Where two lengths of helix, or two sub-units, are in contact, the side chains intermingle and adopt the relative positions which are most favourable for interaction between the side-chain groups. At the surface of the molecule, the exact position of the side-chain groups may depend on conditions in the solution. Thus, for a protein dissolved in distilled water, positive and negative groups at the surface will tend to intermingle:

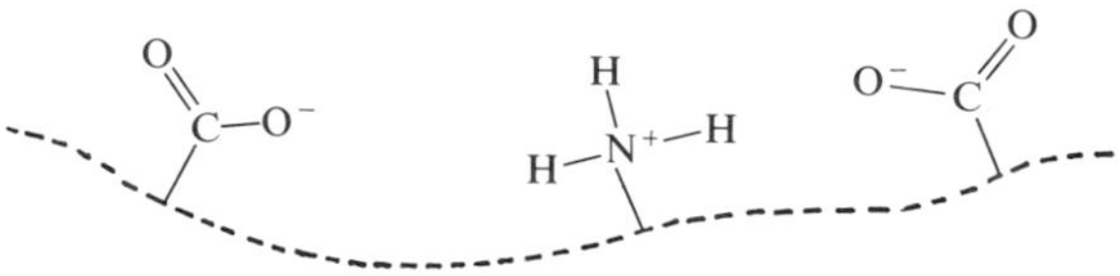

but if Ca^{++} ions are present there may be a rearrangement of surface groups to allow two $—COO^-$ groups to bind the divalent ion:

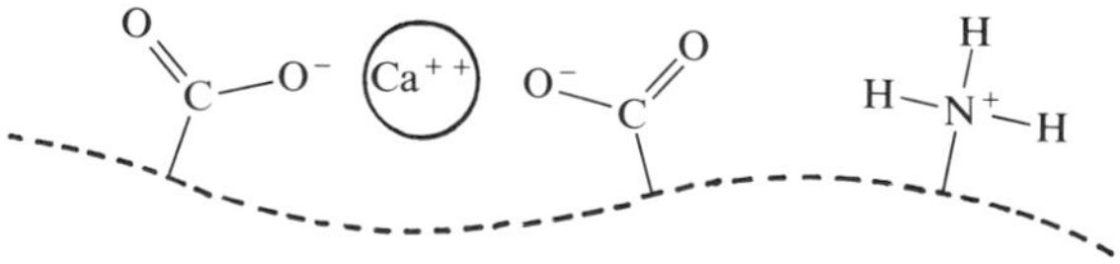

We have to modify slightly our previous picture of a crinkly protein surface studded with a variety of groups, and allow these groups to be jostling and rearranging themselves to some extent. This freedom of movement is important in allowing protein groups to take up the exact three-dimensional arrangement required for binding substrate molecules or antigens.

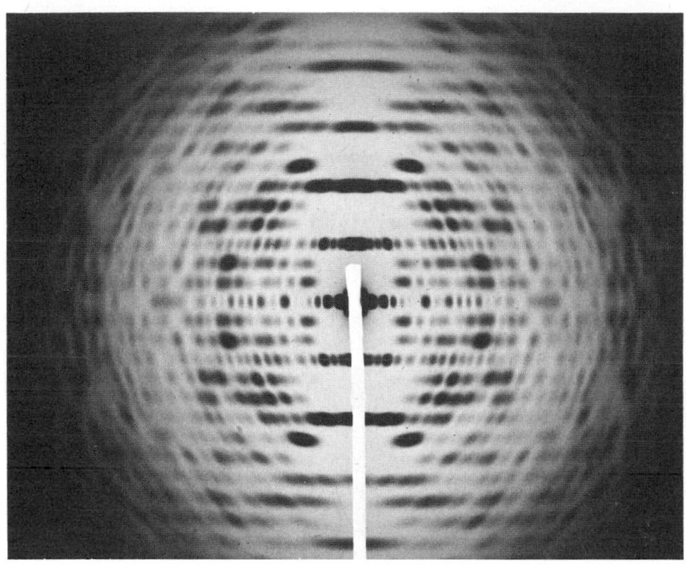

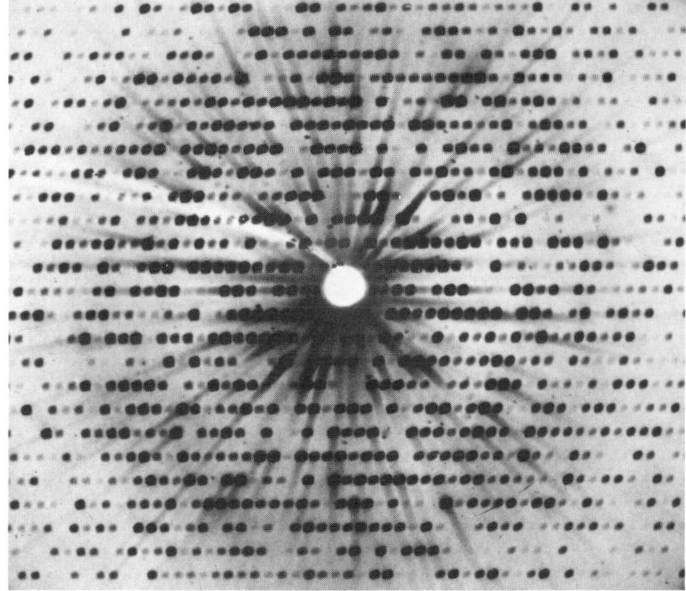

Fig. 2.18 X-ray diffraction patterns produced by a tobacco mosaic virus (TMV) gel, and by a myoglobin crystal.

The unfolding of protein molecules at interfaces

The unfolding of protein molecules at air-water or oil-water interfaces must be briefly described at this point, partly as a contribution to our understanding of protein structure, partly as a background to the discussion in Chapter 4 of the structure of proteins incorporated into cell membranes. At an interface between air and an aqueous protein solution, a surface film of protein forms, in which the protein molecules are denatured to a large extent, unfolding to form a sheet 1 nm thick, with hydrophilic (water-loving) side chains oriented towards the water phase and hydrophobic (water-fearing) side chains towards the air. The same thing happens at an oil-water interface.

It is clear that for globular proteins like myoglobin or haemoglobin this unfolding must involve extensive rearrangement of tertiary and quaternary structure, all segments of the polypeptide chains coming into the 1 nm layer at the interface. We do not know to what extent this denaturation also involves changes in secondary structure, that is in the proportion of the polypeptide chains in α-helical, or other folded form. Lengths of polypeptide chain in an α-helix with alternating hydrophobic and hydrophilic side chains would be expected to unfold completely. Lengths of polypeptide chain in which the side-chain sequence brought most of the hydrophobic groups to one side of an α-helix, and most of the hydrophilic groups to the other, would be expected to remain in α-helical form at the interface. At the interface, as in the globular sub-units, the individual polypeptide chains are probably partly α-helical and partly in other conformations. The unfolding of a globular protein with complete rearrangement of tertiary structure, but little change in secondary structure, is indicated schematically in Fig. 2.19.

Speculation about the forces involved in tertiary and quaternary structure

So far this chapter has been mainly concerned with experimental studies of proteins, and with the type of molecular structure which emerges from these studies. In this final section we turn to discussion and speculation about the way in which the tertiary and quaternary structure of protein molecules is determined by fundamental principles of molecular interaction. This discussion is important because the same principles govern the structure of ribosomes, chromosomes, membranes, and other cell structures to be discussed in later chapters.

It is a well-known observation, discussed at a molecular level in Appendix 2, that oil and water do not mix. Hydrocarbons are hydrophobic. The high proportion of hydrocarbon side chains in most proteins (30–50 per cent, see Table 2.2) makes a long straight α-helix, without tertiary folding, or aggregation, an unfavourable form for a protein to adopt in an aqueous environment. There will be a tendency for polypeptide chains to come together, or for a long helix

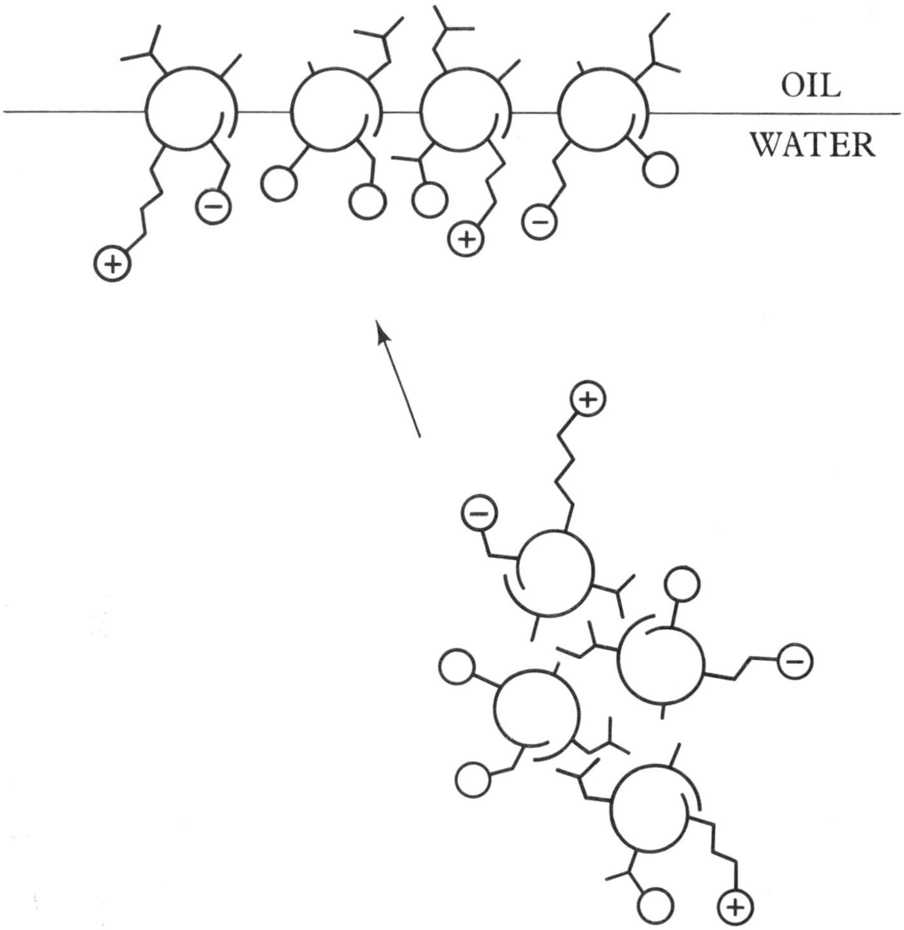

Fig. 2.19 A drawing illustrating schematically the unfolding of protein molecules at oil-water or air-water interfaces. The protein is represented by α-helices (seen end on as in the small drawing of Fig. 2.8) with side chains represented by the small 'shorthand' drawings of Table 2.1. The diagram illustrates the sort of unfolding that must take place to bring the globular protein molecule to a sheet 1 nm thick with hydrophobic (hydrocarbon) groups mainly in the oil or air phase, and hydrophilic groups (—○, —⊕, —⊖) mainly in the aqueous phase. The polypeptide chains in the unfolded molecule are shown in α-helical form on the basis of experiments carried out by Malcolm with polypeptides of restricted amino-acid composition. The extent to which the secondary structure of a globular protein is retained when it unfolds in this way is really not known. For clarity and emphasis the hydrocarbon side chains in this figure are all shown in the oil phase, or in the interior of the globular protein molecule, and the polar groups all in the aqueous phase. In reality the trend towards this decisive separation of the two types of side-chain group would be only imperfectly satisfied.

to fold back on itself, as indicated in Fig. 2.20, thus reducing the hydrophobic area of the molecular surface.

Some sequences of amino acids will make this effect even more pronounced. Thus the sequence of the longer insulin chain between the S–S bonds forms an α-helix in the insulin tertiary structure. This brings most of the hydrocarbon side chains to one side of the helix (Fig. 2.21). Where this happens over certain lengths of a long polypeptide chain it will clearly be energetically very favourable for the chain to undergo some contortion to bring these areas into juxtaposition. This sort of effect seems to be important in determining the secondary and tertiary structure of myoglobin and haemoglobin. When the sequences for the helical regions of these proteins are plotted as in Fig. 2.21 the larger hydrophobic side-chains are found predominantly on one side of the 'helical wheel'. If the sequences of the non-helical regions are plotted in a similar way no such effect is found. Further, the haem group, which is hydrophobic except along one edge (Fig. 9.1) is found to be sitting in a 'pocket' lined by hydrophobic groups. The polar edge of the haem lies outwards. Thus the binding of the haem group to the protein is again a hydrophobic effect. The fact that protein molecules unfold at an air-water or oil-water interface also illustrates the importance of hydrophobic effects in maintaining the native structure of globular proteins in aqueous solution.

Another factor which will be involved to some extent in the determination of tertiary structure is the electrostatic attraction and repulsion between charged side-chain groups. Ionic groups are strongly hydrophilic and hence may be expected to be mainly at the surface, rather than the interior of the molecule. The energetically most favourable position for these groups is achieved if positive and negative groups intermingle over the surface of the molecule. Attraction between groups of opposite charge intermingling in this way will be one of the factors tending to bring the polypeptide chains of a protein into a compact configuration. If the pH of a protein solution is taken to the alkaline or acid side of the isoelectric point the molecules acquire an increasing net charge, positive in acid solution as the $—COO^-$ groups are neutralized, and negative in alkaline solution as the basic groups are neutralized. The repulsive forces arising from this increasing net charge on the molecules are the main factor involved in acid and alkali denaturation of proteins.

Although S–S bonds must be included among the factors holding polypeptide chains in a folded configuration, they are not essential to the formation of a stable tertiary structure. Myoglobin and haemoglobin contain no S–S bonds.

Now that primary and tertiary structures are known for a number of proteins it is becoming feasible to attempt precise evaluation of the side-chain and backbone-group interactions involved, and hence to set up a computer programme for determining an unknown tertiary structure from a known primary structure. The problems are still immense, partly because of the large number of conformations available to a chain as long as 100–150 residues and partly because of the lack of sufficiently precise physico-chemical data applicable to this complex situation. Suppose, for example, that we measure precisely the

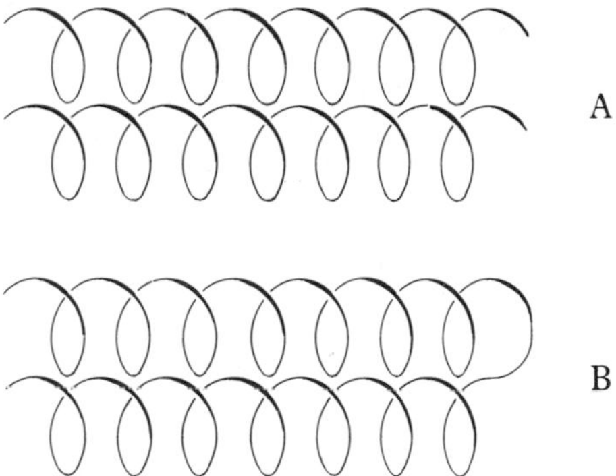

Fig. 2.20 A drawing illustrating the aggregation of two α-helices (A) and an α-helix folding back on itself (B) as a result of side-chain interactions.

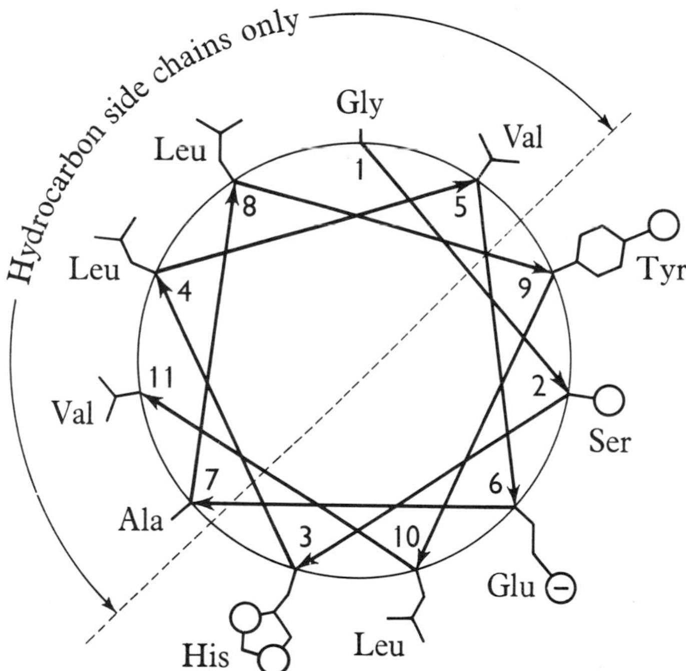

Fig. 2.21 A drawing showing how the side chains are arranged in the 11-residue peptide segment between the S–S bonds in the long chain of insulin (Fig. 2.4) which forms three turns of an α-helix. This is a schematic view along the axis of the helix. The side chains on one side of this length of helix are all hydrophobic. (Side chains represented by the 'shorthand' drawings of Table 2.1.)

free energy change when a leucine side chain moves from a water environment to an environment in which it is surrounded by hydrocarbon chains. Some assumptions still have to be made to estimate what the free energy change will be if the leucine moves from water to a position in the protein structure where it is almost completely surrounded by hydrocarbon chains, but is near the protein surface, and still has water round one of the CH_3 groups.

Model building is now done by calculation or by computer, which makes this approach more flexible than it was in 1950. Instead of using solid models, atomic radii are fed into the calculation and conformations that would be impossible with the model, owing to atoms fouling one another, are represented as inallowable, because of too close an approach of atomic centres. Figure 2.22 shows

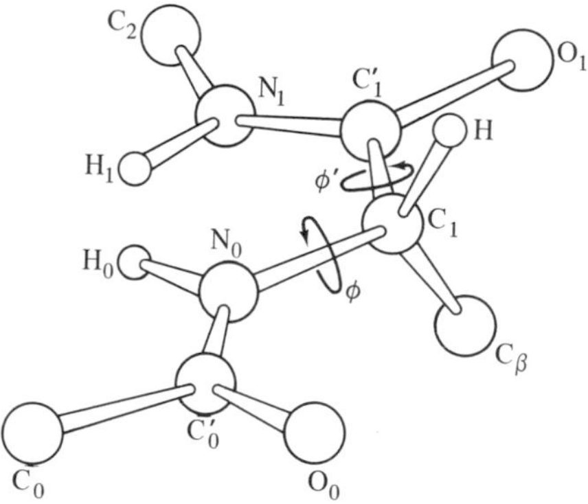

Fig. 2.22 A drawing of two amino-acid residues (numbered 0 and 1) with the atoms C'_1 and N_1 lying in the plane defined by C'_0, N_0 and C_1. The conformation of any polypeptide backbone is fully described if, for each residue, the angles ϕ and ϕ' are specified for rotation about the N_0–C_1 and C_1–C'_1 bonds, measured relative to the planar conformation shown in this drawing. (*From* Ramachandran *et al.*, 1963.)

how the position of a backbone group can be represented by two angles ϕ and ϕ' (this assumes

```
    H
    |
    N
   / \  /
      C
      ‖
      O
```

atoms lie in a plane, see Appendix 2). Figure 2.23 shows, within the areas bounded by full lines, the conformations that are readily allowable, and within the areas bounded by broken lines the conformations available with a bit of 'give' in the atoms, that is, allowing slightly closer approach of atoms than normal solid models would allow (Table 2.4). Shown also in Fig. 2.23 are the actual conformations (ϕ and ϕ' values) for the residues in the non-helical parts of myoglobin. Most fall within the broken line areas, but the amount of 'give' needs to be increased further than that assumed in these calculations to allow formation of the 3_{10}-helix. (For formation of this

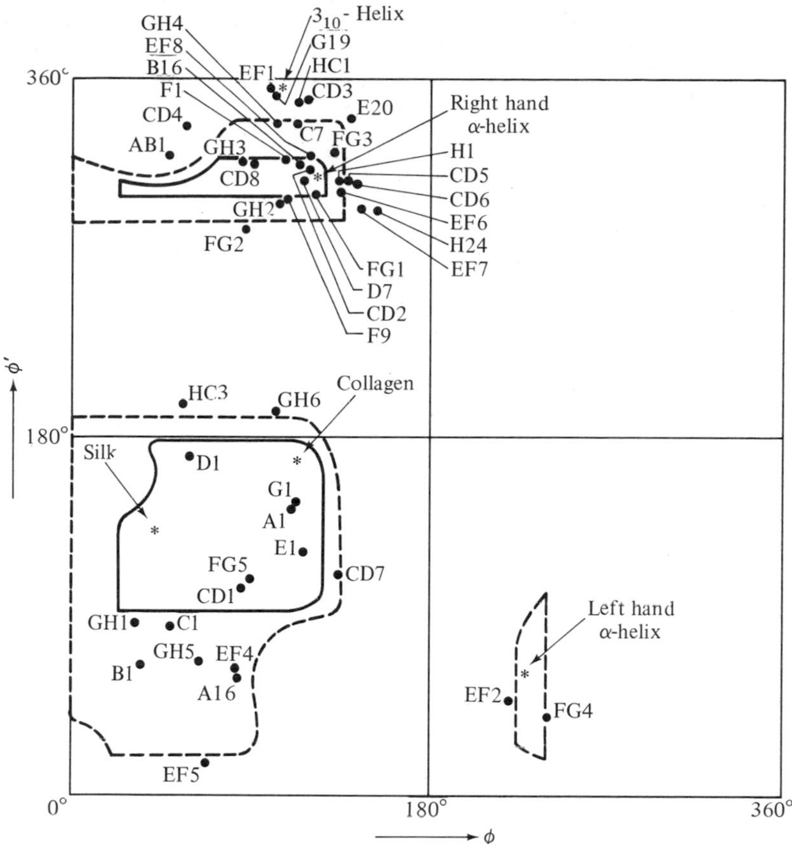

Fig. 2.23 A plot of ϕ' and ϕ (as defined in Fig. 2.22) for the residues of myoglobin. The regular folds such as α-helix, 3_{10}-helix are also shown on the plot. Glycine residues with their short –H side chain have greater freedom than other residues and are not included in the plot. (*From* Davies, 1965.)

helix the N–H distance has to be 0·013 nm less than the value in Table 2.4.) The *left*-handed α-helix comes inside the broken line area, but outside the full line area. Where an unusually close atomic contact is found, this means that there must have been a free energy gain somewhere, as a result of this arrangement, which more than compensates for the energy needed to push the atoms so close together.

The forces which lead to aggregation of protein sub-units (quaternary structure) are again those we have already discussed – the hydrophobic nature of hydrocarbon side chains, and electrostatic forces between charged groups. The form adopted by a polypeptide chain in a small protein molecule, or in the sub-unit of a larger molecule is, after all, a compromise. Not all the hydrocarbon parts of the molecule can be in a purely hydrocarbon environment, not all the backbone intrahelix hydrogen bonds can form, not all the charged groups can

TABLE 2.4 Inter-atomic distances used for plot of Fig. 2.23.

Contact	Normally allowed (nm)	Outer limit (nm)
C . . . C	0·320	0·300
C . . . O	0·280	0·270
C . . . N	0·290	0·280
C . . . H	0·240	0·220
O . . . O	0·280	0·270
O . . . N	0·270	0·260
O . . . H	0·240	0·220
N . . . N	0·270	0·260
N . . . H	0·240	0·220
H . . . H	0·200	0·190

get into juxtaposition to minimize electrostatic potential energy. It is to be expected that if sub-units come together, it will often be possible for the increased interaction between side-chain groups to allow a more satisfactory compromise to be reached. If the sub-unit surface still includes some hydrocarbon side chains then when sub-units make contact over part of their surface this will reduce the area of water-hydrocarbon interface. Charged groups from the sub-units in contact can intermingle. Both factors will be favourable to aggregation. Other factors act the opposite way. For example, thermal agitation, and repulsive forces between protein molecules carrying the same net charge, tend to prevent aggregation. Blood clotting, discussed in Chapter 3, is a result of disturbance of the balance between these competing factors, for the protein fibrinogen.

We have referred in this section to the 'energetically most favourable form' for a polypeptide chain, or aggregate of polypeptide chains, to adopt. As discussed more fully in Appendix 4, the energetically most favourable form, in which the thermodynamic potential energy is a minimum, is the configuration that a molecule, or molecular aggregate, will tend to adopt in equilibrium. The study of living cells is complicated by the fact that their molecules are not always in reversible thermodynamic equilibrium, but there is no reason to expect that living processes involve any departure from the laws of thermodynamics derived from the study of simpler physical and chemical processes. We suppose that the folding and aggregation of polypeptide chains, and indeed the further aggregation of large molecules to form chromosomes, cell membranes, etc., occurs spontaneously in the exact conditions prevailing in the living cell.

We shall see in later chapters that the factors which determine lipid–lipid, lipid–protein, nucleic acid – nucleic acid and nucleic acid–protein interactions in biological structures are primarily the hydrogen bonds, hydrophobic effects, and electrostatic forces, which we have been discussing as responsible for the details of protein structure. Hydrophobic effects are the result of hydrogen bonding between water molecules, and hydrogen bonds are primarily electrostatic in nature. Thus description in simple electrostatic terms gives, on the whole, a very adequate description of the forces between molecules with which we are concerned in this book. The limitations of the electrostatic description are discussed in Appendix 2, where discussion is given also of the way the electron clouds of nitrogen and oxygen atoms are drawn in more closely to the more positive nitrogen and oxygen nuclei than they are to the carbon nucleus, so that these atoms form hydrogen bonds and carbon does not. The possibility of spontaneous formation of proteins under primeval conditions and their further condensation to form molecular aggregates and primitive cells are discussed in Chapter 9. It seems incredible that this should be so, but atoms are apparently of such subtle design that if enough time is allowed, under the right conditions, they condense spontaneously to form stars, planets, large molecules, living cells and . . . men.

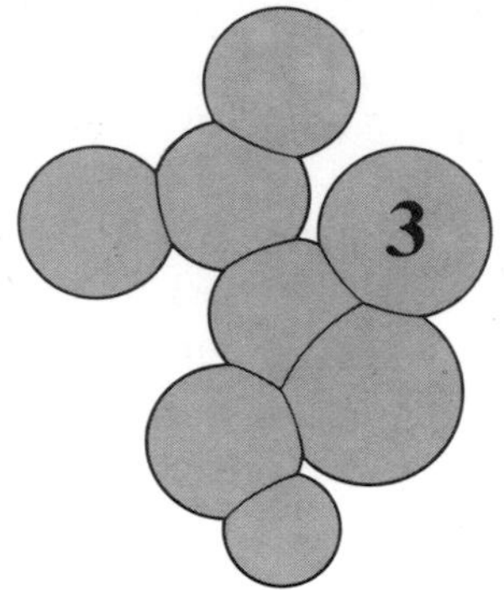

Larger protein structures and the relation of protein structure to function

Chapter 2 has been concerned with the principles of protein structure. We have begun to explore the way in which the primary structure of proteins determines the three-dimensional folding of their polypeptide chains, and the way in which these folded chains aggregate to form larger molecules. In this chapter we first consider more extensive aggregation of sub-units, to form hollow tubes and hollow spherical, or near-spherical, shells. Brief mention is made of the keratins, in which extensive linking of polypeptide chains by S–S bonds leads to the formation of structures such as nails, claws, hooves, hair and wool, and the light ribbed structure of feathers. Certain aspects of protein function are then considered in more detail, for example, the way the polypeptide chains of collagen twist together to form the long fibrous molecules of tendons and cartilage, which act also as condensation sites for the deposition of bone. The mechanism of enzyme action is discussed from various aspects, and a final section is concerned with actin-myosin interaction in muscular contraction.

The protein component of the tobacco mosaic virus

The flow birefringence, and other properties of tobacco mosaic virus (TMV) suspensions, show that the particles are long and thin; in electron micrographs their dimensions can be seen to be approximately 300 nm × 17 nm. The electron microscope pictures show some structural detail in the particles, but we shall describe first the more important contribution made by X-ray diffraction studies. TMV suspensions at sufficient concentration show a curious phenomenon: the suspension separates spontaneously into two phases (as a result of entropy effects, associated with restriction of the freedom of movement of the elongated particles imposed by their neighbours). In the lower phase, which is gel-like and highly concentrated, the particles are oriented in parallel array within localized para-crystalline regions (tactoids). Concentrated TMV gels thus give quite detailed X-ray diffraction patterns (Fig. 2.18), studied by Watson, Rosalind Franklin and others. The radial distribution of electron density, derived from X-ray analysis, shows that the TMV particle is a hollow tube, with a hole down the centre about 4 nm in diameter (Fig. 3.1).

We know from physical and chemical studies that the molecular weight of a

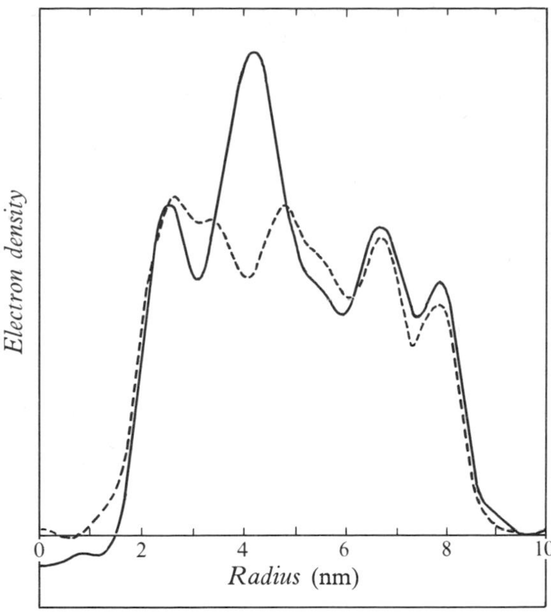

Fig. 3.1 The radial distribution of electron density, derived from X-ray diffraction studies, for the tobacco mosaic virus (TMV) particle (full line), and for TMV protein reaggregated in the absence of the ribonucleic acid component of the virus (broken line). (*From* Franklin and Holmes, 1958.)

TMV particle is about 40×10^6, made up of 94·5 per cent protein and 5·5 per cent ribonucleic acid (RNA). The particles can be broken up, in weakly alkaline solutions, and protein fragments of relatively low molecular weight ($\sim$100 000) can be separated from the RNA. If the pH is now lowered again, these protein fragments reaggregate, even in the absence of RNA, to form hollow tubes of the same internal and external diameter as the intact virus (Fig. 3.1). Detailed X-ray studies have shown that the protein component, both in the intact virus and in the reaggregated form, is made up of sub-units, each approximately $6 \times 2 \times 2$ nm (molecular weight $\sim$17 000), aggregating as shown in Fig. 3.2, to form a helical structure with $16\frac{1}{3}$ sub-units to a turn, or 49 in three turns. In the intact TMV particle about 2200 sub-units are arranged in this way to form a tube 300 nm long. (The fragments of molecular weight $\sim$100 000 from which the tube can be reconstituted must each contain six sub-units.) The TMV protein can also reaggregate in an alternative form as a stack of rings, each ring containing 16 sub-units.

The RNA component of the virus is a long single-stranded nucleic acid chain, and comparison of the radial density distribution patterns for the intact virus and the reaggregated protein (Fig. 3.1) shows that the RNA in the intact virus lies at a distance of 4 nm from the axis of the particle. The RNA chain spirals up between the protein sub-units as indicated in Fig. 3.2.

Fig. 3.2 A drawing of the tobacco mosaic virus particle incorporating the results of X-ray diffraction studies (see text). For clarity a part of the RNA chain is shown without its supporting framework of protein, although it would not in fact retain this helical form if protein sub-units were removed from the virus structure. The individual bases are indicated schematically along the RNA chain. Each of the protein sub-units shown in this figure contains a single tangled peptide chain, with the primary structure of Fig. 8.5 (or a closely similar primary structure, for different strains). (*From* Klug and Caspar, 1960.)

In electron micrographs the tubular form of both the virus and the reaggregated protein can be clearly seen, and in pictures of the reaggregated protein the individual sub-units can almost be distinguished (Fig. 3.3). The reaggregated protein of this preparation is probably in the 'stacked ring' form. Electrons have a very much shorter wavelength than light waves and if electron lenses could be designed as free from aberrations as light microscope lenses, the resolving power of an electron microscope would be a fraction of a nanometer and significant features smaller than 0·1 nm could be seen. With current electron lens design the instrumental resolving power is already 0·2–0·4 nm, which is good enough to resolve individual atoms. However the limits of electron microscopy in biological work are not set by resolution, but by contrast, and by specimen preparative techniques.

Tissue sections are prepared for electron microscopy by methods essentially similar to those used in preparing tissue sections for conventional examination in a light microscope except that, since an electron beam has poor penetrating power, very thin sections are needed and plastics must be used rather than

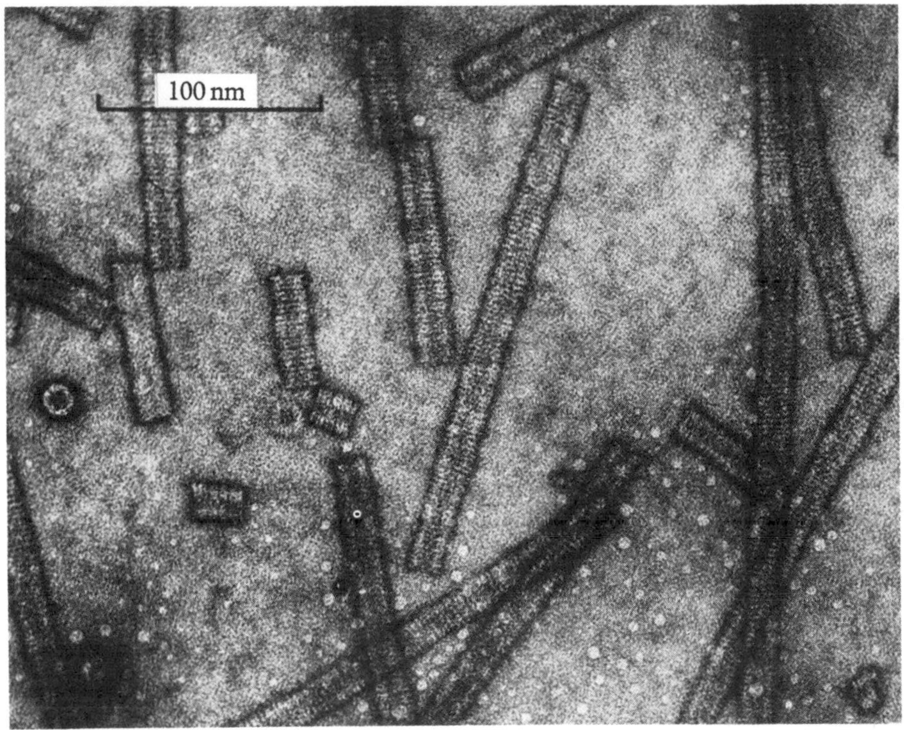

Fig. 3.3 Reaggregated TMV protein by 'negative contrast' (see text). (*From* Nixon and Woods, 1960.)

wax, as an embedding medium. The thin sections are cut with glass or diamond knives. They are mounted on copper mesh 'grids', and observed through the holes of the mesh. Osmium tetroxide and glutaraldehyde are the most satisfactory fixatives for electron microscopy. Osmium tetroxide combines specifically with some of the cellular constituents, and the heavy atoms of osmium increase the contrast of these constituents. Examples of thin-section electron micrographs come later in this chapter (muscle filaments) and in later chapters. An alternative preparative technique, useful for looking at large molecules and viruses, is the 'negative contrast' technique used in the preparation of Fig. 3.3. In the negative contrast technique electron-dense material, such as sodium phosphotungstate, is added to a suspension of virus, or other particles, and a drop of the suspension is dried down on to a thin supporting film stretched across an electron microscope specimen grid. The dense material collects round the particles during drying and some remains, for example, within the hollow tube of TMV protein (Fig. 3.3).

For both types of technique, thin-section or negative contrast, the molecular rearrangement induced by the preparative methods and lack of adequate contrast for viewing very small structural features limit the effective resolution to about 2 nm; individual TMV sub-units in Fig. 3.3 cannot be quite seen, although the repeat pattern along the tubular structure is clear. In interpretation of negative contrast micrographs one has to be rather cautious, for in Fig. 3.3 the images of the upper and lower parts of the tube (that is, the parts against and away from the support film) are superimposed. This can sometimes make a helical structure look like a stack of rings. In spite of these and other limitations of electron microscopy, the electron microscope is a most important tool in modern biology, spanning the gap between structures that can be explored by light microscopy and structures that can be explored by the X-ray diffraction technique.

The TMV sub-units of molecular weight ~ 17000 each contain a single chain of 157 amino acids of known sequence (Fig. 8.5). Each chain of 157 amino acids must be folded into a compact tangle to form the $6 \times 2 \times 2$ nm sub-units revealed by the X-ray diffraction study. X-ray diffraction analysis of TMV has not yet been carried to a point where the folding of the chain within each sub-unit can be seen, though at 1 nm resolution the electron density map shows some features which may be α-helices. The α-helix, if present in TMV protein, is part of the structural detail *within* each sub-unit. It must not be confused with the larger helical arrangement of sub-units of Fig. 3.2, which represents the quaternary structure of the TMV protein (in the sense in which we have defined this term in Chapter 2). This point may be clarified by comparison of the scale markers of Figs. 3.2 and 2.7.

The helix is thus a fundamental feature of macromolecular structure at several levels. We shall see, in later sections of this chapter, that three polypeptide chains twist together in a helix to form the tropocollagen molecule and that α-helices probably twist together like the strands of a rope in keratin and myosin, to form super-helices. Protein sub-units aggregate in a helical way to form hollow tubes (TMV) and fibres (actin – discussed later in this chapter). Nucleic acid chains also form helices (see Chapter 7).

Crane and others have discussed the theoretical and formal reasons for the prevalence of these helical forms. The essential features of their argument can be illustrated by reference back to the chain of Fig. 2.6. The links in this chain are all identical, so that the chain will serve as a model for any molecular chain, like the polypeptide backbone chain, made up from a repeating group. If a chain of this type is going to coil, or fold up, rather than remain in a randomly-folded, constantly changing form, it will do so as a result of forces of attraction between the repeating units. Since the repeating units are all identical the chain will tend to take up a form in which each unit stands in identical spatial relation to its neighbours. For the chain of Fig. 2.6 this corresponds to locking the rotating joints with the angle between the planes defined by adjacent links (θ) the same right along the chain. If $\theta \neq 0$ the chain forms a helix. A helix, formed from a single chain or from a number of intertwined chains, is thus a favourable form for any structure in which there is attraction between repeating units – backbone groups in polypeptide chains, base pairs in the structure of nucleic acids. For similar reasons, a helix is a favourable form of aggregation for identical protein sub-units.

Although nucleic acids and their interaction with proteins are not discussed until Chapter 7, we may conveniently note here the effect of the nucleic acid chain on the aggregation of the TMV sub-units. The tubular structure formed by reaggregation of the protein alone is less stable than the complete virus, and the extent of the aggregation is variable. The presence of the nucleic acid chain apparently increases the cohesion between sub-units in successive turns of the helical quaternary structure, and in the nucleoprotein structure (i.e. the complete virus) the extent of aggregation is controlled by the length of the nucleic acid chain, so that particles are formed of a uniform length of 300 nm.

In a cell infected by a virulent virus, both the nucleic acid and protein components for new virus particles are synthesized; after a certain time lysis occurs, with the release of large numbers of newly synthesized particles from the ruptured cell. During the synthesis of TMV, newly formed protein sub-units will probably be aggregating and breaking apart, both with and without incorporation of nucleic acid chains. But because the complete virus is the most stable form of aggregation for these components, there will be a steady accumulation of complete virus particles, with a decreasing proportion of the material in the incomplete forms. It is in this way, we suppose, that thermodynamic factors discussed in Appendices 3 and 4 lead to the spontaneous formation of complex structures during virus infection, and also in normal synthesis of cell components. The same principle must have been important in biomolecular evolution (see Chapter 9). The structures which form, and persist, are the more stable structures, in which the component molecular chains and sub-units are in a state of lower thermodynamic potential energy.

The protein component of small spherical viruses

Markham and Smith showed, in 1949, that when a preparation of turnip yellow mosaic virus (TYMV) is spun down in a centrifuge it separates into an infectious 'bottom component' and a non-infectious 'top component'. (Separation on centrifugation arises from the fact that two types of particle are present – a quite different phenomenon from the spontaneous separation of TMV suspensions into two phases.) The particles of both top and bottom components of the centrifuged TYMV suspension can be crystallized, and the X-ray diffraction patterns of these crystals show both types of particle to be roughly spherical, with diameter approximately 28 nm. The similarity between the crystals formed by the two types of particle suggests further that the outer surfaces of the two

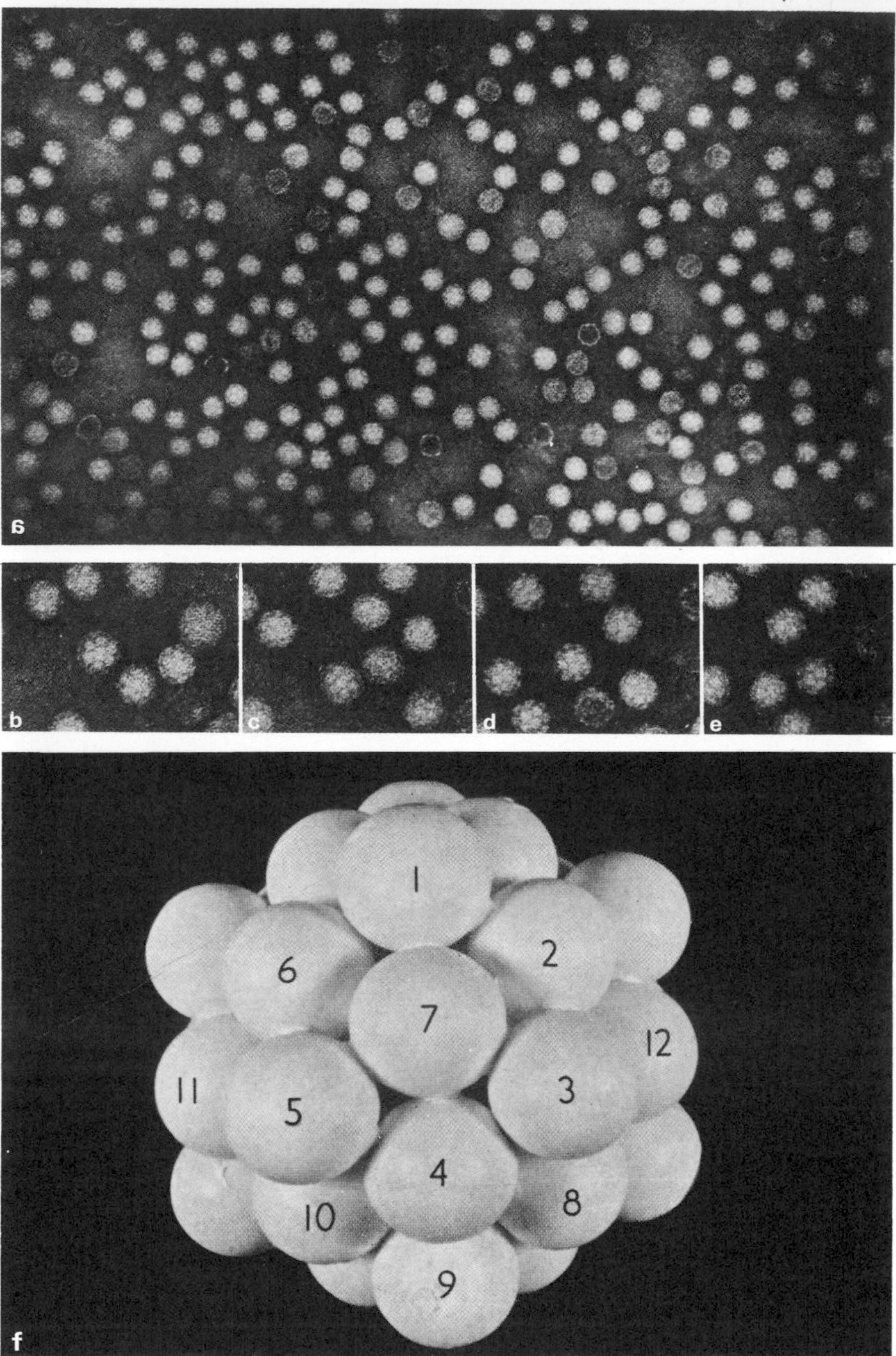

Fig. 3.4 Turnip yellow mosaic virus (TYMV) particles by 'negative contrast'. The occasional rings seen in field (a) are empty protein envelopes. Field (a) is printed at ×110000. Fields (b), (c), (d) and (e) are selected parts of field (a) printed at twice this magnification. At this higher magnification structural detail can be clearly distinguished in the virus particles. The lower picture (f) shows a model of the protein shell of the virus (discussed in the text) based on the detail seen in these electron microscope pictures. The diameter of the virus particle is 28–30 nm and the diameter of each of the 32 capsomeres of which the protein shell, or capsid, is composed is 7–9 nm. (*From* Nixon and Gibbs, 1960.)

types of particle contain similar structural features. Chemical analysis shows the top component to be pure protein, and the bottom component nucleoprotein. It was suspected that the bottom component represented complete virus particles in which the nucleic acid was contained within a spherical, or near-spherical, protein envelope, and that the top component represented non-infectious, empty, protein envelopes. This interpretation has been confirmed by negative contrast electron microscope pictures (Fig. 3.4). (The empty envelopes are probably synthesized to some extent by the infected cell, along with the synthesis of complete particles.) A number of other small spherical viruses have been shown to contain nucleic acid within similar protein envelopes, now called *capsids*.

When the protein component of the tubular TMV virus was found to be made up of a large number of identical sub-units, arranged in a regular way, Watson and Crick suggested that a similar principle might govern the structure of the capsids of the small spherical viruses. It has been found necessary to extend this idea slightly, since some capsids appear to contain more than one type of structural unit or one type of unit in more than one conformation. We will discuss first, however, the possible spherical structures that can be formed from sub-units of a single type, each incorporated into the structure in an identical way. This is actually a simple problem in solid geometry with only two types of solution: structures with four-fold symmetry, and structures with five-fold symmetry. The significance of these symmetry classifications can best be described in relation to model structures of these two types shown in Figs. 3.5 and 3.6. Looking along some axes, the model of Fig. 3.5 is seen to have five-fold symmetry (Fig. 3.5 A), along others three-fold (Fig. 3.5 B), along others two-fold (Fig. 3.5 C). We do not know in general whether the structural units shown up in negative contrast electron micrographs of virus capsids correspond to protein sub-units (as this term is defined in Chapter 2). For some viruses, as we shall see, there is evidence to the contrary: the structural units appear to contain a number of chemical sub-units. The units of the capsid seen in electron micrographs of virus structures have therefore been given the special name of *capsomeres*.

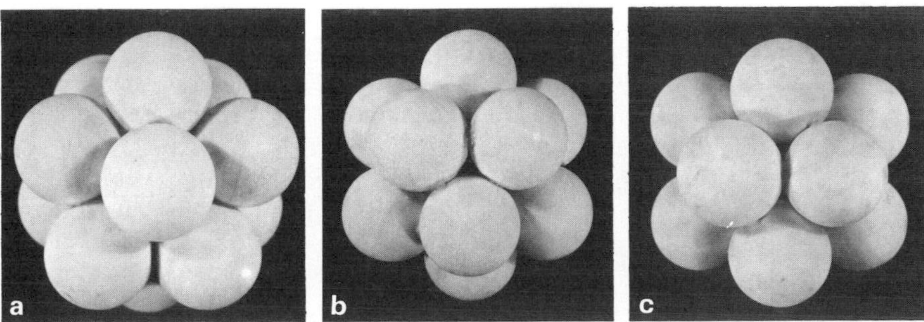

Fig. 3.5 A model made from twelve ping-pong balls illustrating 5—3—2 symmetry. The model is photographed along one of its five-fold axes (A), one of its three-fold axes (B), and one of its two-fold axes (C).

For a structure to have complete 5—3—2 symmetry, rotation about the five-fold axes by 360°/5 = 72°, rotation about the three-fold axes by 360°/3 = 120° and rotation about the two-fold axes by 360°/2 = 180° must in each case lead to a completely identical structure. The ping-pong ball model of Fig. 3.5 has complete 5—3—2 symmetry since the central ping-pong ball of Fig. 3.5 A is a perfect sphere, and if we look away, and then look back at the model, we cannot tell whether it has been meanwhile turned through 72°. But if a virus capsid shows a similar structure in electron micrographs, it may not have true 5—3—2 symmetry, but only pseudo 5—3—2 symmetry (i.e. 5—3—2 symmetry at low resolution only). For if the virus capsid were aligned as in Fig. 3.5 A, and we could improve the resolution to see the detail of the folded polypeptide chains within the central capsomere, we might find that we could then detect rotation by 72°. For complete five-fold symmetry right down to molecular detail, each *capsomere* in this structure must have five-fold symmetry, and the capsid must be made up of 60, or some multiple of 60, sub-units.

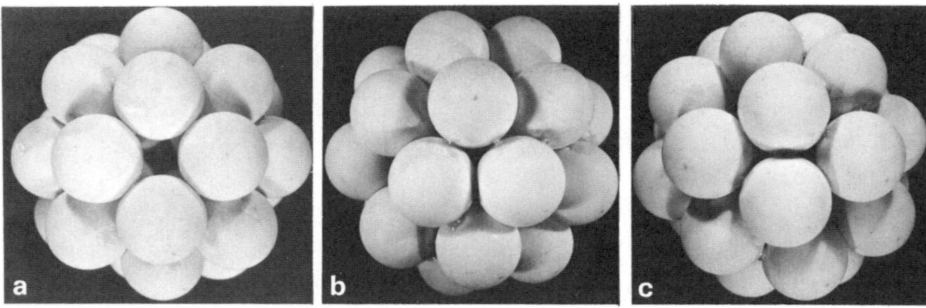

Fig. 3.6 A model made from twenty-four ping-pong balls illustrating 4—3—2 symmetry. The model is photographed along one of its four-fold axes (A), one of its three-fold axes (B), and one of its two-fold axes (C).

For complete 4—3—2 symmetry, a spherical shell of this type must contain 24, or some multiple of 24 identical sub-units (Fig. 3.6). The iron storage protein apoferritin, with diameter ~12 nm, was at one time thought to be a structure of this type, although it now seems more probable that this molecule contains 20 sub-units, arranged with pseudo 5—3—2 symmetry at the corners of a pentagonal dodecahedron. (This is a solid bounded by twelve faces, each face a regular pentagon.) The model of Fig. 3.5 is derived by putting a pin-pong ball on each of the *faces* of a pentagonal dodecahedron.

All the small virus particles which have been studied by either X-ray diffraction or electron microscopy have been found to have 5—3—2 rather than 4—3—2 symmetry. The capsid of TYMV (Fig. 3.4) contains a total of 32 capsomeres, apparently of two different types: twelve with the symmetry of the model of Fig. 3.5, and twenty capsomeres of a different type. Each of the twelve make contact with five neighbouring capsomeres in the capsid (e.g. capsomeres numbered 1, 3, 5, 9 in the model of Fig. 3.4), whereas each of the twenty has six neighbours (e.g. capsomeres numbered 2, 4, 6, 7, 8, 10, 11, 12 in Fig. 3.4). It seems from chemical evidence (Table 7.3) that there is only one type of sub-unit in the TYMV capsid (of molecular weight ~20000) but this has sufficient flexibility in the arrangement of its surface groups to allow aggregation in clusters of five (in the capsomeres with five neighbours) and clusters of six (in the capsomeres with six neighbours). This flexible arrangement may be of wide occurrence in capsid structure accounting for the prevalence of five-fold symmetry. The TYMV capsid thus contains a total of 180 sub-units in true 5—3—2 symmetry. The poliomyelitis virus is thought to have a structure somewhat similar to TYMV.

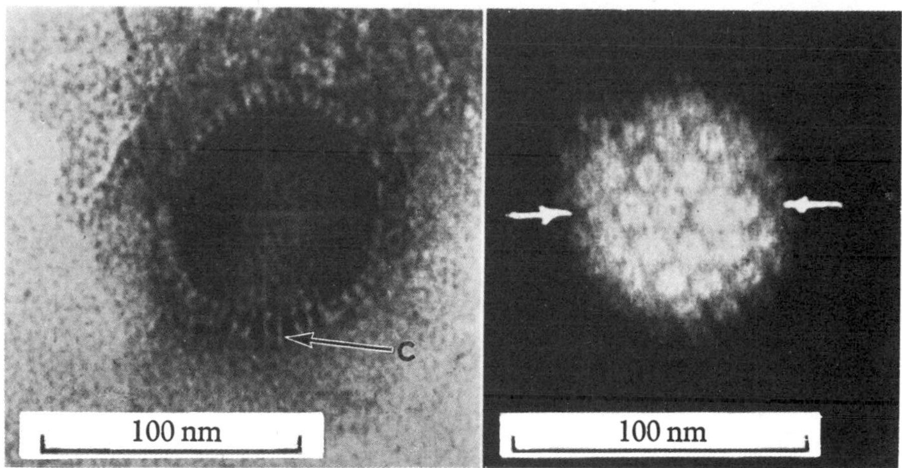

Fig. 3.7 Negative contrast pictures of *Herpes simplex*. The capsomeres in this virus (labelled C in the figure) can be seen to be short tubular structures 13 nm long and 10 nm in diameter. The complete capsid has a diameter of about 100 nm (i.e. more than three times that of TYMV) and is made of 162 capsomeres. The polyoma virus is of similar type, though rather smaller, with a diameter of 45 nm made up of 42 short tubular capsomeres. (*From* Wildy, Russell and Horne, 1960.)

φX 174, a virus infecting *Escherichia coli* bacteria (and hence classified as a bacteriophage) has a diameter of about 25 nm, and is hence a little smaller than TYMV, but is more complex in structure. Negative contrast electron micrographs of φX 174 do not resolve the individual capsomeres, but twelve mushroom-shaped 'spikes' can be seen protruding from the surface at points corresponding to the pentamers (five sub-unit clusters) of the TYMV structure. These spikes (which play a role in adsorption of the phage to *E. coli* prior to infection) can be removed by urea treatment to leave a capsid of 60 identical sub-units, each of molecular weight ~48000. The spike is made of three different types of sub-unit (separated by electrophoresis of the spike fraction on polyacrylamide gels). Each spike contains one sub-unit of molecular weight 36000 and five each of two smaller sub-units of molecular weights 19000 and 5000. The φX 174

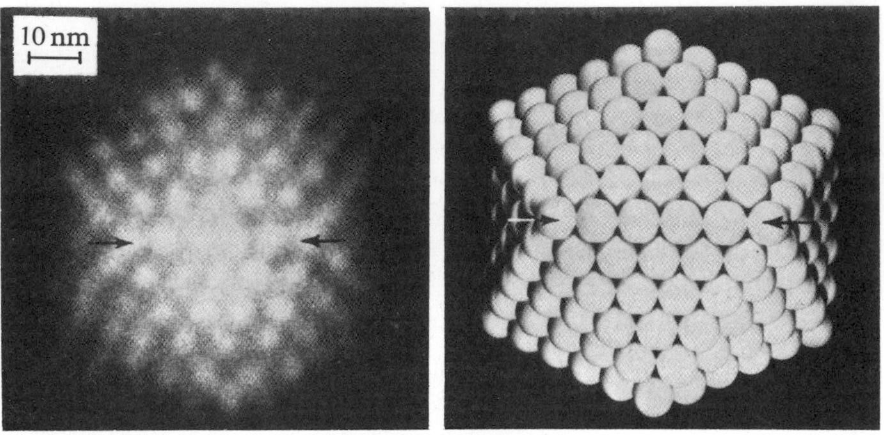

Fig. 3.8 A negative contrast picture of adenovirus type 5, and its interpretation in terms of an icosahedral shell of 252 capsomeres. (*From* Horne *et al.*, 1959.)

particle thus contains a total of 192 sub-units arranged in pseudo 5—3—2 symmetry.

The capsids of the polyoma virus, causing tumours in a variety of animals, and *Herpes simplex* causing lip sores in man, are built to a similar plan (Fig. 3.7), but with a larger number of capsomeres than TYMV. The capsomeres in these viruses appear to be short hollow tubes, and are therefore almost certainly made up of a number of sub-units. In the adenovirus type 5, reponsible for certain throat infections (Fig. 3.8) the capsid is a polyhedral shell 70 nm in diameter with 20 faces, made up of 252 capsomeres, with spikes at the twelve corners similar to those of ϕX 174.

In the total structure of *Herpes simplex* the polyhedral nucleoprotein component is enclosed in a loosely-fitting membrane which is apparently pinched off from the cell membrane as the new virus moves out of the infected cell – an alternative means to cell lysis by which infectious virus particles can be released. In other large viruses such as those of mumps and influenza the nucleoprotein forms long helices – perhaps more flexible versions of the TMV helix – coiled up inside a membrane (Fig. 3.9).

The structure of the T_2 bacteriophage is of some interest, as a variant on the structures we have discussed so far. Also this organism, and the very similar T_4 phage, have been used for modern genetic studies (Chapter 6), for the study of

Fig. 3.9 A negative contrast picture of a partially disrupted mumps virus particle. These large viruses are variable in size (100–600 nm in diameter). Hollow helices, like more flexible versions of the TMV-helix, appear to be coiled up inside a membrane. This membrane shows a palisade of projecting bars on its outer surface each about 9 nm long, spaced 7–8 nm apart. The influenza virus is of similar type but rather smaller (diameter variable around 100 nm). (*From* Horne and Waterson, 1960.)

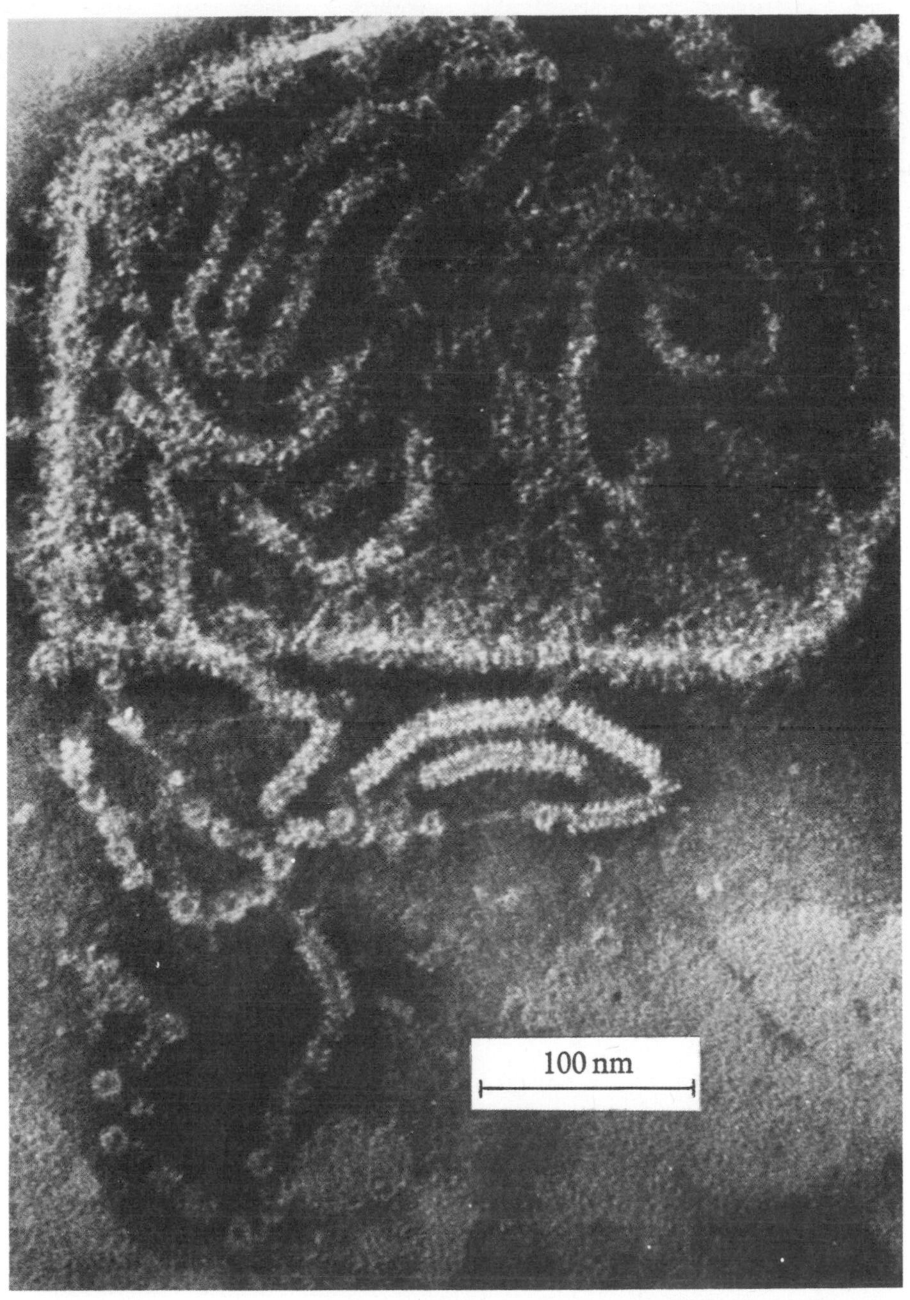

(*See foot of page 84.*)

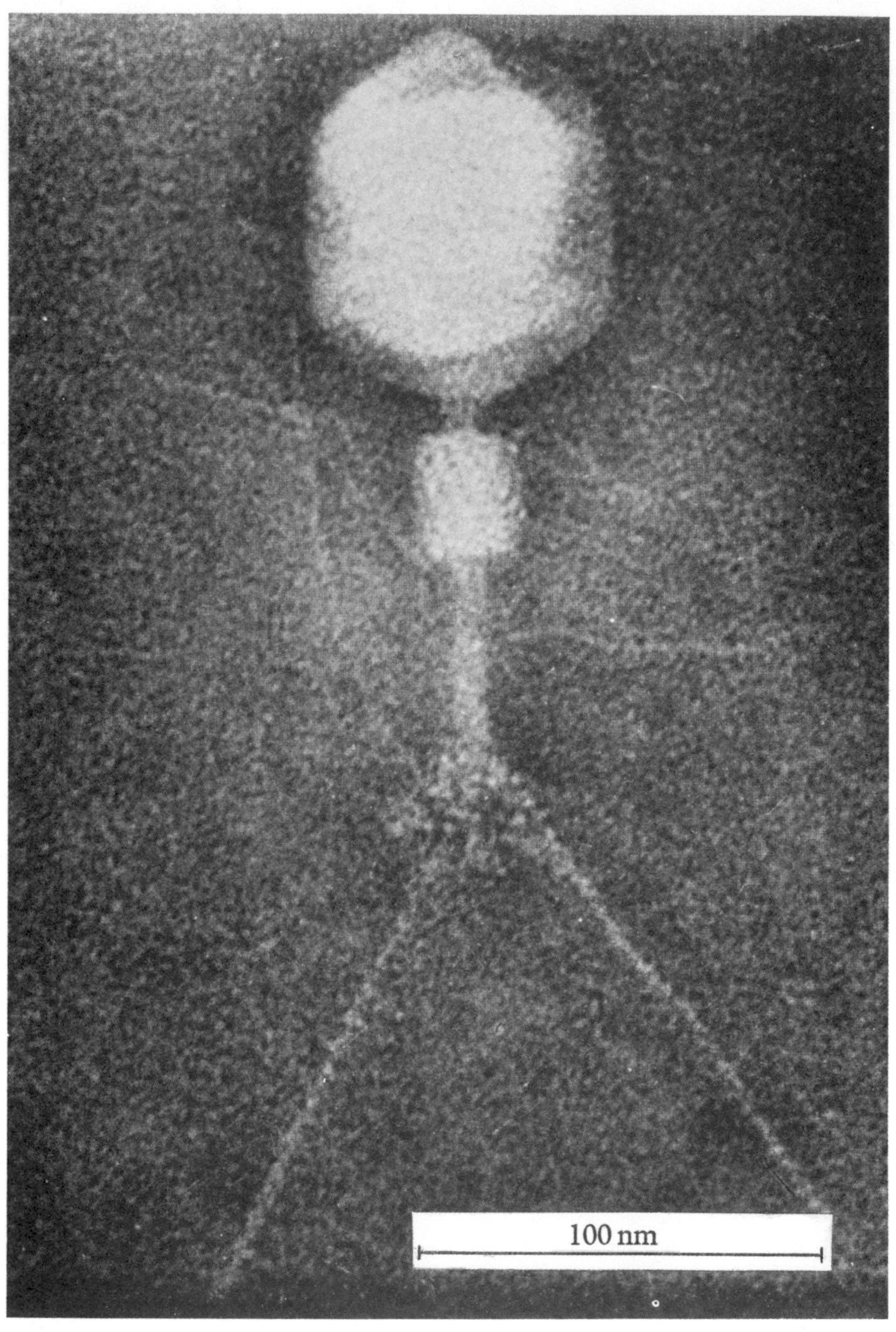
100 nm

interaction between viral and chromosomal nucleic acid, and for the study of 'messenger RNA' formation (Chapter 8).

The 'head' of the T_2 bacteriophage is about the same size as the adenovirus (i.e. about 70 nm in diameter). It consists of a protein envelope containing nucleic acid, in the form of one long double helix of DNA of molecular weight $\sim 120 \times 10^6$ – see Chapter 7 – folded back and forth, or coiled, within the protein envelope. The head appears in electron micrographs as a slightly elongated hexagon (Fig. 3.10) and the protein envelope may be supposed to be made up of capsomeres arranged in some regular way, like those of the adenovirus. Attached to the head is a 'tail', in the form of a hollow tube of protein, and attached to the end of this tube are slender protein fibres (Fig. 3.10). The role of the tail and tail fibres in bacteriophage infection is discussed in Chapter 10.

Microtubules

In the cytoplasm of a wide variety of cells (quite free of virus infection) long tubular structures are found, somewhat similar in construction to the reaggregated protein of TMV. These *microtubules* are found mainly in long slender cell processes like nerve axons and dendrites, and in large cells. In plant cells they tend to lie just inside the plasma membrane. Microtubules appear to have a structural, strengthening role, but may also play some part in determining the shape of elongated cells, and some role in cytoplasmic streaming (see Chapter 1). They form the core structure of cilia and flagella, and form the spindle of dividing cells, as discussed in Chapter 5.

Microtubules are formed by tubular aggregation of protein sub-units. Microtubule protein differs from TMV protein in having no affinity for nucleic acids, so that microtubules are purely protein structures. Their size shows some variation, in cells of different types, with outer diameters ranging from 20 nm to 25 nm. In cilia and flagella, and in the spindle of dividing cells these tubules form part of a contractile system.

Keratins, collagen, deposition of bone

We now turn to a quite different aspect of protein structure: the fibrous proteins. Keratins are the proteins of wool, hair, horns, hooves, nails, claws and feathers. Their common feature is a high cystine content (Table 2.2). Their secondary structure is variable. Hair, wool, porcupine quill and some other keratins, show evidence of α-helical secondary structure, and it has been suggested that in these proteins a number of α-helices twist together to form super-helices (like the

Fig. 3.10 A negative contrast picture of the T_2 bacteriophage. In this preparation, contraction of the outer sheath of the tail has taken place, to form the collar seen near the phage head. In negative contrast pictures of separated tails their non-contractile core can be seen to be a hollow tube, and at the distal end of the tail a hexagonal plate can be distinguished, to which six tail fibres are attached. Two of these fibres are clearly seen in this picture. (*From* Brenner *et al.*, 1959.)

strands of a rope). In feather keratin the polypeptide chains are in a more extended secondary structure.

Cross-linking between polypeptide chains arising from their high cystine content makes the keratins very stable, insoluble proteins. The molecular weight of such an assembly of chemically-linked chains may be very large, but if S–S bonds are broken, a high proportion of feather keratin, for example, is broken

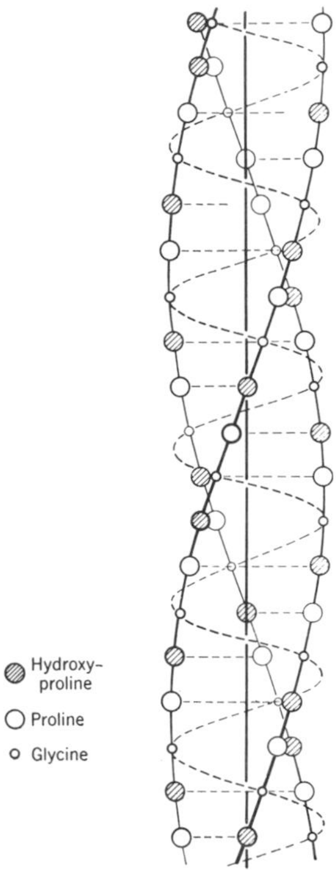

Fig. 3.11 Schematic picture of the helical intertwining of three extended peptide chains in the structure of collagen. (*From* Schmitt, 1959, after Rich.)

up into polypeptide chains of about 100 amino acids. Probably other keratins also are made up of cross-linked chains of a relatively low molecular weight. The problem of how epidermal cells become specialized for keratin synthesis, as they move outwards to the surface of the skin, represents one aspect of the general problem of the control of protein synthesis during cell differentiation, discussed in later chapters.

Collagen is another protein whose unusual amino-acid composition leads to a unique type of secondary structure and to the formation of protein fibres with special properties. These properties are primarily structural in connective tissue, tendons and cartilage, but in the deposition of bone, collagen fibres have the further role of providing condensation centres for the growth of needle-shaped crystals of hydroxyapatite [$Ca_{10}(PO_4)_6(OH)_2$]. One-third of the amino acids of collagen are glycine, and about a quarter are either proline or the closely related residue hydroxyproline which is found only in collagen (Table 2.2).

Proline and hydroxyproline residues cannot be incorporated into an α-helix since they have no N—H group (Table 2.1) and their side chains prevent the close approach of a C=O group to the backbone nitrogen, which is a feature of the α-helix. Proteins with a lower proportion of their polypeptide chains in helical form (Table 2.3) are in general those with relatively high proline content. In collagen, with proline and hydroxyproline content as high as 25 per cent, the chains do not form α-helices, but X-ray diffraction studies show that three chains twist together to give the structure of Fig. 3.11. Glycine residues occupy every third position along each chain, with their N—H groups forming interchain hydrogen bonds. Collagen molecules have a molecular weight ~360 000, each peptide chain having a molecular weight of ~120 000, and containing about 1 000 amino acids. The twisting together of three such chains leads to the formation of a molecule 280 nm long and 1·4 nm in diameter.

The isolated molecules of collagen in this form (molecular weight around 360 000) have been termed 'tropocollagen', and the banding pattern characteristic of native collagen fibres in electron micrographs has been interpreted in terms of a structure in which the tropocollagen molecules aggregate side to side to give alternating close-packed and open regions (Fig. 3.12) which appear as

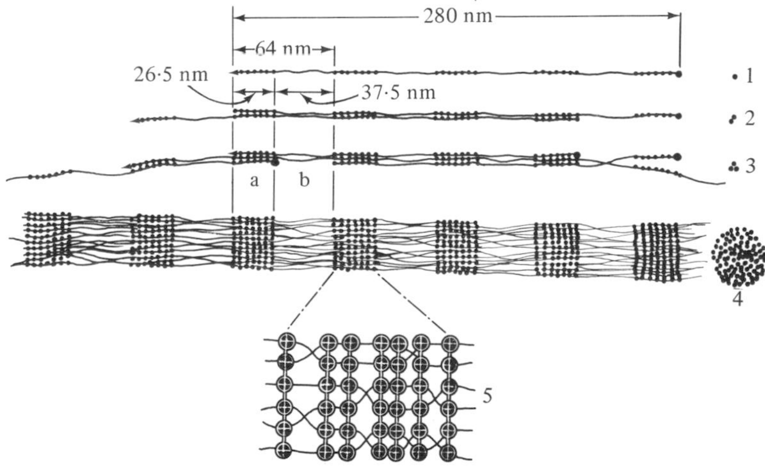

Fig. 3.12 Aggregation of tropocollagen to form collagen. (*From* Grant *et al.*, 1965.)

light and dark bands in negative contrast electron micrographs. Positive staining of collagen fibrils gives a much finer banding due to interaction of metal atoms with specific chemical groups along each tropocollagen molecule. Under *in-vitro* conditions the tropocollagen molecules can be aggregated into a variety of differently ordered structures. The form of the aggregate depends on the conditions, for example, on the salt concentration during aggregation.

In bone, small needle-shaped crystals of hydroxyapatite 20–40 nm long and 1·5–3 nm in diameter are embedded in a matrix of collagen fibres. Electron microscope studies of the early stages of calcification have shown that these apatite crystals begin to form at regular intervals along the fibres (Fig. 3.13) at certain specific sites in the collagen structure. It has further been possible to

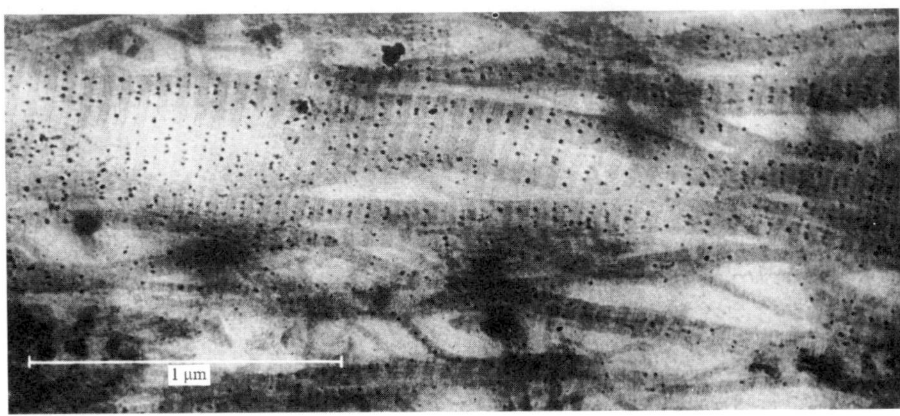

Fig. 3.13 An electron microscope picture of a thin section of avian embryonic bone, at an early stage of calcification. Dense deposits of apatite are forming at specific sites along the collagen fibre. (*From* Fitton-Jackson, 1957.)

study this process *in vitro* by 'seeding' near-saturated solutions of calcium phosphate with collagen fibres. X-ray diffraction studies show that the apatite crystals formed in this way are identical with those formed *in vivo*. An interesting feature of this work is the fact that collagen can be used successfully to seed calcium phosphate solutions, even if it has been previously broken up into its component tropocollagen molecules and reformed *in vitro*. But it is effective as a condensation site only if the tropocollagen molecules have been reaggregated to the native form of collagen. The other more random, or differently ordered, aggregates which can be formed from tropocollagen *in vitro* do not have, apparently, the correct surface sites for the initiation of apatite formation.

Fibrin and the peptic enzymes

Fibrinogen is a protein of molecular weight ~330000, circulating in the blood, and making up, in humans, roughly 4 per cent of the plasma proteins. On chemical evidence the fibrinogen molecule is thought to consist of two identical units, each of molecular weight ~160000, and each containing three separate polypeptide chains. In the electron microscope this two-fold symmetry is not apparent, for fibrinogen appears as an elongated molecule 28 × 6 nm made up of three globular units. These are the dimensions of the molecule at pH ~7. It is slightly shorter at the isoelectric point at pH 5·5, and in alkaline or more acid solution, as it acquires increasing net charge, it becomes more elongated. During blood clotting, fibrinogen is converted to fibrin by the action of the enzyme thrombin. In this conversion four peptides are split off the molecule (Fig. 3.14), two of type A (of nineteen amino acids each) and two of type B (of twenty-one amino acids each). The modified molecules (fibrin) then undergo lateral and end-to-end aggregation to form the fibres of the blood clot.

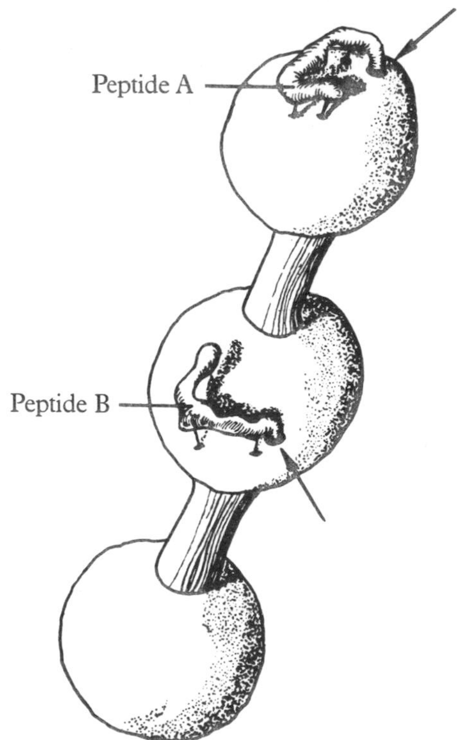

Fig. 3.14 A schematic picture of the 'half-molecule' of fibrinogen (molecular weight ~ 160000). The arrows indicate the sites of action of thrombin. Short threads indicate hydrogen bonds and other physical bonds holding the peptides in their native conformation in the protein surface, prior to the action of thrombin. (*From* Laki, Gladner and Folk, 1960.)

The four peptides contain a preponderance of acidic over basic groups and so carry a net negative charge at pH ~ 7. The modified protein molecules thus carry a different net charge, and the charge distribution over their surface is altered. Also new sites are present at the surface, unmasked by the removal of the peptides or by changes in tertiary structure which may follow their removal. These modifications of the molecular structure and surface charge distribution must lead in some way to a reversal of the previous predominance of intermolecular repulsion forces over forces of attraction. Removal of the A-peptides alone (from the ends of the molecules) favours end-to-end aggregation, while removal of the B-peptides (from the middle of the molecules) favours side-to-side aggregation.

The peptic enzymes pepsin, trypsin and chymotrypsin are formed from pepsinogen, trypsinogen and chymotrypsinogen, as fibrin is formed from fibrinogen, by the hydrolysis of peptide bonds. This hydrolysis converts the peptic enzymes from an inactive to an active form. The conversion does not in every case lead to the splitting-off of a peptide from the precursor molecule. The peptide fragment may remain attached to the protein. The hydrolysis of the peptide bond does result, however, in a rearrangement of tertiary and secondary structure to unmask, or create, the active sites of these proteolytic enzymes. We have noted earlier that insulin is formed from proinsulin in the same sort of way.

The mechanism of enzyme action

As discussed in Chapter 1, enzymic activity involves the formation of a transient complex between a protein molecule (the enzyme) and the substrate molecule (or molecules) taking part in the catalysed reaction. In this complex the substrate molecules are bound to a small area of the enzyme surface, termed the *active site*. For an understanding of the mechanism of enzyme action there are essentially two questions to be answered. First, there is the question of specificity: how is it that a given enzyme will form a complex with only a limited class of substrate molecules? Secondly, there is the question of the exact detail of the catalytic process: how does interaction between substrate and protein groups facilitate the redistribution of electrons within the substrate molecules which underlies the chemical change mediated by the enzyme?

The variety of amino-acid side-chain groups on the surfaces of proteins, and the variety of possible tertiary structure determining their exact three-dimensional arrangement, allows wide variety in the detailed structure of the protein surface. The incorporation of metal ions, coenzymes and other prosthetic groups, increases this variety still further. It is reasonable therefore to suppose that, for a given enzyme, the arrangement of surface groups at the active site will determine the hydrogen bonding, electrostatic and hydrophobic interactions, etc., that are possible between protein and substrate, and rather precisely define the class of molecule that can be adsorbed to the site.

We shall consider in this section the mechanism of action of lysozyme, the first enzyme studied in detail by the X-ray diffraction technique. Lysozyme splits the bonds between amino-sugars in a type of polysaccharide chain found in bacterial cell walls (Fig. 3.15). A substrate is only briefly bound to the protein surface during enzyme action and it is not possible to form a stable complex between lysozyme and the bacterial polysaccharide chain. However stable complexes can be formed between enzymes and inhibitors of their activity which are bound in competition with substrate at the active site. For lysozyme a trisaccharide made from one type of the two amino-sugars which alternate in the bacterial polysaccharide acts as an inhibitor and is almost certainly bound to the same part of the protein surface as the longer chain. Inhibitors can be

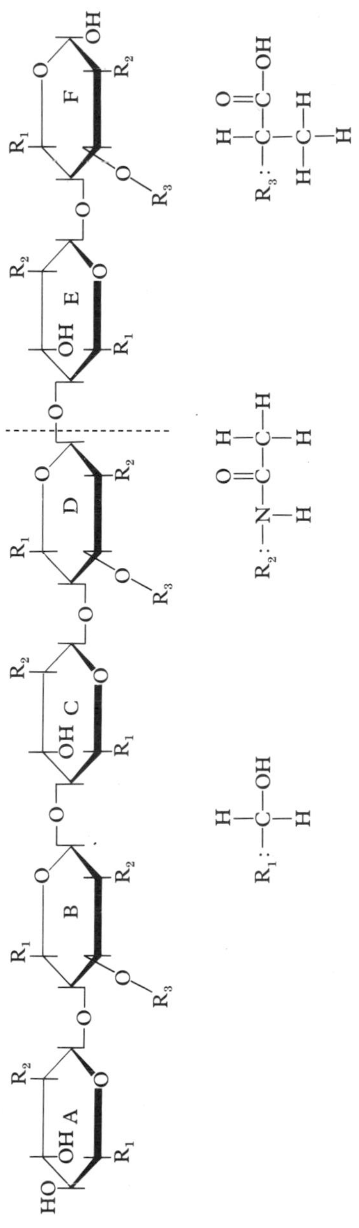

Fig. 3.15 The polysaccharide cleaved by lysozyme.

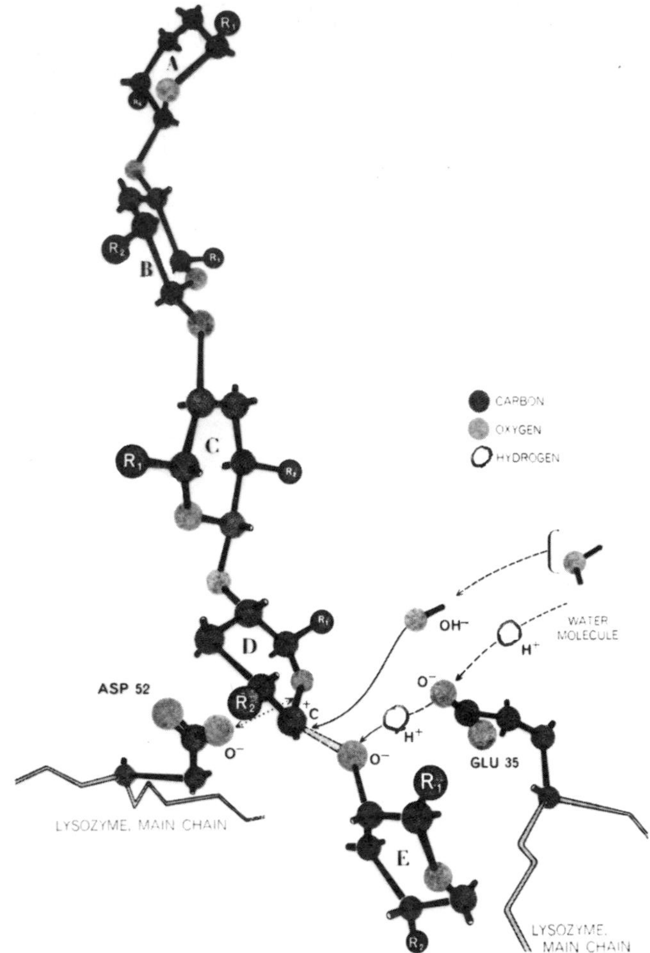

Fig. 3.16 The mode of action of lysozyme. The rings A, B, C, D, E correspond to those of Fig. 3.15. The polysaccharide is cleaved between rings D and E (for details see text). (*From* Phillips, 1966.)

bound to crystallized lysozyme by allowing the small molecules to diffuse into the lysozyme crystals, and the exact position of the inhibitor on the surface of the enzyme can thus be determined from X-ray diffraction analysis.

From studies of this kind Phillips and co-workers have proposed a very plausible model for the detail of lysozyme action which is illustrated in Fig. 3.16. The polysaccharide is held in a groove in the protein surface by hydrogen bonds and areas of hydrophobic interaction. For example, sugar ring C is held by hydrogen bonds from three sugar oxygens to three protein N—H groups (asparagine residue 59 and tryptophane residues 62 and 63, see Fig. 2.14) and from a sugar N—H to a protein backbone oxygen (alanine residue 107). The

—CH_2 group of sugar ring D is in hydrophobic contact with the protein tryptophane residue 108. In binding the substrate, the flexibility of protein structure allows a little 'give', so that the protein side chains can come into the exact position needed to form these hydrogen bonds. However the 'give' is limited, and when the polysaccharide is bound in the groove, sugar ring D (adjacent to the bond to be split by the enzyme) takes some of the strain and is slightly distorted from its normal lowest energy conformation. This distortion together with the nearby presence of the —COO^- group of aspartic acid (residue 52) causes redistribution of electrons in the C—O bond which is to be split by the enzyme. There is some movement of the electron cloud away from the carbon (marked C^+ in Fig. 3.16) towards the oxygen of the E-ring (marked O^- in Fig. 3.16). This electron movement leads to breaking of the bond, for the O^- can attract a proton (H^+) from the nearby glutamic acid

$$\text{—COOH} \rightleftharpoons \text{—COO}^- + H^+$$

residue 35, and the C^+ can attract an OH^- ion from the surrounding solution, so that the

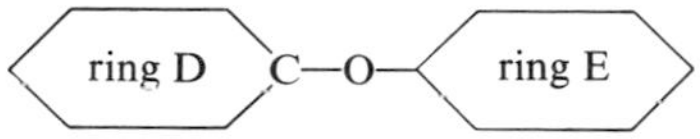

ring D C—O— ring E

bond is broken and

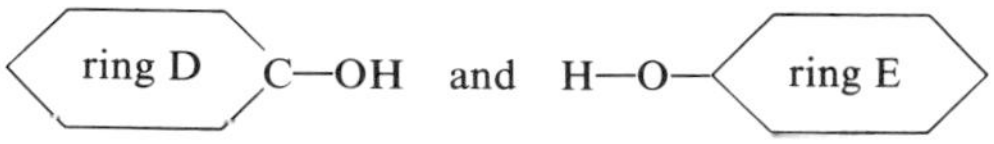

ring D C—OH and H—O— ring E

groups are formed instead.

There is a further interesting point in relation to the role of glutamic acid residue 35. Normally —COOH groups are almost fully dissociated at pH 7, that is to say they spend almost all their time in the —COO^- form, only occasionally and briefly trapping a proton to become —COOH. However the lysozyme —COOH group of residue 35 is situated in a hydrophobic pocket in the enzyme surface and hence is not in a normal aqueous environment. We noted in Chapter 2 and we shall see again in Chapter 4 that ionic groups are hydrophilic; ions are not soluble in lipids, and ionic groups do not want to get down into the hydrophobic interior of a protein. If a —COO^- group is drawn down into a protein molecule it must break quite strong O^- . . . H—O hydrogen bonds to surrounding water molecules, *or* trap a proton and go over to the unionized form —COOH. Thus the glutamic acid residue 35 in its hydrophobic pocket will spend a fair part of its time as —COOH, but the proton is only loosely held, and can move across readily to form the

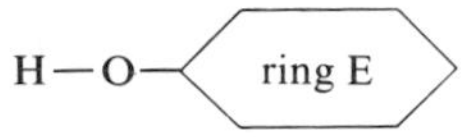

H—O— ring E

bond, with the glutamic acid —COO^- taking up another proton from the surrounding solution.

Feedback inhibition of enzyme activity

Tracing pathways of metabolism, and finding out exactly how enzymes work are two aspects of the problem of understanding the biochemistry of cells. But we need to know also how enzymes are made and how movement along the various metabolic pathways is controlled and co-ordinated. We shall see in later chapters that enzymes are made under control of the genes, and that to some extent co-ordination of metabolic activity is under genetic control – more enzyme is made if its substrate accumulates, or if the product of its activity is in short supply. However, apart from control of enzyme *synthesis* there is also control of the *activity* of the enzymes present in a cell at any one time. It is this latter control that we consider in this section.

At the simplest level some control over enzyme activity is effected by the relative affinity of sites on different enzymes for the same substrate or the same coenzyme, or by other direct effects such as inhibition of the activity of an enzyme by the product of its action if this accumulates to high concentration. A more indirect control is exerted by feedback inhibition. Figure 3.17 shows in outline some of the pathways through which the amino acids lysine, methionine and threonine can be synthesized in *E. coli* from aspartate. Cytidine-triphosphate (CTP), one of the building blocks from which nucleic acid chains are made (Chapter 7), can also be formed from aspartate. All these pathways show the phenomenon of feedback inhibition. Lysine inhibits the activity of one of the aspartate kinases, threonine inhibits the activity of the other. Threonine inhibits also the homoserine dehydrogenase, and isoleucine inhibits the threonine deaminase. Further, the aspartate transcarbamylase, which catalyses the first step in the synthesis of the pyrimidine ring of cytidine from aspartate, is inhibited by CTP. We shall now discuss this enzyme in more detail.

Aspartate transcarbamylase is a multi-subunit enzyme of total molecular weight around 310000. It contains six sub-units of molecular weight ~33000 which bind aspartate, and carry the active (catalytic) site of the enzyme, and six sub-units of molecular weight ~17000 which bind CTP, the regulator of enzyme activity. In some way, binding of CTP to the smaller sub-unit modifies the activity of the active sites on the larger sub-units. Further, binding of aspartate to the enzyme shows evidence of interaction between the large sub-units; when aspartate is bound to some of the sub-units this increases the affinity for aspartate at the empty sites. This effect is very similar to an effect observed in the binding of O_2 to haemoglobin that will be discussed in detail in Chapter 9. In the case of haemoglobin the detailed changes in the conformation of the molecule which accompany binding of O_2 can be studied by X-ray diffraction. Haemoglobin has therefore been taken as a model for the sort of changes of conformation, called *allosteric* effects, that may take place when a multi-subunit protein binds substrate and regulator molecules. The binding of regulator is supposed to bring about a small change in the tertiary structure of the regulator sub-unit. This might have a variety of effects, but one simple possibility is that the regulator sub-units may block access to the active sites when binding CTP, but

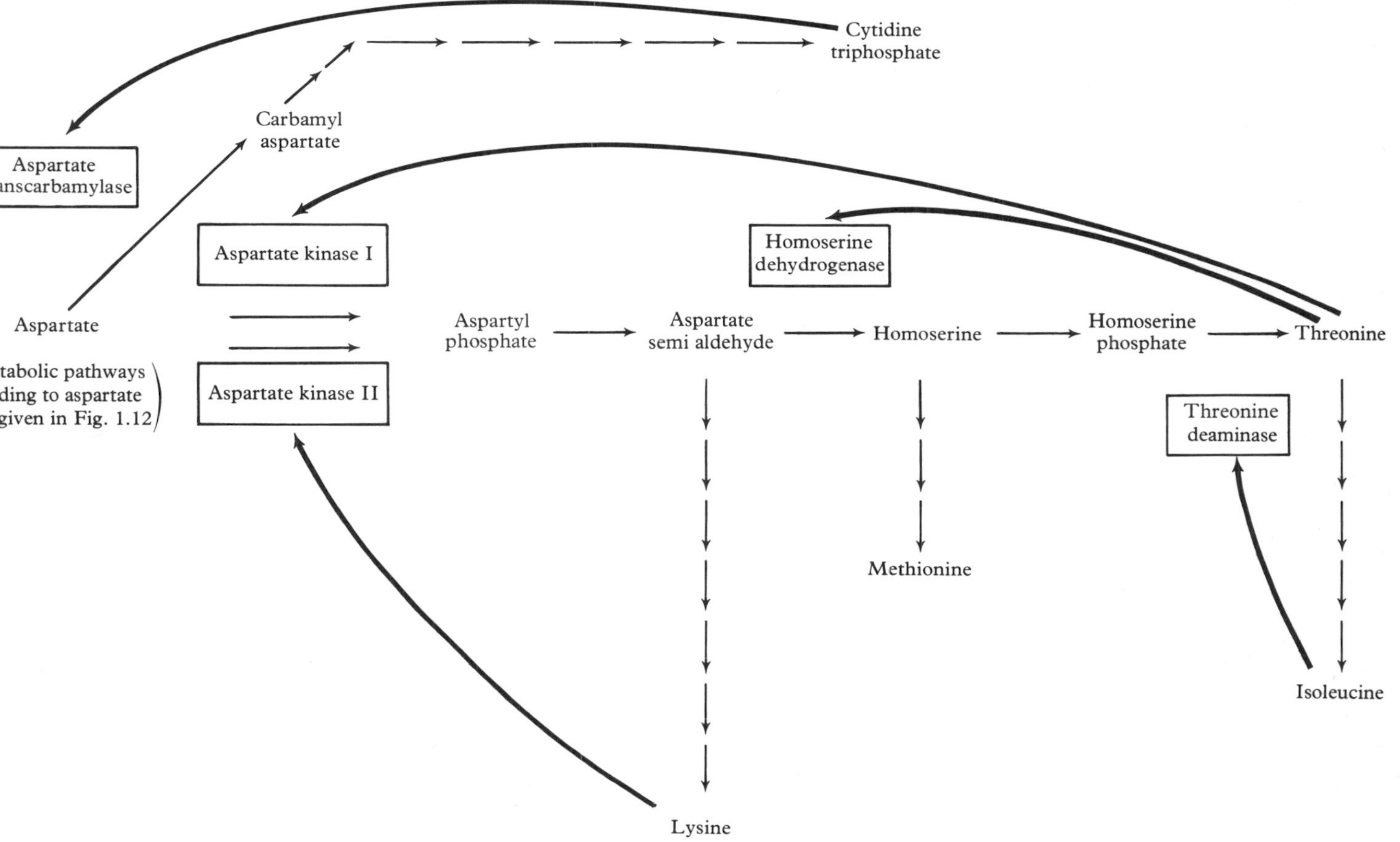

Fig. 3.17 An outline of the pathways of amino acid and pyrimidine synthesis from aspartate in *E. coli*. The heavy arrows indicate feedback inhibition of enzyme activity. (*From* Monod *et al.*, 1963.)

be partially withdrawn, and allow access, when no CTP is bound. Another possibility is that conformational change in one sub-unit can lead to slight re-arrangement of groups at the part of the sub-unit surface where it interacts with the neighbouring sub-units, inducing in them a conformational change to maintain a good 'fit' at the interacting surfaces.

In allosteric inhibition, as distinct from simple inhibition by competitive binding of substrate and inhibitor at the active site, the regulator can be a quite different type of molecule from the substrate, i.e. without chemical groups common to substrate and inhibitor, so that allosteric inhibition can act many steps back along a metabolic pathway. Allosteric processes can lead to activation as well as inhibition. For aspartate transcarbamylase, succinate acts as a competitive inhibitor, CTP as an allosteric inhibitor and ATP as an allosteric activator. The aspartate transcarbamylase, aspartate kinase and other feedback controls of Fig. 3.17 thus act to maintain a balanced supply of the different types of amino acids and nucleotides needed for protein and nucleic acid synthesis.

Allosteric enzyme control illustrates a point made in the last chapter, namely that protein molecules are not completely rigid structures. If they were, they would probably not act efficiently as enzymes. We noted in the last section that a small conformation change accompanies the binding of substrate to lysozyme. If no 'give' were possible, and to bind substrate the enzyme surface had to precisely fit the substrate molecule, this would require a precision of enzyme design that could probably not be achieved by folded molecular chains. Small conformational changes also allow the allosteric effects involved in control of enzyme activity.

There is no reason, in principle, why a single-subunit enzyme like lysozyme should not have one site on its surface binding substrate and another binding an inhibitor with allosteric interaction, but all enzymes so far found to show allosteric regulatory effects seem to contain more than one sub-unit. The significance of this will perhaps become clear when allosteric effects are understood in more detail. It may be that the conformational change within one sub-unit (change in tertiary structure) that can be brought about by binding of the regulator molecule is quite limited, and allosteric regulatory effects are achieved mainly by the rearrangement of quaternary structure that can follow from a small tertiary structure change.

Multi-enzyme complexes

We have been discussing enzymes catalysing a single chemical reaction, but made up of more than one type of sub-unit. We turn now to consider some examples of a related type of structure – macromolecular complexes made up of many sub-units catalysing not one, but a number of sequentially related steps along a metabolic pathway.

Lynen and his co-workers at Munich have studied the enzymes responsible for the cycle of steps by which saturated hydrocarbon chains are synthesized

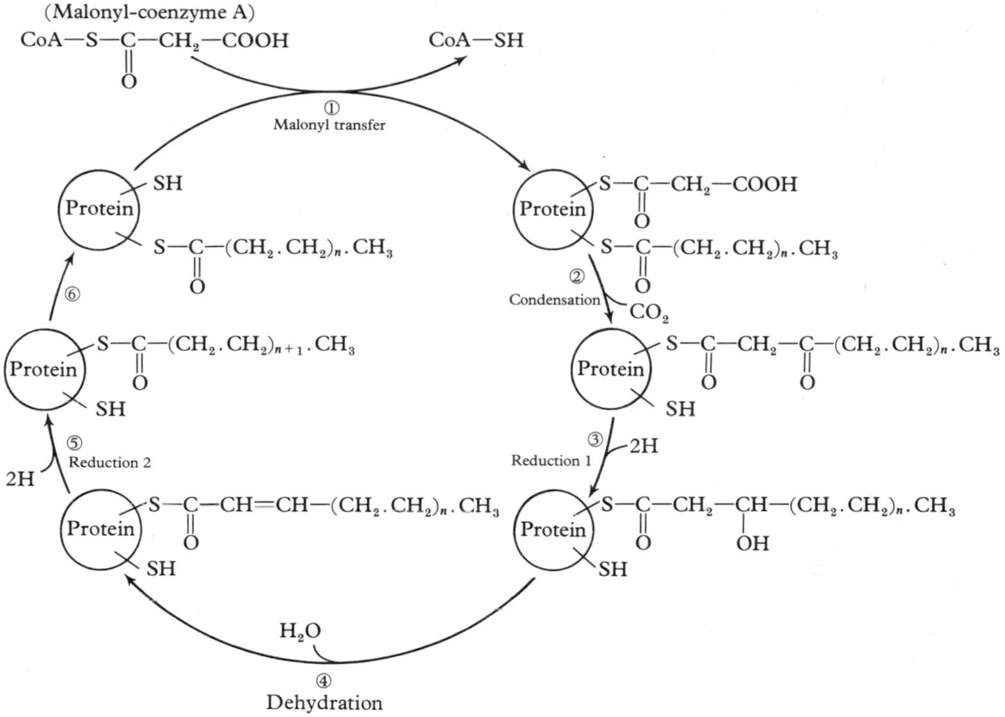

(Malonyl-coenzyme A)

$CoA{-}S{-}\underset{\|\atop O}{C}{-}CH_2{-}COOH$ → $CoA{-}SH$

① Malonyl transfer

(Protein)–SH, (Protein)–$S{-}\underset{\|\atop O}{C}{-}(CH_2.CH_2)_n.CH_3$

(Protein)–$S{-}\underset{\|\atop O}{C}{-}CH_2{-}COOH$, (Protein)–$S{-}\underset{\|\atop O}{C}{-}(CH_2.CH_2)_n.CH_3$

② Condensation → CO_2

(Protein)–$S{-}\underset{\|\atop O}{C}{-}CH_2{-}\underset{\|\atop O}{C}{-}(CH_2.CH_2)_n.CH_3$, (Protein)–SH

③ Reduction 1 ← 2H

(Protein)–$S{-}\underset{\|\atop O}{C}{-}CH_2{-}\underset{|\atop OH}{CH}{-}(CH_2.CH_2)_n.CH_3$, (Protein)–SH

④ Dehydration → H_2O

(Protein)–$S{-}\underset{\|\atop O}{C}{-}CH{=}CH{-}(CH_2.CH_2)_n.CH_3$, (Protein)–SH

⑤ Reduction 2 ← 2H

(Protein)–$S{-}\underset{\|\atop O}{C}{-}(CH_2.CH_2)_{n+1}.CH_3$, (Protein)–SH

⑥

Fig. 3.18 Synthesis of hydrocarbon chains. Two further enzymes (acyl transfer and palmityl transfer) initiate and terminate this cyclical process and the former also mediates recycling of the growing hydrocarbon chain by transferring it from one SH group to the other (step 6)

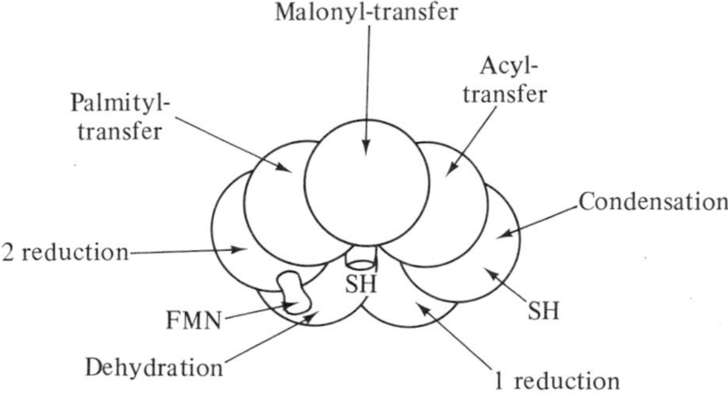

Fig. 3.19 The multi-enzyme complex responsible for the reactions of Fig. 3.18. (*From* Lynen, 1964.)

(Fig. 3.18). Starting from malonyl coenzyme A (whose ties with other metabolic pathways are indicated in Fig. 1.12) the hydrocarbon chain is lengthened by two CH_2 groups for each turn of the cycle, as indicated in Fig. 3.18. Five of the enzymes involved are indicated in this figure and further enzymes are required to initiate and terminate the process, making seven in all. Lynen's work suggests that all these enzymes are joined in a single complex which he represents (purely schematically) as in Fig. 3.19. The growing chain is supposed to remain covalently linked to the central —SH group as it swings round the active sites of the complex. Electron micrographs (Fig. 3.20) suggest that the complex is perhaps 'barrel-shaped' and made up of three rings. The complex looks like a ring when seen end-on, but side views show a three-layer structure. Since newly synthesized hydrocarbon chains are mainly destined for incorporation into membranes, it will not be surprising if this enzyme complex is found to be bound to cell membranes *in vivo*.

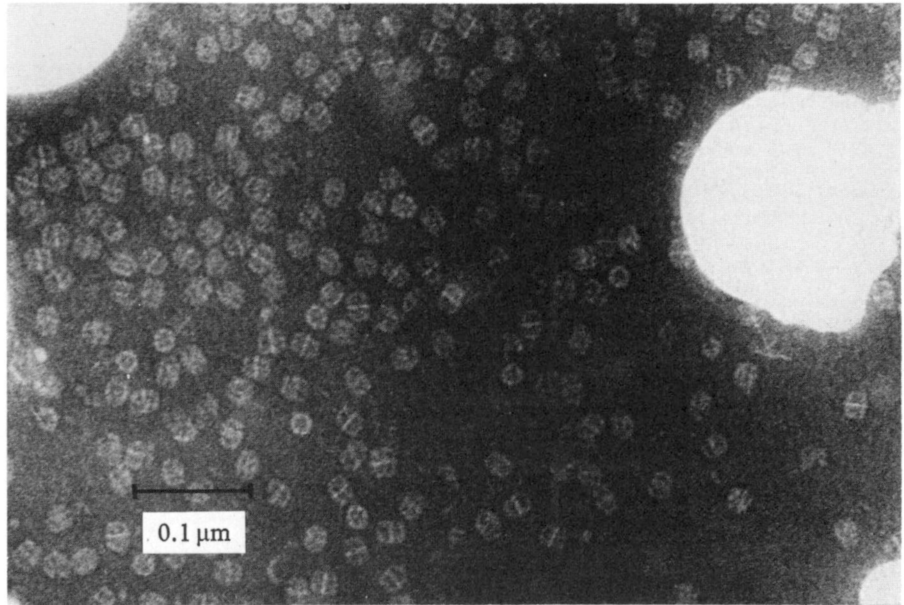

Fig. 3.20 Electron micrograph of the multi-enzyme complex which is represented schematically in Fig. 3.19. (*From* Lynen, 1964.)

Another example of a multi-subunit enzyme is pyruvate dehydrogenase, studied by Reed and his co-workers. The series of reactions catalysed by this complex are represented in Fig. 1.12 by a single arrow from pyruvate to acetyl-coenzyme A, but in fact a number of separate steps are involved as indicated in Fig. 3.21:

(1) Pyruvate decarboxylase
(2) Dihydrolipoyl transacetylase
(3) Dihydrolipoyl dehydrogenase

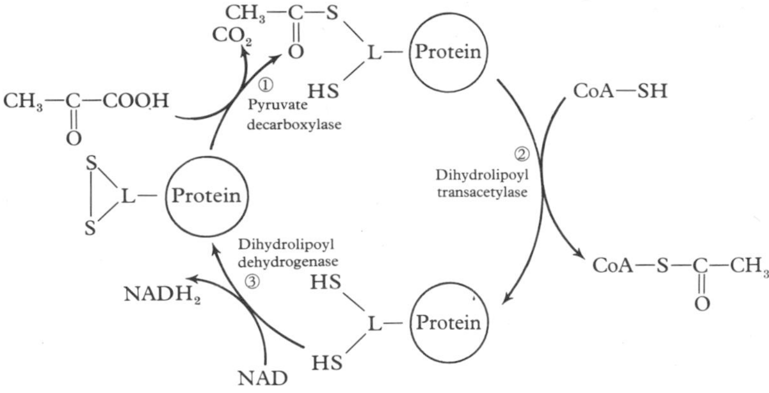

Fig. 3.21 The reactions of the pyruvate dehydrogenase enzyme complex

$\begin{matrix}S\\|\\S\end{matrix}\!\!>$L, lipoic acid, has the formula $\begin{matrix}S\text{———}S\\ |\qquad\quad |\\ CH_2.CH_2.CH.(CH_2)_4.COOH\end{matrix}$

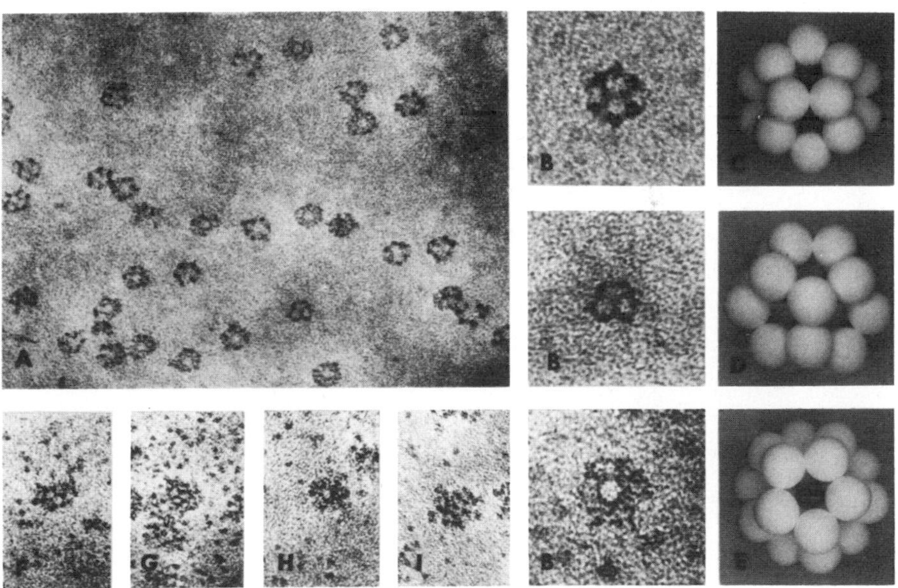

Fig. 3.22 A Electron micrograph of mammalian dihydrolipoyl transacetylase.

B Selected images from fields such as shown in A.

C, D, E Twenty-sphere model viewed down 2-, 3- and 5-fold axes, corresponding to the three different views of the complex seen in B.

F to I Electron micrographs of the complete mammalian pyruvate dehydrogenase complex. (*From* Ishikawa *et al.*, 1966.)

This is still a simplified representation since step (1) includes a number of separate reactions, but it is sufficient for our discussion.

All the enzymes of steps (1), (2) and (3) are grouped in a single pyruvate dehydrogenase complex. In the complex isolated from *E. coli* it seems that eight molecules of the enzyme catalysing step (2) aggregate together at the corners of a cube, then twelve of the enzyme clusters of step (1) aggregate to the edges of this cube, and six of the enzymes of step (3) to the faces. The whole structure thus has pseudo 4—3—2 symmetry (as defined earlier in this chapter). In the pyruvate dehydrogenase isolated from mammalian cells, 20 of the enzymes of step (2) aggregate, like the sub-units of ferritin, at the corners of a pentagonal dodecahedron (Fig. 3.22). Thirty enzyme clusters of step (1) then aggregate to the edges and twelve enzymes of step (3) to the faces of this dodecahedron. The mammalian pyruvate dehydrogenase is thus a structure with pseudo 5—3—2 symmetry.

The enzymes which catalyse steps (2) and (3), and the various enzymes of the step (1) reactions may themselves be made of two, four or more sub-units, so that the total structure may contain more than 200 sub-units.

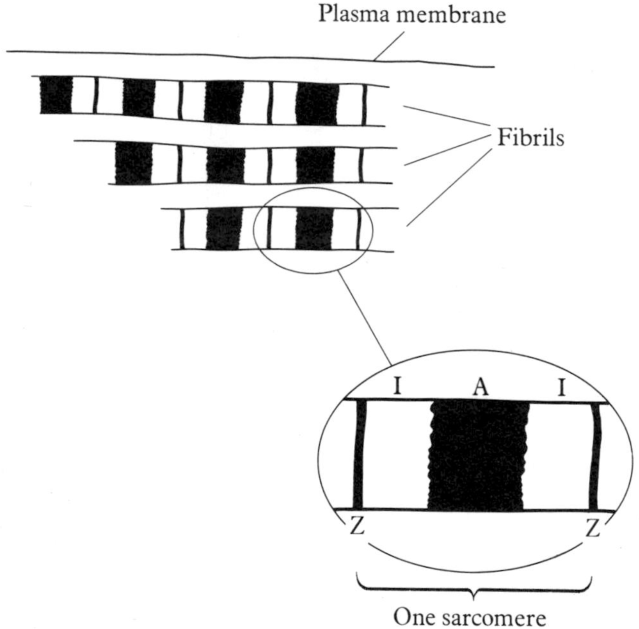

Fig. 3.23 The main features of a striated muscle cell as seen in fixed preparation or under a polarizing light microscope. The bands of the individual fibrils are in register to a large extent across the width of the cell. The I- and A-bands are optically isotropic and anisotropic respectively. The segment from one Z-band to the next, along a fibril, is termed a sarcomere.

Myosin, actin and muscular contraction

Under phase-contrast microscopy, or in polarized light, or in fixed and stained preparations, vertebrate skeletal and heart muscle cells show the lateral striations indicated in Fig. 3.23. These striations are a feature of the individual fibrils within the muscle cell, and it is because these fibrils are in register to a large extent, that the striations are seen, at relatively low magnification, as bands running right across the cell. In light microscope preparations, and more clearly in electron micrographs (Fig. 4.6) mitochondria can be seen between the fibrils, and also a membranous reticulum which will be described in Chapter 5. The mitochondria are mobilizing, in the form of ATP, the energy required for muscular contraction (see Chapter 1).

We are concerned in this section with the muscle proteins, and the detailed structure of the fibrils. The inset of Fig. 3.23 shows the main features of the striation pattern: the prominent Z-bands, and the I- and A-bands, so called because they appear isotropic and anisotropic respectively in polarized light. In the electron microscope the A-band is seen to contain an array of rod-like filaments each about 1·6μm in length and 10 nm in diameter, lying parallel to the fibril axis, spaced 30–40 nm apart (Figs. 3.24 and 3.25). These filaments lie in register, and their length defines the length of the A-bands. In transverse section (Fig. 3.26) they are seen to be arranged in a hexagonal pattern. Thinner, less rigid filaments, about 8 nm in diameter, run through the I-bands, and some distance into the A-band between the thick filaments (Figs. 3.24, 3.25 and 3.26). They are apparently attached to a plate-like structure lying across each fibril at the Z-band.

During normal contraction of striated muscle (to about 60 per cent of the muscle rest length) the A-band remains constant in length, and the shortening, which occurs uniformly along each fibril, involves a reduction in the length of the I-bands. Stretching of the muscle beyond the rest length involves only an increase in the length of the I-bands. During stretching, or during contraction there is movement of the thin filaments further out from, or further into, the A-band, but no change in the length of either filament (Fig. 3.27). The region of overlap between thick and thin filaments becomes less during stretch and greater during contraction. At 85 per cent of the rest length the ends of the thin filaments from adjacent Z-bands meet at the centre of the A-band, and during further contraction from 85 per cent to 60 per cent of the rest length, they begin to overlap at the centre of the Z-band (Fig. 3.27).

This 'sliding filament' mechanism was originally proposed by H. E. Huxley and Jean Hanson from X-ray and electron microscope studies and from phase-contrast light microscope studies of isolated fibrils. It was supported by A. F. Huxley and R. Niedergerke's careful interference microscope studies on intact muscle cells, which showed that the A-band dimension remains constant during stretch and mild contraction, and later by A. F. Huxley's analysis of the tension that can be developed by a muscle at different stages of stretch and contraction. The tension is found to depend on the extent of overlap of the two types of

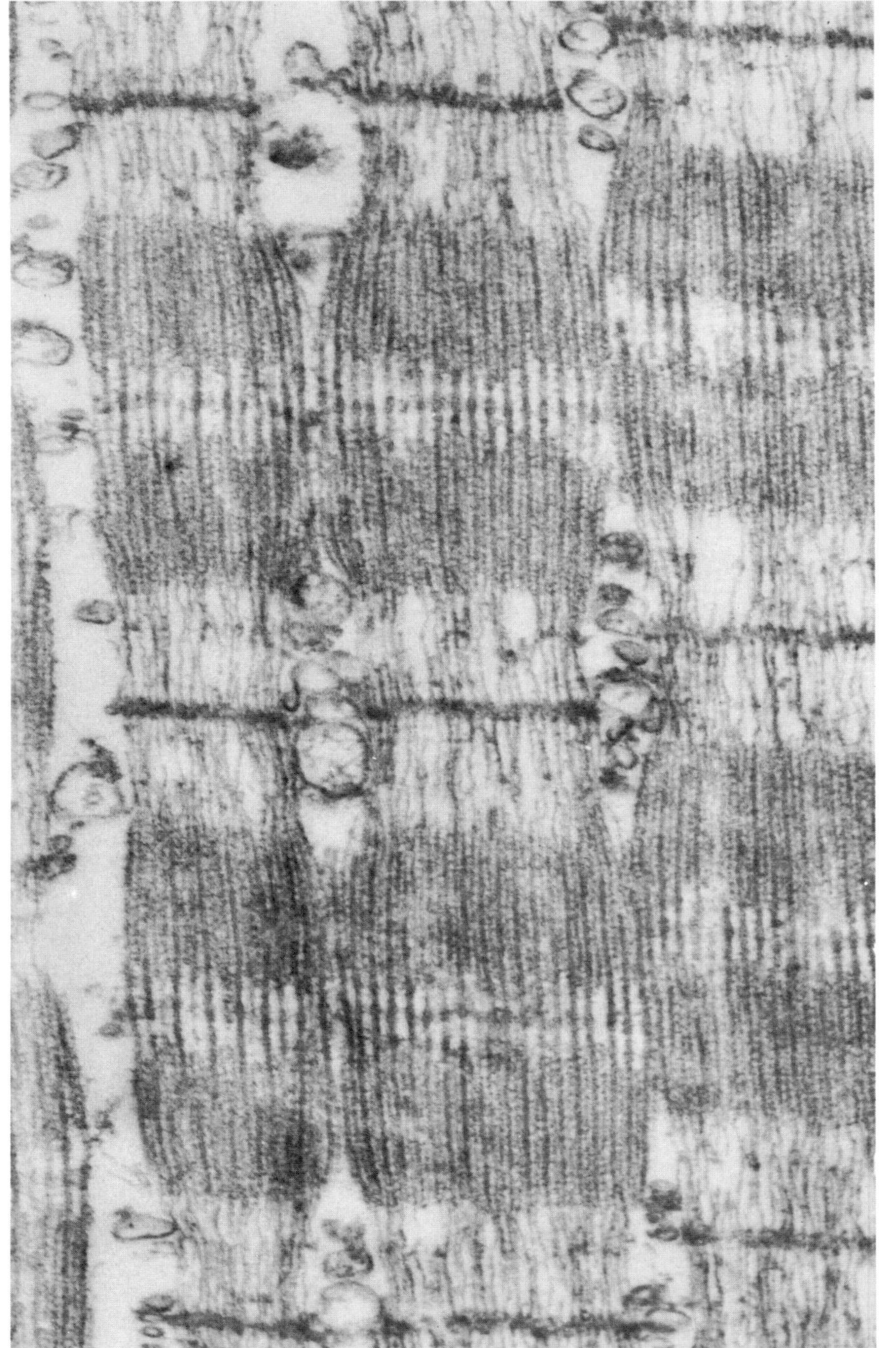

Fig. 3.24 Electron microscope picture of a longitudinal section of rabbit psoas muscle (glycerol-extracted) printed at a magnification of $\times 80000$. Thick filaments lie in register in the A-band. Thin filaments run from the Z-bands, through the I-bands and part way into the A-bands. (*From* Huxley, 1957.)

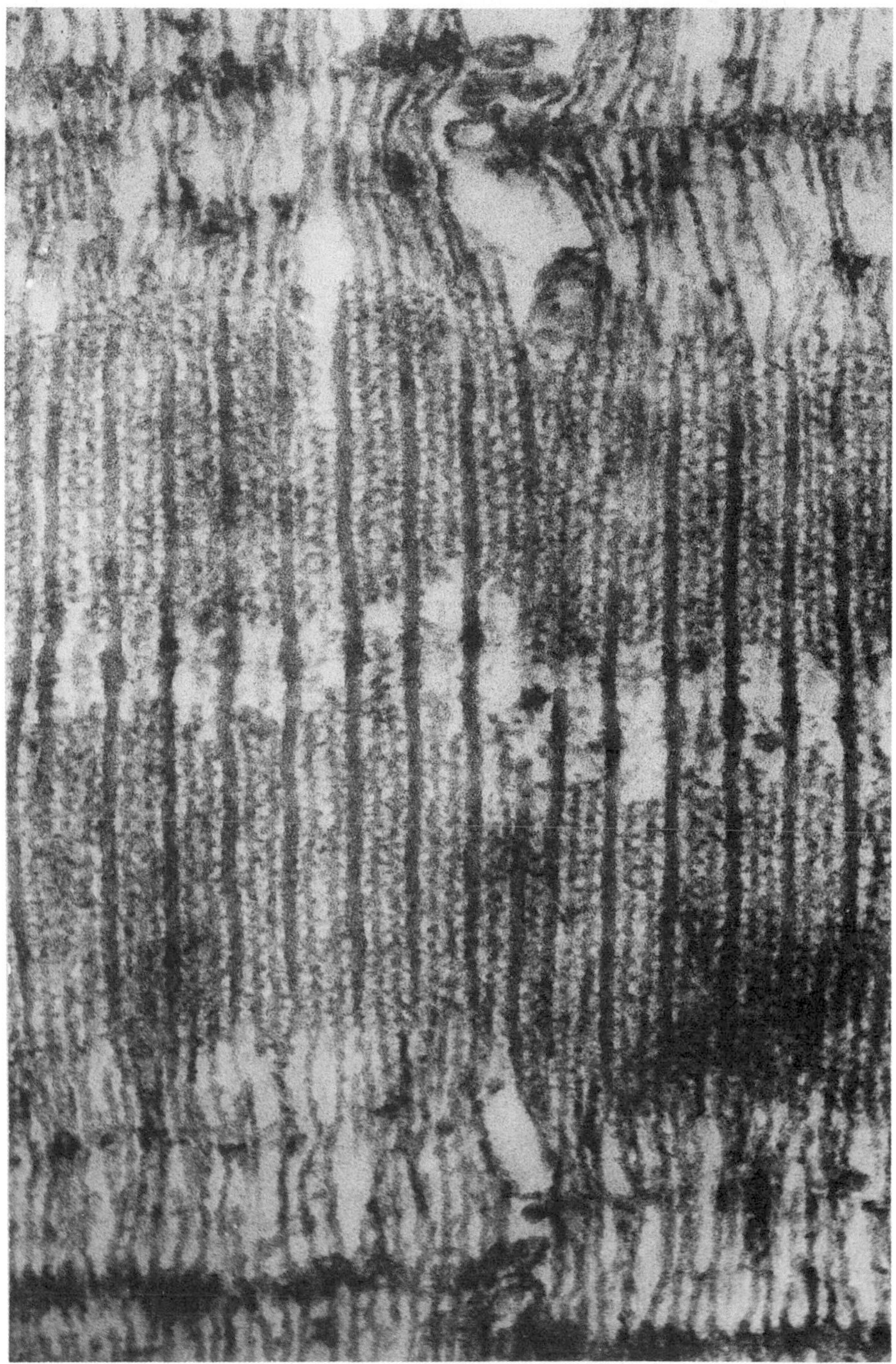

Fig. 3.25 Detail of one sarcomere of a preparation similar to that of Fig. 3.24 enlarged to twice the magnification (i.e. total magnification ×160000). Note the cross-bridges between the thick and thin filaments. (*From* Huxley, 1957.)

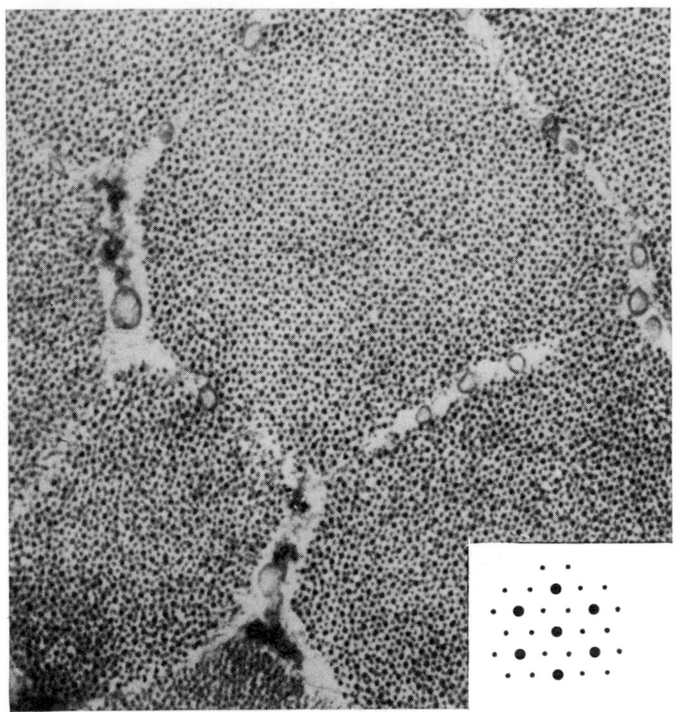

Fig. 3.26 An electron microscope picture of a transverse section of rabbit psoas muscle (glycerol-extracted) through the region of the A-band where thick and thin filaments interdigitate. Magnification ×67000. The inset shows the hexagonal arrangement of the thick and thin filaments. Each thick filament is immediately surrounded by six thin filaments. Each thin filament is immediately surrounded by three thin and three thick filaments. (*From* Huxley and Hanson, 1960.)

filament, which is the result to be expected from the 'sliding filament' model if tension is generated by some sort of interaction between the different types of filament in the overlap region. Some muscles, in barnacles for example, are capable of contraction to less than 60 per cent of their rest length, with spaces opening up in Z-band structure to allow the thick filaments to penetrate the Z-band and overlap the thick filaments of adjacent sarcomeres.

A variety of different proteins can be extracted from muscle. Myosin and actin together make up 55 per cent of the total protein of striated muscle; other proteins present in smaller amount are tropomyosin, troponin, α-actinin, β-actinin and myoglobin. Szent-Györgyi showed, in 1942, that mixed protein threads formed from myosin and actin can be caused to contract on addition of ATP. It therefore seems that these are the two more important proteins in the contractile structure of the muscle fibrils.

Myosin is a protein of molecular weight ~500000, making up about 40 per cent of the total muscle protein (in vertebrate striated muscle). It is a long thin molecule about 200 nm long by 2–3 nm in diameter, made up of two α-helices

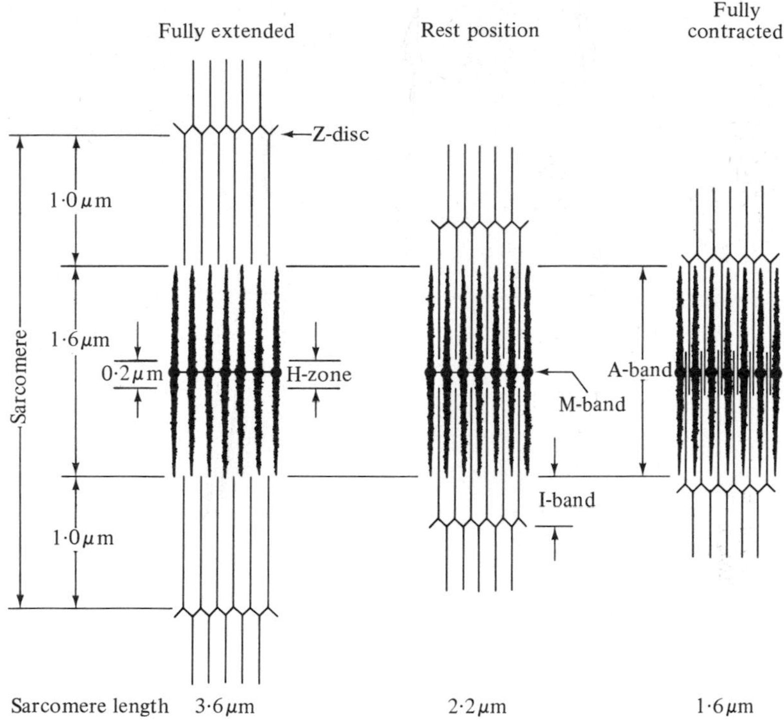

Fig. 3.27 Schematic representation of contraction of striated muscle. (*From* Muir, 1960).

twisted together like the strands of a rope, with a globular thickening or protrusion at one end of the molecule. This protrusion carries an ATPase site, splitting the high-energy phosphate group of ATP. Myosin molecules readily form elongated aggregates *in vitro* under physiological conditions of pH and salt concentration, and H. E. Huxley has suggested that both in these aggregates and in the thick filaments of the muscle cell, the molecules are arranged as in Fig. 3.29 with the protrusions forming the crossbridges seen in thin sections (Fig. 3.25). X-ray diffraction patterns of the type shown in Fig. 2.16 show that these protrusions are arranged in a helical way about the thick filaments with a repeat distance of 42·9 nm (Fig. 3.30).

Actin is a protein of molecular weight ~60 000, making up about 15 per cent of the total muscle protein. Its molecules readily aggregate to form fibres *in vitro*, and the details of this aggregation are of some interest, for actin molecules incorporate ATP (or ADP) as a firmly bound prosthetic group, and this bound ATP is split during aggregation. (This process may therefore have some relevance to the mechanism of muscle contraction.) In solutions of low salt concentration actin molecules exist as globular units of molecular weight ~60 000. The globular molecules aggregate into fibres as the salt concentration is brought up

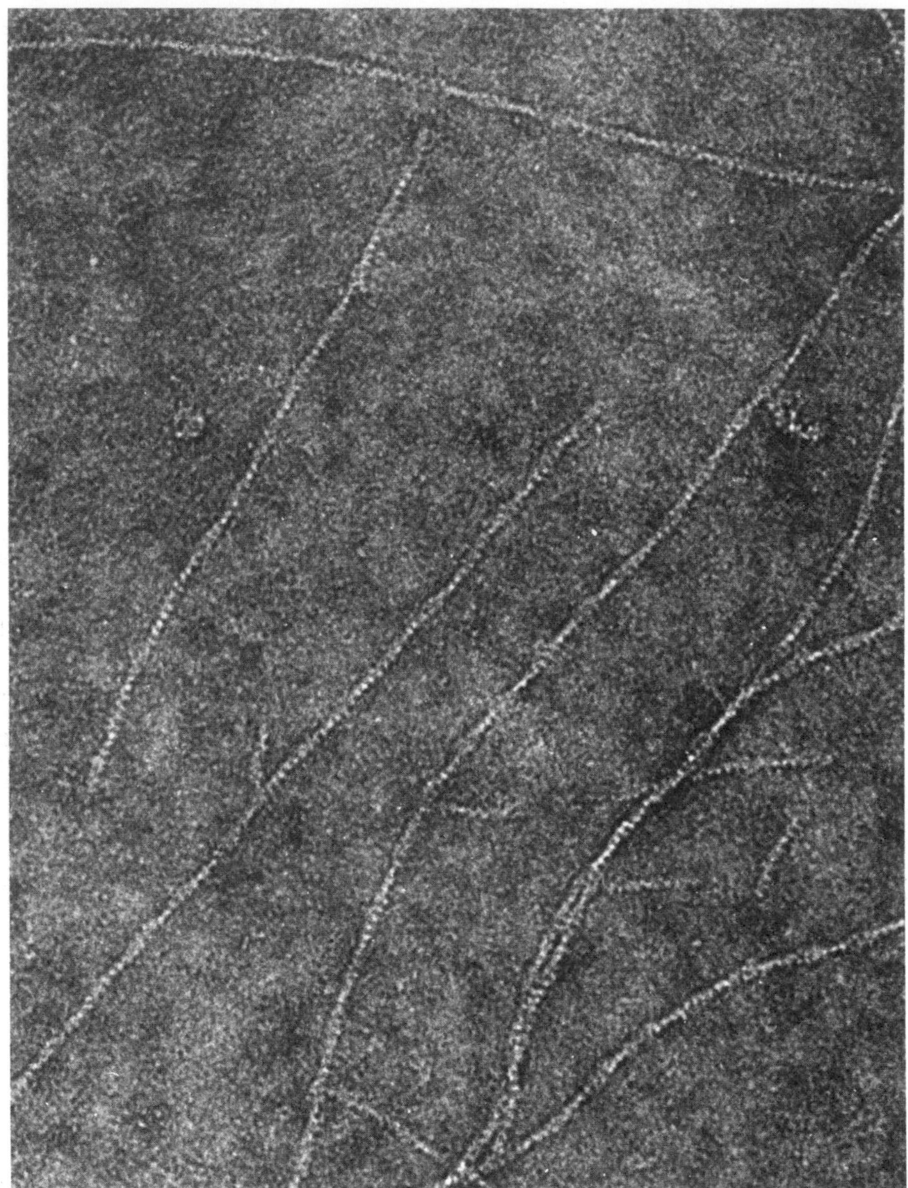

Fig. 3.28 F-actin threads viewed by negative contrast. (*From* Hanson and Lowy, 1963.)

to physiological levels, provided Mg^{++} ions are present, and provided the prosthetic group is in the triphosphorylated form (ATP). During fibre formation the terminal phosphate is split off the bound ATP. ADP and Mg^{++} ions remain bound to the actin and become incorporated into the fibre. The actin fibre is a helical structure which, in negative contrast electron micrographs, looks like

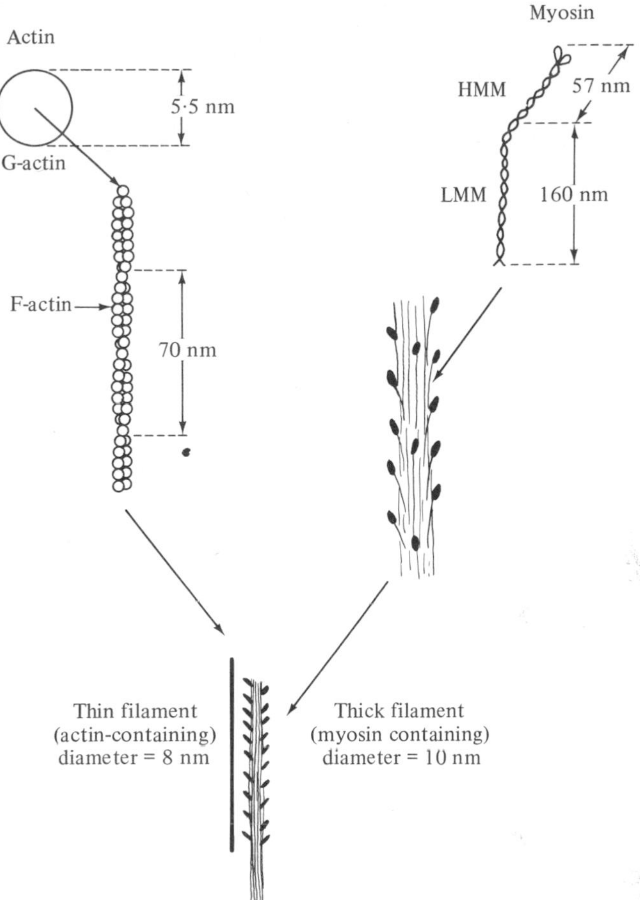

Fig. 3.29 The aggregation of actin and myosin molecules. (*From* Muir, 1971.)

two strings of beads twisted together (Figs. 3.28 and 3.29). Each 'bead' has a diameter of 5–6 nm and is presumably a 60 000 molecular weight unit. Each complete turn of the double helix is 70–80 nm in length and contains about thirteen pairs of beads.

Study of muscle fibrils by light microscopy after specific extraction procedures shows that the thick filaments of the A-band are predominantly myosin, and the thin filaments predominantly actin. *In vitro* aggregation of either myosin or actin sub-units leads to the formation of filaments of variable length, so that there must be some component in the filaments of the muscle cell, or some other factor as yet unknown, which controls their length *in vivo* (compare the way in which the nucleic acid strands of TMV limit the aggregation of sub-units in the virus, and see discussion below). Each of the thick filaments seen in the electron microscope, which are about 1·6 μm long and 10 nm in diameter, must contain

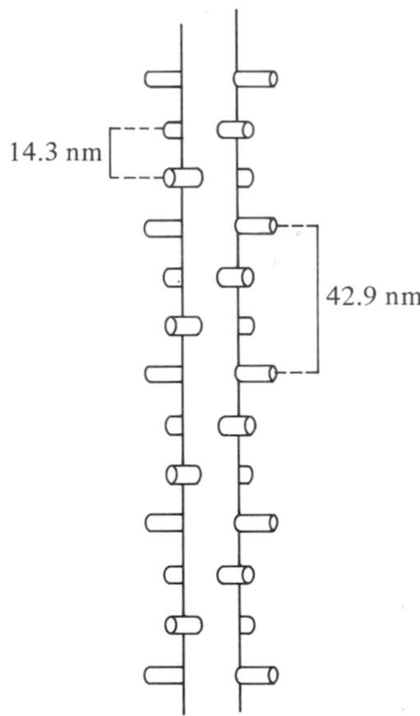

Fig. 3.30 Interpretation of the X-ray diffraction pattern of muscle in Fig. 2.16. The protrusions from the thick filaments, which form the cross-bridges, are in helical arrangement (due to the way myosin aggregates) with a repeat of 42·9 nm. (*From* Huxley, 1969.)

100–200 myosin units of molecular weight ~500 000. Cross bars can be seen in electron micrographs linking the thick filaments at the centre of the A-band.

When solutions of purified myosin and actin are mixed together, at relatively high salt concentration, and the salt concentration then lowered to physiological values, the solution gels, as a result of cross-linking between the molecules of actin and myosin. By suitable treatment, the protein molecules in these actomyosin gels can be oriented to form threads. In the absence of ATP the threads are relatively inextensible, but if ATP is added, with conditions so chosen that the myosin ATPase activity is inhibited, they become readily extensible. The myosin ATPase activity may be inhibited by suitable adjustment of Ca^{++} and Mg^{++} concentrations, and when the enzyme activity is inhibited the presence of ATP tends to break the cross-links between actin and myosin. If conditions are now changed (by varying the concentration of Ca^{++}, Mg^{++}) to remove the ATPase inhibition, the threads contract and develop tension. The behaviour of the actomyosin threads parallels, in all these respects, the behaviour of glycerol-extracted muscle cells, from which nearly all components but the fibril proteins have been removed or inactivated.

In living, relaxed, muscle cells the rate of breakdown of ATP is relatively low, so that the myosin ATPase activity is to a large extent inhibited *in vivo*, under resting conditions. Probably ATP is bound to the myosin ATPase site, without being split, and prevents cross-links forming between these myosin sites and the actin threads, thus accounting for the extensibility of muscle cells in their relaxed state. (Rigor in muscle fatigue, or rigor mortis, may be supposed to result from cross-linking between actin and myosin associated with a lowering of the intracellular concentration of ATP.)

Contraction of the muscle cell seems to be associated with temporary removal of the inhibition of the myosin ATPase activity and may be brought about by intracellular injection of Ca^{++} or by the passage of an action potential along the muscle cell plasma membrane. We shall return in Chapter 5 to a discussion of how the passage of this action potential might remove the myosin ATPase inhibition, and how this inhibition can be reimposed during the relaxation phase of the contraction-relaxation cycle. Confining discussion here to the contractile process itself, it seems that the central problem is that of understanding how removal of the inhibition of myosin ATPase activity, the subsequent splitting of ATP at the myosin active site, and the interaction with the actin threads at these sites, which we know to occur if the ATP comes off the myosin, can lead to a unidirectional movement of the actin thread along the myosin. A. F. Huxley and Simmons, and H. E. Huxley have suggested detailed models for the way in which the cross-bridges seen in electron microscope pictures (Fig. 3.25) might take part in this way in the contractile process. In other schemes that have been proposed, no direct contact is assumed between actin and myosin, during contraction, but only long-range electrostatic interaction between charged groups on the two types of filament. The fact that detailed ideas such as these can now be proposed, and tested quantitatively against muscle contraction data, is a measure of the progress which has been made towards localizing the sites in the muscle structure at which the chemical energy of ATP is converted to the mechanical energy of contraction. The problem of how this conversion is made remains unsolved. The work of Ebashi and his co-workers in Tokyo suggests that it may be in fact a rather complex process with troponin involved as well as actin and myosin (Fig. 3.31).

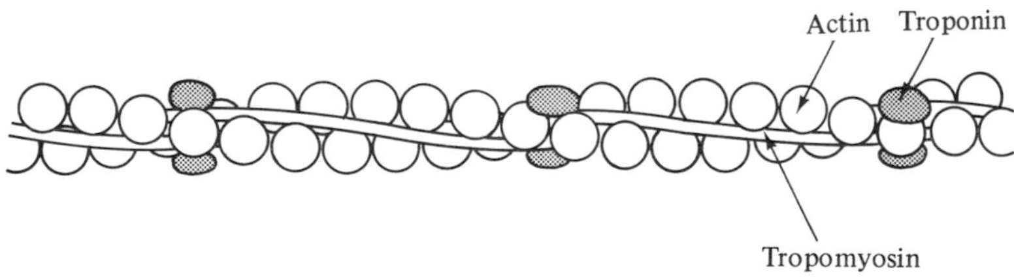

Fig. 3.31 A model showing the way in which tropomyosin and troponin may co-aggregate with actin in the thin filaments of muscle. (*From* Ebashi *et al.*, 1969.)

Apart from the question of how exactly muscle cells contract, there is growing interest in the further question of self-assembly of the muscle fibrils and how these proteins can come together spontaneously to form a structure of this complexity. Careful examination of thin sections shows that the Z-band has the structure shown in Fig. 3.32. This lattice structure is perhaps made up of tropo-

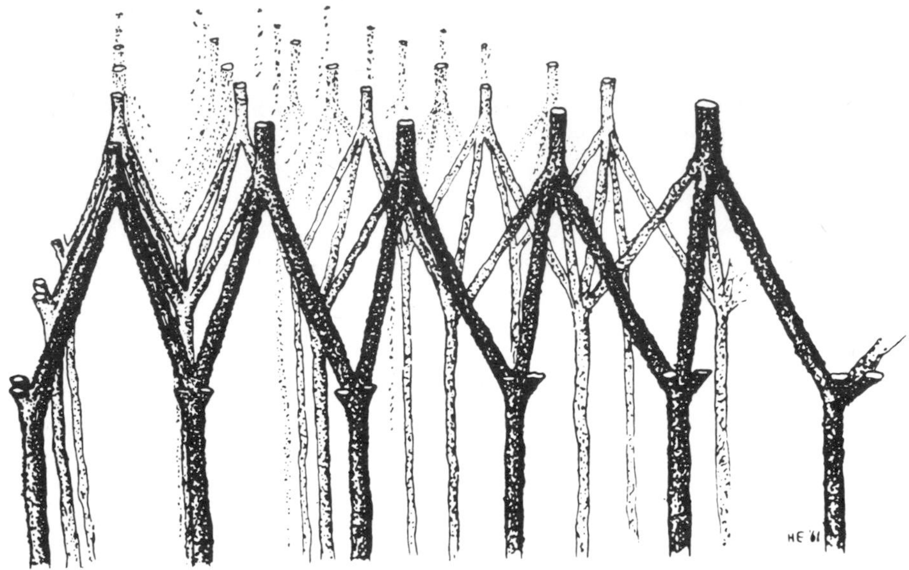

Fig. 3.32 A drawing showing the structure of the Z-bands of skeletal muscle. (*From* Knappeis and Carlsen, 1962.)

myosin or α-actinin. Troponin, β-actinin and tropomyosin appear to coaggregate with actin in the thin filaments, the coaggregation of several components determining the length of the filaments as indicated in Fig. 3.33. Either the Z-band lattice must form first, with the thin filaments growing out from it on either side, or perhaps thin filaments form, of 1 μm length, with one 'frayed' end and these frayed ends then come together to form the Z-band lattice. These questions are being tackled by study of the earliest formed structures that can be seen in thin-section electron micrographs of developing muscle cells, but it is difficult to study in this way the initial stages of aggregation, before the proteins have taken up a form clearly identifiable as part of a developing fibril.

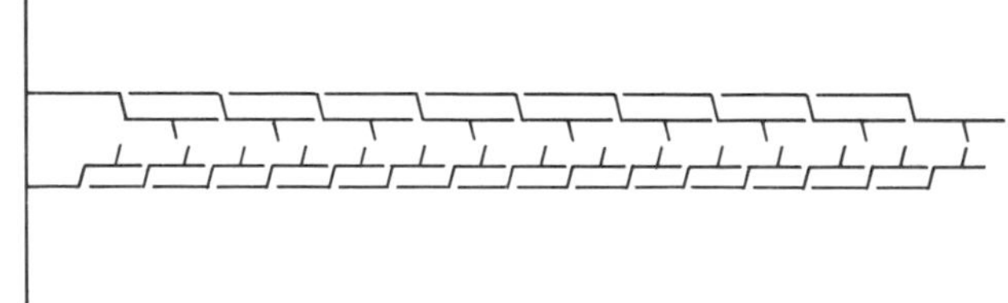

Fig. 3.33 A drawing showing how coaggregation of two components, to form a fibre, can lead to a stable structure of determined length, due to a stabilizing attraction between specific parts of the two components which only come into register after growth to this length. (Modified *from* Huxley, 1967.)

Similar problems arise in the electron microscope study of smooth muscle, which seem to be a more primitive, less organized, form of contractile, cell. Smooth muscle cells appear to contain thin filaments similar to those of striated muscle, but the myosin molecules are less extensively aggregated, and perhaps in some smooth muscle form dimers only, which cross-link and draw together the net of actin filaments during contraction.

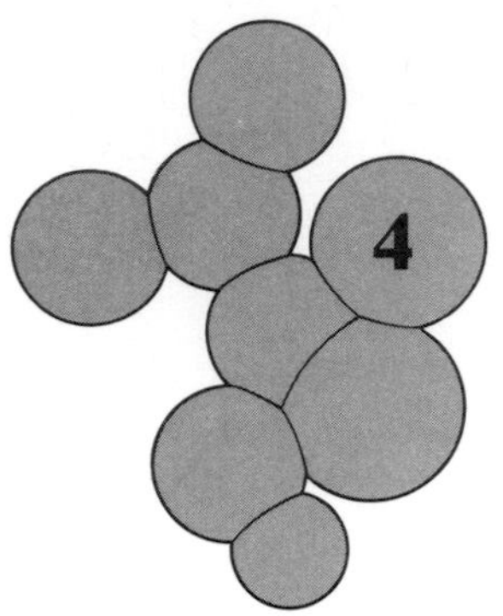

4 The cell surface

The plasma membrane

In thin-section electron micrographs, a common feature is found at the surface of all cells. A lamellar structure about 8 nm thick, taking up osmium or permanganate during fixation to a depth of 2–3 nm on either side, apparently defines the outer limit of the cytoplasm. This structure is seen at low magnification as a single dense line and at high magnification as a pair of electron dense lines 2–3 nm thick, separated by a less dense line of similar width (Figs. 4.1 and 4.2). In the red blood cell this is the only structure observed between the haemoglobin-laden cytoplasm and the blood serum. It will be convenient to refer to this feature in electron micrographs as the plasma membrane, for we shall see that there is very good evidence for equating it with the plasma membrane as defined in Chapter 1, that is the membrane which acts as the main barrier to diffusion into and out of the cell. Since the membrane is about 8 nm thick, and sections for electron microscopy are about 50 nm thick, the unstained central layer is seen clearly only where the membrane lies perpendicular to the plane of the section, as for example in the transverse section through microvilli shown in Fig. 4.4. Although a plasma membrane is apparently present at the surface of all cells, there is some variation in its structure in different cells and even in different parts of the same cell. The outer dense layer often appears thinner than the inner dense layer. Sometimes the outer layer may only show after fixation in potassium permanganate so that the osmium-fixed membrane appears even at high magnification as a single layer of osmiophilic material.

Epithelial layers of cells are usually strengthened by an extracellular layer of fibrous material, termed a basal lamina. Plant cells are surrounded by extracellular walls of varied composition. For bacteria also, walls and membranes are often present outside the plasma membrane, so that the total structure at the bacterial cell surface can be quite complex. It is usually possible however to identify the outer limit of the cytoplasm, to recognize the plasma membrane, and hence to classify all structures outside this membrane as extracellular. For bacteria the plasma membrane sometimes appears rather less than 8 nm in thickness (5–7 nm).

For most cells, even where no clearly defined extracellular wall or basal

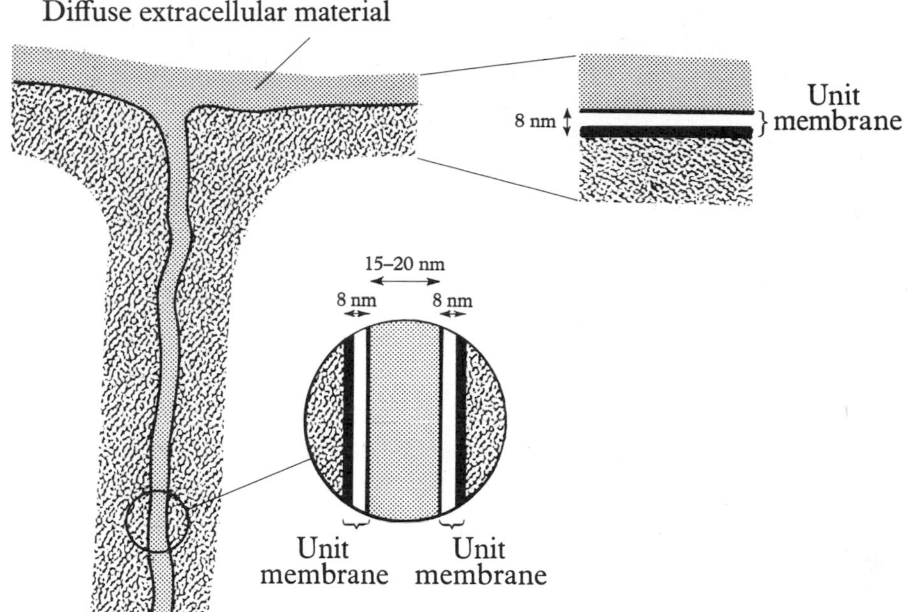

Fig. 4.1 Diagrammatic representation of the fine structure at the surface of cells. Two tissue cells are shown with part of their surfaces in apposition, and part exposed to extracellular space. The plasma membranes of these cells are represented by a single dense line, as they would appear at low magnification. Their high magnification appearance and dimensions are shown in the insets.

lamina is present, it is observed in electron micrographs that the external surface of the plasma membrane is covered with diffuse material, which has no sharp outer boundary. Occasionally, as at the surface of the cells lining the intestine, this material extends from the plasma membrane as far as 0·5 μm into the lumen, and gives a furry appearance to the surface (Fig. 4.3). When two cells are in contact (Fig. 4.1), similar diffuse material is found to fill a gap of 15–20 nm between the plasma membranes of the two cells. This gap is often found to be greater than 20 nm, but it is remarkable that it is not usually less than 15 nm. Hypertonic solutions reduce the dimension of the gap, which suggests that the diffuse material apparently filling the gap is perhaps a hydrated gel. This gel would not be expected to much affect the free diffusion of small molecules or ions through the 15–20 nm channels between cells, although it might restrict the movement of large protein molecules.

It is difficult to decide how closely the dimensions of these channels seen in fixed preparations correspond to their dimensions *in vivo*, but some experiments on giant squid axons suggest that there are aqueous channels of about these dimensions between cells in living tissues. Squid axons are surrounded by Schwann cells which abut against each other, and electron micrographs show a typical 15–20 nm gap between their plasma membranes; a similar gap is seen

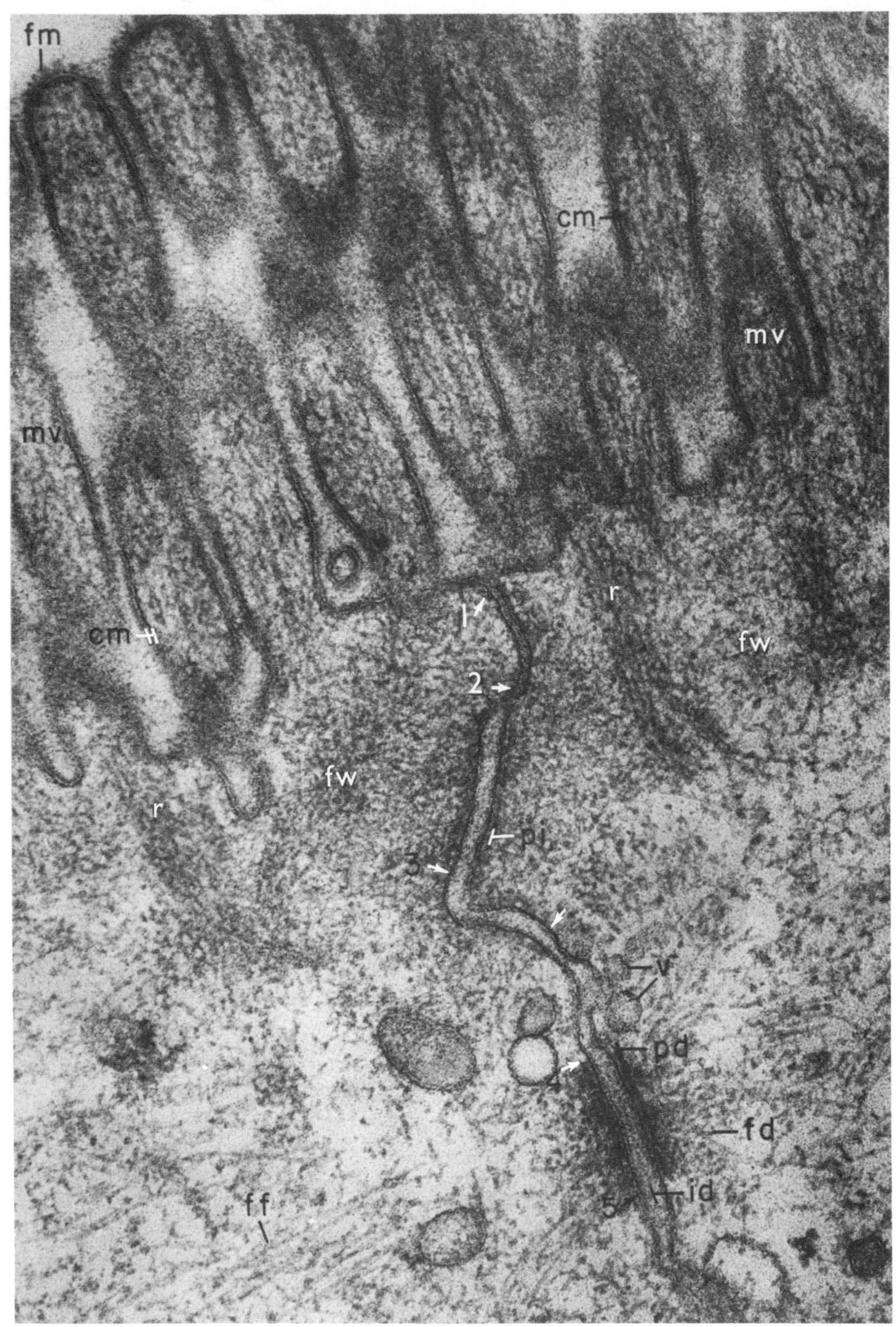
fm
cm
mv
mv
cm
r
1
fw
2
fw
r
pi
3
v
pd
fd
ff
5
id

between the axon and Schwann cell plasma membranes. If the axon is stimulated electrically to conduct a nerve impulse, sodium ions flow into, and potassium ions flow out from, the axon. The flow of ions from the external medium to the outer surface of the axon plasma membrane is very rapid and demonstrates the ready diffusion of ions through the intercellular channels between Schwann cells. With frequently repeated stimulation, however, the diffusion of ions is not rapid enough to prevent accumulation of potassium in the gap between axon and Schwann cell plasma membranes. The extent of this accumulation, determined from the change in the resting axon membrane potential, suggests that this gap is about 30 nm *in vivo*.

Other evidence, indicating that these narrow gaps between cells might be relatively free diffusion channels, is provided by electron microscope studies of the intestinal epithelial cell during fat absorption. Fat globules can be seen in the channels between the cells appearing to push the apposed plasma membranes apart as the globules move down the channels. When the fat has gone, the usual 15–20 nm gap is restored. This evidence suggests therefore that globules, or other bodies, larger than 20 nm can move along the channels. Other evidence is provided by the kidney tubular cell: here the area of the basal surface of the cells is greatly increased by complex infoldings, a specialization probably associated with the active transport of ions across the plasma membranes of these cells during tubular reabsorption. The infoldings are separated from one another by gaps of 15–20 nm. This infolding would not increase the effective area of the plasma membrane unless there were free diffusion in the gaps between the membranes. The closely packed microvilli of intestinal epithelial cells (Figs. 4.3 and 4.4) provide a mechanism for increasing the surface area of these cells, presumably to increase the rate at which material can be taken into the cells from the intestinal lumen. A similar 'brush border' lines the lumen of kidney tubules and the surface of the placenta. In all these situations, increase in effective area of membrane implies free diffusion of ions and small molecules in the spaces between microvilli.

When a basal lamina is formed to support an epithelium (Fig. 4.5), it is made up of an interwoven meshwork of fine fibres embedded in the diffuse material of the layer (free of fibres) just outside the plasma membrane. In Fig.

Fig. 4.2 The junction between two intestinal epithelial cells of the rat. (*From* Farquar and Palade, 1963.) Osmium tetroxide fixation. Section stained with lead hydroxide. Magnification ×85 000. mv = microvilli, cm = plasma membrane, fm = osmiophilic material coating the outer surface of the plasma membrane round the tips of the microvilli, r = fibrils running from the microvilli down into the cytoplasm spreading out and interlacing to form a web (fw). Between arrows 1 and 2 there is an occlusion zone; between arrows 2 and 3 fibrils from the web are attached to the plasma membranes (pi = attachment plate). Between arrows 4 and 5 there is a desmosome or adhesion disk. id = line of osmiophilic extracellular material midway between the plasma membranes, pd = attachment plate for the tonofibrils (fd and ff) running into the cytoplasm from the desmosome, v = vacuoles; the unnumbered arrow (between arrows 3 and 4) indicates a region where the plasma membrane at the junction can be seen to take up stain more densely on the cytoplasmic side than on its outer surface.

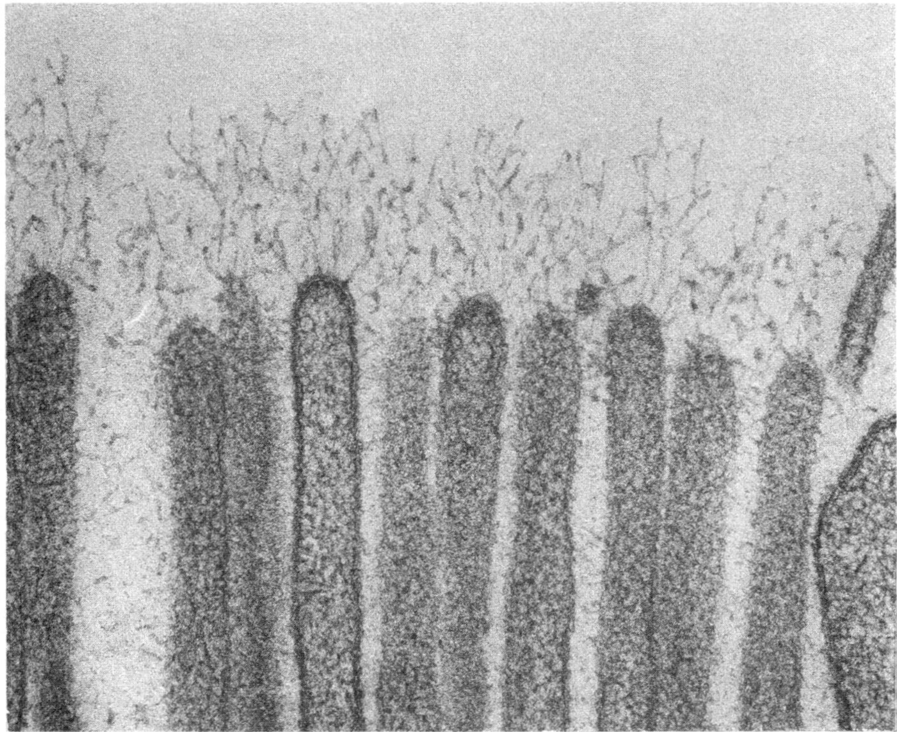

Fig. 4.3 Longitudinal section of intestinal microvilli. Magnification ×100 000. (*From* Ito, 1965.)

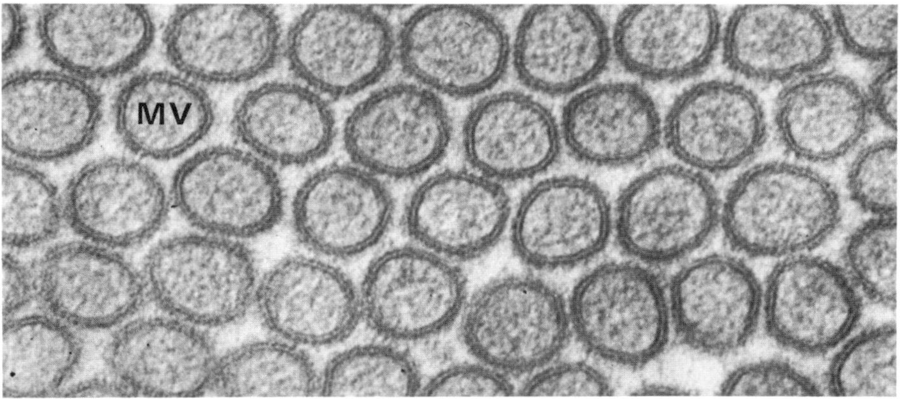

Fig. 4.4 A transverse section through the intestinal brush border. The individual microvilli (MV) are seen to be packed in a regular hexagonal array. Each microvillus is surrounded by a membrane which can be resolved into two osmiophilic layers separated by a clear layer. This is the high magnification appearance of the plasma membrane, seen particularly clearly in this situation because the membranes surrounding the microvilli all lie perpendicular to the plane of the section. Magnification ×130 000.

4.5 collagen fibres, cut in longitudinal and transverse section, are seen embedded in the diffuse material outside the basal lamina. It seems that the diffuse material found surrounding cells, and in the channels between cells, may be connective tissue ground substance, termed glycocalyx, known from light microscope histochemical studies to contain hyaluronic acid and other non-sulphated mucopolysaccharides.

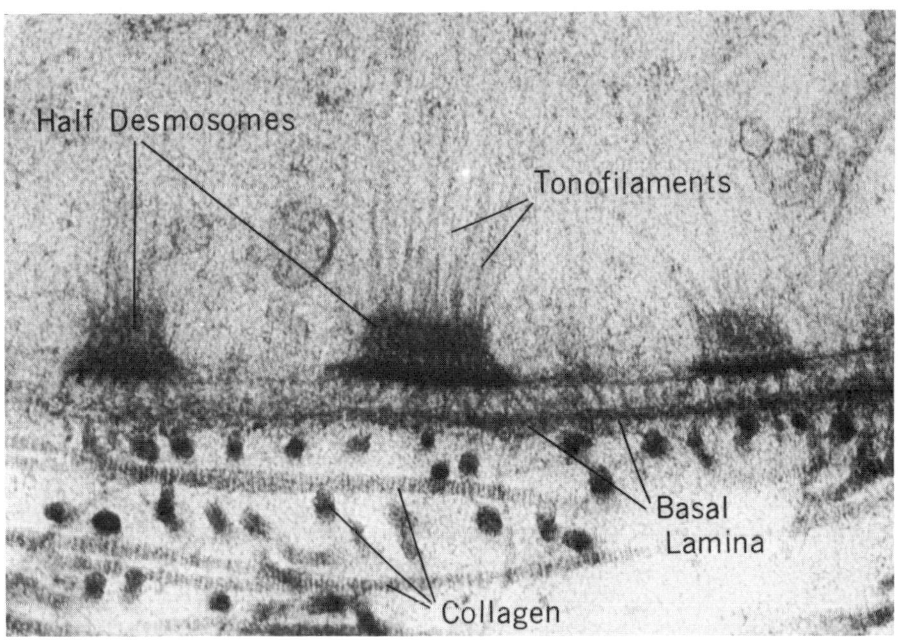

Fig. 4.5 Basal surface of cell in epidermis of a larval *Amblystoma*. Magnification × 50 000. (Micrograph by Elizabeth Hay, *from* Fawcett, 1969.)

In bone, cartilage, or white fibrous tissue, additional extracellular material confers on the tissue its characteristic properties. In cartilage the extracellular material contains sulphated mucopolysaccharides, in white fibrous tissue it contains large amounts of oriented collagen fibres, and in bone apatite crystals are arranged in a regular pattern along collagen fibres (see Chapter 3). In all these cases the pericellular glycocalyx blends into the intercellular structure. In plants and bacteria, polysaccharides and proteins form thick walls and capsules outside the plasma membrane.

The synovial fluid of skeletal joints contains sulphated mucopolysaccharides. The many negatively charged $—COO^-$ and $—SO_3^-$ groups and the absence of any large hydrophobic groups cause these chains to adopt a very open random coil (as compared with protein and nucleic acid chains) with water interspersed between the folds of the chain. This type of molecule acts as a good lubricant for the joint.

Intercellular adhesions

The distensibility of the usual 15–20 nm space between cells, mentioned above, suggests that little force is required to separate membranes apposed in this way. In plants, and in vertebrate bone and cartilage, strengthening of the tissues is accomplished by the rigid nature of the intercellular substance. Epithelial layers are strengthened by the underlying basal lamina. But in epithelia, and between the branching muscle cells of the heart, which are subject to shearing forces, direct methods of adhesion between cells would seem to be necessary. Two types of structure are seen in electron micrographs which appear to execute this function, termed maculae and zonulae adhaerens.

A *macula adhaerens*, also known as a desmosome, is a limited area of contact which appears as a modification of the apposed surfaces of two cells (Fig. 4.2). The two plasma membranes run parallel to each other with a separation of 13 nm between their outer surfaces; this space is bisected by a dense line, so that at a macula adhaerens a series of five parallel dense layers is observed with four intervening lighter layers. Cells which possess maculae adhaerentes always contain cytoplasmic fibrous elements, termed tonofibrils, and these converge on to, and appear to be attached to, the inner surface of the plasma membrane at the macula. The region of attachment of the fibrils forms a dense plaque on the cytoplasmic side of the membrane, and the opposed pair of dense plaques is the characteristic feature of a macula adhaerens. In a cell adhering to adjacent connective tissue, only one plaque is present but the outer surface of the plasma membrane shows some specialization to transmit tension to the basal lamina and collagen (Fig. 4.5). The structure of these adhesion points shows that the intercellular material and adjacent cytoplasm are modified, and experimental shrinkage of the cells proves that these are points of adhesion. Further, the fact that maculae adhaerentes are found mainly on those cells, such as skin epithelium, which are subject to shearing forces gives support to this interpretation of their function.

A *zonula adhaerens* differs from a macula mainly in that it surrounds a whole cell like a hoop or girdle and is not an isolated 'spot-weld'. Zonulae adhaerentes bind cells into flat sheets to form the endothelia of blood and lymphatic vessels. Zonulae adhaerentes are also seen at the inner border of intestinal epithelial cells (Fig. 4.2, between arrows 2 and 3). As in maculae, the plasma membranes run parallel to each other and there is a similar separation of about 13 nm between the membranes, but the outer layer of the plasma membrane disappears in the zonula and no bisecting line is present, so a clear space of about 30 nm lies between the inner layers of the two plasma membranes. The structure of the zonula adhaerens thus appears in electron micrographs as considerably simpler than that of a macula. Zonulae adhaerentes are widely distributed: between the liver cells where they form a bile canaliculus, at the inner margin of all secreting gland cells, and on cells of columnar epithelia.

Myofibrils at the end of a heart muscle cell are attached to dense plaques through which the force is transmitted to similar plaques and myofibrils in the

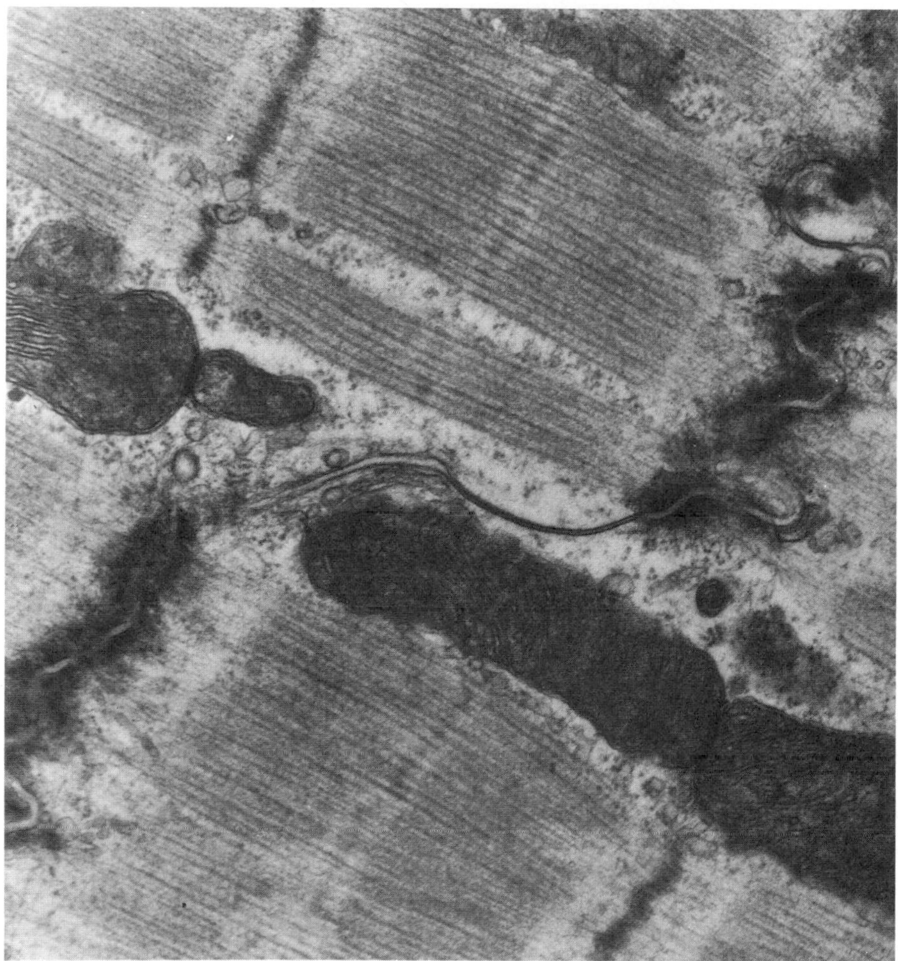

Fig. 4.6 Junction between heart muscle cells. An electrotonic synapse can be seen near the centre of the micrograph. At these contacts the membranes still appear to be separated by a 2 nm gap and the junctions are sometimes called 'gap-junctions'. The hexagonal array of 8·5 nm discs lies in this gap (see text). The discs do not stain with osmium. Magnification ×35000. (Micrograph by A. R. Muir.)

next cell (Fig. 4.6). These paired plaques resemble a macula adhaerens but there is no bisecting line. At the end of a skeletal muscle cell, similar plaques transfer the force to the extracellular collagen of tendon. Presumably at maculae and zonulae adhaerentes, and at the paired plaques of heart muscle, some material must be present joining the outer surfaces of the apposed plasma membranes, similar to the material that appears to join the plasma membrane to the basal lamina in Fig. 4.5, although it may not take up stain, or may show varying staining patterns after osmium fixation in these different situations.

Sometimes the plasma membranes of adjacent cells appear to be in direct contact, with the space between the cells occluded. If such a relationship prevails around a whole cell the contact is known as a *zonula occludens* and these have been demonstrated between all epithelial cells (Fig. 4.2). Since the usual space between cells is probably a channel permeable to water and ions, the zipping together of the adjacent plasma membranes at these occlusion contacts may be concerned with restricting diffusion along the intercellular gaps. Experimental proof of the sealing at least against the passage of protein molecules has been provided for the cells of kidney tubules: mouse kidney sections were examined in the electron microscope after intraperitoneal injection of ox haemoglobin, and the haemoglobin was seen to be penetrating between cells only as far as the point where the membranes were in contact. For capillary endothelia it seems that the zonula occludens does not usually completely surround each cell. There are small regions where there is slight parting of the closely apposed membranes to allow some diffusion of water, ions, small molecules and even protein molecules through the intercellular channels from the blood to the extracellular spaces of the tissues.

Close contact between plasma membranes can apparently serve a further purpose, apart from restricting intercellular diffusion, and this is to facilitate the *transcellular* passage of ions and small molecules. Smooth muscle, cardiac muscle and certain nerve cells in invertebrates are coupled together by synapses (i.e. functional contacts which transmit electrical impulses) which differ from the usual chemical transmitter synapses (such as the neuromuscular junction discussed in Chapter 5) in that they have no delay, they are reversible, and the electrical resistance between the cytoplasm of the two cells is lower than between either cell and the surrounding fluid. At these electrotonic synapses, the plasma membranes of the two cells form close contact and such a discrete area of fusion should be called a *macula occludens* (Fig. 4.6). Close contact between apposed membranes apparently, in these contact regions, modifies the molecular structure of each membrane, making the membranes more permeable to ions. Loewenstein and co-workers have demonstrated that there are specialized cell contacts of this type, which allow rapid diffusion of ions and small molecules, between the cytoplasm of adjacent cells in mammalian liver. At maculae occludentes the two apposed membranes appear to be cemented together by small discs about 2 nm thick and 8·5 nm in diameter. Colloidal lanthanum will penetrate between the membranes in these contact regions and can thus be used as a 'negative stain' in thin sections to show that the discs are arranged in a hexagonal pattern with centre-to-centre distance 9–10 nm. There may be a hole through the centre of each disc to allow cytoplasmic contact between the two cells.

Phagocytosis and pinocytosis

Phagocytosis and pinocytosis in amoebae, and cultured cells *in vitro*, as seen by light microscopy, have been discussed in Chapter 1. In electron micrographs very many types of cell, such as muscle cells, nerve cells and endothelial cells, are seen to contain small vacuoles bounded by a membrane of appearance and thickness similar to those of plasma membranes. Sometimes an infolding of the plasma membrane is seen which apparently represents the formation of such a vacuole. These vacuoles are regarded as pinocytotic vesicles. Although the dynamic process cannot be observed in the electron microscope, a variety of electron-opaque particles such as colloidal gold, mercuric sulphide, iron-dextran and ferritin can be seen in these vesicles in electron micrographs and appear to be entering the cells by pinocytosis. There seem to be three types of pinocytotic

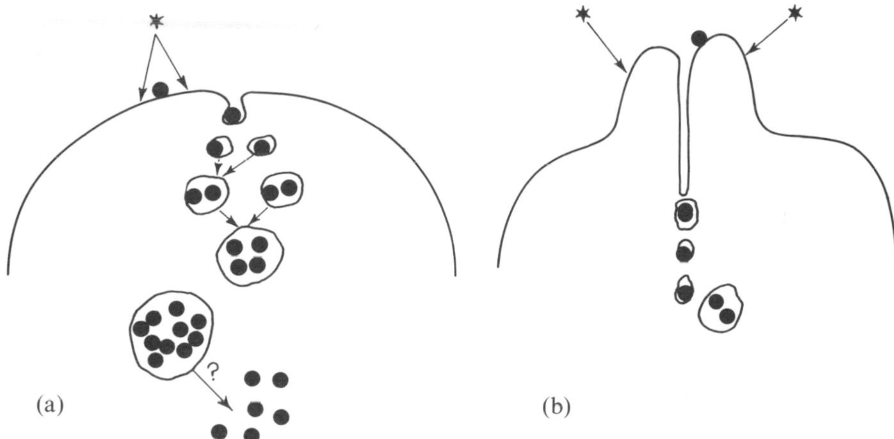

Fig. 4.7 Diagram showing hypothetical mechanisms for pinocytosis based on electron microscope studies. In each case the first step is supposed to be attachment of a particle to the cell surface. Then in (*a*) synthesis or flow of the membrane at the points marked with an asterisk causes the particle to be invaginated in a pinocytotic vesicle. Fusion of the lips of the invagination produces an isolated vesicle. Fusion of this vesicle with other similar vesicles produces a large absorption vesicle. Possibly the final step is the enzymatic removal of the limiting membrane to release the absorbed material into the cytoplasmic matrix. A slight modification of this process is shown in (*b*). In this case synthesis or flow of the membrane at the points marked with asterisks produces a tubular invagination of membrane carrying attached particles deep into the cell. Vesicles break off from the end of this tube.

mechanism. In two types, the first step is adhesion of the particles to the cell surface; then either the part of the surface bearing the particles is invaginated to form a vesicle at the edge of the cell (a in Fig. 4.7), or the particle is carried down a deep pit into the cell with a string of vesicles forming at the bottom of the pit (b in Fig. 4.7). In the third type of pinocytotic mechanism, waving protrusions of plasma membrane adhere and fuse to enclose a part of the extracellular fluid.

The constitution of the medium can alter the rate of pinocytosis, and there is probably, in the first two mechanisms, a specific interaction between the molecules in solution and the cell surface. This initial stage of adhesion of molecules to the surface may be important in achieving selective absorption, so that the vesicle, when formed, contains a higher proportion of the adhering material than does the surrounding medium.

In electron micrographs two types of pinocytotic vesicle can be defined, one is a simple invagination of the plasma membrane while the other is larger and shows an organized frill attached to the cytoplasmic surface of the vesicle. The latter type is found in those situations where protein uptake is known to be important, for example, in the ovum during the accumulation of yolk. When the small vesicles formed near the surface move towards the middle of the cell, they fuse together, fluid is withdrawn and a much higher concentration of the absorbed material is found in the central vesicles. Finally the wall of the vesicle may disappear so that the ingested material is liberated into the cytoplasmic matrix. In electron micrographs small vesicles have also been seen apparently budding off large vesicles in the cytoplasm. The disappearance of a vesicle in the light microscope may thus sometimes represent a parcelling into vesicles too small to be seen in the light microscope.

The electrical properties of plasma membranes, and direct permeability studies with radioactive isotopes, show that these membranes are restricting passage of molecules and ions as small as potassium ions, about 0·3 nm in diameter. Yet studies with protein molecules labelled with fluorescein, in the ultraviolet microscope (see Chapter 1) or with ferritin in the electron microscope, show that large protein molecules can pass into the interior of the cell. Although we might allow that chemical transport mechanisms, or small pores, could control the permeability of the membrane to small molecules and ions, it is difficult to visualize a pore structure which could allow the entry of even a 4 nm protein molecule and still retain a selective permeability to ions. Pinocytosis provides a method of segregating larger molecules in packets of plasma membrane, so that they are taken into the cell without the continuity of the plasma membrane being at any time effectively interrupted. Once in the cytoplasm, the wall of the vesicle can be destroyed and its contents incorporated without any loss of cytoplasmic materials. We shall see in the next chapter that a reverse of pinocytosis is used to transport specific substances out of the cell, with little loss of the more freely diffusible small molecules. The crew escape from a submerged submarine in a rather similar way, by means of an air-lock which allows each man to get out, without the interior of the submarine being flooded. The realization that pinocytosis is a widespread phenomenon demands that consideration of permeability be divided into the total permeability of the whole cell, including pinocytosis, and the permeability of the plasma membrane itself.

Nerve myelin

The sheaths surrounding myelinated nerve cells are formed by extensive infolding of plasma membranes. These structures merit special attention, not only because such specialization is unusual and related to the function of the cell, but also because they provide large, regularly arranged, masses of plasma membrane for chemical and physical analysis. Later we shall discuss the evidence which has been derived from these studies and consider the molecular structure of the component membranes. Here we shall merely describe the morphology of the myelin sheath of nerves.

The nerve cells in the anterior horn of grey matter in the spinal cord of a large mammal are about 80 μm in diameter. The cylindrical axons of these cells, which extend down the limb to innervate a muscle in the foot, are only about 10 μm in diameter and they are often more than a metre long. Much of the structure of a whole nerve is assigned to providing these slender processes with physical strength and electrical insulation. All axons are insulated from the surrounding tissues by Schwann cells or glial cells. In unmyelinated nerves many axons are enveloped by a single Schwann cell (Fig. 4.8*a*). The area where the lips of a Schwann cell come into apposition around one or more axons is called the mesaxon (because this resembles the arrangement of the peritoneum forming the mesentery around the gut tube). Where a myelin sheath is to be formed only one axon is included within the Schwann cell (Fig. 4.8*b*), and during development the mesaxon elongates and wraps itself around the axon many times, to form a spiral (Fig. 4.8*c*). Early in development this is a loosely wound spiral, with the cytoplasm of the Schwann cell separating successive turns. Later a compact mass of spirally wound plasma membrane is formed – this spiral structure being the myelin sheath as seen in Fig. 4.8*d*. The outer layer of each turn of the plasma membrane comes into contact with another outer layer of the membrane in the next turn; similarly the inner layers come into contact. The opposed inner surfaces give the major repeating dense line in the myelin structure (D in Fig. 4.8*e*) while the outer surfaces produce a fainter line, called the intraperiod line (I in Fig. 4.8*e*). The myelin sheaths in a large nerve may be up to 2–3 μ thick, and over 200 turns are used to produce such a sheath. The actual mechanism causing the spiralling of myelin is not clear; the axon could rotate, or membrane could be synthesized at the edge of the cell marked with an asterisk in Fig. 4.8*b*, with the whole mass of the myelin rotating in a stationary Schwann cell, around a stationary axon.

When a single axon and its surrounding sheath are teased out from a myelinated nerve, the myelin sheath is seen under a hand-lens to be divided into segments, each about 1·5 mm long. The gaps between the segments are known as the nodes of Ranvier. Only one Schwann cell nucleus is usually present in each internodal segment of myelin, so that each segment corresponds to a single cell wound in a complex spiral. At the nodes there is considerable interdigitation of processes from adjacent Schwann cells, but the adjacent Schwann cell plasma membranes are separated by the usual gap of 15–20 nm. It is presumably this gap which forms the aqueous channel at the nodes, from the axon membrane to the fluid surrounding the Schwann cell, essential to the theory of saltatory conduction. This theory also requires the lengths of axon membrane under the internode segments of the myelin sheath to be electrically insulated from the fluid surrounding the Schwann cell. Electron microscope pictures have in fact

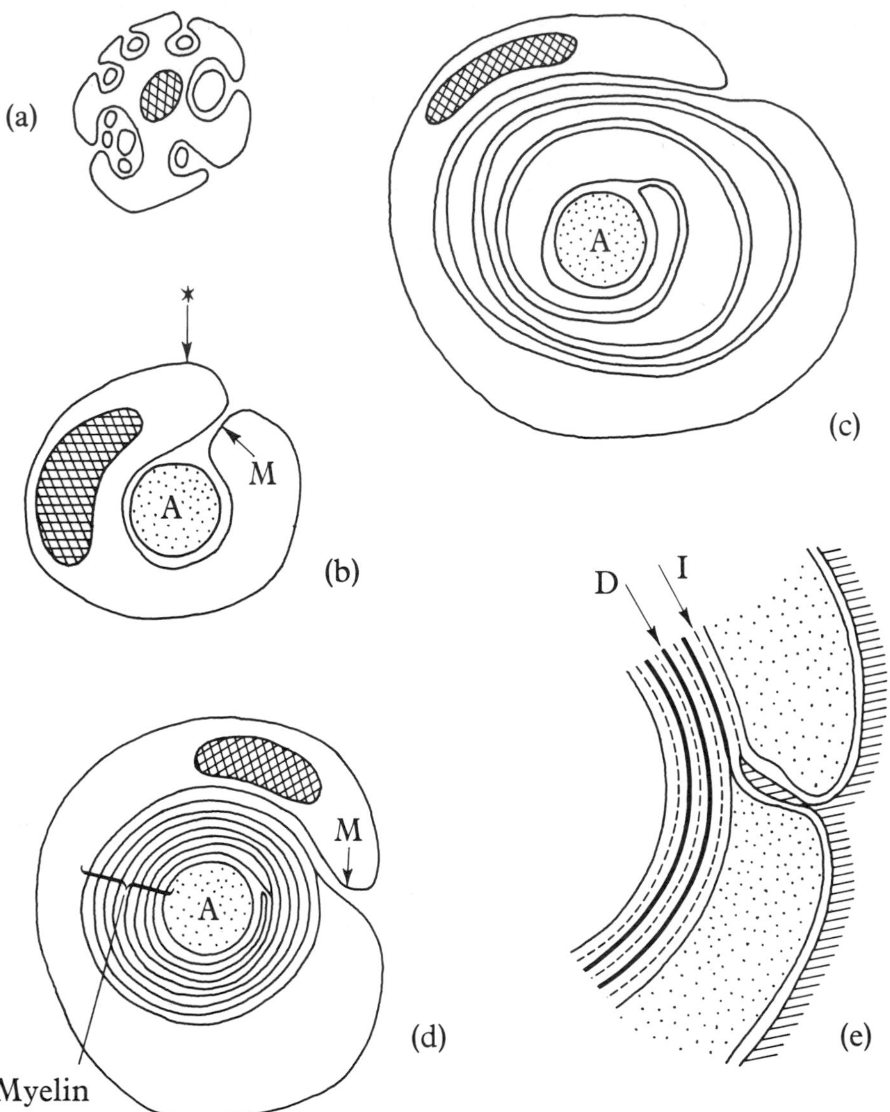

Fig. 4.8 Diagrams illustrating the formation of the myelin sheath. In unmyelinated nerve numerous axons are often enclosed by a single Schwann cell (*a*). Embryonic nerves have this structure, and before myelination begins, the Schwann cells proliferate, so that a single axon (A) is enclosed within each Schwann cell as in (*b*). The infolded cell membrane forms the mesaxon (M). The first step in myelination is a lengthening of the mesaxon, and its spiral wrapping around the axon (*c*). The cytoplasm between the turns of the spiral later disappears so that a compact spiral mass of membrane material is formed. This is the myelin sheath shown in (*d*). A higher power view of the mesaxon in compact myelin is shown in the diagram (*e*). Where the external surfaces of the infolded plasma membrane come into contact a faint intraperiod line is formed (I). Where the cytoplasmic surfaces of the plasma membrane come into contact a dark line is formed, known as the major dense line (D). A section through adult myelin thus shows a series of alternating dense and intraperiod lines, the repeating period from one dense line to the next representing a pair of Schwann cell plasma membranes.

shown a fairly intimate contact between axon and Schwann cell membranes near the nodes. Although these membrane contacts are not as close as in zonulae occludentes they could nevertheless markedly restrict ionic diffusion along the gap between axon and Schwann cell in these regions.

To summarize some of the main points in these last four sections: we have seen that when plasma membranes come into apposition they are normally kept 15–30 nm apart (perhaps repulsive forces arise from charged groups on the membrane surface). In specialized areas, however, membranes can be cemented together, or come into close contact, sometimes with local modification of permeability properties. Specialized areas of the plasma membrane can also infold, to form a pinocytosis vesicle, or proliferate, as in the formation of nerve myelin.

Permeability properties of cell membranes

We have discussed so far the appearance of the plasma membrane and various associated structures in electron micrographs. We consider now evidence from other types of work – permeability measurements, chemical analysis, etc., which bears on the question of the nature and properties of the membrane at the surface of cells. We can then take a more critical look at the electron micrographs, and discuss the significance of these thin-section profiles of dehydrated, fixed material in relation to the hydrated, dynamic structure present in life, and discuss the molecular events which underlie and govern the properties of these membranes.

In the light microscope, which is the only means we have for direct study of living cells, no definite structure can be seen at the boundary of animal cells (Chapter 1) or at the cytoplasmic boundary of bacterial and plant cells separated from their walls. The presence of a membrane at these boundaries must be inferred indirectly, or from electron micrographs. Certain phenomena which seem at first sight to suggest the presence of a boundary membrane, such as the shrinking of cells in hypertonic solutions, or the ionic concentration differences which are found between the interior and exterior of cells, or the electrical potential which is found between cytoplasm and extracellular fluid in nerve and muscle cells, do not in fact provide conclusive evidence for a definite membrane structure at the cell boundary. Similar effects can be produced by blobs of protein gel, without boundary structure.

The best evidence, apart from electron microscope studies, that a definite membrane is present at the boundary of cells is the fact that ions and molecules do not move freely into and out of cells. Study of this movement leads to the operational definition of the 'plasma membrane' given in Chapter 1: that component of the total structure at the boundary of cells which is mainly responsible for limitation, or control, of the free diffusion of ions or molecules from the external medium to the cytoplasm. In this section we shall consider what properties of the plasma membrane may be deduced from its permeability properties. We shall see that its low permeability to ions and high permeability to lipid-soluble substances suggest that it may be a layer of lipid or lipoprotein. (Lipids are the chemical class which includes all fats and oils. Lipid-soluble substances are substances with a relatively high solubility in oils. Lipoproteins are molecular aggregates containing both lipid and protein.)

The relation which is found between permeability and lipid solubility is illustrated by the results of Fig. 4.9. The oil/water partition coefficient is a measure of the relative solubility of a substance in oil and water. It is defined as the ratio

$$\frac{\text{concentration in oil}}{\text{concentration in the aqueous phase}}$$

for a solute in equilibrium across an oil-water interface. The partition coefficient of a substance depends on the capacity of its molecules to form hydrogen bonds (see Appendix 2).

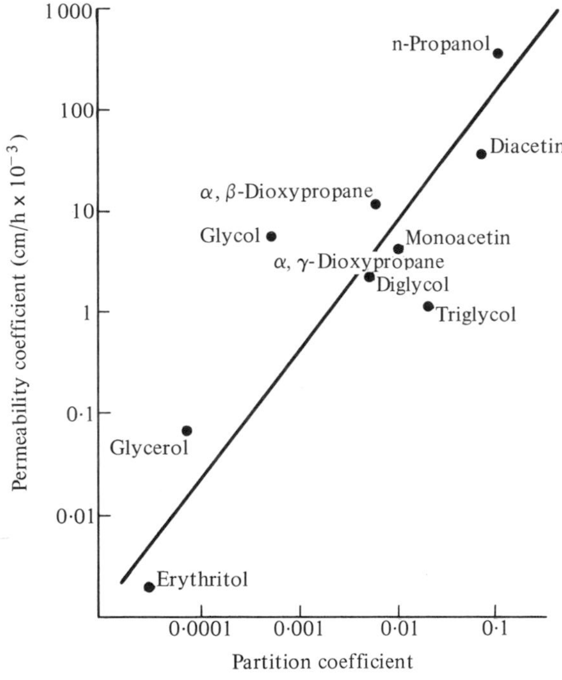

Fig. 4.9 Permeability of bovine red cells plotted against partition coefficient for a variety of solutes. (Data *from* Jacobs *et al.*, 1935.)

The permeability coefficient of a membrane, for a given solute, is defined as the net number of moles of solute which cross each square centimetre of membrane in unit time, when the concentration difference across the membrane is one mole/litre. Permeability coefficients are determined, in the experiments represented in Fig. 4.9, by suspending the cells in an isotonic solution of the solute under investigation and measuring the time for haemolysis (arbitrarily chosen as the time taken for 75 per cent of the cells to haemolyse, i.e. burst). This time is directly related to solute permeability since movement of solute into the cell disturbs osmotic equilibrium, and leads to inward movement of water, and eventual lysis.

So far as possible, in this and the following sections, discussion will start with the plasma membranes of mammalian red cells, since these membranes have been extensively studied by a wide variety of methods. Permeability and chemical determinations are at their simplest for these cells, since electron micrographs and chemical evidence, to be discussed later, suggest that only a thin membrane, less than 10 nm thick, separates the cytoplasm from the exterior medium, and the cytoplasm is an almost pure solution of haemoglobin. Outer layers of polysaccharide are minimal and there are no membrane-bound compartments within the cytoplasm – two factors which complicate interpretation of permeability and chemical data. However, to draw conclusions of more general application to the plasma membranes of other cell types, we must in each section extend the discussion as far as possible after the preliminary discussion of red cells.

A series of permeability measurements similar to those of Fig. 4.9, were made by Collander on the large unicellular alga, *Chara ceratophylla*, with similar correlation between permeability and partition coefficient, and pioneer studies which established this sort of relation were made by Overton on plant and muscle cells around 1899–1902. For plant and muscle cells, the presence of internal membranes and outer walls and basal laminae make it less certain that the main permeability barrier under study in permeability experiments is in fact the 7–10 nm membrane seen in electron micrographs, though the constancy of this feature in micrographs of different cell types, with outer walls of varying structure and with varying types of internal membrane system, makes this interpretation reasonable. Allowing for these difficulties, it seems generally true that for movement of solutes across cell membranes there is a correlation between permeability and oil/water partition coefficient. This suggests, as Overton appreciated, that cell membranes must be predominantly of lipid or lipoprotein composition. The absolute values of the permeability coefficients are in general much lower than values that would be expected for free diffusion of solute molecules into and out of the cells, so that membranes must be present which impose a barrier to free diffusion, though this barrier is not so marked for solutes of high lipid solubility.

Movement of ions across membranes

When there is a solute concentration difference across a permeable membrane, molecular diffusion (i.e. random thermal movement) leads to a net flow of solute molecules across the membrane, from the high to the low concentration side. In the study of the permeability properties of cell membranes the reverse effect is also found – a movement of solute across the membrane against the concentration gradient. Movement against the concentration gradient or, more generally, movement of the transported molecule or ion to a state of higher free energy, is termed *active transport*.

Active transport must always require a supply of energy, derived from the metabolic activity of the cell. There is evidence we shall discuss in this section for active transport of ions, and evidence that the energy for this process is derived from the ubiquitous ATP. Other examples of active transport are found in the uptake of amino acids and sugars into cells. The term 'active transport' is however, in a certain sense, a misnomer, for any transport process is made up of a series of passive steps in each of which the system as a whole, including membrane components, ATP, and the transported solute or ion, moves to a state of lower free energy. In active transport there is a gain in the free energy of the transported solute or ion, at the expense of a loss of energy in some other part of the total system.

In discussing the movement of ions across a membrane we have to consider the interrelation of concentration differences and electrical potential differences across the membrane. Diffusion of ions across a membrane will normally generate a potential difference, as a result of the different rates of diffusion of positive and negative ions. This potential difference interferes with the simple diffusion arising from concentration differences. Since ions tend to move down a concentration gradient and also down a potential gradient, an equilibrium situation is possible in which there is a stable concentration difference across the membrane, exactly balanced by a potential difference. It may be shown by thermodynamic arguments (Appendix 4) that under such equilibrium conditions, the potential difference (E, measured in millivolts) and the concentration difference for any ion type (e.g. K^+) to which the membrane is at all permeable, are related by the equations:

$$E = 58 \log_{10}\left(\frac{\text{concentration outside}}{\text{concentration inside}}\right) \text{ for univalent positive ions at } 20^\circ\text{C}$$

and

$$E = 58 \log_{10}\left(\frac{\text{concentration inside}}{\text{concentration outside}}\right) \text{ for univalent negative ions at } 20^\circ\text{C}.$$

A variety of cells, such as muscle cells, the giant axons of squid, and certain plant cells, are large enough for the insertion of electrodes to measure the potential across the total structure at the cell boundary. If the potential across a membrane can be measured and the ionic concentrations on the two sides determined, it is possible to compare the calculated equilibrium potential for any given ion type with the measured potential. Results of such measurements for squid axons are given in Table 4.1.

If there were agreement, for any given ion type, between the calculated and measured potentials, this would indicate that ions of this type were in equilibrium across the membrane, and were crossing only by a simple or facilitated diffusion process. But this condition is not satisfied for Na^+, K^+ or Cl^- ions across the squid axon membrane. K^+ ions and Cl^- ions show some departure from equi-

TABLE 4.1 Membrane potentials and salt concentrations for squid giant axons.

Ion	Internal concentration c_i (mM/litre)	External concentration c_o (mM/litre)	Calculated equilibrium potential (mV) $\left(E = \pm 58 \log \frac{c_o}{c_i}\right)$	Measured membrane potential (mV)
K^+	400	20	−75	
Cl^-	110	560	−41	−63 to −72
Na^+	50	440	+55	

(Data from HODGKIN, A. L. (1958) *Proc. Roy. Soc.* **148B**, 1, and KEYNES, R. D. (1962) *J. Physiol.* **163**, 19P.)

librium, and the concentration difference for Na^+ ions shows a very large departure from equilibrium conditions.

There is reason to believe that the bulk concentrations of Na^+, K^+ and Cl^- ions found for squid axon cytoplasm are not very different from the *free* concentrations of these ions. The rate of diffusion of K^+ ions through the cytoplasm, for example, is not very different from their diffusion rate in free solution. It seems probable in fact that univalent ions are not bound to any great extent in cells, and that a large difference between the measured potential and the equilibrium potential, for a given univalent ion type, must indicate that there is active transport of this ion across the membrane.

All cells, except certain mammalian red cells, show a ratio of internal to external Na^+ very much lower than the equilibrium value. There must be in the membrane of most cells an 'ion pump' actively transporting Na^+ ions across the membrane out of the cell; it seems that the ion pump is often an exchange pump, bringing in one K^+ ion for each Na^+ ion carried out (or exchanging K^+ and Na^+ ions in a ratio other than 1 : 1; for red cells 3 Na^+ ions are carried out for each 2 K^+ ions). It has been shown for squid axons and red cells that the energy for this active transport is derived from the breakdown of ATP, and that the pump is inactivated by certain enzyme inhibitors. The molecular mechanisms that may be involved are discussed in later sections.

Superimposed on any active pumping, there is some diffusion of K^+, Na^+ and Cl^- ions across the membranes, which continues when the pump is inactivated. In the case of the resting squid axon the passive permeability coefficients for K^+ and Cl^- ions are of the order of 100 times larger than the passive permeability coefficient for Na^+ ions. Although both Na^+ and K^+ ions are involved in the activity of the ion-exchange pump it is the relatively low permeability of the resting membrane to Na^+ which leads to the greater departure from equilibrium conditions for the Na^+ ions. Low ionic permeability coefficients provide further support for the idea that the diffusion barrier is made up of lipid or lipoprotein. Small ions are very insoluble in oil; to move into the oil they must either shed their hydration layer of water molecules (held in the strong electrostatic field near the surface of a small ion) or carry water of hydration into the oil and break a number of hydrogen bonds between the water of hydration and the surrounding molecules of the aqueous phase.

We conclude this section with a brief discussion of the changes in ionic permeability associated with the activity of nerve and muscle cells. The electrical impulses which travel along the membranes of nerve and skeletal muscle cells during activity are termed *action potentials*. During the passage of an action potential, past a particular point on the membrane, the approaching electrical disturbance first reduces the magnitude of the membrane potential slightly. This leads to a transient increase in permeability to Na^+ ions and a flow of Na^+ ions into the cytoplasm, which reduces the potential more, and further increases permeability to Na^+ ions. Potential and permeability changes thus act cooperatively: a small initial disturbance builds up to a large disturbance in which the local potential across the membrane is reversed, and approaches the equilibrium potential corresponding to the Na^+ concentration difference across the membrane (Table 4.1). There appears to be some flow of Ca^{++} ions into the cytoplasm, as well as Na^+ ions. This phase is followed, within about 1 millisecond for squid axons, by a return to normal Na^+ permeability. But there is meanwhile an increase in K^+ permeability; the resulting outflow of K^+ ions brings the membrane back to the resting potential. For squid axons (at 18°C) the potential and K^+ permeability return to their resting values within about 4 milliseconds. The transient local depolarization (or reversal of polarization) spreads to produce a slight depolarization ahead of the advancing action potential. The process we have described is thus repeated a little further along the membrane, and a wave of transient depolarization (the action potential) spreads along the membrane.

These ionic permeability changes associated with the action potential have been studied in squid axons and frog muscle cells by Hodgkin, Huxley and others with the very elegant 'voltage clamp' technique of Marmont and Cole. The voltage across the membrane is reduced and held to a series of different values by currents supplied by fine electrodes inserted into the cells, and the current flow across the membrane determined from the electrode current. The ions carrying the current are identified by study of the effects of changes in the external salt concentrations, and by studies with radioactive isotopes of Na^+ and K^+. It is more difficult to apply these techniques to heart muscle cells, but the permeability changes associated with the action potential in these cells appear to be similar to those of skeletal muscle. However, differences in the magnitude and time relations of the permeability changes lead to longer time intervals before the return to resting conditions (1 second or more for frog heart at 18°C).

The transmission of an action potential along a skeletal muscle cell is initiated by the action of acetylcholine on the muscle cell membrane at the neuromuscular junctions. Acetylcholine is released in 'packets' from the terminal points of the axons which innervate the muscle (the packets possibly corresponding to the vesicles seen in electron micrographs, as discussed in Chapter 5). In frog muscle, which has been intensively studied by Katz, the acetylcholine apparently produces a local increase in the permeability of the muscle cell membrane to both Na^+ and K^+ ions. The resulting flow of ions, mainly inflow of Na^+, reduces the membrane potential and initiates the electrical disturbance which then spreads along the muscle cell membrane. A different effect is found for the membranes of muscle cells in the 'pace-maker' region of frog heart. Here acetylcholine apparently produces a specific increase in K^+ permeability alone.

A question of considerable interest is how far these electric properties of nerve and muscle reflect general properties of cell membranes, or how far they are a special development to meet a functional need. The postsynaptic membrane area of a neuromuscular junction certainly seems to contain specialized acetylcholine 'receptors', since the remaining area of the muscle cell is not sensitive to depolarization by acetylcholine (unless the nerve to the muscle is destroyed, and then the receptors seem to become more uniformly spread over the muscle surface). The property of electrical excitability might be a general property of cell membranes, though detectable only in cells large enough and long enough to allow study of the spread of an action potential with fine electrodes. *Nitella* show slow-spreading action potentials, and quite simple artificial membranes of a type to be discussed in later sections, show electrical excitability.

Chemical composition of membranes

We consider now a quite different approach to the study of the properties and structure of membranes based on the following viewpoint: since it is suspected from permeability studies that lipids may be involved in membrane structure, let us extract the lipids from suitable cells, study their properties and see whether these suggest what the structure of the membrane might be.

The red cell provides particularly favourable material for this type of analysis. The cytoplasm of these cells contains little else but haemoglobin, and the haemoglobin can be leached out by haemolysis to leave a ghost which appears, under phase contrast, as a tenuous envelope with approximately the same diameter as the original cell. Moreover, the ghost retains all the lipid, and certain of the permeability properties of the original cell – if prepared by the gentle procedure of successive haemolysis in solutions of decreasing tonicity, the ghost still retains the ability to concentrate K^+ and pump out Na^+ in the presence of ATP. It seems reasonable to assume that the ghost membrane is closely similar to the membrane of the original cell, although its detailed structure may have been modified by the passage of haemoglobin molecules (6 nm in diameter) through a membrane normally very impermeable to sodium ions (hydrated diameter 0·5 nm). The fact that permeability to Na^+ ions returns to a fairly low value, after mild haemolysis, incidentally represents an additional experimental

TABLE 4.2 Chemical composition of membranes (all figures given as % of total dry weight)

	Cell type	Phospholipids				Glyco-lipids (ceras-ine, etc.)	Choles-terol	Protein
		Phospha-tidyl ethanol-amine	Phospha-tidyl choline	Phospha-tidyl serine	Others (sphingo-myelin, phospha-tidic acid, etc.)			
PLASMA MEMBRANES	Human red blood cell	3·4	6·7	2·4	11·5	Trace	9·2	60
	Liver cells	–	8	–	18	–	13·0	60
	Bovine myelin	11·7	7·5	7·1	6·7	22·0	17·0	22
	Gram-positive bacteria	5–10	–	–	10–15	Some present	–	60–75
	Bovine heart mitochondria	8·4	9·3	Trace	4·8	Trace	0·2	76
	Rabbit sarcoplasmic reticulum	1·4	8·3	Trace	1·3	–	–	54
	Guinea pig brain synaptic vesicles	4·2	11·5	3·3	9	Trace	5·6	66

Data from DEWEY, M. M. and BARR, L. (1970) *Current Topics in Membranes and Transport* **1**, 1. New York, Academic Press, except for gram-positive bacteria figures which are from OP DEN KAMP, J. A. F. *et al.* (1965) *Biochim. Biophys. Acta* **106**, 438 and SALTON, M. R. J. and FREER, J. H. (1965) *Biochim. Biophys. Acta* **107**, 531.

feature of the behaviour of the membrane of red cells, which must somehow be incorporated into our final picture of its structure.

The chemical composition of ghosts from human red cells is shown in Table 4.2. The amount of lipid in the ghost (and in the original red cell) is sufficient to form a layer 3 nm thick over the cell surface. The amount of protein in the ghost is sufficient to form a surface layer of comparable thickness. This is simply an estimate of the amount of lipid and protein present. Whether these components are structurally arranged as uniform layers remains an open question to be discussed in later sections.

There are two main further sources from which plasma membrane material may be obtained in sufficient amount, and in sufficiently pure form, for useful chemical analysis: bacterial protoplasts and nerve myelin. For many bacteria, the outer walls and capsules can be digested away by enzymes to leave a bacterial protoplast (bacterial cell minus outer wall and capsule) bounded by a membrane which still retains the main permeability properties of the original bacterium. Bacterial protoplast membrane material may be obtained, as for red cell ghosts, by bursting open the protoplasts in hypotonic solution and centrifuging down the membrane material. The composition of this material is also included in Table 4.2. The amount of lipid present in the protoplast membranes of *Staphylococcus aureus* is sufficient to form a layer 1–2 nm thick, and the protein and carbohydrate a layer 3–4 nm thick.

The myelin sheaths of nerve are formed, as discussed in an earlier section, by proliferation of Schwann cell membranes. We may expect therefore some similarity in chemical composition between myelin sheath material, and the membrane material from red cell and bacterial protoplast ghosts. The same components: phospholipids, protein, carbohydrate, are in fact found in all three types of material, although with some variation in their relative proportions (Table 4.2). For all these membranes there is evidence that a certain amount of carbohydrate material may be incorporated in the membrane structure, probably mainly at the outer boundary.

Methods more recently developed for isolating membrane fractions from homogenized tissues, and for separating the plasma membranes and different internal membranes by density flotation, are giving more extensive data on lipid and protein composition for membranes of various types (Table 4.2).

Properties of phospholipids and other lipids

The molecular structure of the lipids of Table 4.2 is illustrated in Figs. 4.10 and 4.11 and Tables 4.3, 4.4 and 4.5. A typical phospholipid, phosphatidylethanolamine, is shown in Fig. 4.12. The natural phospholipids usually contain one saturated and one unsaturated hydrocarbon chain, as indicated in this figure.

Phospholipids and cerebrosides are *polar lipids*. Their molecules contain long (lipid) hydrocarbon chains which are strongly hydrophobic, and polar groups, i.e. groups capable of forming hydrogen bonds (see Appendix 2) which are

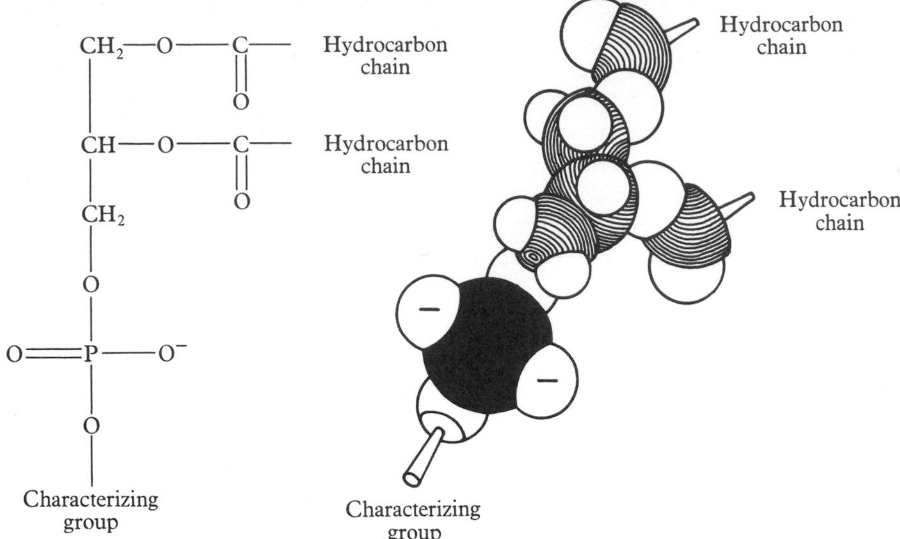

Fig. 4.10 Chemical formula, and drawing of a molecular model, of the group in phosphatidic acid, phosphatidyl-ethanolamine, phosphatidyl-choline and phosphatidyl-serine linking the appropriate group of Table 4.3 to two hydrocarbon chains of the type shown in Table 4.4. (In all these drawings of molecular models phosphorus atoms are shown black.)

TABLE 4.3 The characterizing groups of the more common phospholipids (note 'short-hand' drawings which will be used in later figures)

PHOSPHOLIPID	CHARACTERIZING GROUP
Phosphatidic acid	$—H$
Phosphatidyl - ethanolamine	$—CH_2—CH_2—\overset{+}{N}H_3$
Phosphatidyl - chloline (lecithin)	$—CH_2—CH_2—\overset{+}{N}(CH_3)_3$
Phosphatidyl - serine	$—CH_2—CH(\overset{+}{N}H_3)—C(=O)O^-$

CH_3 CH_2 CH_2 CH_3
CH_3 CH CH_2 CH
CH_2 CH_3
CH_2 C CH
CH_3
CH_2 CH CH CH_2
CH_2 C CH CH_2
CH C CH_2
HO CH_2 CH

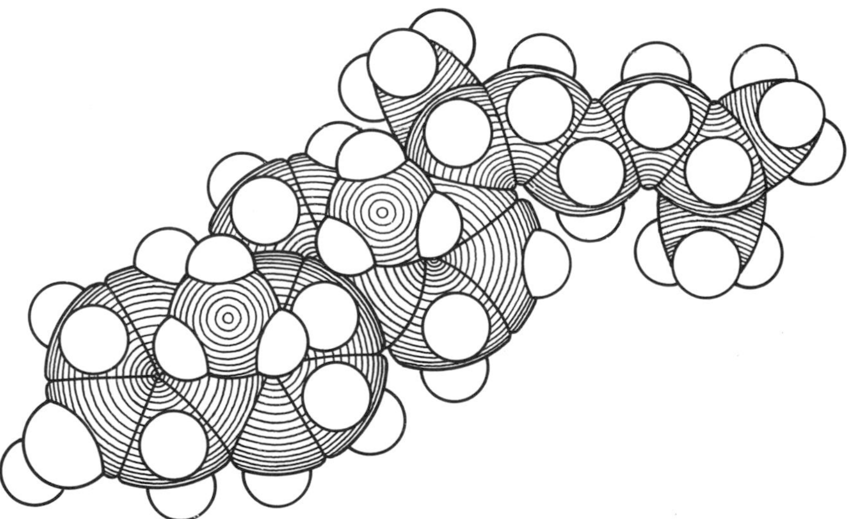

Fig. 4.11 Chemical formula, and drawing of a molecular model, of cholesterol.

TABLE 4.4 The more common phospholipid hydrocarbon chains

Saturated	palmitic	$-CH_2-CH_2-CH_2-CH_2-CH_2-CH_2-CH_2-CH_2-CH_2-CH_2-CH_2-CH_2-CH_2-CH_2-CH_3$
	stearic	$-CH_2-CH_2-CH_2-CH_2-CH_2-CH_2-CH_2-CH_2-CH_2-CH_2-CH_2-CH_2-CH_2-CH_2-CH_2-CH_2-CH_3$
Unsaturated	oleic	$-CH_2-CH_2-CH_2-CH_2-CH_2-CH_2-CH_2-CH{=}CH-CH_2-CH_2-CH_2-CH_2-CH_2-CH_2-CH_2-CH_3$
	linoleic	$-CH_2-CH_2-CH_2-CH_2-CH_2-CH_2-CH_2-CH{=}CH-CH_2-CH{=}CH-CH_2-CH_2-CH_2-CH_2-CH_3$
	linolenic	$-CH_2-CH_2-CH_2-CH_2-CH_2-CH_2-CH_2-CH{=}CH-CH_2-CH{=}CH-CH_2-CH{=}CH-CH_2-CH_3$
	arachidonic	$-CH_2-CH_2-CH_2-CH{=}CH-CH_2-CH{=}CH-CH_2-CH{=}CH-CH_2-CH{=}CH-CH_2-CH_2-CH_2-CH_2-CH_3$

TABLE 4.5 Formulae for the phospholipid sphingomyelin, and the cerebroside cerasine

Sphingomyelin

```
                                   NH—C—(CH2)22—CH3
                                   |  ||
                                   |  O
                                   CH—CH—CH=CH—(CH2)12—CH3
                         O-        |  |
           +             |         |  OH
(CH3)3—N—CH2—CH2—O—P—O—CH2
                         ||
                         O
```

Other hydrocarbon chains, e.g. those of Table 4.4, are also found in place of $—(CH_2)_{22}—CH_3$

Cerasine

```
                        NH—C—(CH2)22—CH3
                        |  ||
                        |  O
                        CH—CH—CH=CH—(CH2)12—CH3
                        |  |
        CH2OH           |  OH
         |              |
    H    C——O    O—CH2
    |   /|    \  |
    C  H  OH   C
    |\ H  |   /|
   OH  C——C    H
       |  |
      OH  H
```

In other cerebrosides, other hydrocarbon and hydroxyl-hydrocarbon chains are found in place of $—(CH_2)_{22}—CH_3$

hydrophilic. In the case of the phospholipids, the charged phosphate group and the ionic groups of Table 4.3, make this part of the molecule strongly hydrophilic. Cholesterol can also be classed as a polar lipid, although its polar group (the single —OH) is only weakly hydrophilic.

Polar lipids with strong hydrophilic groups are rather insoluble in either water or oil, as isolated molecules. They are in an energetically favourable situation, however, at an oil-water interface, and also in the oil or water phase, if they come together to form *micelles* (Fig. 4.13). Dissolved in water or oil, at sufficient concentration, polar lipids spontaneously form micelles. If an oil-water or air-water interface is available they concentrate at the interface.

In aqueous solution at high concentrations, and under appropriate conditions of pH, etc., a more favourable type of micelle is often a flat sheet, or bimolecular layer, as indicated in Fig. 4.14. Over certain ranges of concentration these layers become spaced out regularly through the solution (alternating with layers of water) and under these circumstances, or in the hydrated solid state (corresponding to almost complete removal of the water between the bimolecular layers) it becomes possible to study their dimensions and molecular structure in some detail by X-ray diffraction and other techniques. For mixed lipids of the composition found for natural membranes, in the temperature range 20–37°C, the bimolecular layers are probably liquid layers. Hydrocarbon chains have the

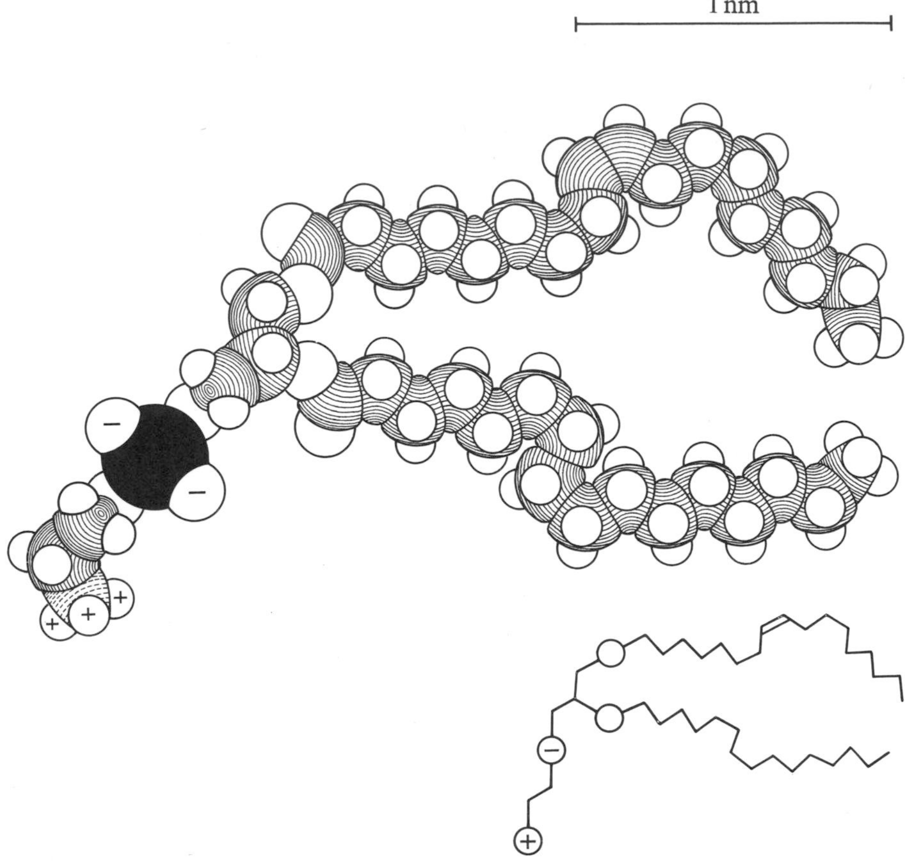

Fig. 4.12 A drawing of a molecular model of a complete phospholipid: a phosphotidyl-ethanolamine molecule with one oleic acid hydrocarbon chain, and one palmitic acid chain. Note the 'shorthand' skeleton representation, which will be used in later illustrations:

— carbon–carbon single bonds
= carbon–carbon double bonds
—o hydrogen-bonding (i.e. polar) groups.

same sort of flexibility as peptide chains (Chapter 2, Fig. 2.6). In a liquid lipid these chains are in constant movement, like a basket of lively snakes, although in a bimolecular layer of polar lipids the movement of any one molecule will be restricted by the need for its polar group to move for the most part only in the planes of water-lipid interfaces. An attempt has been made to indicate the liquid nature of the lipid layer in Fig. 4.14 by drawing the hydrocarbon chains in random configurations. However there is still debate about the question of how much movement is possible for mixed-lipid hydrocarbon chains in a bilayer. It is possible that complexes may form of the type indicated in Fig. 4.15, to allow

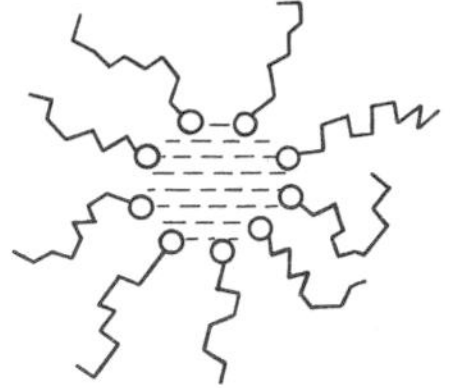

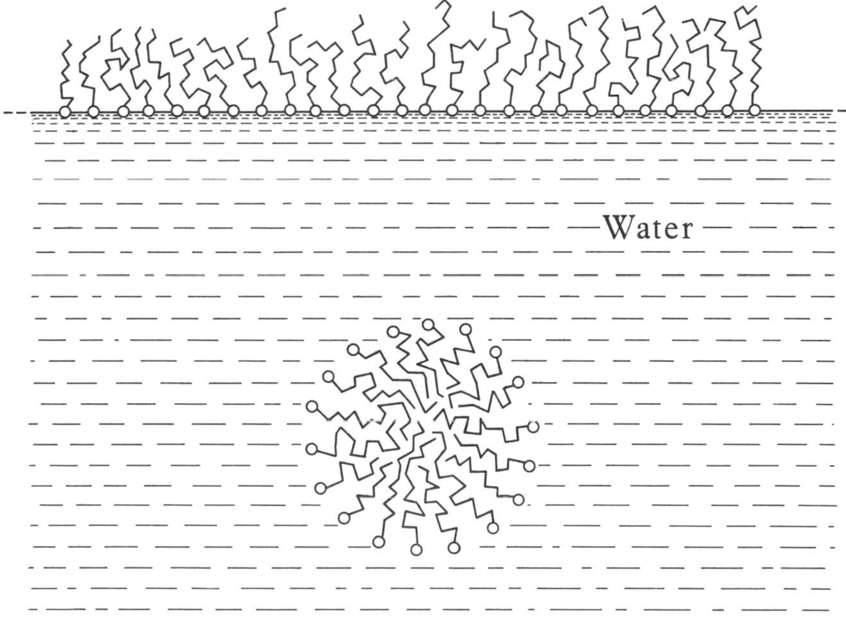

Fig. 4.13 A schematic illustration of polar lipids arranged in a monomolecular layer at an oil-water interface, and also forming spherical micelles in the oil and aqueous phases. randomly folded hydrocarbon chains, —o polar groups. (*From* Luzzati, Mustacchi and Skoulios, 1958.)

close packing of cholesterol molecules with phospholipids with all the bonds of the hydrocarbon chains in the lowest-energy fully-extended conformation. The free energy difference for a CH_2—CH_2 bond fully extended as in Fig. 4.15 or rotated through 120° to alternative low-energy positions is 0·6 kcal/mole. This is not a large energy step compared with the energy of thermal motion (see Appendix 4) so that complexing with cholesterol or other steroids may critically affect the fluidity of phospholipids and control permeability properties.

Occasionally, as a result of random thermal movement, the polar end of a molecule may be drawn down among the hydrocarbon chains, particularly the

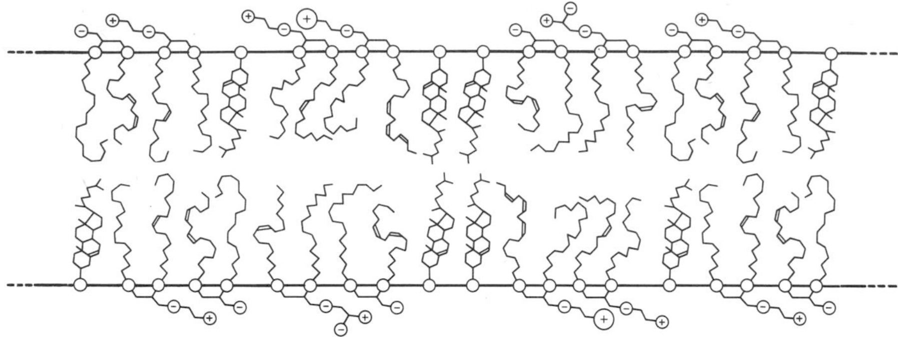

Fig. 4.14 A schematic illustration of a bimolecular layer of mixed phospholipids and cholesterol (phospholipid to cholesterol ratio 2:1). For full molecular formulae, and models, of molecules represented here by 'shorthand' drawings, see Figs. 4.11 and 4.12, and Tables 4.3 and 4.4.

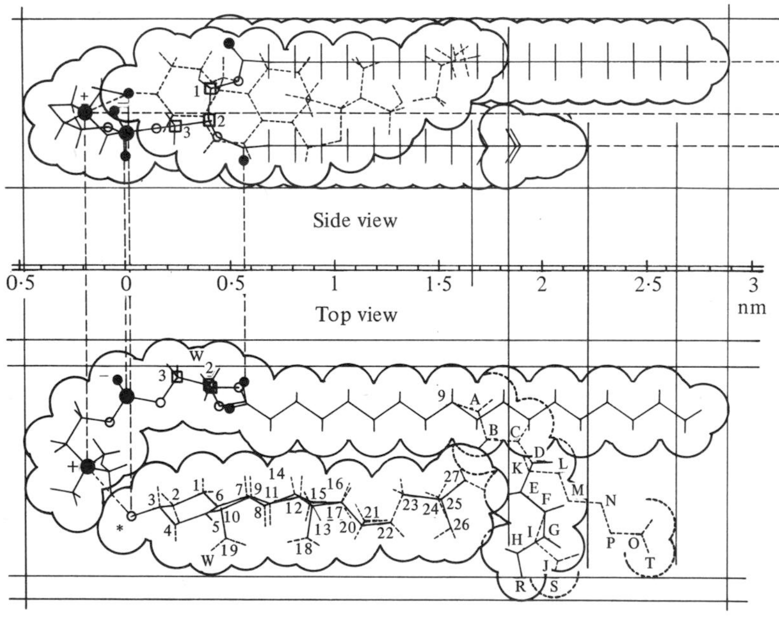

Fig. 4.15 Drawing of a molecular complex of a cholesterol molecule and a phospholipid with one extended saturated hydrocarbon chain and one extended unsaturated chain (compare Fig. 4.12). (*From* Vandenheuvel, 1965.)

cholesterol molecules with their single O—H polar group. The polar end may then surface again on either side of the bimolecular layer, so that there is the possibility of molecules moving across from one side of the layer to the other.

Molecular models for membrane structure suggested by chemical composition and permeability properties

The first really detailed ideas on the possible molecular structure of cell membranes were based on an experiment of the most elegant simplicity carried out by Gortner and Grendel in 1925, using techniques developed around 1917 by Langmuir and others.

Gortner and Grendel extracted the lipids from a sample of blood of known cell number, and spread these lipids from benzene solutions on to the air-water interface of a shallow trough part filled with water. In an experiment of this type the benzene evaporates to leave polar lipid molecules spread about the surface. If a thread is drawn across the surface from the edge of the trough, this surface layer is compressed, until at a certain quite sharply defined area, the surface layer begins to offer resistance to further compression. At this point the polar lipids form a close-packed monomolecular layer with their hydrocarbon chains sticking up into the air, and their polar groups in the water. Resistance to further compression arises from the fact that such compression must either push some hydrocarbon chains down into the water, or bring some polar groups out of the aqueous phase.

Gortner and Grendel measured the area of the close-packed monomolecular layer formed by their lipid sample. From an estimate of the surface area of each red cell, and the known number of cells in their sample, they calculated that the amount of lipid was just sufficient to form a bimolecular layer over the cell surface (i.e. area of spread monolayer = twice total area of cells). They suggested that a bimolecular layer of the type shown in Fig. 4.14 might form the basic element of membrane structure.

A bimolecular lipid layer of this type would, of course, account in a satisfactory way for the main permeability properties of the cell membranes. It would provide a diffusion barrier to ions and water soluble substances, while remaining relatively permeable to lipid-soluble substances, which could dissolve in, and diffuse through the lipid layer. We now know, from modern measurements, that Gortner and Grendel failed to extract all the lipid from their red cells and underestimated the area of membrane per red cell, the two errors cancelling out to some extent. The true area of monomolecular layer per cell ranges from 1·2 to 2 depending on the surface pressure chosen. *Thus a close-packed bilayer could only occupy just over half the red cell surface area.* If a bilayer covers the whole surface it must be at low surface pressure, or there must be a tendency for the phospholipid polar groups to be separated from one another by inserted protein molecules.

Danielli and Davson retained the bimolecular lipid layer as a basic element in the membrane structure they proposed in 1935. They pointed out that the elastic properties of cell membranes make it probable that protein must be adsorbed to the surfaces of the bimolecular layer. Later Danielli suggested that protein might also be incorporated to interrupt the lipid layer in places, forming protein-lined aqueous pores. We shall find it convenient to refer to a structure of either of these types as a 'Danielli-type' structure.

Danielli suggested that the first layer of protein adsorbed to the bimolecular lipid layer might be unfolded, as protein is known to unfold at an interface between water and *non-polar* lipids (Fig. 2.19). However, at the surface of a bimolecular layer of *polar* lipids, with the molecules oriented with their polar groups towards the aqueous phase, this type of unfolding is in no sense obligatory. In a later section we shall consider the interaction between proteins and polar lipids in more detail, and we shall see that speculation based on the known properties of proteins and polar lipids is extremely difficult. In interaction with *polar* lipids we do not know how much the protein tertiary or quarternary structure will be modified. Further, the hydrophobic effects which bring about the formation of a bimolecular layer of polar lipids in the absence of protein may be completely modified when there is protein present.

The study of the electrical capacity of cell membranes seems, at first sight, to provide support for the Danielli-type structure. This capacity has been determined for red cells, sea urchin eggs and bacteria from the dielectric constant of suspensions, and for plant, muscle and nerve cells from measurements with electrodes. The high frequency membrane capacity is surprisingly similar in all these cells at a value about 1 $\mu F/cm^2$. (The low frequency capacity shows more variation, but it is the high frequency value which is of more significance in relation to the main structure of the membrane.) A capacity of 1 $\mu F/cm^2$ corresponds to the capacity of a layer of lipid (of dielectric constant 3) about 3 nm thick (capacity in $\mu F/cm^2 = \dfrac{1\,000 \text{ dielectric constant}}{36\pi \quad \text{thickness in nm}}$). This is exactly what would be expected for the Danielli-type structure assuming that the protein pores, if present, were not too numerous or extensive in area. However, the capacity measurements unfortunately only set a *lower limit* to the membrane thickness. They would be equally compatible with a structure of mixed lipid and protein, of thickness $3x$ nm and dielectric constant $3x$, where x remains a completely unknown quantity. The importance of capacity measurements lies primarily in the fact that they do set a finite lower limit to membrane thickness of about 2 nm (since the dielectric constant of the material cannot be less than about 2). For these cells (red cells, sea urchin eggs, bacteria, nerve, muscle, plant cells) this represents very good evidence, distinct from permeability evidence, that there is a definite membrane structure at the cell boundary of material of low dielectric constant and low ionic conductivity relative to cytoplasm or extracellular fluid.

Evidence that plasma membranes seen in electron micrographs are permeability barriers

The most direct evidence that the plasma membranes seen in electron micrographs correspond to the main lipoprotein barriers that limits diffusion into and out of cells comes from the study of red cell ghosts. After extensive washing, to remove all the haemoglobin, the amount of lipid and protein left is only sufficient to form a membrane of thickness about 6 nm unhydrated, 7·5–9 nm hydrated. Yet ghosts show a typical 'plasma membrane' profile in thin sections. It follows that there cannot, in this case, be any significant outer or inner layers of membrane unrevealed by staining for electron microscopy. The thin-section profile must represent the protein-phospholipid-cholesterol permeability barrier.

A further most important result of electron microscopy is the demonstration that the myelin sheath is made up of Schwann cell membrane material (see earlier section of this chapter). This provides direct evidence that the osmiophilic layer seen at the Schwann cell boundary is a layer of lipoprotein. Further, detailed studies of the composition and structure of the myelin sheath become immediately relevant to the study of membranes.

The myelin sheath is a laminated structure suitable for study by X-ray diffraction (Fig. 2.16). Since X-ray diffraction patterns can be taken on fresh, wet, unfixed nerve, the dimensions of the myelin lamellae can be compared in electron micrographs and in freshly isolated nerve. In fresh frog nerve, a dimension of 9 nm is found corresponding to the 7–8 nm dimension in the electron micrographs. Changes produced by the electron microscopc preparative procedures may be followed by X-ray diffraction; there is considerable shrinkage during fixation, and during dehydration in alcohol solutions (9 nm → 6·5 nm), due to loss of water, and presumably to extraction of lipids. During the penetration of the dehydrated structure by methacrylate monomer and polymerization, the original dimension is almost restored (6·5 nm → 8 nm), possibly because methacrylate molecules move into the structure to replace the extracted lipid molecules. The *in vivo* thickness of membranes seen in the electron micrographs as 7–8 nm profiles may therefore be nearer 9 nm.

The details of the X-ray diffraction pattern of myelin have been extensively studied by Finean. Some time before the electron microscopical demonstration of the connection between the myelin lamellae and the Schwann cell membrane, he suggested from the X-ray diffraction evidence a structure for the myelin lamella essentially similar to that proposed by Danielli for cell membranes. This seems at first sight good support for the Danielli structure; but unfortunately Finean's proposals for the detailed arrangement of molecules in the myelin structure are almost as much a matter of speculation as Danielli's, and moreover, a similar sort of speculation, based on chemical composition and the known arrangement of polar lipids in lamellae without protein present. As we have noted, for polar lipids interacting with proteins, speculation of this sort becomes very tentative.

The myelin sheath does not, in fact, give very detailed X-ray diffraction patterns (Fig. 2.16). The main results which come directly out of the early X-ray diffraction studies on which Finean's model is based, are the lamellar thickness, and a periodicity within the lamellae suggesting close-packed hydrocarbon chains lying predominantly perpendicular to the plane of the lamellae. The X-ray studies are in this respect confirming a conclusion reached earlier from studies of myelinated nerve in polarized light. More recently Finean, and Wilkins and co-workers, have refined the X-ray diffraction studies to show two main regions of high electron density in each lamella. The separation of these regions suggests that they represent the electron-dense phosphate groups of a bimolecular lipid layer.

A further contribution to chemical identification of plasma membranes seen in electron micrographs comes from study of sheets of pure phospholipid and lipoprotein prepared and fixed *in vitro*. Laminar micelles of pure phospholipid give an electron micrograph appearance, after osmium or permanganate fixation, embedding and thin sectioning, essentially similar to that of the membrane at the boundary of cells. Adsorption of protein to these pure phospholipid lamellae *in vitro* prior to fixation leads to a slight thickening of the outer lines in the electron micrographs (Fig. 4.16). Thin sheets of protein alone give a single dense line.

Osmium tetroxide has been shown, by chemical tests, to interact with a variety of groups in proteins, also with phospholipid polar groups, and with the double bonds of unsaturated hydrocarbon chains (Table 4.4). Potassium permanganate is also a relatively non-specific 'electron stain', though it seems to interact less strongly with protein than does osmium tetroxide. We cannot tell from these studies which component of a lipoprotein structure will be more strongly osmiophilic. We know with some certainty that in the lamellae of pure phospholipid, the molecules must be oriented with their polar groups towards the aqueous phase. It would seem from the electron microscope appearance of these lamellae, and from studies of other water-lipid systems, that the polar groups provide a greater number of osmium-binding sites than the hydrocarbon chains. But this is by no means certain. Hydrocarbon chains may become reoriented to some extent during fixation, to bring their unsaturated groups to the surface of the lamella, with the osmium forming cross-links between them, and between polar groups, at the aqueous interface. In fact, if reaction with osmium takes place primarily at the aqueous interface, a layer of lipoprotein may produce a similar profile whatever the arrangement of lipid and protein within the layer.

We may sum up the arguments of this section as follows: the evidence for believing the electron micrograph 7–8 nm membrane found at the boundary of cells to be a layer of lipoprotein, and for supposing this layer to be the main barrier to free diffusion into and out of the cell, seems very convincing. The hydrated, *in vivo* thickness of the layer may be 9 nm. How the protein and lipid are arranged in this layer is much less certain. The similarity in electron micrograph appearance between plasma membranes of cells, and the bimolecular lamellae of phospholipid with adsorbed protein, may be considered quite good evidence for what we have called a Danielli-type structure, the clear region between the dark lines in the electron micrographs corresponding to predominantly hydrocarbon material. In the myelin sheath, at least, and in the pure phospholipid bilayers, the hydrocarbon chains are oriented mainly perpendicular to the surface of the lamellae.

The electron micrographs and the X-ray study of myelin certainly do not rule out possible variants on the Danielli-type structure: small areas of pure protein

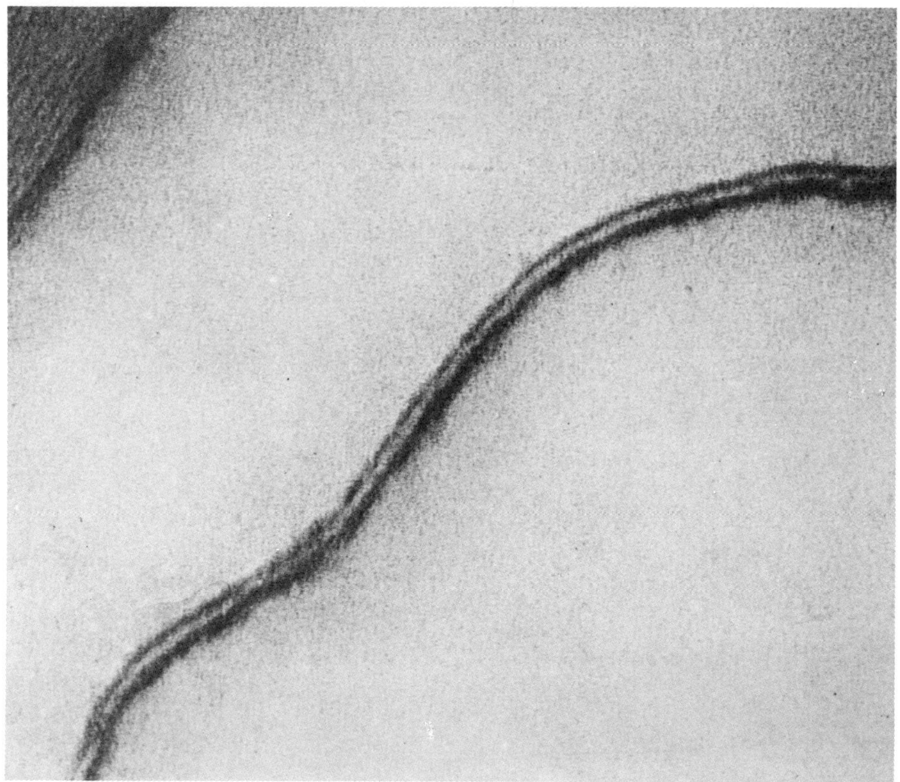

Fig. 4.16 Electron micrograph of the smallest lamellar structure seen in aqueous suspension of phospholipid and protein after osmium fixation. Magnification ×530000. (*From* Stoeckenius, 1962.)

interrupting an otherwise continuous lipid sheet, or a net-like membrane of protein fibrils with lipid plaques in the meshes. In both the X-ray diffraction study of myelin, and in electron micrographs of thin sections, we are studying an effect produced by those membrane components which are organized in a more regular way (see Appendix 1 for a discussion of this technical limitation to X-ray diffraction studies). We learn nothing by these methods of more irregularly arranged material which might compose up to 30 per cent of the membrane structure.

Stability of lipid bilayers and their interaction with proteins

Since we are so ignorant of the detailed molecular structure of membranes, particularly for small or transient areas which may be important in electrical or active transport phenomena, we must at present keep an open mind on different possibilities.

In the first place, the lipid may not everywhere be arranged as a bilayer; where it is so arranged it may be of variable fluidity, which will affect the readiness with which molecules can cross from one side to the other. There is the further possibility, shown up rather strikingly by negative contrast examination of mixtures of saponin, lecithin and cholesterol, that lipid mixtures can sometimes form lamellar aggregates of spherical micelles. Thus, for example, limited areas of a lipid bilayer might under certain circumstances undergo transient or more permanent transformation to a lamellar aggregate of spherical micelles, allowing transient or permanent aqueous pores to form between the micelles.

For interaction between lipids and proteins in membranes there are, again, a number of possibilities, illustrated in Figs. 4.17 to 4.19. If, for simplicity, we take the lipid to be arranged as a bilayer, protein molecules might be adsorbed to the bilayer without any change in the structure of the protein molecules (Fig. 4.17). There is secondly the possibility (4.18) of incorporation of protein

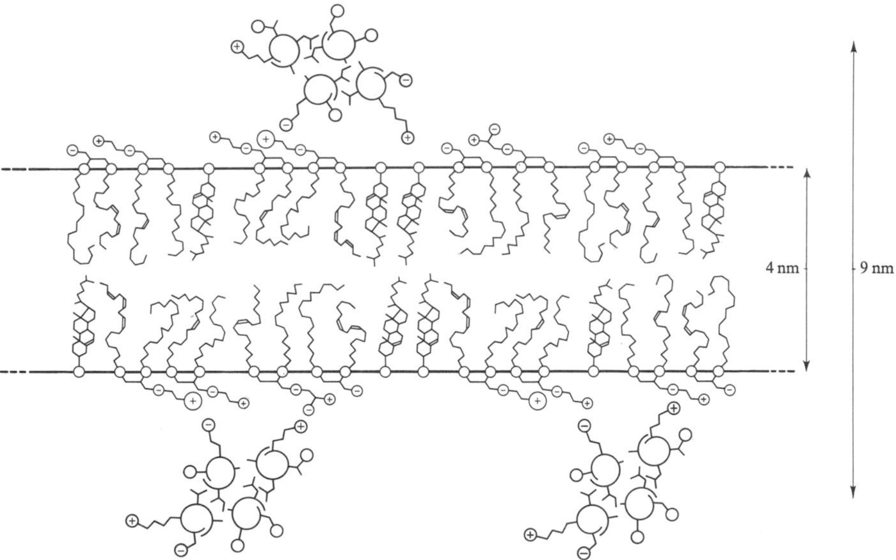

Fig. 4.17 A schematic illustration of the adsorption of a globular protein molecule to a bimolecular layer of polar lipids, without change in the secondary or tertiary structure of the protein. The bimolecular layer is represented as in Fig. 4.14. The globular protein is represented as in Fig. 2.19. Electrostatic attraction between oppositely charged groups may be expected to play some part in binding the protein to the lipid polar groups, as indicated in this figure. Divalent positive ions can also link negatively charged groups.

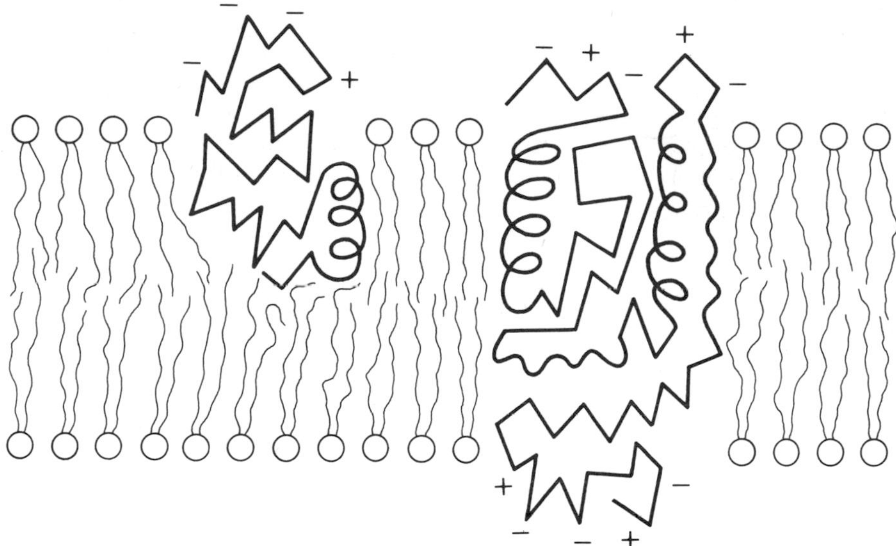

Fig. 4.18 Protein molecules embedded in a bimolecular layer of polar lipids. The proteins are shown with parts of their chains in helical conformation and parts in more extended secondary structure, as typically found for globular proteins in aqueous solution (e.g. lysozyme, Fig. 2.14). However, for membrane proteins, incorporated into the lipid bilayer as indicated here, it must be supposed that the secondary and tertiary structure adopted by the polypeptide chains brings non-polar amino-acid side chains in contact with the hydrocarbon chains of the lipids, and polar amino-acid side chains to the aqueous interface (indicated in this figure by + and − signs on the pro tein). (*From* Singer, 1972.)

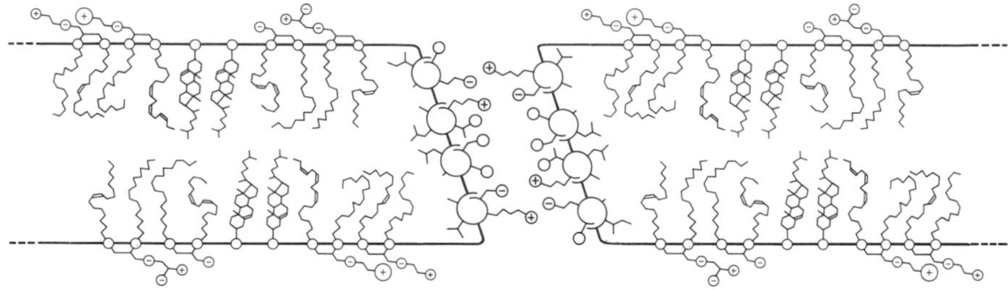

Fig. 4.19 A schematic illustration of the incorporation of protein molecules into a bimolecular layer of polar lipids to form an aqueous channel through the layer lined by protein polar groups. The protein is supposed to undergo some modification in secondary and tertiary structure, as it unfolds to stabilize the hydrocarbon-water interfaces at the break in the bimolecular layer. The protein molecules are therefore represented by the unfolded drawing of Fig. 2.19. The bimolecular layer of polar lipids is represented as in Fig. 4.14. Note the interaction between oppositely charged protein side-chain groups tending to reduce the pore to a series of small water-filled crevices between the protein groups. Divalent positive ions (e.g. Ca^{++}) can also link negatively charged groups. Low Ca^{++} concentrations may be expected to loosen the pore structure and make the membrane more permeable to water and ions.

into the bimolecular layer, with a secondary and tertiary structure which allowed hydrophobic contact with the lipids. An enzyme incorporated in this way might retain sufficient three-dimensional structure for its active site, and hence its enzymic activity, to be little modified by incorporation into the membrane, or the enzyme active site might only be formed in interaction with lipids. The concept that globular proteins may be rather deeply embedded in the membrane surface in this way receives support from 'freeze-etch' examination of membranes. In this technique frozen tissue is fractured and a replica taken from the fracture surface for electron microscope examination. In the fracture, membranes tend to cleave along their weakest plane, which in a Danielli-type structure is the mid-plane of the membrane, between the ends of the hydrocarbon chains (Fig. 4.14). For myelin this cleave-surface is smooth, suggesting a simple bilayer structure for the lipid, but most other membranes show particles 5–10 nm in diameter in this plane which probably represent globular proteins embedded in the lipid bilayer.

A third possibility is that a protein might be incorporated into the surface with complete loss of secondary and higher structure, so that the polypeptide chains lay threaded amongst the phospholipid polar groups in a random, constantly changing form, with their hydrocarbon side chains oriented towards the lipid, and their polar side chains towards the aqueous phase. Proteins in this situation would presumably have no enzymic activity and probably could have no functional role in active transport. They might however play an important part in stabilizing the membrane structure. Further possibilities are the incorporation of sufficient protein to form aqueous channels (Fig. 4.18), and the stabilization by protein of a lipid monolayer. In the situations of Figs. 4.18 and 4.19, there is always the possibility of further adsorption of protein (as in Fig. 4.17).

Speculation about molecular mechanisms underlying permeability and active transport phenomena

Since we know little of the true arrangement of lipid and protein in plasma membranes we consider in the final section of this chapter what are the simplest modifications of the Gortner and Grendel lipid bilayer model that would be necessary to account for the observed properties of plasma membranes. The main experimental approach, in this field, lies in study of the permeability properties of pure lipid-phospholipid bilayers of defined composition. Phospholipids dispersed in aqueous solution by ultrasonic agitation form small 'liposomes' 50–500 nm in diameter, each made up of a number of concentric spherical bilayer shells. Bangham and co-workers have formed liposomes in solutions containing radioactive ions, and studied the movement of these ions out across the bilayers when the liposomes were suspended in unlabelled medium. Another method, pioneered by Mueller, involves forming a lipid bilayer across a hole in the wall separating two aqueous compartments. A droplet of mixed lipids, dissolved in a water-miscible solvent, is spread across the hole. If the amount of lipid is suitably chosen, the solvent dissolves in the water to leave a bilayer across the main area of the hole, with some unspread lipid round the edge. There are difficulties in determining whether the whole main area is a bilayer, or whether part of the area is perhaps several bilayers thick. Also natural lipid mixtures have not yet been spread successfully, unnatural lipids like tetradecane have to be added; further the composition of the bilayer areas may not be the same as the overall composition of the lipid drop spread across the hole. When the difficulties have been overcome for permeability studies of artificial membranes of this kind, or for liposome measurements (where the problem is to determine the total area of liposome surface in the preparation under investigation) then it should be possible to find out whether the movement of water and the passive movement of ions across plasma membranes can be adequately explained in terms of the lipid component of the membrane alone. On present evidence the water and passive ionic permeability properties of natural membranes seem to require either permanent or transient pores of some kind or movement of protein across the membrane. If pores penetrate a lipid bilayer they could be stabilized by protein (Fig. 4.19) or could arise by spherical micelle formation in the bilayer (discussed earlier), or possibly by a sort of incipient micelle formation whereby polar groups of several phospholipids come together to form a cleft in the membrane surface, infolding a very small aqueous droplet that then moves across the membrane. In the limit, a single phospholipid molecule might move from one side of the membrane to the other (as discussed above), binding ions and water molecules. Any of these processes could allow ions, which are very insoluble in lipids, to move across the membrane.

Yet another possibility for formation of permanent or transient pores in bilayers has arisen from the study of the action of antibiotics like gramicidin and valinomycin on natural and artificial membranes. Antibiotics of this kind are ring-shaped molecules with hydrophilic groups on the inside and hydrophobic groups on the outside. Some, like gramicidin A, can stack as cylinders and penetrate into a bilayer to open up an aqueous pore of highly selective ionic permeability. Others, like valinomycin, act as carriers binding ions and taking them across through the lipid. The valinomycin rings have an internal pore diameter of 0·7 nm, allowing hydrated K^+ ions to be bound within the ring but not hydrated Na^+ ions. The question as to whether natural membranes contain molecules of this kind, or any lipids like saponin that might favour local micellation of lipid bilayers, is at present conjectural.

For active transport, proteins must be involved at some point, linking transport to the splitting of ATP. For the Na^+ and K^+ exchange pump the ion-translocator membrane site seems to be closely linked to an enzyme ATPase site. Fragments prepared from red cell ghosts, and from other membrane sources show ATPase activity sensitive to Na^+ and K^+ levels, and like the ion transport system, sensitive to Ca^{++}, Mg^{++} levels and to the inhibitor ouabain. The simplest models that have been proposed to account for active transport processes are 'carrier' models in which a lipid-soluble carrier is supposed to move from one side of the membrane to the other binding the transported ion or solute. Thus for the Na^+—K^+ exchange pump the carrier may be supposed to take two forms, binding Na^+ and K^+ respectively, and be converted from the K^+ to the Na^+ binding form on the inside of the membrane by the action of ATP, and back to the K^+ binding form at the outer surface of the membrane. However, no lipid soluble carriers with the required properties have yet been isolated from membranes, nor have any proteins yet been described with the property of being able to 'turn themselves inside out' (so that their polar groups move to the interior of the protein molecule and non-polar groups to the outside).

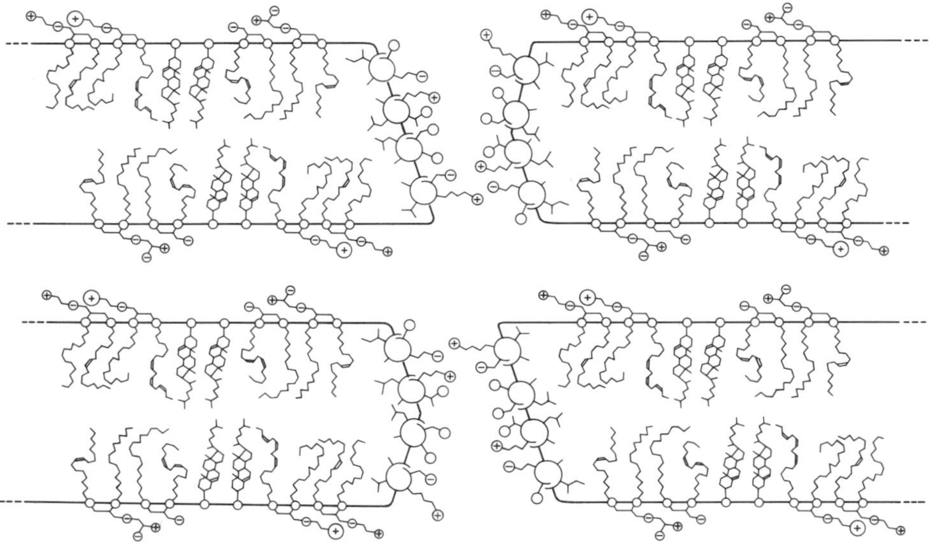

Fig. 4.20 A figure showing how protein groups in a membrane pore might be exposed alternately to the media inside and outside the membrane.

However, perhaps it is not necessary for a protein to undergo such drastic conformational change to cross a membrane. It may be that a globular protein rather deeply embedded in the bilayer on one side can cross to the other side occasionally in random thermal movement with relatively minor change in tertiary structure in the process. The properties of the Na^+—K^+ pump noted above seem to require that a protein site should be presented alternately to the inside and outside surface of the membrane. One possibility would be a protein lined crevice like that of Fig. 4.19, with enough freedom of movement to allow the groups near the centre of the crevice to be presented alternatively to the interior and exterior, as indicated in Fig. 4.20. In active transport of sugars and amino acids a translocator protein site again seems to be involved.

The sudden increase in ion permeability that accompanies depolarization of a nerve or muscle cell is not understood at all at the molecular level. If the membrane structure is at all like the model depicted in Figs. 4.19 and 4.20, and if the crevice is normally held closed partly by Ca^{++} ions linking negative groups, one could understand that depolarization might lead to a movement of Ca^{++} ions from the membrane into the cytoplasm, with opening of the crevices.

If the reader glances back thoughtfully at this point to the early sections of this chapter it will become clear how little is yet known about the structure and function of plasma membranes. Apart from lack of knowledge of membrane protein and possible pore structure we do not know how the 15–30 nm gap between apposed membranes is maintained, how pinocytosis is brought about, or how specificity of cell adhesion is controlled (Chapter 1). We do not know how synthesis of new membrane is controlled to allow membrane proliferation to form microvilli, or myelin sheathes, or how the enzymes are controlled which synthesize outer polysaccharide coats and walls, or how tonofibrils and myofibrils become attached to membranes at certain sites. The field of membrane structure and function remains a large area of molecular biology awaiting means for detailed exploration. Mainly perhaps we need better methods to separate and characterize membrane proteins, while keeping them in their natural conformation. Unfortunately the detergent-type molecules used to separate membrane proteins from lipids are very likely to modify protein conformation, which depends so much on hydrophobic effects (see Chapter 2). Some progress has been made in characterizing a number of membrane proteins, at least as regards molecular weight and amino-acid composition. They tend to have a higher proportion of non-polar amino acids than most soluble proteins, indicating hydrophobic interaction with lipids in the natural state (see *Staph. aureus* membranc phosphokinase, Table 2.2). Preliminary amino-acid sequence studies for the main glycoprotein of the red cell membrane show short carbohydrate chains linked to amino acids towards one end of the chain, and a non-polar region in the central part of the sequence. Labelling of this protein by interaction of radioactive iodine first with intact red cells, and then with ghosts, shows that the carbohydrate-binding end of the chain is accessible to the outside, and the other end of the chain to the inside, suggesting this protein lies right across the membrane with its non-polar sequence spanning the hydrophobic membrane interior.

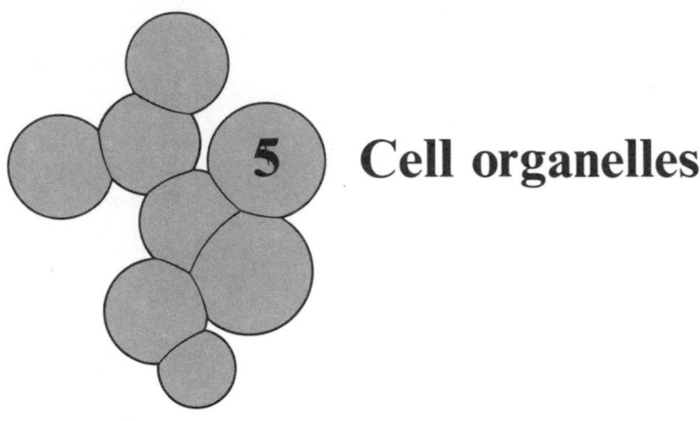

Cell organelles

In previous chapters we have explored the types of structure that can be formed from protein alone and the functional roles of structures of this kind, then moved on to consider the interaction of proteins with lipids at the plasma membrane. In this chapter we continue to explore the structure and function of lipoprotein membranes in mitochondria, chloroplasts and other membrane systems and cell organelles. We include in this chapter also discussion of another rather fascinating type of cytoplasmic structure – the centriole, found at the poles of dividing cells and at the bases of cila and flagella.

Mitochondria

All cellular activity requires energy, usually obtained from step-wise oxidation of energy-rich materials such as glucose or fatty acids. Early observations indicated that the mitochondria might be involved in cellular oxidations, and the more recent studies of the enzymes present in different cell fractions prepared by the methods of differential centrifugation discussed in Chapter 1 have confirmed that mitochondria are directly, and possibly exclusively, concerned with important steps in these processes. Of all cells, only mature red cells, which do not expend very much internal energy, and bacteria, are devoid of mitochondria. There is great variation in the shape, size, and number of mitochondria in different cells; usually they are long slender threads – hence their name – but they can also be short stout rods, or spherical globules. Mitochondria were originally recognized in the light microscope by their affinity for the dye Janus Green, but the demonstration in electron micrographs of a unique internal membrane structure has provided a valuable criterion for their identification, and shown their widespread occurrence.

In thin sections of material fixed in either osmium tetroxide or permanganate, each mitochondrion has an external envelope 4–6 nm thick. This is separated from a similar internal membrane by a clear gap of about 10 nm. The central mitochondrial compartment within the internal membrane is filled with a matrix containing small granules. The inner membrane projects as folds, or tubes, into the matrix to form *cristae*, each crista appearing in a thin section as a pair of internal membranes with a 10 nm separation (Fig. 5.8). Serial sectioning

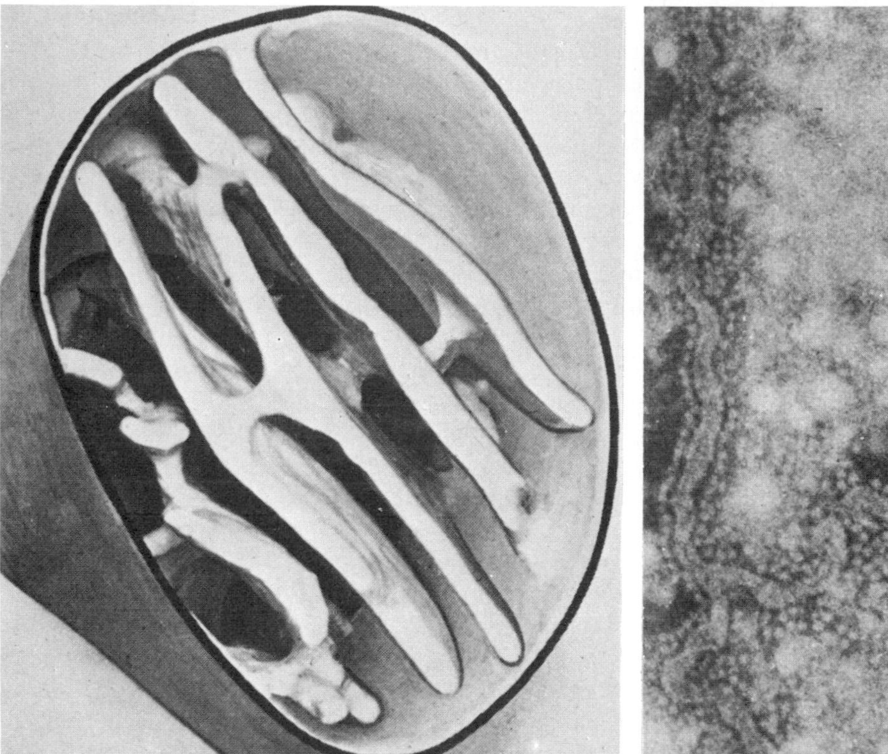

Fig. 5.1 *a*. Photograph of a model showing the three-dimensional structure of a mitochondrion. Magnification ×280000. This model was made from serial sections of a mitochondrion in an axon at a nerve-muscle junction. An external membrane surrounds the whole organelle. An internal membrane runs parallel to the external membrane, and also projects as folds, or cristae, into the central compartment. These cristae are irregular and anastomose with each other. The spaces between the cristae would in life be filled with the mitochondrial matrix. (*From* Andersson-Cedergren, 1959.)

b. Detail of structure of cristae revealed by negative contrast studies of fragmented mitochondria. Magnification ×150000. Particles 8–9 nm in diameter, on stalks 4–5 nm long, appear to line the membranes of the cristae, and the inner membrane at the boundary of the mitochondrion, on the sides facing the mitochondrial matrix. (*From* Stoeckenius, 1963.)

and reconstruction give the three-dimensional structure of a mitochondrion (Fig. 5.1). The fold of membrane forming a crista is often only continuous with the inner membrane at the surface of the mitochondrion over a small area, so that the flat plate of the crista abuts against, but does not join, the inner membrane around much of the crista circumference. The cristae do not form complete partitions and it is possible by a devious route to travel through the matrix from one end of a mitochondrion to the other.

There is some variation in mitochondrial internal structure, as there is in the shape and size of the whole organelle, in different tissues. The mitochondria of early embryonic animal cells and plant cells contain just a few short cristae, while in cells of the adrenal cortex the cristae are replaced by tubular invaginations, a pattern also seen in the gill epithelium of fishes. In muscle cells (Fig. 4.6) flat cristae are seen; here there is little matrix, and the entire inside of the mitochondrion is filled with close packed cristae. These variations are only superficial; the underlying structure of an outer envelope with a contorted internal membrane is constant, and the identification of mitochondria is never in doubt in an electron micrograph. After tissue homogenization and centrifugation, the mitochondrial structure is preserved, and the pellet which contains them can be identified, and its purity assessed, by thin-sectioning of the osmium-fixed embedded pellet.

Certain specialized cells have high energy requirements. Muscle cells are an obvious example, but the secretion of material against an osmotic gradient is equally demanding; the nasal gland of the duck secretes a strong salt solution, the oxyntic cells in the stomach form a 0·1 N HCl solution, and the tubular cells of the kidney produce a large increase in the concentration of the glomerular filtrate. In all these cells a large proportion of the cytoplasm is occupied by mitochondria, and each of these has more profuse infoldings of their internal membrane than is found in the less numerous mitochondria of other glandular, nerve or plant cells. Observations on the structure of mitochondria indicate that the energy expenditure of a cell is related to the total area of internal mitochondrial membrane. Bacteria do not contain mitochondria – the size of a mitochondrion is in fact about the same as that of a bacterium – but part of the surface of the bacterial plasma membrane is specialized to carry out the energy-producing oxidation reactions associated with the internal mitochrondrial membrane in eukaryotic cells (cells with nuclei).

The main pathways of the mitochondrial enzyme systems are summarized in Fig. 5.2. There are further pathways not shown in this figure leading to the electron transfer chain from intermediate compounds produced during the breakdown of fatty acids. In the oxidation reactions of the tricarboxylic acid cycle, and of fatty acid breakdown, hydrogen atoms are removed from the intermediate compounds; after a number of further steps, an equivalent number of hydrogen atoms combine with oxygen to form water. In these further steps the hydrogen electrons and hydrogen nuclei are separated from one another, and the electrons move along a chain of carriers: the electron transfer chain. This movement is coupled to the phosphorylation of ADP to ATP, and under aerobic conditions provides the main means by which the cell regenerates the ATP which is broken down (to ADP) during muscular contraction, active transport and other energy-requiring metabolic reactions. This oxidative-phosphorylation system remains functionally active *in vitro* if mitochondria are extracted and maintained in suitable media. Also, small vesicles with a single bounding membrane derived from the inner mitochondrial membrane by disruptive treatments such as ultrasonic agitation or swelling in hypotonic solutions, are still capable of transferring electrons along the electron transport chain and coupling this transfer to phosphorylation of ADP.

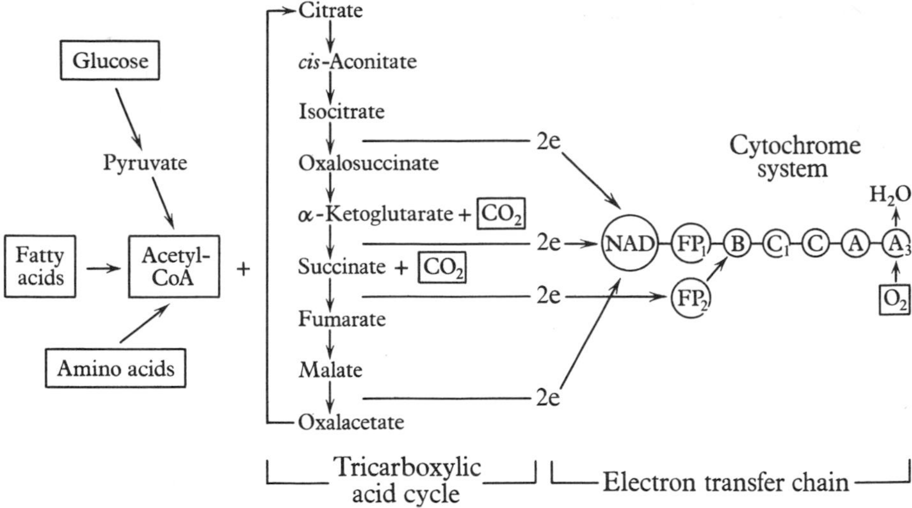

Fig. 5.2 The intermediate compounds of the tricarboxylic acid cycle and the carriers of the electron transfer chain (cytochrome system).

NAD = nicotinamide-adenine dinucleotide
FP = flavoprotein
B, C, etc. = cytochrome b, cytochrome c, etc.

(Modified *from* Lehninger, 1959.)

Many of the enzymes of the tricarboxylic acid cycle are readily brought into aqueous solution if the mitochondria are fragmented by ultrasonics, suggesting that these enzymes exist in solution in one or other compartment of the mitochondria, or at least are not very firmly bound to the mitochondrial membranes. The enzymes and carriers of the electron transfer chain, on the other hand, remain firmly attached to the small vesicles formed on disruption. It is possible by mild swelling of mitochondria to burst the outer membranes, and subsequently to separate them from the inner membranes. It can be shown in this way that the enzymes and carriers of electron transfer and phosphorylation of ADP form part of the structure of the inner mitochondrial membrane. Transfer along the electron chain and the coupling of this transfer to the phosphorylation of ATP are apparently achieved by incorporation of the enzymes and carriers of the transfer chain into this inner mitochondrial membrane.

In Fig. 5.3 the successive energy steps associated with the movement of electrons to different levels of reduction-oxidation (redox) potential along the electron transfer chain are compared with the free energy required for the formation of ATP from ADP. The movement of a pair of electrons down the transfer chain can lead to the phosphorylation of two molecules of ADP for the chain leading from succinate, and three molecules of ADP for the chains leading through NAD from the other intermediate compounds of the tricarboxylic acid cycle (Fig. 5.2). One phosphorylation appears to be associated with electron movement across the span NAD to flavoprotein, another with movement across the span flavoprotein to cytochrome c, and a third with movement from cytochrome c to oxygen. The electrons from succinate pass directly to a flavoprotein rather than through NAD. Electrons from the intermediate compounds of fatty acid breakdown pass in some cases through NAD, and in other cases direct to flavoproteins.

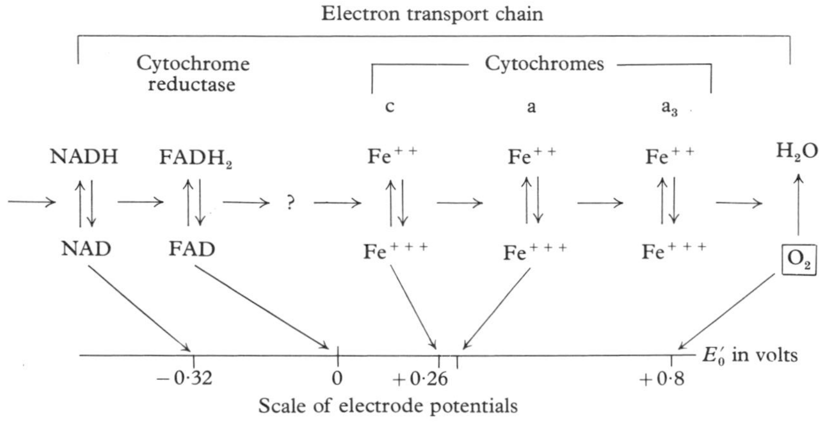

Fig. 5.3 A schematic picture of the electron transfer chain, showing the approximate reduction-oxidation potentials (E_0') of the electron carriers (at pH ~ 7). The free energy change associated with the ADP + P ⇌ ATP reaction is generated by a pair of electrons moving down a reduction-oxidation potential step of 0·25 V. (NAD = nicotinamide-adenine dinucleotide, FAD = flavin-adenine dinucleotide.) (*From* Lehninger, 1954.)

Lipids make up 30–40 per cent of the dry weight of mitochondria, and with certain exceptions (for example, much less cholesterol) the mitochondrial lipid composition is broadly similar to that of plasma membranes (Table 4.2). It is reasonable to conclude, therefore, that the inner and outer mitochondrial membranes seen in thin-section electron micrographs represent sheets of lipoprotein which impose barriers to free diffusion between, firstly, the inner matrix and the space between the inner and outer membranes, and, secondly, between this space and the cytoplasmic matrix. The osmotic behaviour of isolated mitochondria (their swelling, for example, in hypotonic solution) is also consistent with the view that they are bounded by a membrane or by membranes impermeable to ions and small solute molecules. However the permeabilities of the inner and outer membranes appear to be very different, at least after isolation of the mitochondria from cells. The inner membrane represents a permeability barrier comparable to plasma membranes, *but the outer membrane seems to be quite permeable to ions and to small solute molecules such as sugars.* Yet the lipid composition and thin-section appearance of the two membranes are quite similar. This shows that the apparent uniformity of different membranes in thin-section electron micrographs is deceptive, if taken to imply uniformity at the molecular level.

There are differences in the enzymes present in the two mitochondrial membranes; the inner membrane contains the enzymes and carriers of oxidative phosphorylation, the outer membrane carries a different set of enzymes and electron carriers (cytochrome b_5, etc.) not linked to phosphorylation reactions.

Leaving to one side the problem of the permeability properties and role of the outer membrane, and concentrating attention on the structure and function of the inner membrane, the main problem is to understand how energy available in transfer of hydrogen atoms and electrons, from one carrier to another of higher hydrogen or electron affinity (higher redox potential) can be used in a phosphorylation reaction. Negative contrast electron micrographs of inner mitochondrial membranes show a series of knobs on stalks on the side facing the inner matrix (Fig. 5.1). During negative contrast drying there is adsorption of phosphotungtate ions to the membrane and to the knobs, which may introduce electrostatic repulsion forces which push out the knob as a partial artefact. However, it seems probable that the knob structure corresponds to some component lying at the membrane surface *in vivo*, for ultrasonic treatment detaches from the membrane particles which look like isolated knobs in electron micrographs. The cytochromes remain bound to the membrane. The isolated knob preparation shows ATPase activity, and the knobs probably carry the phosphorylation site of mitochondrial oxidative phosphorylation. This can act to add phosphate to ADP during electron flow, and reversibly, to split the terminal phosphate off ATP, if ATP is present, when oxidation is prevented (no electron flow) or when the knob carrying the phosphorylation site is detached from the cytochrome chain on the main part of the membrane.

An interesting line of work following from studies of isolated mitochondria, and later of sub-mitochondrial vesicles formed by fragmentation of the inner membrane, was the discovery of active transport of ions across this inner membrane. Flow of electrons down the cytochrome chain can lead *either* to phosphorylation of ADP *or*, if no ADP is present, to movement of K^+ or Ca^{++} ions across the membrane. Further, if electron flow is stopped (by cutting off supplies of oxygen, or tricarboxylic acid cycle substrates like succinate), then ATP split by the phosphorylation site can provide energy for ion movement. Thus electron flow, phosphorylation and ion movement are linked.

There are two main theories as to how this linking might be achieved. On the classical view it is supposed that electron transfer from one carrier to the next brings some intermediate compound to a 'high energy' form, and that this high energy compound mediates either phosphorylation or active ion transport. However support has grown for an alternative explanation with the merit, forcefully argued by Peter Mitchell for a number of years, that it avoids postulating a 'high energy' intermediate compound (for which there is no direct evidence) and that it neatly shows why the whole carrier-enzyme system must be situated in a membrane to function effectively. On this *chemo-osmotic* hypothesis the hydrogen and electron carriers are supposed to be situated in the membrane structure in such a way that flow of hydrogen atoms and electrons down the transfer chain leads to release of H^+ ions on one side of the membrane and uptake of H^+ ions on the other, so that hydrogen and electron flow generates both a potential and pH difference across the membrane (Fig. 5.4). This potential and pH difference could then provide the driving force which disturbs ionic

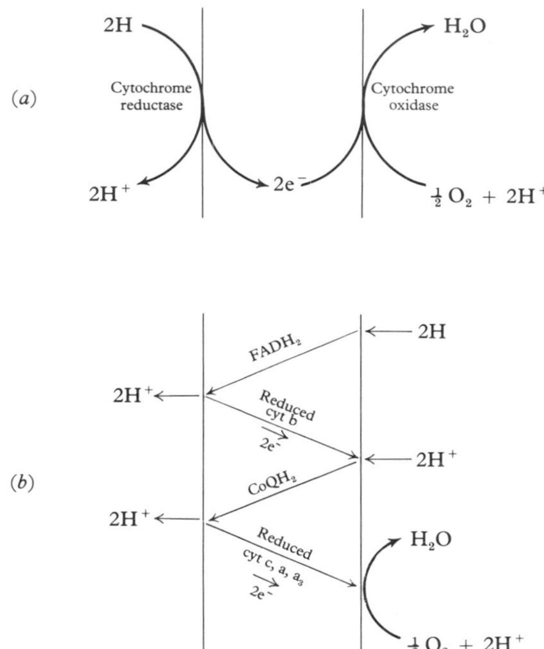

Fig. 5.4*a* H^+ pump of the simplest kind. It is supposed that the carriers of the electron transfer chain might be so positioned in the membrane that transfer of electrons to cytochromes with release of H^+ takes place on one side, and withdrawal of electrons from cytochromes with uptake of H^+ takes place on the other (*after* Lundegardh, 1945).

b A scheme for H^+ pumping at the inner mitochondrial membrane that may come closer to the real situation. Alternation of hydrogen and electron carriers, with coenzyme Q inserted as an additional hydrogen barrier between cyt b and cyt c, leads to pumping of two H^+ ions across the membrane for each electron travelling from FAD to molecular oxygen. (*After* Mitchell, 1967.)

equilibrium across the membrane and disturbs equilibrium at the phosphorylation site. It may be noted, in this context, that splitting of ATP to ADP+P is a hydrolysis (cf. discussion of lysozyme action in Chapter 3). If the ATPase site is suitably situated in the membrane the splitting of ATP could lead to uptake of an H^+ ion from one side of the membrane and uptake of an OH^- ion from the other. Splitting of ATP would then cause pumping of H^+ ions across the membrane and conversely, if the ATPase site acted reversibly, an H^+ gradient would lead to phosphorylation of ADP (Fig. 5.5). If these views are correct the ATPase of the inner mitochondrial membrane would differ from that of the plasma membrane in that for the former, phosphorylation would be directly linked to H^+ movement, indirectly affecting K^+ and Ca^{++} transport, while for the latter phosphorylation would be directly linked to K^+ and Na^+ movement.

Peter Mitchell and others have shown that oxidation of tricarboxylic acid cycle substrates by mitochondria does appear to be linked to H^+ movement across the bounding membranes, the suspending medium becoming more acid, which corresponds to outward movement of H^+ ions, during electron flow.

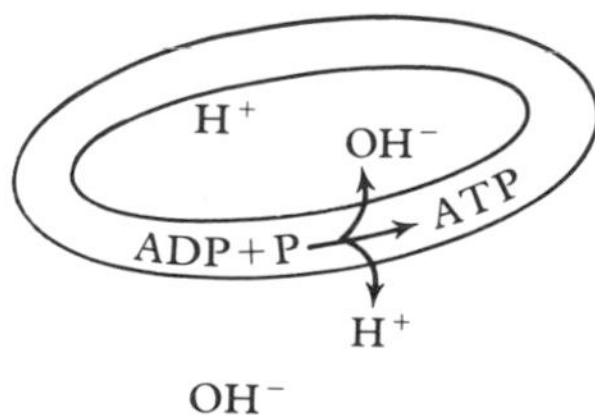

Fig. 5.5 Reversible ATPase site so situated in a membrane that phosphorylation of ADP can be driven by an H^+ gradient across the membrane. (*From* Jagendorf and Uribe, 1967.)

However there are difficulties in measuring the internal pH and potential and hence the pH and potential differences across the membranes, and in showing that the arrangement of carriers is compatible with the scheme of Fig. 5.4*b*. The chemo-osmotic hypothesis is not entirely satisfactory in its present form. It is nevertheless a stimulating concept and suggests simple and feasible molecular mechanisms by which hydrogen and electron transfer might be coupled to phosphorylation and ion pumping.

Considerable progress has been made in the task of chemically isolating and characterizing the proteins of mitochondrial membranes. Unfortunately most membrane proteins are complexed with lipids by hydrophobic interaction in the natural state, and to separate these proteins from lipid, or to separate specific lipoproteins from the total membrane structure, detergent-type molecules must be used. These are polar lipids, similar to phospholipids in that they stabilize hydrocarbon-water interfaces. When membranes are broken apart with detergents there is the possibility of protein denaturation and extensive rearrangement of lipid–lipid and lipid–protein interactions, since detergents at high concentration tend to interfere indiscriminately with all types of hydrophobic interactions. Some detergents at lower concentrations appear to break up membranes at specific points, without disrupting other parts of the membrane structure or causing total protein denaturation. Using deoxycholate (a detergent-type molecule found in bile, which emulsifies fat) Green and his co-workers have shown that it is possible to separate different parts of the cytochrome chain, with the proteins and lipids remaining complexed together as lipoprotein aggregates. It is possible to isolate a 'structural' protein fraction from mitochondrial membranes by treatment with a mixture of cholate, deoxycholate and sodium dodecylsulphate (another detergent) and subsequent fractionation by ammonium sulphate precipitation, followed by alcohol extraction of lipid. This fraction makes up 40–50 per cent of the total mitochondrial membrane protein; it is insoluble at pH ~ 7, but soluble at pH ~ 11. Some of the proteins of this fraction might have a true structural role. They will form complexes *in vitro* with phospholipids and with cytochromes a, b and c_1 and may link these components together in the inner mitochondrial membrane. However, we have noted above the problems of structural rearrangement associated with use of detergents in the study of membranes and although many drawings can be found in text books and in semi-popular scientific articles suggesting possible arrangements for the components of the inner mitochondrial membrane the true arrangement of these proteins and lipids still remains unknown.

Chloroplasts

The chloroplasts of green plants are cytoplasmic organelles with many structural and biochemical features rather similar to those of mitochondria. Chloroplasts are bounded by an outer and an inner membrane (Fig. 5.6). In mitochondria, cristae membranes appear to be continuous infoldings of the inner surface membrane; in chloroplasts internal flattened sacs are seen, similar to cristae, but, in the fully developed chloroplast at least, these do not appear to be attached to, or continuous with, the inner surface membrane. In chloroplasts the internal flattened sacs are termed *lamellae* (or thylakoids), and in contrast to mitochondrial cristae, the membranes of these lamellae appear in close contact, both at their inner surface (i.e. at the surface inside the flattened sac) and in parts of their outer surface, where the lamellae form packed stacks termed *grana*. The grana are large enough to be seen by light microscopy, but the arrangement of the membranes in the grana and in intergrana regions is, of course, only seen in electron micrographs (Fig. 5.6). Lipids make up about 40 per cent of the dry weight of chloroplasts, and include phospholipids as well as other lipid components characteristic of chloroplasts. The membranes seen in thin-section electron micrographs are thus again lipoprotein structures and, presumably, permeability barriers separating the interior of the chloroplast into different compartments. The main volume of the chloroplast, the *stroma*, lies outside the lamellae and inside the inner bounding membrane (Fig. 5.6), and there is a small separate space between the inner and outer bounding membranes. However, both bounding membranes appear to be quite permeable to ions and small solute molecules, with only the membranes of the lamellae forming a permeability barrier comparable with the plasma membrane and the inner mitochondrial membrane. The compartments within the lamellae appear to be of small volume, though these spaces may be larger *in vivo* than they appear in fixed preparations, and may vary in size in different states of chloroplast activity (see below).

In chloroplasts, light energy is absorbed by chlorophyll molecules which

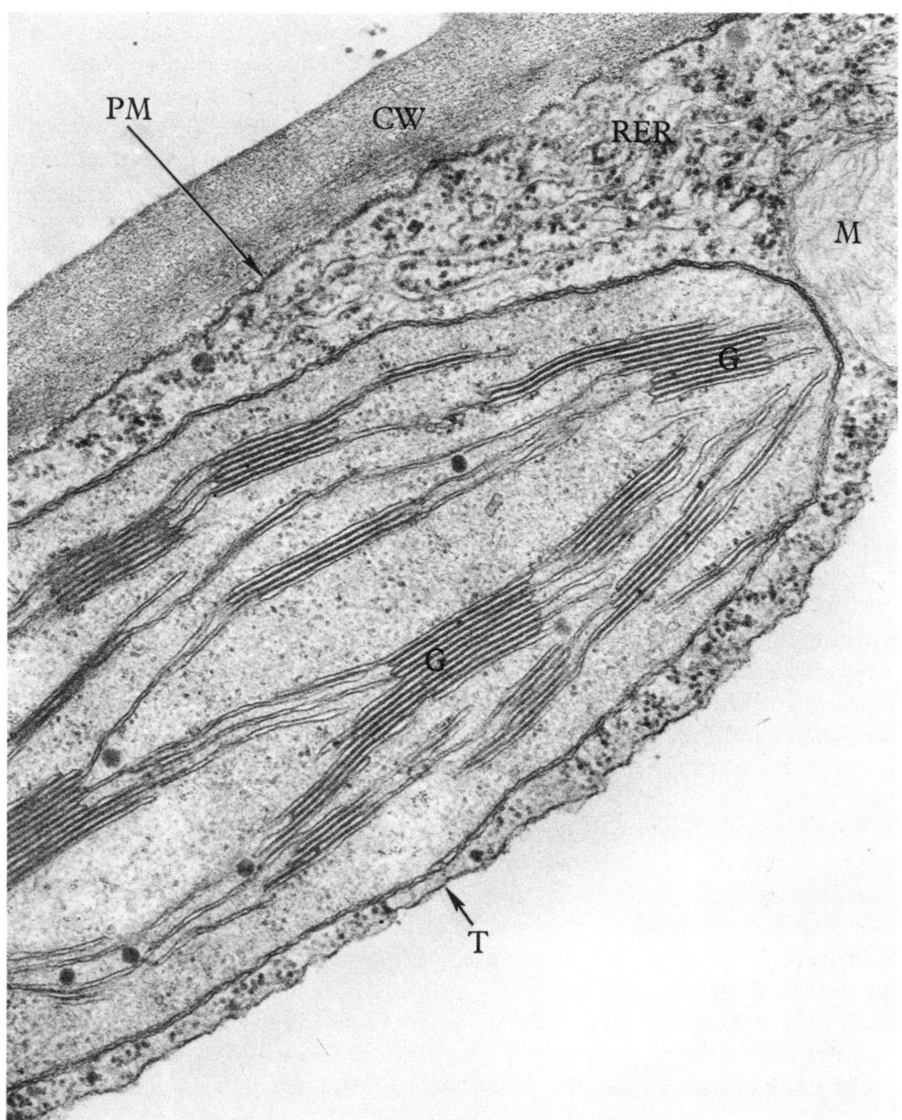

Fig. 5.6 Chloroplast of *Pisum sativum*. Part of a chloroplast extends from the left margin of the micrograph towards the upper right corner bounded by a double membrane. Within the chloroplast flattened sacs (thylakoids) can be seen stacked together in part of their area to form the grana (G) and extending out also into the inter-grana regions. Part of a mitochondrion (M) can be seen upper right (also bounded by a double membrane). The chloroplast lies in a thin layer of cytoplasm which is separated from the cell wall (CW) by the plasma membrane (PM) and separated from the vacuole by the tonoplast (T). RER = rough endoplasmic recticulum. (Micrograph by A. Massalski.) (Magnification × 50 000.)

have a structure rather similar to the haem group of haemoglobin (Fig. 9.1) except that Mg replaces Fe at the centre, and an additional 20-carbon hydrocarbon chain is linked by an ester bond to one of the acidic groups, conferring increased lipid solubility. There are further minor differences between these two molecules, but this seems to be an example of the way in which, through evolution, a molecule of a certain kind can be modified to fill quite different roles. The chlorophyll molecules form part of the structure of the lamellar membranes.

The main synthetic activity of chloroplasts, through which atmospheric CO_2 becomes built into small organic molecules, takes place in a cycle of reactions (Calvin cycle) which are different from, but produce a similar effect to, the mitochondrial tricarboxylic acid cycle acting in reverse. In the tricarboxylic acid cycle small organic molecules are broken down to CO_2, and hydrogen atoms are withdrawn at various points to pass down a chain of hydrogen and electron carriers to atmospheric O_2. In the Calvin cycle, CO_2 is built into small organic molecules, and hydrogen atoms are *supplied* by a chain of hydrogen and electron carriers, which draw electrons from water molecules, releasing O_2. Electron flow in mitochondria releases energy, but electron flow in chloroplasts requires energy at certain steps, and this comes from the absorbed light energy. The enzymes of the Calvin cycle, like the enzymes of the mitochondrial tricarboxylic acid cycle, are for the most part present as soluble enzymes in the stroma of the chloroplast. However the carriers of the hydrogen and electron chain form part of the membrane structure of the chloroplast lamellae.

The redox potentials of the electron and hydrogen carriers of chloroplasts are shown in Fig. 5.7. Part of the electron flow, from cytochrome b_6 to cytochrome f, for example, is from carriers of low electron affinity to carriers of higher electron affinity (higher redox potential). This part of the electron flow is 'downhill' (i.e. energy yielding) and closely similar to electron flow in mitochondria. It is coupled in chloroplasts to phosphorylation reactions forming ATP from ADP (Fig. 5.7). Other parts of the electron flow require energy: system I is linked to one type of chlorophyll molecule (chlorophyll a) and system II is linked to chlorophyll b. (The two chlorophylls differ only in that in chlorophyll b a formyl group replaces one of the pyrrole methyl groups.) Cycling of electrons powered by light adsorption in system I, generates ATP from ADP, independent of *net* electron flow from H_2O through to NADP which leads to donation of hydrogens to intermediate compounds of the Calvin cycle.

When a chloroplast suspension is illuminated, under appropriate conditions, there is an increase in the pH of the suspending medium. Chloroplasts also show swelling, or shrinkage under different conditions, in response to either illumination or to change in the pH of the suspending medium. Experiments of this kind, combined with thin-section electron microscopy of swollen chloroplasts have led to the view that illumination causes H^+ transport across the lamellae membranes from the stroma into the internal spaces of the lamellae. This H^+ movement is apparently mediated by electron flow induced by light

(a)

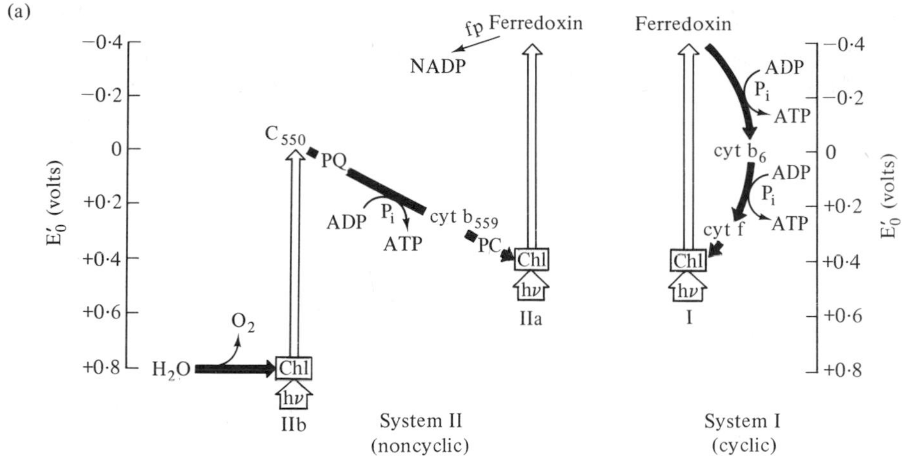

(b)

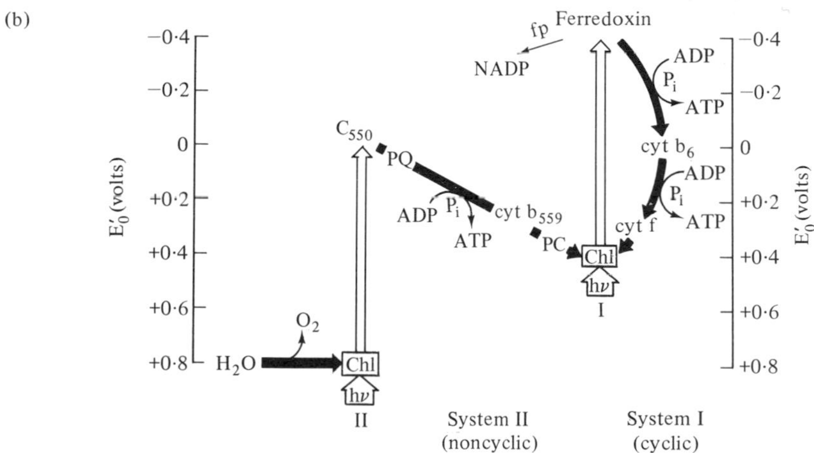

Fig. 5.7 Alternative schemes proposed for electron flow in chloroplasts. (*a*) Cyclic electron flow from ferredoxin (System I) totally separate from non-cyclic flow from water to NADP (System II). (*From* Knaff and Arnon, 1969.)

(*b*) The more popular view, with system II electrons joining system I somewhere around cytochrome f and plastocyanin.

cyt f, etc. = cytochrome f, etc.
PC = plastocyanin
PQ = plastoquinone
Chl = chlorophyll
C550 = component 550
fp = ferredoxin – NADP reductase
hv = absorbed light quantum
P_i = inorganic phosphate

absorption. Illumination can also lead to active transport of K^+ and Mg^{++} across the lamellar membranes.

Swelling is due to movement of water into the internal spaces of the lamellae. For mitochondria also, a variety of swelling and shrinkage effects can be produced under different conditions. It is probable in both cases that the water movement follows passively from the disturbance of osmotic equilibrium resulting from ion movement. Thus for chloroplasts, as for mitochondria, electron transport, phosphorylation of ADP, H^+ and water movement across membranes, and active transport of K^+ and other positive ions are all interrelated processes. The chemo-osmotic hypothesis developed for mitochondria (Fig. 5.4) can be adapted to account for the interaction between these different aspects of the activity of chloroplast membranes. As for mitochondria there are difficulties in accounting for some experimental data and in checking whether the predictions of this hypothesis are quantitatively correct, but it is clear, at the least, that very similar molecular mechanisms are operative in both mitochondria and chloroplasts. For chloroplasts there is, of course, the separate problem of how the adsorption of light by a chlorophyll molecule mediates electron flow from one carrier to another, and perhaps electron flow across the lamellar membranes. After light absorption the chlorophyll molecule is in a high-energy state, and spectroscopic evidence shows that for a short while this energy can be transferred from one chlorophyll molecule to another. The chlorophyll molecules are perhaps arranged side by side at the membrane surface, close enough for energy transfer. After a short while this energy comes to a chlorophyll molecule different from its neighbours in that it forms part of an energy or electron transfer complex. This complex now becomes altered in some way and the energy is channelled irreversibly to electron flow through system I or system II. In the linking of chlorophyll to a lipoprotein membrane we have an example of a biological energy-conversion system quite different from those of muscle and of the ion pumps of membranes – a 'light-energy to thermodynamic-energy' converter as compared with the chemical-mechanical converter of the muscle fibril or the chemical-thermodynamic converter of ion pumps. The examination of chloroplast membranes by the freeze-etch technique (described in the previous chapter) shows a variety of particles 10–20 nm in diameter on, and in, the lamellar membranes, with system I (Fig. 5.7) associated perhaps with particles ~11 nm in diameter, and system II associated with particles ~17.5 nm in diameter within the membrane.

Lysosomes

During differential centrifugation studies on the liver, De Duve noticed a fraction which was lighter than, and both enzymatically and structurally distinct from mitochondria. As it contained hydrolytic enzymes it was called the lysosomal fraction. Subsequent studies have shown that this fraction contains a wide variety of enzymes including acid phosphatase, β-glucuronidase, cathepsins, ribonuclease and deoxyribonuclease. A variety of organelles are found in this fraction, all seen in thin-section micrographs to be bounded by a phospholipid membrane; this is a constant feature and is perhaps not a surprising one since the contained enzymes, if free, would digest other cellular constituents. Some lysosomes are spherical and small (<1 μm in diameter) with a uniform finely-granular matrix, others contain coarse granules within this matrix, others are large and irregular containing numerous granules of varying density, and also ferritin and phospholipid myelin figures.

Since the development of histochemical techniques for electron microscopy, the term lysosome has come to mean a membrane-limited organelle showing acid phosphatase activity. Within this definition all the above particles are accommodated; but the small lysosomes with the finely granular matrix are now considered as primary lysosomes. These fuse with vesicles containing ingested material, in phagocytic cells, to form secondary lysosomes, in which the lysosomal enzymes digest the contents. This process accounts for the heterogeneity of organelles found in the lysosome fraction, since the content of the vesicle depends on the stage of its digestion.

As would be expected, lysosomes are particularly common in phagocytic cells, but as other cells probably digest components of their own cytoplasm as part of the constant process of break-down and renewal of cell structure, lysosomes are widely distributed. In this break-down, cytoplasmic components are segregated in autophagic vacuoles, which then fuse with lysosomes. One type of mammalian cell, the neutrophil polymorphonuclear leucocyte, has a cytoplasm studded with small vesicles. Each of these is a primary lysosome and this cell is found at sites of inflammation discharging the contents of its lysosomes to digest the invading bacteria. The subsequent necrosis of bacteria, leucocytes and other tissue produces the debris known as pus.

The endoplasmic reticulum in a protein-secreting cell

So far the organelles described are discrete, in that they can be isolated from homogenized cells in their intact form and their morphology and chemical content indicate a particular function. In the cytoplasm, there is also a pervading system of spaces, lined by phospholipid membranes, which is known as the endoplasmic reticulum. It is not always a network (reticulum) or even restricted to the central part of the cytoplasm (endoplasm), but in some form it is present in almost all cells. Its pattern is very varied and it serves a wide variety of different functions in different situations. Hence it is impossible to describe it except by relating it to particular functions, such as protein secretion, striated muscle contraction and cell division. This we shall do in these next three sections.

There is now abundant evidence (Chapter 8) that the process of assembly of amino acids to form a protein is associated in some way with cytoplasmic particles containing ribonucleic acid. These particles, known as *ribosomes*, may be seen in thin sections as electron-opaque bodies 15 nm in diameter. They are also identifiable in tissue homogenates, where they have been shown to contain ribonucleic acid. Most of the protein synthesis in a cell occurs at these ribosomes.

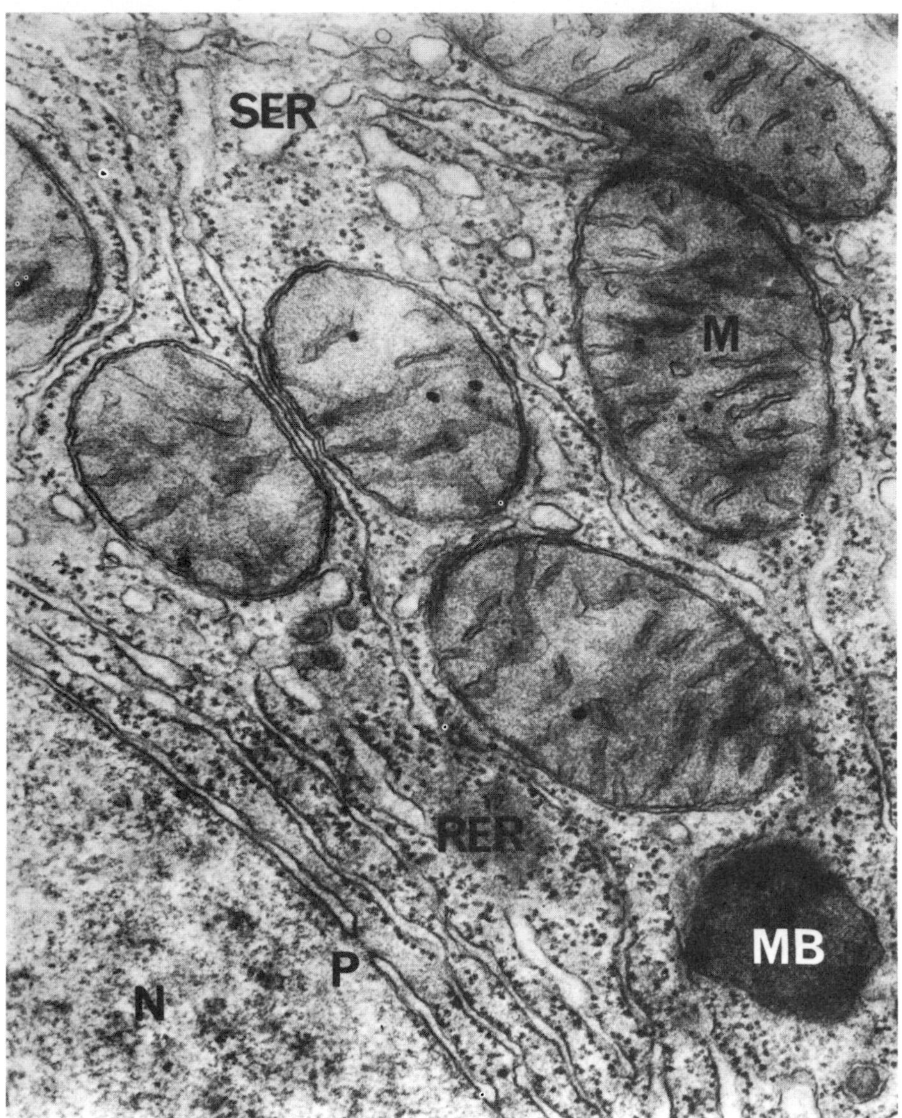

Fig. 5.8 Electron micrograph of rat liver cell, selected to show most of the common organelles. The nucleoplasm (N) in the lower left corner is bounded by its envelope with an occluded pore (P). Numerous free ribosomes are present and some are attached to rough endoplasmic reticulum (RER). Mitochondria (M), microbody (MB) and smooth endoplasmic reticulum (SER) are also present. Osmium fixation, Araldite embedding. Magnification ×50 000. (Micrograph by A. R. Muir.)

Some protein synthesis occurs on slightly smaller ribosomes (Chapter 7) found within mitochondria and chloroplasts. The cytoplasmic proteins can be divided into two groups, those required for the internal composition of the cell and those synthesized for export into the surrounding medium.

The myofibrils in muscle, and haemoglobin in the red cell, are outstanding examples of the first group, but all cells require new proteins during growth and division, and are subsequently constantly resynthesizing their enzymes and other proteins. Ribosomes are in fact found in the cytoplasm of all cells, except the mature red cell. A fully developed cell in an adult may have comparatively few, but cells preparing for a cell division, or yeast and bacteria during their exponential growth phase, have large numbers of ribosomes throughout their cytoplasm. In cells of this type there are few cytoplasmic membranes.

The production of protein for export, by contrast, seems to be associated with a characteristic arrangement of ribosomes on the outer surface of an elaborate series of flattened membranous sacs; these together form the ergastoplasm or rough-surfaced endoplasmic reticulum (Fig. 5.8). Examples of this organization are found in the plasma cell forming antibodies, the fibroblast forming collagen, the pancreatic cell secreting digestive enzymes, and the liver cell forming plasma proteins. The membranous sacs communicate with each other to form a three-dimensional continuous network which is surrounded by the cytoplasmic matrix. The interior of the cell is thus divided into two distinct phases – inside and outside the endoplasmic reticular membranes. Not all the ribosomes in such a cell are attached to the surface of the endoplasmic reticulum, but it must be remembered that even a cell specialized for protein secretion is also involved in maintaining its own integrity. The ribosomes found in the cytoplasmic matrix may be supposed to be serving this purpose. Cells such as liver cells which produce proteins for export, but are also metabolically very active, apparently have about half their ribosomes free and half attached to membranes.

If protein synthesis is a property of the ribosomes, what role can be assigned to the elaborate cytoplasmic membranes in a protein-excreting cell? This question has been investigated by choosing cells whose products are recognizable, and in which the cycle of secretion can be controlled. The exocrine cells of the guinea-pig pancreas fulfil these criteria, for they commence secretion if a meal follows a period of starvation. The relative uniformity of the pancreas allows a chemical analysis of homogenized gland to be related to the fine structure of the glandular cells.

A resting pancreatic cell is shown in Fig. 5.9; pyramidal in shape, its apex points towards the lumen of the gland, and its broader basal zone contains the nucleus. The rough-surfaced endoplasmic reticulum is mainly restricted to this basal zone, where the communicating membranous sacs carry ribosomes on their outer surfaces and the whole network is embedded in cytoplasmic matrix containing mitochondria and free ribosomes. In the apical zone, spherical dense bodies with a clear zone around them are contained within smooth membranes; these bodies are the zymogen granules seen in the light microscope.

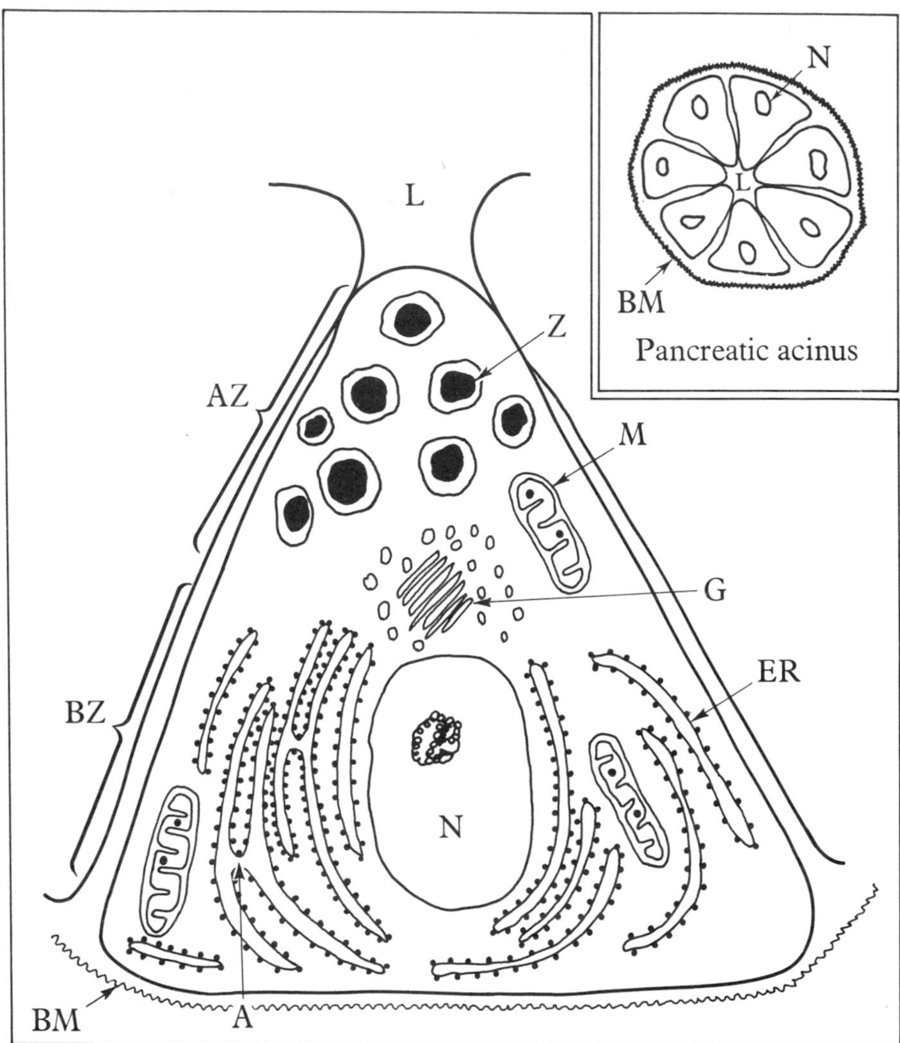

Fig. 5.9 Diagram showing the fine structure of the resting exocrine pancreatic cell, with an inset drawing showing the arrangement of these pyramid-shaped secretory cells around the lumen (L) of the pancreatic acinus. The apical zone of each cell (AZ) contains zymogen granules (Z) surrounded by smooth membranes. The nucleus (N), containing a nucleolus, occupies the centre of the basal zone (BZ) and membranous sacs and vesicles on the apical side of the nucleus form the Golgi region (G). The cytoplasm of the basal zone contains an extensive array of parallel lamellar sacs forming the endoplasmic reticulum (ER). The external surfaces of the membranes forming the endoplasmic reticulum carry large numbers of small dense granules, the ribosomes. The endoplasmic reticular cavities communicate with each other through anastomoses (A). Mitochondria (M) are distributed throughout the whole cytoplasm. The whole acinus is surrounded by a basal lamina (BM).

Finally, in this simplified description, a region between the nucleus and the apical zone contains apparently empty vesicles, and piles of flattened sacs, all with limiting membranes free of attached ribosomes; this is the Golgi region.

If such a resting pancreas is homogenized and centrifuged according to the method outlined in Chapter 1, various fractions are obtained. The fractions which are of interest in protein synthesis are the zymogen granule fraction, the microsomes and the supernatant solution containing free ribosomes. The microsomal fraction is really a homogenization artefact since it consists of small spherical membranous vesicles with ribosomes attached to their outer surface; these are derived from the long flattened sacs of rough-surfaced endoplasmic reticulum fragmented during homogenization. When the reticulum is disrupted the free ends of broken membranes join to form discrete vesicles. The term microsome is useful when discussing the biochemical activity of these fractions, but it should be remembered that microsomes are merely remnants of an originally continuous reticulum.

Starvation for forty-eight hours produces little change in the resting gland structure, there is merely an increase in the size and number of the apical zymogen granules. Starvation does, however, put all the glandular cells in the same functional state, and one hour after the first meal following starvation significant changes are seen in a high proportion of cells. The endoplasmic reticular cavities of the basal zone are not now so regularly arranged, they are distended and contain small granules (intracisternal granules) similar in density and texture to the contents of zymogen granules. During the second hour after feeding, these granules disappear from the endoplasmic reticulum of the basal zone, but similar granules are now found in the smooth-membrane vesicles of the Golgi region. These move towards the apical region of the cell. At the end of the second hour, some of the apical zymogen granules are seen near the surface of the cell, their smooth limiting membranes continuous with the plasma membrane, an appearance likely to be due to the discharge of the contents of the zymogen granules into the lumen of the gland.

This whole series of images can be interpreted as in the secretory sequence shown in Fig. 5.10. This interpretation receives some support from the occasional observation of continuity between smooth and rough-surfaced membranous sacs, but much more reliable support comes from the analysis of cellular fractions obtained at various times during a synchronized secretory cycle. The proteolytic activities of the microsomal fraction are twice as great, in the pancreas one hour after feeding, as in the starved animal. If the microsome membranes are disrupted with bile salts, the resulting suspension is, as might be expected, very rich in enzyme activity; it consists mainly of isolated intracisternal granules. But a challenging additional observation is that the isolated ribosomes have a proteolytic activity almost equal to that of the intracisternal granules. This could be merely contamination of the ribosome fraction, but it could be a demonstration of newly formed protein, at its site of synthesis, before it is passed into the cavity of the endoplasmic reticulum. Labelling the amino-acid pool with ^{14}C leucine has confirmed the latter interpretation, since at

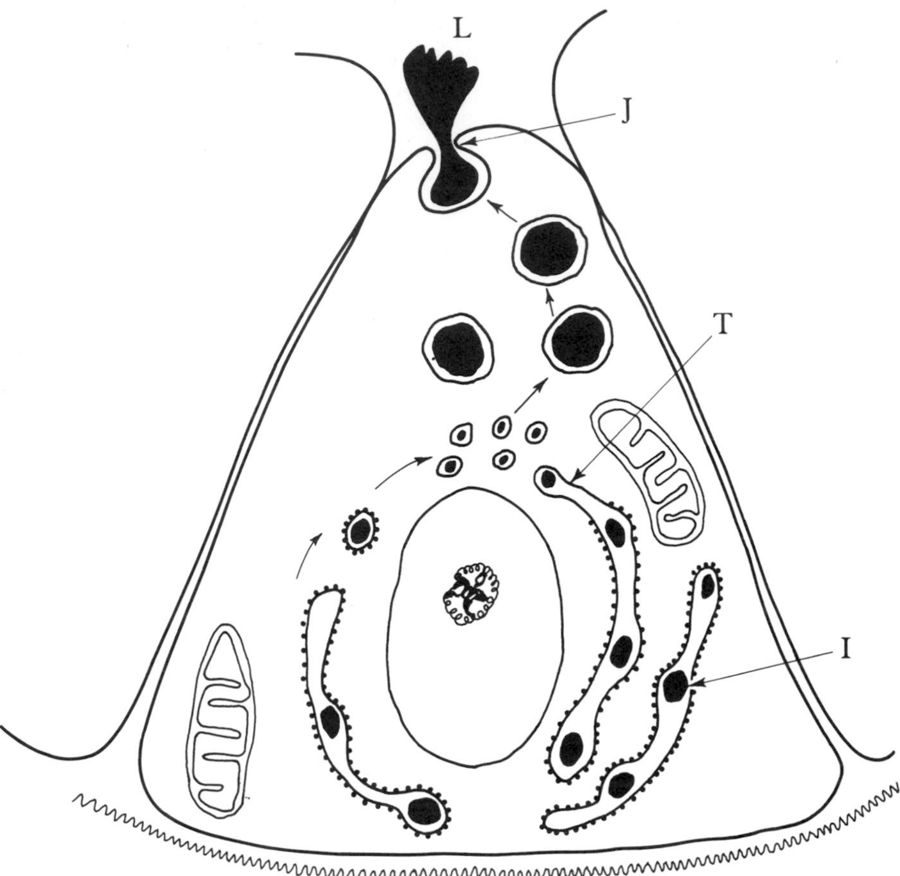

Fig. 5.10 Diagram showing a possible secretion mechanism for the exocrine pancreatic cell. During secretion the membranes bounding the zymogen granules fuse with the surface membrane (J) so liberating their contents into the lumen of the acinus (L). Smaller zymogen granules can be observed in the cavities of the endoplasmic reticulum (intracisternal granules I). Transitions (T) between the membranes of the endoplasmic reticulum, carrying ribosomes, and the smooth-surfaced membranes of the Golgi region have been observed. (See text for fuller discussion of this secretory process.)

periods of less than ten minutes after injection of the labelled leucine, the maximum radioactivity is found in the ribosomes, to be followed, in time, by labelling of the intracisternal and then the zymogen granules.

A variety of possible mechanisms may be suggested for the movement of the synthesized product across the pancreatic cell (Fig. 5.10). The intracisternal granule, enclosed in membrane, could break off from the lamellar endoplasmic reticulum, shed the attached ribosomes and migrate into the Golgi region to produce the small smooth vesicles found above the nucleus, these vesicles finally fusing to produce the zymogen granules. Alternatively the endoplasmic reticulum could remain intact and its products be passed along its cavities into the Golgi region, past the observed transitions between rough- and smooth-surfaced membranes. Membrane flow is another possibility, the synthesis of membrane in the reticulum, and the acquisition of ribosomes, then preceding the formation of intracisternal granules. Continued synthesis of membrane would then carry the formed granule towards the Golgi and apical regions.

Segregation of the finished products, the pancreatic enzymes, may be necessary to allow their synthesis to proceed, or segregation may simply be a device that allows the basal regions of the cell to be synthetically active, although only a limited area of the apical surface of the cell is available for the discharge into the lumen. Or segregation into small vesicles could be a mechanism for the removal of material from the cell cytoplasm, so that on fusion of the vesicles with the plasma membrane other constituents of the cytoplasm are not lost.

These studies strongly suggest that the internal cell membranes of cells of this type are concerned with the segregation, transport and discharge of an end-product, the actual synthesis being a function of the ribosomes.

There is some evidence that acetylcholine is discharged from nerve endings by a similar process of fusion of vesicles with the plasma membrane. More precisely this evidence is that acetylcholine is known (from physiological studies of the muscle cell membrane potential at the neuro-muscular junction) to be released in 'packets', and that large numbers of vesicles are seen, in electron micrographs, in the axon cytoplasm near its terminal synaptic or neuro-muscular junction. Mucus is apparently discharged from goblet cells by a similar process, and this fusion of vesicles with the plasma membrane, a reverse of pinocytosis, is perhaps a fundamental method of cellular excretion. Endothelial cells in capillary walls contain vesicles, some of which appear to fuse with the plasma membranes on either side of the cell. Electron microscopic studies of the transport of electron-opaque particles across these thin cells, suggest that this is effected by pinocytosis, followed by vesicular migration and discharge of the contents by fusion with the outer plasma membrane. Plant cells contain numerous Golgi regions (termed *dictyosomes* in plants) and, as in animal cells, vesicles from these regions move to the plasma membrane, discharging their contents into the extra-cellular space. These mechanisms all imply that the vesicular and external membranes have a sufficiently similar molecular structure to allow their continuity.

The endoplasmic reticulum in striated muscle

The contractile myofibrils of a muscle cell run the length of the cell, which may be many centimetres. Each myofibril consists of thousands of contractile units in series, the sarcomeres, which extend for 1·6 to 3·6 μm between Z discs along the myofibril (Chapter 3). The middle portion of each sarcomere is the A-band containing the 1·6 μm long, thick, myosin-containing filaments. The thin, actin-containing filaments attached to the Z discs form half of an I-band before they interdigitate with the A-band filaments.

Because the cytoplasm of a muscle cell is known as sarcoplasm, its system of

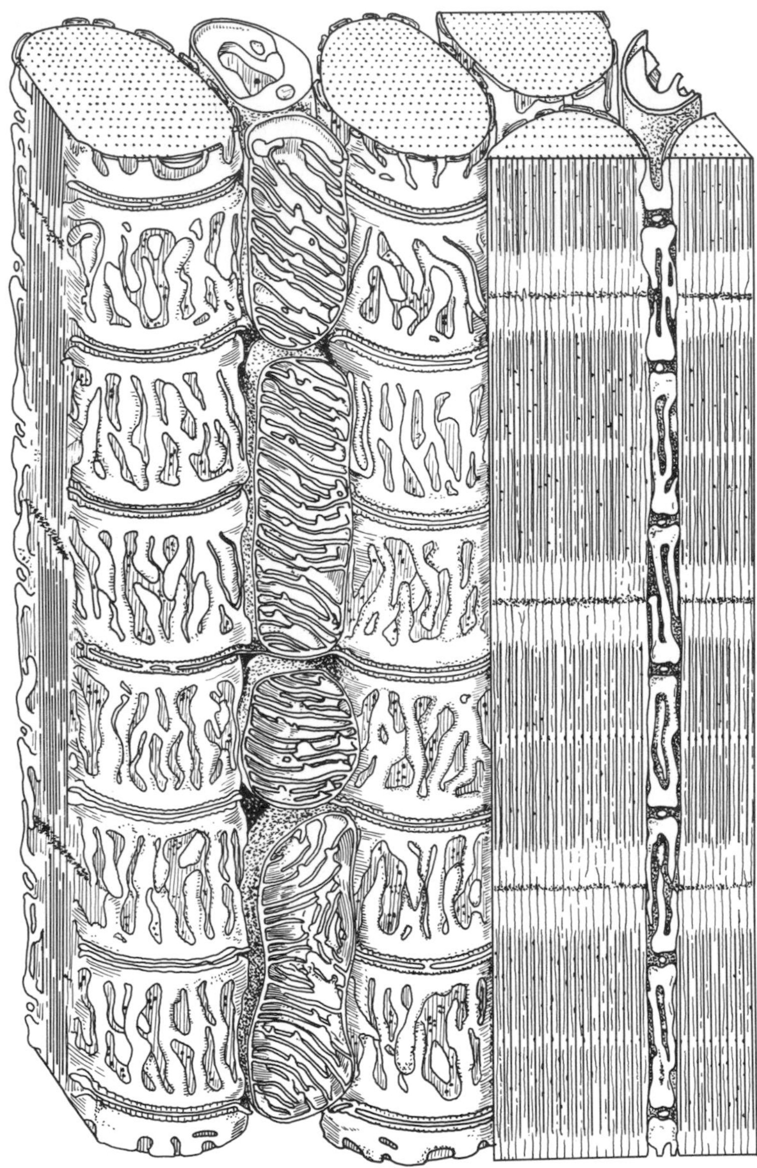

Fig. 5.11 Diagram showing the arrangement of the sarcoplasmic reticulum in skeletal muscle, based on electron micrographs of the toadfish swim-bladder, and bat crico-thyroid muscles of the type shown in Fig. 5.12. The transverse tubules running across the cell near the junction between A- and I-bands are the T-tubules (see text). These tubules are sandwiched between the apposed elements of the sarcoplasmic reticulum to give a characteristic appearance in thin sections known as a 'triad' (Fig. 5.12). (*From* Revel, 1962.)

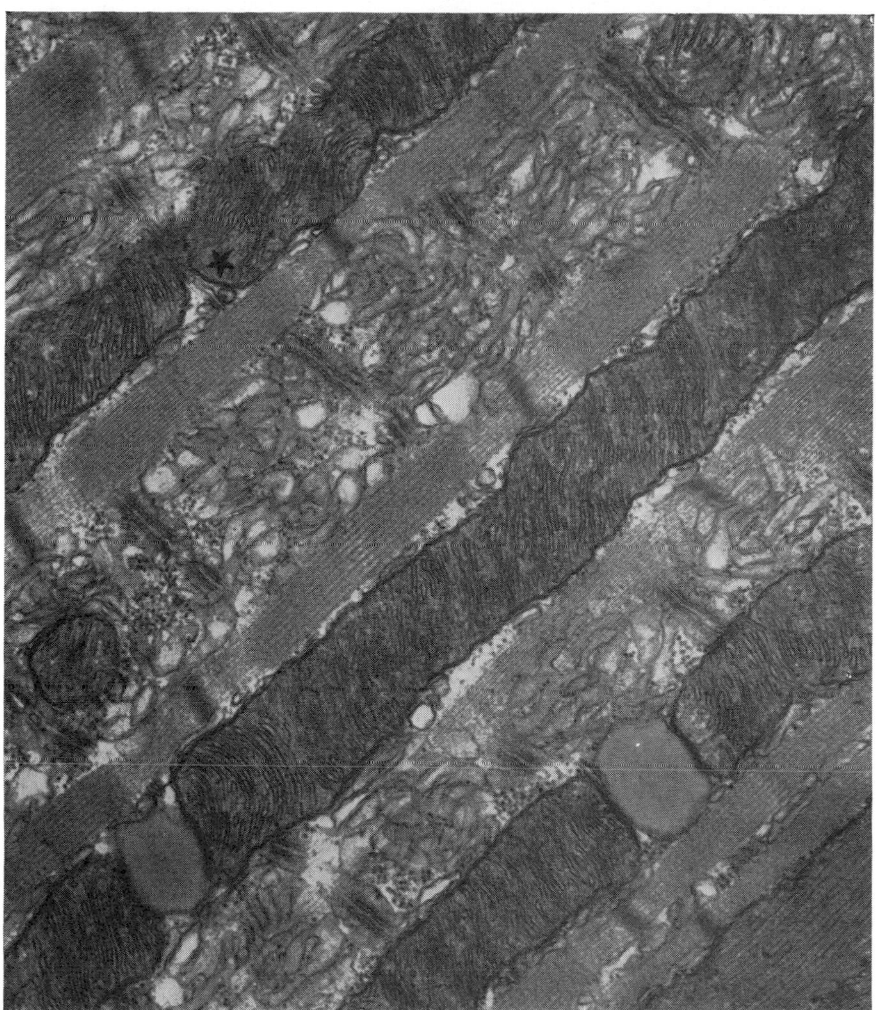

Fig. 5.12 Electron micrograph showing the sarcoplasmic reticulum in the bat crico-thyroid muscle (magnification ×23000). The section passes along the surface of myofibrils which lie diagonally across the picture. Triads can be seen near the junction between the A- and I-bands. Longitudinal tubules of the sarcoplasmic reticulum connect the triads at opposite ends of an A-band. At the asterisk a T-tubule leaves the triad, passes between two mitochondria, and becomes continuous with another triad in the adjoining myofibril (a feature included in the diagram of Fig. 5.11). Note the closely packed cristae in the mitochondria. (*From* Revel, 1962.)

membranous tubules is called the sarcoplasmic reticulum. In all striated muscles, the sarcoplasmic reticulum forms a cylindrical basket of tubules surrounding each of the many myofibrils in the cell (Figs. 5.11 and 5.12). The main function of the sarcoplasmic reticulum appears to be the regulation of calcium ion concentration in the sarcoplasm bathing the myofilaments. Calcium ions are necessary for the enzymic release of energy from ATP by the myosin ATPase and control of calcium concentration can control this release and hence the contractile process. Release of calcium ions from the lumen of the sarcoplasmic reticulum into the sarcoplasm causes contraction, and their active transport by the sarcoplasmic reticular membranes back into the lumen permits relaxation, by lowering the sarcoplasmic Ca^{++} concentration. This can be shown by isolation of sarcoplasmic reticular membranes by homogenization and centrifugation. Their relaxing effect on myofibrils *in vitro* can then be demonstrated, as can their ability to concentrate Ca^{++} by a factor of a few thousand over that of the suspending medium. The active transport of Ca^{++} is dependent on ATP, and the sarcoplasmic reticular membranes possess the necessary ATPase.

If the sarcoplasmic reticulum acts in this way as the controller of contraction, how is its behaviour coupled to the known stimulus of the action potential travelling along the plasma membrane, or sarcolemma? In unusually slender muscle cells with diameters of 1 or 2 μm, each myofibril is adjacent to the sarcolemma and, in such cells, extensions of the sarcoplasmic reticulum are applied as flattened sacs to the cytoplasmic surface of the sarcolemma. At these points of apposition, the ionic disturbance of an action potential could be transmitted, in some form, to the sarcoplasmic reticulum and so initiate Ca^{++} release and contraction.

An intriguing structural adaptation of the sarcolemma is present in the more common muscle cells whose diameters are more than 50 μm. At particular levels opposite each sarcomere, the sarcolemma forms tubular invaginations which pentrate deeply into the cell in a transverse plane. These tubules are known as the T-tubules and their lumina are continuous with the extracellular space; so, when an action potential passes the orifice of the tubule, an electrotonic impulse of some kind could pass rapidly deep into the cell along the T-tubules (Fig. 5.13). In these normal thick muscle cells, the cisternae of the sarcoplasmic reticulum are applied to the T-tubules throughout the whole muscle cell, and at these points of apposition the electronic impulse passing down the T-tubule membrane could be transmitted in some form to the sarcoplasmic reticulum, to initiate Ca^{++} release. The T-tubule system would thus cause the central myofibrils to contract at almost the same time as those close to the peripheral sarcolemma.

Proof of the role of the T-tubule in intracellular conduction comes from stimulating the sarcolemma with sub-threshold stimuli which do not propagate along the cell. Instead of a total contraction, a local contraction involving only the sarcomeres at the stimulated level is seen to spread into the cell. Comparative studies show that the T-tubules are at the level of the Z discs in amphibian skeletal and mammalian cardiac muscle, but at the level of the A–I junction in

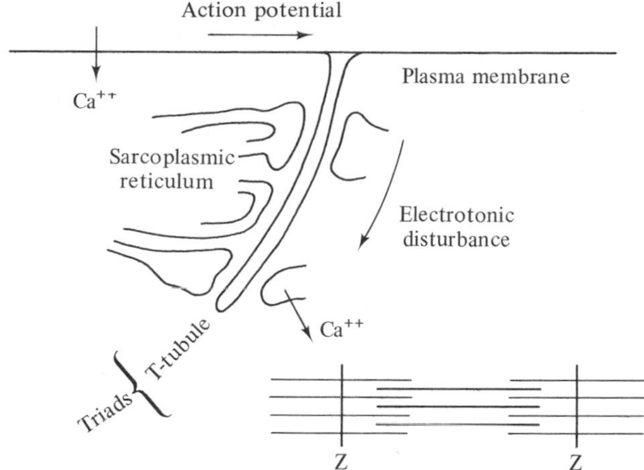

Fig. 5.13 A schematic illustration of the role of the sarcoplasmic reticulum, and of Ca^{++} ions, in striated muscle cells. It is supposed that the T-tubules play some role in conducting an electrical impulse down into the cytoplasm, following the passage of an action potential along the plasma membrane. The passage both of the action potential and of the impulse travelling along the T-tubules, transmitted in some way to the sarcoplasmic reticulum at the triads, is supposed to release Ca^{++} ions into the cytoplasm to provide the direct stimulus for contraction of the actin-myosin system.

mammalian skeletal muscle. The site of maximum sensitivity to surface stimulation follows these species differences.

Thus in muscle, the sarcoplasmic reticulum is involved in excitation-contraction coupling. It is correspondingly more evident in muscles which contract at high frequency; it is very impressive in the laryngeal muscles of the bat which produce a sound of supersonic frequency to guide the animal in the dark.

The endoplasmic reticulum in dividing cells

Because of the optical and chemical differences between nucleoplasm and cytoplasm, the boundary of the nucleus is clearly seen in all forms of light microscopy. At this boundary the electron microscope shows the nuclear envelope (Fig. 5.14). In an interphase cell, the nuclear envelope consists of two membranes with an intervening perinuclear cisterna which forms a shell around the whole nucleus. The inner membrane is applied directly to the nucleoplasm while the outer one often carries ribosomes on its cytoplasmic surface.

At numerous points on the surface of the nucleus the inner and outer membranes fuse with each other to form circular pores, 50 nm in diameter, between the nucleus and the cytoplasm. Although these pores may provide communications between nucleoplasm and cytoplasm they are not unrestricted openings. An occluding substance is seen by electron microscopy within the pores and electrical resistance measurements show that there is not free ionic passage into and out of the nucleus.

The fate of the nuclear envelope during cell division has been studied in rapidly dividing animal tumour cells and in the growing plant root tip. No change is observed in electron micrographs during prophase, but during

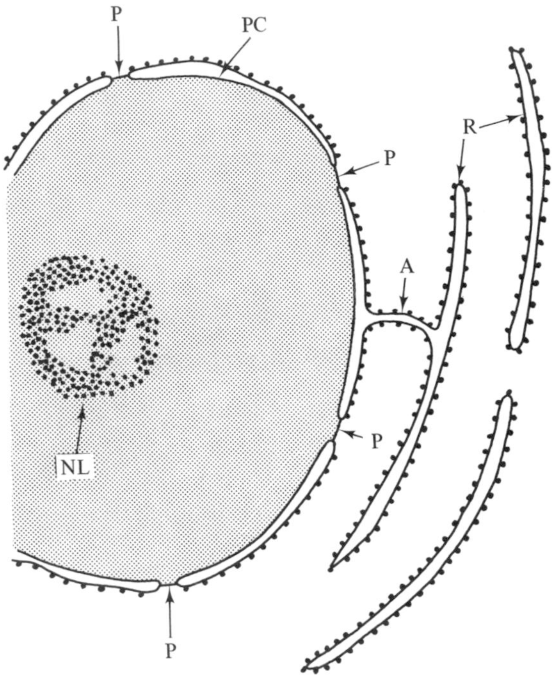

Fig. 5.14 Diagram showing the fine structure of the nucleus. The ovoid section of the nucleus almost fills the picture, with a small part of the cytoplasm included on the right. The nuclear envelope is a flattened sac. Its interior space forms the perinuclear cisterna (PC). The inner membrane of the cisterna is smooth but the outer membrane carries ribosomes (R). The cavity of the perinuclear cisterna is sometimes observed to be continuous with other endoplasmic reticular cavities (A). Circular pores (P) between the nucleoplasm and the cytoplasm are formed by the fusion of the inner and outer membranes of the perinuclear cisterna. Osmiophilic material is often condensed against the nuclear side of the inner membrane of the perinuclear cisterna, and branching channels penetrate this nuclear material, running in from the pores. The nucleolus (NL) appears as a spherical aggregate of small dense granules.

metaphase the perinuclear cisterna apparently breaks up, and its fragments round up to form vesicles, which later may be seen in the cytoplasm, thus accounting for the 'disappearance' of the nuclear membrane observed at this stage in living cells under phase contrast. The sequence of events suggested by electron microscope studies is that when the chromosomes separate during anaphase, these vesicles move into the daughter cells; they then aggregate around the chromosome mass, and flatten out around the telophase nucleus to create a new nuclear envelope. In plant cells, the membranous vesicles invade the actual substance of the spindle during anaphase and follow the line of the spindle fibres to reach the separating chromosomes. Some of the vesicles then invest the chromosomes, and form a new nuclear envelope, while others pass towards the division plane between the two daughter cells. At the site of separation of the two plant cells a new cell plate forms. Here the vesicles of the endoplasmic reticulum flatten out along the line of separation where the new plasma membranes and extracellular wall are formed.

The whole process is remarkably similar in the animal and plant cell, and a process similar to the invasion of the spindle seen in plant cells may be taking place during the division of animal cells also, but only have been revealed, so far, in plants, where the spindle axes are always oriented parallel to the axis of the root.

Centrioles

All the organelles discussed so far in this chapter are formed predominantly of phospholipid membranes, as bounding layers or as part of the internal organization. In the cytoplasmic matrix between these membranous structures comparatively few features are discerned with present electron microscope techniques; microtubules have been mentioned in Chapter 3, and muscle filaments; in Chapter 4 the fine filaments which radiate through the matrix from their attachment to maculae and zonulae adhaerentes have been described; glycogen and ferritin molecules may be present in the cytoplasmic matrix, and ribosomes unattached to endoplasmic reticulum. In this relatively homogeneous medium a pair of highly-organized, non-membranous structures are present in most animal cells; these are the *centrioles*. They lie close together, near the nucleus, often near the Golgi region. The pair of centrioles is sometimes termed a *diplosome*. A centriole resembles a barrel, 0·5 μm long and about 0·2 μm in diameter, and one end of the barrel is closed. Its wall consists of nine imbricated staves, each stave being composed of three slender cylindrical rods (Fig. 5.15(*a*)). It is not known whether these are solid rods or hollow tubes, since the dark periphery in electron micrographs could be produced by the stain or fixative failing to penetrate a compact rod, or by the staining of a tube around an aqueous core. The outer rod of each triplet forming a stave is linked to the next triplet (Fig. 5.15(*a*)), so giving strength to the organelle. Cylindrical aggregations of dense material often surround the centriole and these are called centriolar satellites.

During interphase, two centrioles are present and they lie with their axes at right angles to each other. Early in prophase the centrioles replicate, perhaps as a result of a new barrel growing out from the side of each centriole so that two such pairs are formed. These migrate to opposite poles of the cell, and spindle fibres (structures identical to microtubules) run to both pairs of centrioles. Whether the spindle fibres are attached to the centrioles is not certain

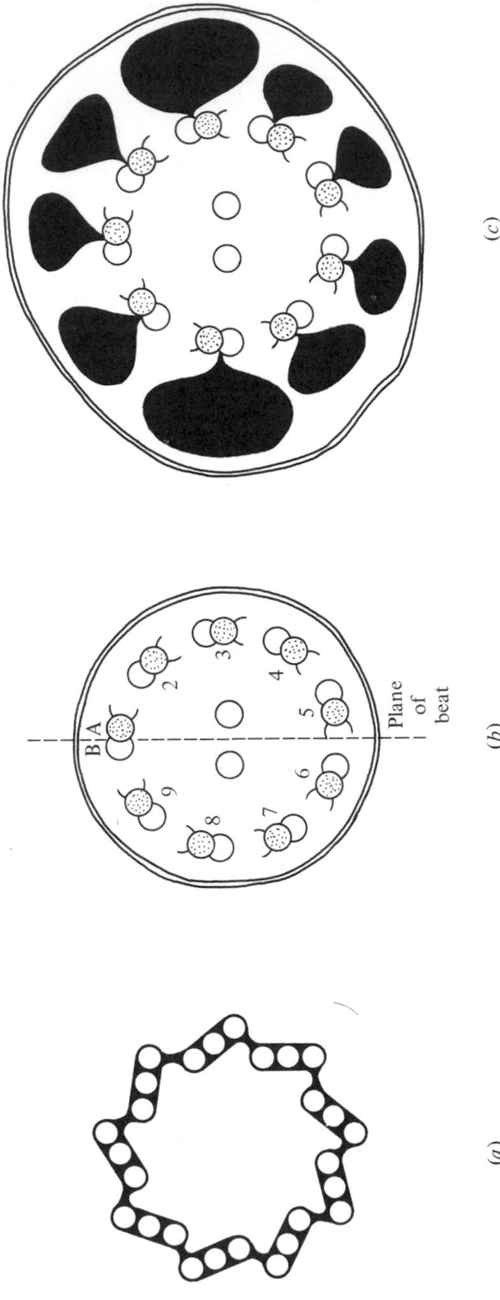

Fig. 5.15 Diagram showing the appearance of transverse section through (*a*) centriole, (*b*) cilium and (*c*) the proximal part of a flagellum.

but it seems likely that they end in the centriolar satellites of one of the centrioles in each diplosome. In telophase one pair of centrioles moves into each daughter cell. The position of the centrioles at the poles of a dividing cell and the fact, discussed in the next section, that similar bodies play some role in generating the microtubules of flagella makes it seem probable that the centrioles play some controlling role in spindle formation. But before assigning any significant role to the centrioles during mitotis and meiosis, account must be taken of the absence of these elaborate structures in dividing plant cells!

Cilia and flagella

Structures similar to centrioles are found at the base of all motile cell processes. These processes can be long (> 50 μm) whip-like flagella, which undulate in one or more planes, or shorter cilia which are present in large numbers to beat in unison with a forceful stroke in one direction. Both are widely distributed, from protozoa to mammals, and as their structure varies little a general description is justified.

In the cytoplasm under the process, the basal body, similar to a centriole, has its closed end facing the cell surface. Particularly in invertebrates the basal bodies are anchored in the cytoplasm by fibrils showing a cross-striated periodicity. From the centre of the closing plate of the basal body, a pair of tubules, with a structure the same as microtubules, extend along the axis of the motile process. From each of the nine staves in the basal body further pairs of tubules extend along the process. These pairs are joined, so that in thin section the cilium has the appearance shown in Fig. 5.15(*b*). A tubular extension of the plasma membrane covers the cilium. At high magnification, one of the paired tubules (A), which has a darker core than its partner, is seen to possess short arms which extend towards the tubule (B) of the next pair.

The assembly of these details in Fig. 5.15(*b*) shows that it is possible to number the outer pairs of tubules: number 1 being chosen in the plane bisecting the two central tubules and numbers 2–9 taken in the direction of the short arms. If the direction of this reading is clockwise, then the cilium, in section, is being examined from its cellular end.

It would be satisfying to relate all this fine structure to the function of these processes, but this is not yet possible. The short arms have been shown to possess ATPase activity and these could interact with the adjacent tubules in the manner described for the cross bridges of myofilaments in Chapter 3. The tubules, like the microtubules of the cytoplasm, appear to be made up of globular sub-units. Small dislocations in the attachment of the short arms to adjacent tubules could produce bending of the cilium. One clear fact has been established in cilia and some flagella, namely that the plane of beat is the bisecting plane which goes between the central tubules and through the pair of tubules number 1. The design does confer an asymmetry, as pairs 3 and 8 are in line with the central pair, thus leaving three pairs on one side (9, 1, 2) and four pairs on the other side (4, 5, 6, 7), and this may be related to the asymmetric action of a cilium. The

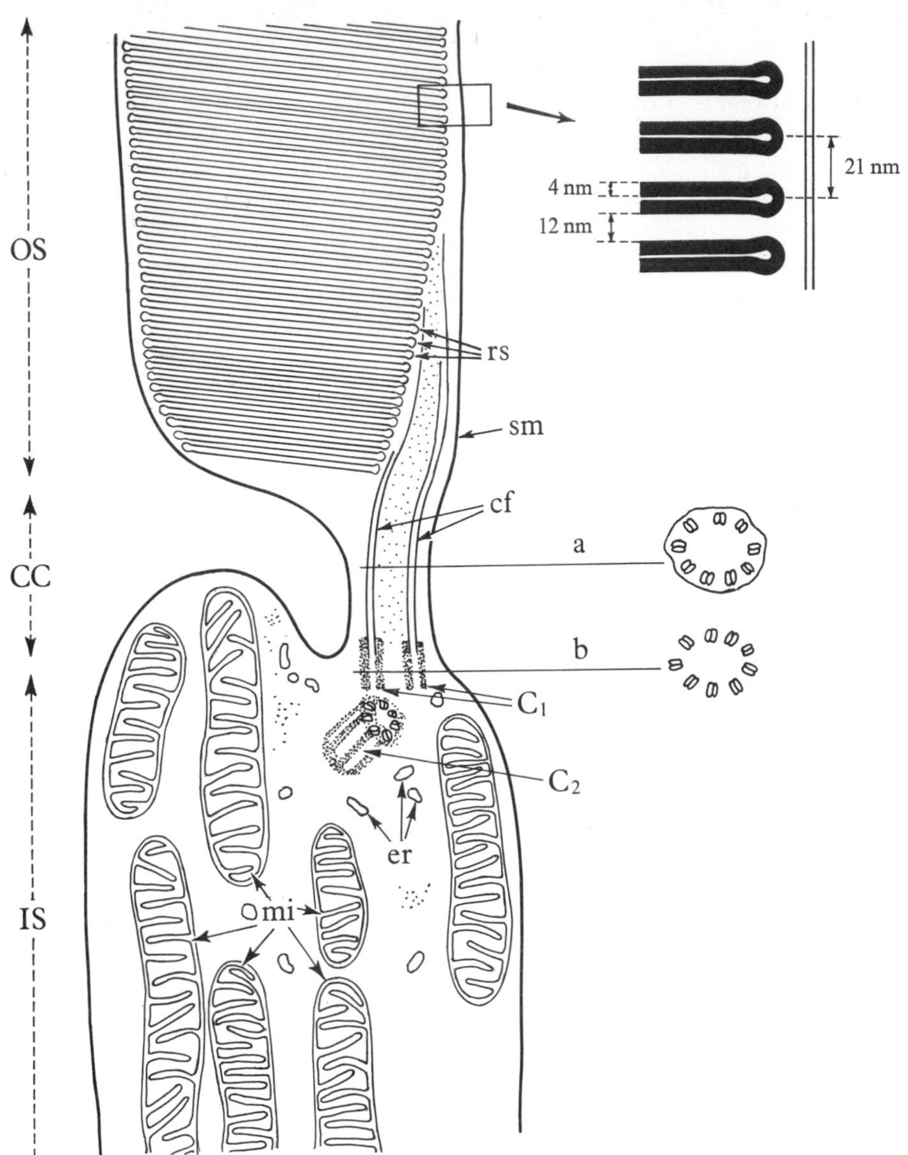

Fig. 5.16 Diagram based on electron microscope observations of retinal rod cells of mammals. At the left are indicated the outer segment (OS), the inner segment (IS) and the connecting stalk, or cilium (CC). In cross-section (a on the right) this stalk shows a circular arrangement of nine pairs of filaments typical of cilia, although in normal cilia an additional central pair of filaments are present. It seems therefore that the outer segment of the retinal rod cell represents a variation on normal cilium development. This diagram also shows the electron microscopic appearance of centrioles (C_1, C_2). The centrioles, which in dividing cells act as centres for aster formation (see text), can apparently also act as a centre for the growth of cilial filaments (cf.). Other features in this picture are the rod sacs (rs) and their dimensions (upper right), plasma membrane (pm), mitochondria (mi) and tubular elements of the endoplasmic reticulum (er). Electron microscope studies of developing retinal rod cells suggest that the rod sacs are formed by invagination of the plasma membrane. (*From* De Robertis, 1960.)

idea of a supporting role for pairs 3 and 8 which lie in the 'hinge plane' receives confirmation in certain flagella where thick condensations of cytoplasmic material are found applied to the outer side of these two pairs along the length of the process. The proximal part of a flagellum contains such condensations of material on the outer side of all nine peripheral pairs but only those accompanying pairs 3 and 8 continue far along the process, and then only when the flagellum beats in one plane (Fig. 5.15(*c*)). Some flagella have a more complex action, beating in many planes, and in these the central tubules may be reduced to one, $(9 \times 2)+1$, and the outer condensations remain uniformly distributed or even join to form a cylindrical shell.

Long processes sometimes protrude from secretory cells which are concerned with discharge of material. These structures, known as stereocilia, are non-motile and do not contain any internal tubular structures. But some cells, in sense organs, have processes which detect movement, or chemical changes, or electrical potential changes in the bathing fluid. These processes resemble cilia, but they usually lack the central pair of tubules, which gives them a $(9 \times 2)+0$ organization. Examples can be seen on the cells of the Organ of Corti and in the ampullae of the semicircular canals in the inner ear where, by detecting movement, they respond to sound vibrations or positional movements of the head. The bases of these cells make synaptic contacts with nerve cells, so that depolarization of their plasma membrane will transmit an impulse to the central nervous system.

Even cellular processes which detect changes in illumination have this $(9 \times 2)+0$ structure, although such receptors contain further an elaborate pattern of internal membranes in the cytoplasm along the side of the cilial structure (Fig. 5.16). These membranes form flattened sacs, perpendicular to the direction of incident light, carrying light sensitive pigments. The sacs trap the light and, in some way, initiate action potentials in the plasma membrane. The first step in this process is the channelling of absorbed light energy to retinal molecules (Fig. 5.17). In the dark, 11-*cis* retinal is covalently bound to the pro-

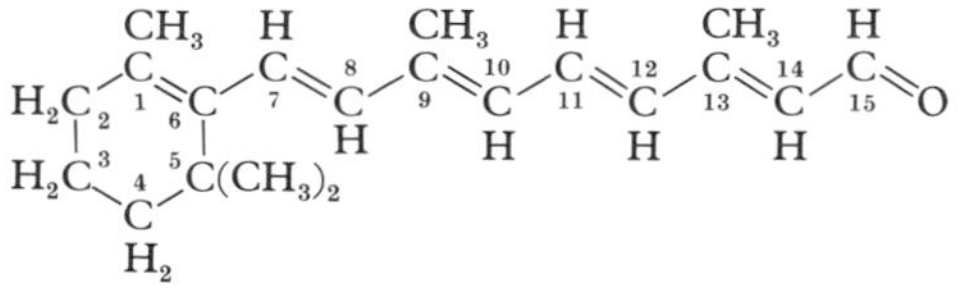

all-*trans* retinal$_1$

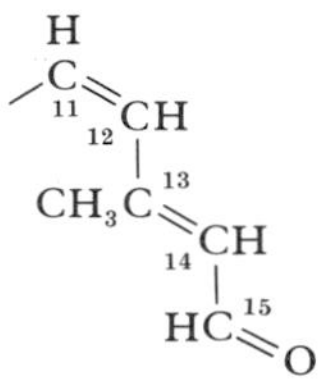

11-*cis* retinal

Fig. 5.17 The structure of retinal.

tein opsin to form rhodopsin, a component of the sac membranes. In the light, the retinal energy level is raised and the molecule is converted to the all-*trans* form and released from opsin. It is not certain whether the all-*trans* retinal can perhaps move to the plasma membrane, directly affecting the ionic permeability of this membrane, or whether this effect is achieved indirectly through release of some other component from the sac membrane (perhaps Ca^{++}) which diffuses to the plasma membrane to produce a change in its Na^{+} permeability.

The fact that such different but fundamental activities as cell division, motility and sensitivity, are associated with intricate nine-fold arrangements of fine tubules suggests a phylogenetic linkage of great ancestry.

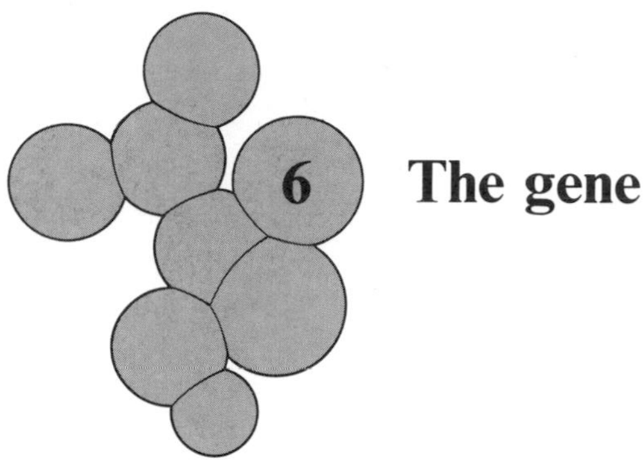

6 The gene

The particulate nature of inheritance

The methods of genetics provide an approach to the study of certain aspects of cellular activity which is quite independent of those of microscopy and biochemistry; these three different approaches have now converged to provide a unified view of inheritance at the molecular level. To appreciate the contribution of modern genetics we must briefly trace the progress made through this century in refinement of the geneticist's fundamental technique: analysis of the distribution of parental traits among the offspring of controlled crosses (i.e. controlled mating or fertilization).

At the simplest level the theory of the gene does not go beyond two simple assertions: (1) that it is necessary to distinguish between the assemblage of outwardly recognizable traits by which an individual is defined, known as its *phenotype*, and the assemblage of inherited factors which determine these traits, known as its *genotype*, and (2) that inheritance is 'particulate' rather than 'blending', i.e. that the factors transmitted by the parents are combined in the offspring not in the manner that black paint may be combined with white to make grey, but in a way more nearly resembling a mixing of black and white marbles, which do not contaminate each other in the mixing and which can in principle be segregated cleanly again into their constituent classes of black and white.

The blending theory was still current in Darwin's day. Darwin himself partially accepted it, but he also gave due weight to facts of common observation with which it is in conflict. Thus he was much struck by the occurrence of throw-backs, when obvious resemblances to some member of the grandparental or earlier generation suddenly make their appearance without having been outwardly manifest in the parents. Such phenomena are easier to interpret in terms of marbles than in terms of paints, and suggest that the hereditary units are capable of being transmitted in a dormant state without loss of identity or potency. The well-known circumstance that a pair of brothers, or sisters, although having the same parentage as each other, do not exhibit the complete matching of traits which is observed in identical twins, also argues against a blending type of inheritance.

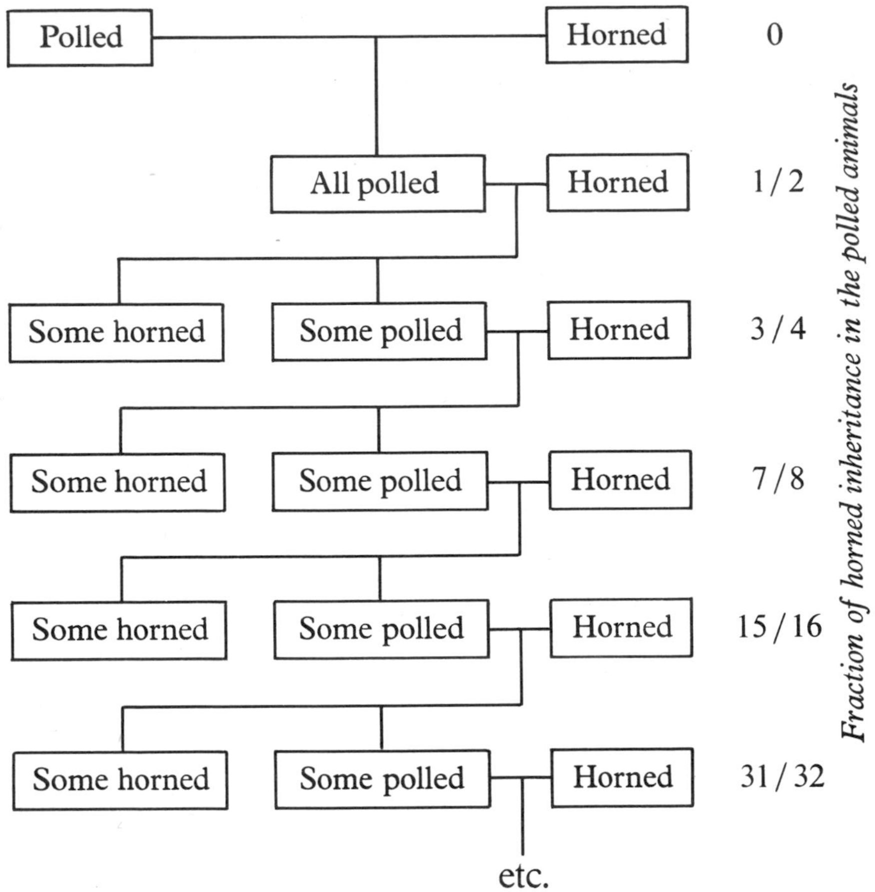

Fig. 6.1 Plan of a repeated crossing experiment in cattle. However far the series is taken, the polled character is never diluted out.

A most convincing proof that there are units or particles of heredity which can indefinitely survive admixture is obtained if one tries to 'dilute out' an inherited trait by the method of repeated crossing. Our example will be taken from cattle-breeding, where the method has been utilized for practical purposes. The cattle of some breeds are without horns. This 'polled' condition, as it is called, is valued by the farmer because its possessors are more manageable than horned cattle in the large herds used in modern farming. If a bull from a polled breed is crossed to a horned cow the first generation progeny are polled. The character is therefore said to be 'dominant'. When the polled offspring of the initial cross are crossed to the horned breed two types of offspring result, polled and horned. Yet the polled cattle of this generation now have three-quarters of their ancestors from the horned breed. This offers a method, which is illustrated in Fig. 6.1, whereby the farmer can superimpose the polled character upon a horned breed, while preserving all the other characters of the latter, e.g. milk yield, etc. He can select a polled bull from the second generation progeny, make a second cross into the horned breed, reselect, and repeat the process as many times as he likes. Examination of the numerical series in the right-hand column of Fig. 6.1 shows that by continuing this process for ten generations the breeder would obtain cattle possessing the heredity of the horned breed to the extent of 1023 parts in 1024; but the residual part contains unchanged the determinant of the desired polled character.

A theory of inheritance which allowed any degree of blending would predict that this breeding plan would be thwarted by attenuation of the polled character under the brunt of continued crossing. Not only does this *not* happen but there is also no reduction in the frequency of the trait in successive cross generations. This frequency is found to remain constant – one-half, on average, of all the animals bred in each generation is polled.

The case which we have taken is a specially selected one, since the same clear-cut result will not be obtained with a trait which expresses the cumulative action of many genes of individually small effect (polygenic inheritance). The general run of traits, which vary on a continuous rather than an 'all-or-none' scale, are found to have a polygenic basis and hence to present a superficial appearance of blending. More detailed analysis invariably shows this appearance to be only superficial. The inherited determinants themselves, whether these be one, few, or many, in any particular case, are transmitted as units, like marbles which remain unblended and uncontaminated after being shaken up with other marbles.

Mendel's discoveries

The essential features of particulate inheritance were first discovered by the Moravian monk Gregor Mendel, whose epoch-making experiments were published in 1866. Yet the Mendelian epoch failed to open until long after his death. In 1900 his neglected work independently attracted the attention of three different workers, de Vries, Correns and Tschermak.

Mendel's paper on inheritance in peas stands to this day as a model of experimental design and scientific writing. The text is widely available in English translation, so that both opportunity and incentive exist to read the paper for oneself. In the quest for laws governing the distribution of traits among the descendants of an initial cross, Mendel pointed out that his predecessors had neglected 'to determine the number of different forms under which the offspring of hybrids appear, or to arrange these forms with certainty according to their separate generations, or definitely to ascertain their statistical relations'. This, then, is the Mendelian method: to make a controlled cross, to sort out the offspring into classes, to count these classes and to count the number of individuals in each class. Today, almost a century later, procedures which amount to nothing more than this are being used to promote the latest advances in analysing the fine structure of the gene.

Mendel chose the garden pea as experimental material partly because it is a self-fertilizing plant which normally can only be crossed artificially, and partly because many true-breeding varieties of different colour, shape, etc., are known. From thirty-four seedmen's varieties he selected twenty-two which bred entirely true to type during a two-year trial. These showed clear-cut differences in characters involving the seeds, the cotyledons, the seed coat, the seed pods, the flowers and the stem.

For each character-difference Mendel began by making a cross between two contrasting varieties, for example between a tall and a short variety. In every case he found that progeny of the first generation (F_1 hybrids in modern terminology) uniformly resembled one of the two parental types, regardless of which type had acted as the male and which as the female parent. The parental character which appeared in the F_1 generation he termed 'dominant' and the one which disappeared 'recessive'. Dominance is not a universal phenomenon, as Mendel himself was aware; for certain characters in every species dominance is incomplete. Such cases do not, however, introduce anything new in principle, and will not be further considered here.

He now allowed the F_1 plants to breed by self-fertilization. The second (F_2) generation thus obtained was *not* uniform, but showed a mixture of dominant and recessive forms in a ratio of 3:1. The meaning of this orderly *segregation* of characters was made clear on raising a further generation by continued self-fertilization. It turned out that the dominant forms, which had comprised three parts in four of the F_2 generation, actually consisted of two distinct types:

1. Pure-breeding for the dominant character, like the plants of one of the original parental varieties.

2. Giving dominant and recessive offspring in the ratio 3:1, like the plants of the F_1 generation.

Dominants of 'mixed' constitution proved to be twice as numerous in the F_2 generation as the 'pure' dominants. The final breakdown for the F_2 generation was therefore 1 pure-breeding dominant: 2 'mixed' dominants: 1 pure-breeding recessive. In modern terminology the pure breeding condition is called *homozygous* (Greek roots: homo- = same, zygo- = join) and the 'mixed' condition is called *heterozygous* (hetero = different). The individual plant or animal is known as a homozygote or heterozygote respectively.

Mendel's result follows neatly from the simplest possible model of particulate inheritance. Consider the offspring of a tall × short cross. Tall is dominant to short in peas, so that all the offspring are tall. Since the recessive character, short, reappears in the F_2 generation, we know that the determinants of tallness and of shortness do not blend in the F_1 generation. Let us suppose, then, that when the F_1 generation produces gametes (pollen cells or ovules) half of these gametes receive the determinant for tallness and the other half the determinant for shortness. The entire sequence of events can now be represented as in Fig. 6.2, writing T for the dominant and t for the recessive determinant. Homozygotes contain two determinants of the same type, heterozygotes two different determinants.

The frequencies predicted from these simple considerations of probability are those which Mendel actually observed. It is worth emphasizing that the regularities *are* based on probability, and hence cannot be expected to find exact fulfilment in observational data, any more than the 50–50 chance governing the toss of a coin guarantees that exactly 50 heads will be obtained in a hundred throws. The expectation of 50 per cent is simply a limit towards which the empirical proportion tends as the number of throws is increased. The Mendelian ratios of 3:1 and 1:2:1 are also limits towards which the empirical ratios tend as the number of progeny bred is increased. This result may be precisely paralleled by an experiment in which two marbles are drawn from two different bags, each of which contains a mixture of black and white marbles in equal proportions.

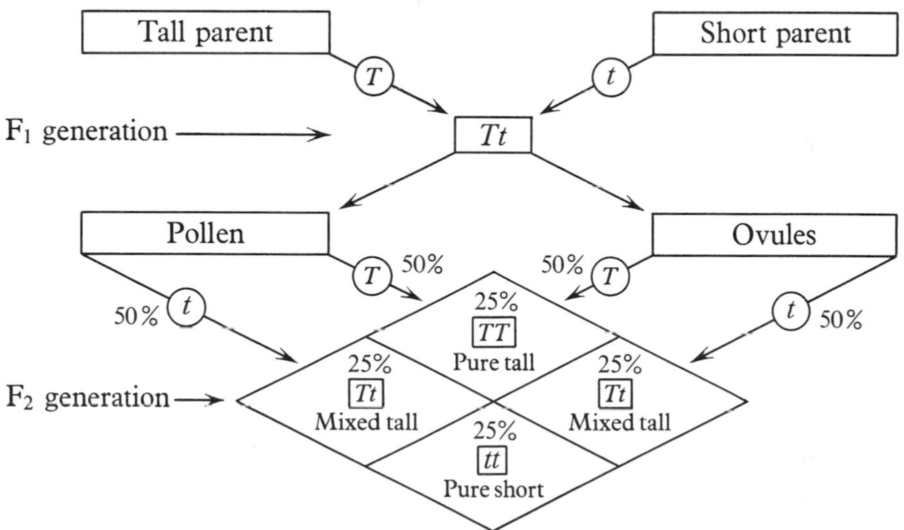

Fig. 6.2 Results of crossing tall with short peas and breeding from the offspring. Mendel's 1:2:1 ratio follows naturally from a simple hypothesis concerning gamete-formation in the F_1 generation (see text).

We can now understand in more precise terms why a Mendelian trait cannot be 'diluted out' by repeated back-crossing. Writing *P* for the dominant hereditary factor determining the polled character and *p* for its recessive allelomorph (Greek roots: allelo-=other, morph-=form), the cattle-breeding experiment can be diagrammed as in Fig. 6.3. The initial cross gives a crop of heterozygous polled animals, and thereafter each generation contains heterozygous polled animals and homozygous horned animals in equal proportions.

Mendel's next step was to ask what happens when segregation occurs for two characters simultaneously. Suppose, to take Mendel's own example, an F_1 plant has received the smooth-seed factor *and* the yellow-cotyledon factor from one parent, but is of wrinkled-seed and green-cotyledon descent on the other side. Do the factors which have been received together tend to be transmitted together to future generations, or does reshuffling occur?

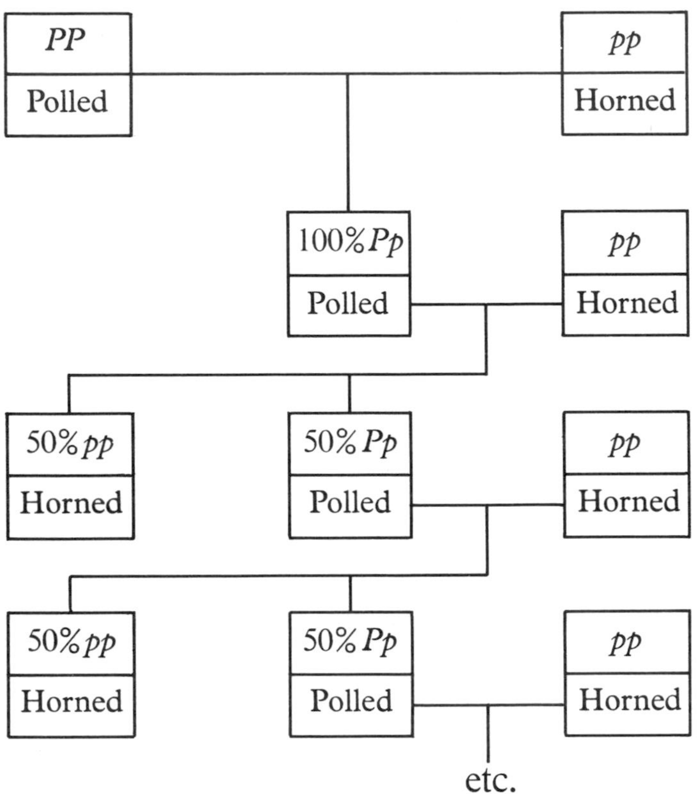

Fig. 6.3 Mendelian interpretation of the cattle-breeding experiment illustrated in Fig. 6.1. *P* represents the dominant factor responsible for the polled character and *p* represents its recessive allelomorph.

We can assign a Mendelian formula to the F_1 plant considered above, and see at once that in principle four distinct kinds of gamete can be produced by it (Fig. 6.4). Note that in writing the genetic formula of the F_1 plant in this figure we have grouped the factors received from one parent on one side of an oblique stroke and those received from the other parent on the other side. This convention is useful in enabling a distinction to be made between two forms of double heterozygote, namely one receiving both dominants from the same parent, written AB/ab, and a heterozygote which has received one dominant from one parent and the other from the other parent, written Ab/aB.

As a means of counting the frequencies with which the four classes of gametes shown in the figure were produced, Mendel test-crossed the double heterozygotes to doubly recessive plants, obtaining the four corresponding phenotypic

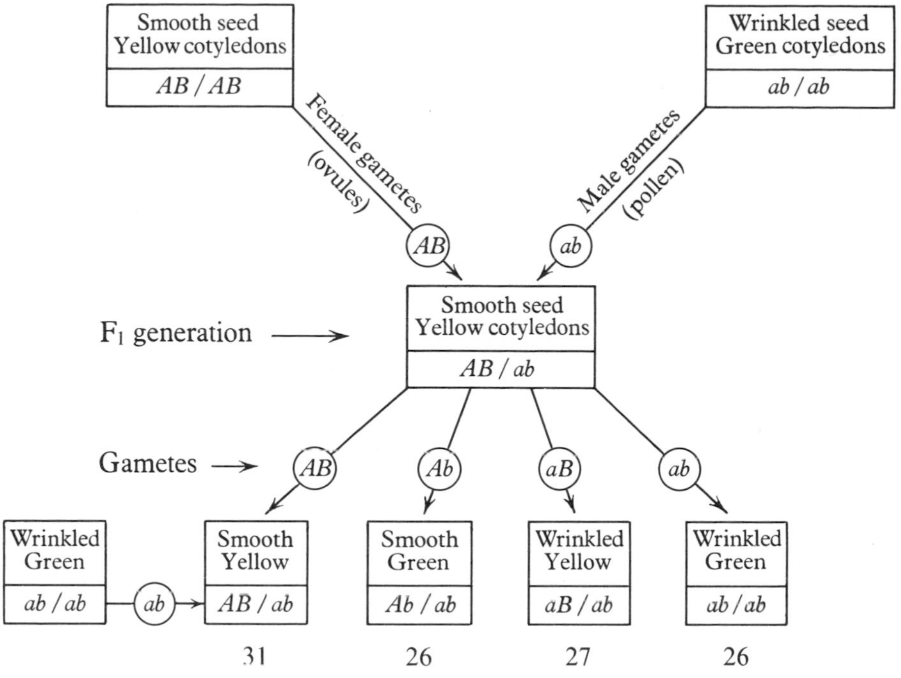

Number of old combinations: smooth yellow plus wrinkled green = 31+26
Number of new combinations: smooth green plus wrinkled yellow = 26+27

Fig. 6.4 Test to determine the relative frequencies of the four types of gamete produced by a double heterozygote. A cross is made to a double recessive and a count is made of the numbers of offspring falling into the four possible phenotypic classes. The numbers shown are those actually observed by Mendel in a typical experiment.

A = factor for smooth seed
a = factor for winkled seed
B = factor for yellow cotyledons
b = factor for green cotyledons.

classes. In all cases he observed approximate equality of new and old combinations. The factors which have been inherited together in these experiments show no tendency to preserve their association when transmitted to the progeny, but are reshuffled at random.

This conclusion, sometimes termed the Law of Independent Assortment, has subsequently been found to be a common but not an invariable result of experiments of this kind. In other cases genetic *linkage* is observed, that is, a tendency for the genetic factors to preserve their previous combinations in inheritance rather than to form new ones. It was the discovery and analysis of linkage which enabled a later generation of geneticists to tie Mendel's purely formal scheme to definite intracellular structures, the chromosomes.

Linkage groups, chromosomes and genetic maps

The idea that the genetic determinants, or *genes* as they came to be called, might be located in the chromosomes was put forward in 1903 by W. S. Sutton, and independently by Th. Boveri. It follows very naturally from the facts of meiotic and mitotic cell-division outlined in Chapter 1. Consider first Mendel's law of segregation, as illustrated in Fig. 6.2, according to which the F_1 heterozygote *Tt* forms gametes of two kinds in equal proportions, containing either *T* or *t*.

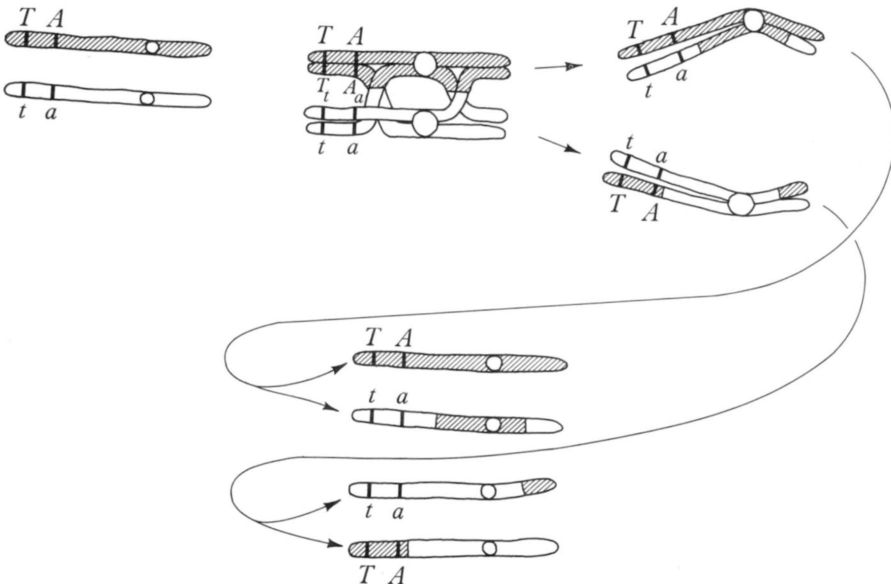

Fig. 6.5 The diagram of Fig. 1.6 is here reproduced with the addition of two allelomorphic gene pairs, *T*, *t* and *A*, *a*, to illustrate the phenomenon of linkage.

Now suppose that these allelomorphic genes, *T* and *t*, represent alternative structures of a particular segment in the homologous chromosomes of a particular chromosome pair, as shown in Fig. 6.5. Following the fate of this chromosome as far as the gamete stage we find that exactly half the gametes resulting from meiosis will contain the *T* form of the relevant chromosomal segment and half will contain the *t* form. Mendel's 1:1 ratio thus follows directly from, and is explained by, the behaviour of the chromosomes at meiosis.

Let us now consider the transmission of two separate pairs of allelomorphic genes. If the two pairs are located in two different chromosome-pairs, we can deduce that they will be reshuffled at random ('independent assortment') during gamete formation, since the movement of maternal and paternal chromosomal elements towards one or the other pole of the dividing cell is a chance phenomenon. We now know that this explains Mendel's observation of independent assortment in peas – the particular characters which he chose to study happened to be determined by genes located in different chromosomes. But what will be observed in a case of two pairs of genes fairly close to one another on the same chromosome such as *A*, *a* and *T*, *t* in Fig. 6.5? Clearly such pairs will preserve their original combinations (*A* with *T* and *a* with *t* in the figure) except in the relatively rare event of a chiasma happening to form, during meiosis, in the intercept between them (see Chapter 1). Two such genes therefore do not show independent assortment, and are said to exhibit *linkage*.

If a chiasma does form in the intercept there will be a genetic cross-over allowing the combinations *At* and *aT* to be formed. These are termed *recombinants* – an unfortunate and confusing term, since they are in fact new combinations. The frequency with which such new combinations are found among the progeny in crossing experiments is known as the recombination frequency. Independent assortment corresponds to a recombination frequency of 50 per cent. When the observed frequency is found to be significantly less than 50 per cent, the genes are linked and can be assumed to lie on the same chromosome. An illustrative case of linkage in the mouse is reproduced in Table 6.1. It is evident from inspection of Fig. 6.5 that the shorter the distance between two genes along the chromosome, the rarer will be the occurrence of recombination between them. It follows that the observed frequency of recombinant types in the progeny obtained in crossing experiments can be used as a measure of the chromosomal distance between any two linked genes. Using this method, T. H. Morgan and his school were able in the second decade of this century to build up a detailed map of the four chromosomes of the fruit fly *Drosophila melanogaster*. As a consequence of their work and that of other investigators, some thousand genes were located in linear order on the map, one map unit of distance being defined as the distance corresponding to a recombination frequency of one per cent. Map units are additive over short distances so that if a gene has neighbours on each side lying at x and y units from it respectively, then the distance between the two outlying genes is approximately $x+y$.

The genes of *Drosophila melanogaster* can be mapped in four distinct linkage groups, using the property that each gene shows 50 per cent recombination with all genes of other groups, but shows linkage with the genes of its own group. The haploid number of chromosomes in *Drosophila* is observed cytologically to be four. The fact that the number of linkage groups identified by the purely genetic criterion agrees with the haploid number of chromosomes observed with the microscope (not only in *Drosophila* but in every organism of which sufficently extensive linkage studies have been made) is perhaps one of the most striking examples known of the experimental corroboration of a theory – in this case the chromosome theory of heredity.

TABLE 6.1 A cross showing simultaneous segregation of three linked genes in the mouse.

Note that 'leaden' (ln) *and 'fuzzy'* (fz) *are recessive, only showing their effects when present in double dose, whereas the effect of 'Splotch'* (Sp) *is recognizable in single dose. From the observed frequencies it is possible to infer that the order of the loci is* ln Sp fz, *and that* Sp *is much closer to* ln *than it is to* fz.

	Genotypes	Phenotypes	Numbers of offspring observed	Interpretation in chromosomal terms
Father	+ *Sp* + / *ln* + *fz*	Splotch marking		
Mother	*ln* + *fz* / *ln* + *fz*	leaden colour and fuzzy fur		
Offspring	+ *Sp* + / *ln* + *fz*	Splotch marking	107	no cross-over
	ln + *fz* / *ln* + *fz*	leaden colour and fuzzy fur	92	
	+ + *fz* / *ln* + *fz*	fuzzy fur	12	cross-over between *ln* and *Sp*
	ln *Sp* + / *ln* + *fz*	leaden colour and Splotch marking	3	
	+ *Sp* *fz* / *ln* + *fz*	Splotch marking and fuzzy fur	52	cross-over between *Sp* and *fz*
	ln + + / *ln* + *fz*	leaden colour	44	
	+ + + / *ln* + *fz*	wild-type	3	double cross-over
	ln *Sp* *fz* / *ln* + *fz*	leaden colour, Splotch marking and fuzzy fur	0	
			313	

(Data from SNELL, G. D., DICKIE, M. M., SMITH, P. and KELTON, D. E. (1954) *Heredity* **8**, 271.)

The distances in genetic maps must not be interpreted as exactly proportional to the physical distances along the chromosome strand, since the statistical density of cross-overs varies, some regions of the chromosome being more prone to chiasma formation than others. The essential fact, however, is that methods based exclusively on the classification and counting of the offspring of controlled crosses have been successfully used to assign the genes to their stations in the chromosomes and to arrange them in a definite linear order.

Refinement of genetic analysis

In modern work directed towards increasingly detailed mapping of genes, micro-organisms are used, rather than mice or *Drosophila*. Mapping of two points very close together on a chromosome requires measurement of very low recombination frequencies since chiasma formation in the short region between these points may occur only once in hundreds, thousands or millions of cell divisions. To study such events therefore requires the rearing, classification and counting of literally millions of individuals, a heroic undertaking for a drosophilist. With micro-organisms, however, the raising of progenies numbered in millions presents no problem, and the rare aberrant individuals can be readily isolated from the mass by the use of suitable selective procedures.

The second advantage offered by micro-organisms is the relative precision with which the traits available for study can be defined in biochemical terms. Although in *Drosophila* we can relate definite points on the linkage map with, say, the shape of the wings or the colour of the eyes, we have no exact information on the nature of the gene-product from which the effects on the wings or eyes arise. Nor do we know much about the chemical and physical steps which intervene between the (unknown) defective – or missing – gene-product and the defective phenotype. With a fungus, such as *Aspergillus nidulans*, we are in a more favourable position. The characters with which the microbial geneticist works are for the most part simple nutritional requirements. Thus the normal *Aspergillus* is able to synthesize adenine from simple precursors absorbed from the environment. A great number of forms have, however, been isolated which lack this ability. Adenine is the end result of a sequence of chemical reactions, each catalysed by a separate and specific enzyme. Lack of any of these enzymes blocks the sequence, and so gives rise to the 'adenineless' (i.e. adenine requiring) phenotype. Successful growth may sometimes be achieved by addition to the environment of an intermediate metabolite, thus establishing the point at which the sequence is blocked. The microbial geneticist can thus in favourable cases work with strains with very specific defects, a given strain lacking a defined enzyme.

As a very rare occurrence a gene may suffer a sudden change to a new stable state, and the new form is then reproduced in place of the old in succeeding generations. This process is known as *mutation*. The frequency with which this happens spontaneously is of the order of one occurrence per 10^6 cell divisions, although the frequency can be greatly increased by various physical and chemical agents. The term mutation embraces stable chromosomal variations of any and every kind, including total loss of an interstitial part or even the whole of a chromosome (deletion), the turning of an interstitial segment back to front (inversion), and the exchange of a broken-off bit of one chromosome for a broken-off bit of another (translocation). A less drastic type of alteration is known as *point-mutation.* This term implies the conversion of a chromosomal segment from one stable state to another without positional change.

Mutations provide the mechanism of biological variability, and together with natural selection form the basis of evolutionary change. In Mendel's work, naturally occurring mutants provided the strains used in cross-breeding experiments. For a naturally occurring mutant form to survive in competition with the non-mutant or other mutant forms, the mutation must confer some short or long term advantage, or must, at least, not be disadvantageous. Many mutations are lethal and the organism does not survive, or does not survive long enough to produce progeny. In an artificial environment, however, it is possible to breed strains with serious defects. In the case of *Aspergillus nidulans* a mutation leading to lack of adenine would be lethal in the wild state, for adenine is essential to form ATP and as a building block in nucleic acid synthesis (see Chapter 7). However, adenineless mutants can be bred artificially by growing *Aspergillus* in an adenine-containing medium and selecting the occasional adenine-requiring cells that crop up through random mutation, and breeding from them. The adenine requirement is detected by checking for inability to grow in an adenine-free medium. Let us trace, step by step, the genetic analysis of two *Aspergillus* adenineless mutants, ad_8 and ad_9.

In higher organisms the gametes are haploid, the somatic cells are diploid. In fungi, on the other hand, the diploid state is transistory. Spores of different types can fuse together to give diploid cells in a process similar to the fusion of gametes in higher organisms, but the diploid cells rapidly undergo meiotic division and the cells of vegetative fungal growth are haploid. The geneticist is thus able to classify the products of meiosis in a very direct fashion, for in haploid organisms the genotype is fully expressed in the phenotype, without the complications introduced by the phenomena of dominance and recessiveness encountered in diploids. It is also possible in *Aspergillus nidulans* to cause the formation, as a rare event, of stable diploids, and Pontecorvo and his school have been able to exploit, as an alternative means of genetic mapping, the aberrant occurrence in these diploids of mitotic crossing-over.

Each mutant form is separately analysed by essentially the same procedure as that applied by Mendel to the case of tall and short peas. The difference in the case of *Aspergillus* is that we can identify the phenotype of cells equivalent to gametes (the haploid cells of the vegetative organism) as well as the phenotypes of diploid cells. It turns out that the mutations ad_8 and ad_9 are both recessive, as shown by the facts set out in Table 6.2. We adopt here the convention in which the symbols $+$ and $-$ are used as superscripts denoting the normal and mutant forms respectively. Symbols such as ad_8 and ad_9 can also be used without superscripts if we wish to refer non-committally to the chromosomal segments which are altered in these mutations.

Thus ad_8^+/ad_8^+ and ad_9^+/ad_9^+ represent normal, wild-type diploid *Aspergillus* cells with the normal genetic segments at points ad_8 and ad_9 in both homologous chromosomes, while, for example, ad_8^+/ad_8^- is a diploid cell with a normal segment on one chromosome and mutant ad_8 segment on the other. The haploid cells ad_8^- and ad_9^- are, of course, adenineless. The phenomenon observed here (Table 6.2), that the mutants are recessive, is fairly generally found. In an ad^-

TABLE 6.2 Determination of the dominance relations of two independently arisen mutations in *Aspergillus nidulans.*

Each mutant has as its phenotypic expression the inability to synthesize adenine. Completion of the analysis is achieved by demonstrating in each case that the heterozygote segregates in a ratio of 1 : 1.

	Diploids carrying the ad_8 mutant		Diploids carrying the ad_9 mutant	
	Genotype	Phenotype	Genotype	Phenotype
Normal homozygote	ad_8^+/ad_8^+	normal	ad_9^+/ad_9^+	normal
Heterozygote	ad_8^+/ad_8^-	normal	ad_9^+/ad_9^-	normal
Mutant homozygote	ad_8^-/ad_8^-	adenineless	ad_9^-/ad_9^-	adenineless
Conclusion	ad_8^- is recessive to ad_8^+		ad_9^- is recessive to ad_9^+	

mutant the block in the adenine metabolic pathway means that one of the necessary enzymes is not made, or is otherwise defective. The diploid cell ad_8^+/ad_8^- thus carries on one chromosome a defective gene for one of the enzymes of adenine synthesis, but on the other chromosome this segment is normal and makes normal enzyme. A diploid cell can usually get by, under these conditions, with a reduced amount of the enzyme, and show normal phenotype.

We now consider the purely genetic study that can be made of the question of whether ad_8 and ad_9 represent mutations at differing points in the genetic segment for one enzyme, or mutations affecting two different enzymes. Since we are considering here the purely genetic approach let us talk of 'functional units' rather than enzymes and consider a little later how these genetically defined functional units are related to enzymes and sub-units of enzymes. We assume, for the genetic approach, that synthesis of adenine requires a number of functional units each specified by a genetic segment. We suppose that to show normal phenotype the diploid cell must have a normal segment for each of the functional units of adenine synthesis on one or other of the two homologous chromosomes. The two possible results of analysis of ad_8^-/ad_9^- diploids are illustrated in Figs. 6.6 and 6.7. In fact for these two *Aspergillus* mutants the result found experimentally is the one represented in Fig. 6.6. The ad_8^-/ad_9^- diploid shows normal phenotype, so that the ad_8 and ad_9 mutations are affecting different functional units.

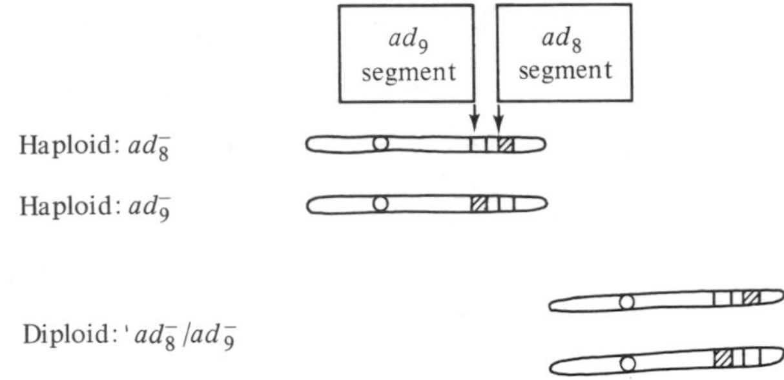

Fig. 6.6 *Aspergillus nidulans* experiment. First possibility: ad_8 and ad_9 are located in different segments of the same chromosome. The doubly heterozygous diploid possesses one normal (+) version of each gene and hence is phenotypically normal. The same would be true if ad_8 and ad_9 were located on different chromosomes. For graphic purposes the ad_8 and ad_9 segments are drawn in this and the following figure too large by many orders of magnitude relative to the remainder of the chromosome.

Breeding behaviour in *Aspergillus* can be tested by methods similar to those used to study the simultaneous segregation of two different gene-pairs in higher organisms. In the case of ad_8 and ad_9 it is found that new combinations are formed with a frequency of only about 16 per cent, the remaining 84 per cent of the progeny carrying the parental combinations. Hence the ad_8 and ad_9 segments lie on the same chromosome (Fig. 6.6) separated by approximately sixteen map units.

We shall now turn to a third adenineless mutation, ad_{16}, and consider what happens when its relation to ad_8 is subjected to the same analytical procedure. Persistent diploids obtained by combining strain ad_8^- and strain ad_{16}^- are found to be adenineless like their parents. These two mutations must therefore affect the same functional unit.

We have not so far considered the methods used to establish that two adenineless mutant forms under study are not the same mutant strain arising on two different occasions, but are in fact two mutants that have suffered slightly different genetic alterations. Where the defects turn out

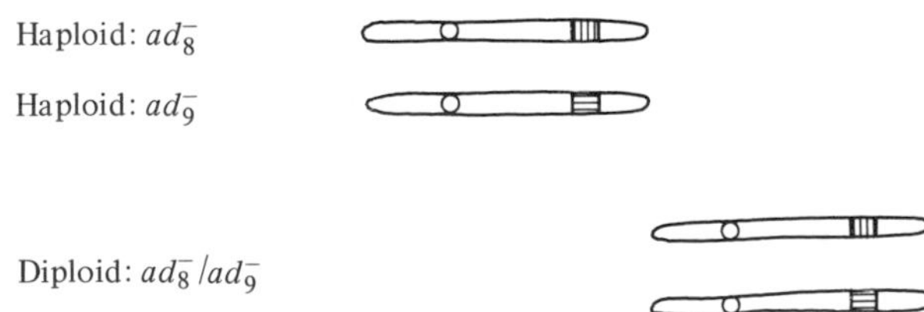

Fig. 6.7 *Aspergillus nidulans* experiment. Second possibility: ad_8^+ and ad_9^- represent different defective states of the same functional gene. The diploid heterozygote would possess no normal version of the gene but only two defective ones. Hence the phenotype would be adenineless.

to be in different genetic segments, as for ad_8 and ad_9, they clearly represent two different mutations, but more refined analysis is needed to establish this point for ad_8 and ad_{16}. When two adenineless strains with mutations affecting the same functional gene are crossed with each other, the natural expectation is that the haploid cells produced by their transient F_1 will all be adenineless. This expectation is in fact borne out in the overwhelming majority of cases. But the methods of microbial genetics make it possible to search through literally millions of cases to identify a few aberrant cells. Exposure to a medium entirely lacking in adeninc and related compounds, but otherwise adequate, causes the elimination of all the adenineless cells in the culture, but supports the growth and multiplication of any rare adenine-synthesizers which may be present. Such

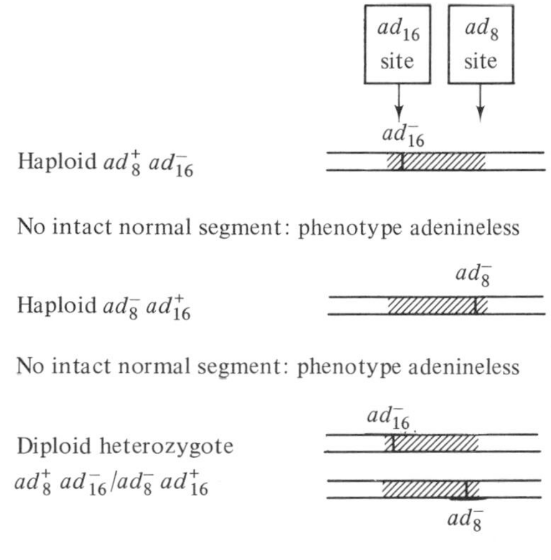

Fig. 6.8 *Aspergillus nidulans*. The ad_8 and ad_{16} mutations pictured as affecting different sites in the same functional gene, cross-hatched in the diagram.

cells do in fact come to light, in the case of the ad_8 and ad_{16} mutations with a frequency of a few per ten thousand. Refined techniques of analysis have shown that the great majority have not arisen by any further mutation. This shows that we are in fact dealing with two different mutants, since cells with the normal adenine-synthesizing ability have reappeared. But how has this come about?

The mystery is entirely resolved by the conclusion that the two mutations, ad_8 and ad_{16}, represent alterations to different *parts* of the same functional unit. Let us postulate that the whole unit must be intact in its original form if it is to permit the synthesis of the required enzyme. Then both mutant forms will be inactive, and so will be the persistent heterozygous diploids formed by combination of the two haploid strains. This can be represented schematically as in Fig. 6.8. When a transient F_1 is left to segregate the resultant classes of haploid cells will in general simply be those of the first and second parent in 1:1 ratio. But so far we have reckoned without the possibility of crossing-over. Suppose that genetic exchange by this mechanism can occur not only between the chromosomal segments underlying different functional units but also *within* them. More particularly, what will be the consequence of a cross-over in the region *between* the two sites affected by the ad_8 and ad_{16} mutations? Such an event is depicted schematically in Fig. 6.9. It results in one strand which is doubly defective and

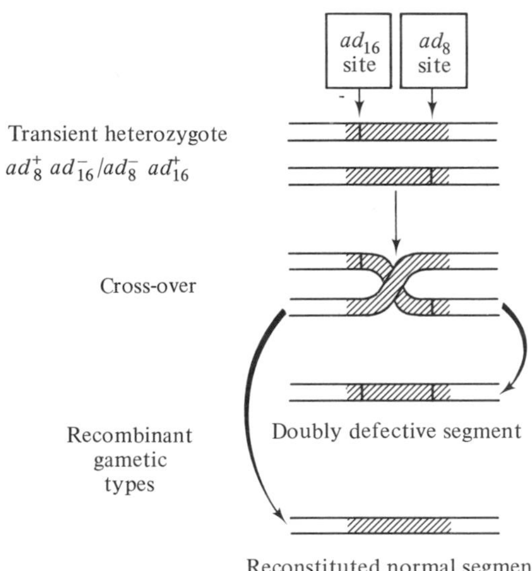

Fig. 6.9 *Aspergillus nidulans*. Crossing-over *within* a length of chromosome (shown shaded) corresponding to one functional unit. For simplicity only one chromatid of each of the paired chromosomes is drawn.

another which now contains no defects at all and is consequently capable of supervising the synthesis of the functional unit. Offspring which have inherited a cross-over chromosome of the latter kind consequently show phenotypical normality. It is these which account for the minute fraction of adenine-independent haploid cells produced by the transient F_1 of a cross between the two haploid strains ad_8^- and ad_{16}^-.

In classical genetic thought the genes were considered, like beads on a string, units of function, mutation and recombination. Each gene was thought to control a specific aspect of function, a point mutation was thought to affect the whole gene and it was supposed that crossing-over in chiasma formation could only take place between genes.

We have seen from our discussion of the *Aspergillus ad* mutants that the units of mutation and recombination must in fact be smaller than the genes, regarding these now more precisely as genetic segments controlling units of function.

Point mutations can occur at different points within a gene, and crossing-over can occur between these points.

The earliest evidence for these new ideas was obtained in 1949 with *Drosophila* by M. M. Green and K. C. Green, but it was misinterpreted. Soon, however, these results, together with analogous observations made by J. A. Roper in 1950, were given a clear interpretation by G. Pontecorvo which in its essentials has been fully borne out by subsequent experimental analysis. By studying bacterial viruses, with generation times even shorter than those of fungi, Seymore Benzer has mapped mutations down to the molecular level, in ways to be described in a later section.

The genetic test we have described to determine whether or not two mutations occur within a genetic segment controlling one unit of function is termed the *cis-trans* test. When two mutant sites are on different chromosomes (upper part of Fig. 6.9) they are said to be in the *trans* (Latin for 'across') arrangement, and when they are on the same chromosome (lower part of Fig. 6.9) in the *cis* (Latin for 'on this side of') arrangement. If a phenotypic difference is observed between the *trans* and *cis* heterozygotes, then the mutations both lie within a genetic segment controlling one unit of function. Benzer has termed the genetic segment defined this way a *cistron*. The occurrence of normal phenotype in the situation shown in Fig. 6.6 is termed *complementation*.

The sex life of bacteria and bacterial viruses

Genetic studies of bacteria and bacterial viruses (called bacteriophages or phages) have revealed a variety of bizarre effects, analogous to chiasma formation and crossing-over in higher organisms. Bacteria and viruses do not contain at any stage of their division cycle thick strands of nuclear material visible in the light microscope, so we shall use the term *genome* rather than chromosome for these organisms, to describe the physical entity that must be assumed to exist from genetic experiments, and can be seen in electron micrographs. The viral and bacterial genomes are in fact much thinner than the condensed strands which form chromosomes visible in the light microscope.

In phage infection the viral genome enters the host cell. The result may be virulent infection or lysogeny. In the former case the viral genome duplicates and reduplicates many times in the host cell, new virus particles are formed and eventually the cell lyses, releasing a large number of virus particles identical with the original infecting particle. Genetic experiments are possible if a cell is doubly infected with virus particles of two closely related kinds. Then something analogous to crossing-over can occur during duplication of the two genomes, to give virus particles which contain genetic segments from both types (Fig. 6.10).

In lysogeny the viral genome does not duplicate in an uncontrolled way. It enters into some close relationship with the host cell genome, possibly as indicated in Fig. 6.10, and duplicates only when the host genome duplicates. The virus in this latent state is termed prophage. After many divisions the viral segment may be released from this control and duplicate as in virulent infection,

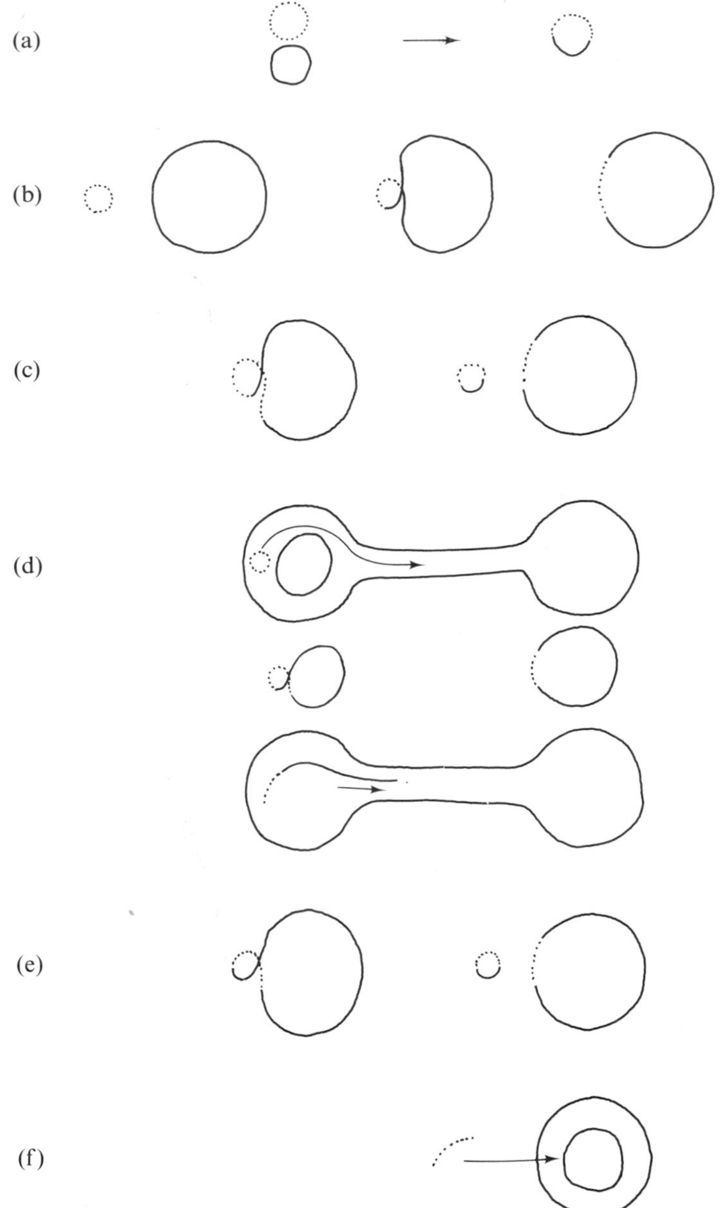

Fig. 6.10 *a*. Formation of a viral genome of mixed ancestry during double infection with virulent viruses. The viruses whose genetics have been mainly studied are the *E. coli* bacteriophages. Genetic mapping and microscopy show that *E. coli* and some, at least, of the phages have circular genomes. The phage genomes are therefore represented by circles in this figure. The mixed viral genome is probably formed by some sort of crossing over occurring while both parental genomes are in course of duplication.

b. Lysogenic viral infection. Incorporation of the viral genome into the bacterial with subsequent lysis of the cell. The virus particles so released sometimes carry part of the host genome (Fig. 6.10).

In bacterial conjugation a bridge is formed between donor (male) and recipient (female) cell through which segments of genome can pass. This process has been mainly studied in *E. coli.* The formation of the bridge for these bacteria is controlled by a small genetic segment, the sex factor, about one fiftieth of the length of the main bacterial genome. The presence of this factor in an *E. coli* bacterium confers the property of maleness, and the capacity to initiate conjugation. The sex factor can exist in two alternative states, either as a small separate genetic segment which duplicates in synchrony with the main genome or, alternatively, incorporated like a lysogenic virus into the main bacterial genome, the sex factor probably adopting a circular form and becoming incorporated by crossing-over during the duplication of the genome (Fig. 6.10*b*). When the sex factor is incorporated in this way conjugation leads to transfer of a segment of the main bacterial genome. This process can be studied by artificially interrupting conjugation after varying times. The more prolonged the conjugation, the more of the bacterial genome is transferred, and after sufficient time, last of all, the sex factor is transferred. The genome is thus transferred as a linear structure, the circle having been broken at a junction with the sex factor, and the broken end of the main genome enters first. In subsequent duplication there can be genetic exchange between the transferred segment and the female genome.

Another aspect of bacterial sexual activity, sexduction, is illustrated in Fig. 6.10*e*, and also the further phenomenon of bacterial transformation. Wherever a full or partial diploid state exists, for example after transfer of part of the main bacterial genome in conjugation or sexduction, there is the possibility of genetic recombination and mapping. Frequency of recombination determines map distance as in higher organisms.

Some common molecular mechanism similar to chiasma formation between homologous chromosomes of higher organisms seems to lie behind all these phenomena. There seems to be a tendency for homologous lengths of genome

---

genome probably occurs by some sort of crossing over, with both in circular form, while both genomes are undergoing duplication.

c. Bacterial transduction. Release of the lysogenic viral genome can occur by a process which is the reverse of incorporation. The released virus can carry away part of the host genome. When this virus infects another bacterium the latter may acquire heritable characteristics of the original host.

d. Bacterial conjugation. The upper drawing shows transfer of free sex factor (the circular form may break to be transferred as a linear structure). The middle drawing shows incorporation of the sex factor into the main bacterial genome, and the lower drawing transfer of the bacterial genome initiated by incorporated sex factor.

e. Bacterial sexduction. After a process which is the reverse of incorporation of sex factor, part of the main bacterial genome may be carried by free sex factor. The free sex factor may then transfer this part to a recipient cell.

f. Bacterial transformation. Part of the genome of fragmented bacteria can be taken up by a recipient bacterium of a different strain, which subsequently acquires some heritable characteristics of the donor strain.

to come together so that some sort of switch or crossing-over occurs during duplication of the genome, and the efficiency of this process apparently depends on the closeness of the homology (homology implies some similarity between genetic segments, and close homology implies close similarity). For example, in male cells carrying free sex factor incorporation of the factor into the main bacterial genome is a very rare event. This is thought to reflect a lack of any close homology between the sex factor, and the main genome. After sexduction, however, incorporation of the sex factor carrying a segment of bacterial genome is very efficient, presumably because crossing-over takes place very readily between this segment and the homologous region of the recipient's genome. In a similar way closeness of homology between regions of viral and host genomes may determine whether a virus infection is virulent or temperate, i.e. leads to lysogeny. We return to discussion of crossing-over and the molecular mechanism involved in the next chapter, and we return to the question of incorporation of a viral genome in discussing λ-phage and virus-induced tumours in Chapter 10.

Of these various aspects of bacterial and viral activity, we shall be considering in more detail some of the extensive experimental studies that have been made of two genetic segments – the RII region of T_4 bacteriophage and the *lac* region of the *E. coli* bacterial genome. The RII region will be discussed in the next section, as an example of genetic mapping now carried down to the molecular level, and the *lac* region in Chapter 10, in discussion of enzyme induction.

The limits of genetic mapping

Benzer's experimental study of recombination in bacterial viruses (phages) has had an important influence on modern genetics. The traits of which he made an intensive investigation are those which determine the phage's ability to multiply in one rather than another strain of the host bacterium, *E. coli.* Although it is impossible to see the phage particle except with the electron microscope, its presence can be spotted macroscopically in the following manner. A culture of *E. coli* is grown on agar in a glass dish until it covers the agar surface in a thin film. A dilute suspension of phage particles is then spread on the film and the culture is incubated further. Whenever a bacterial cell is penetrated by a virulent phage, its biochemical machinery passes under the control of the invader's genome and is used to manufacture fresh replicas of the original phage. After a thousand or so have been made, the cell bursts, releasing new phages to infect neighbouring cells in which the cycle is repeated. The sequence of events initiated by each infecting particle thus leaves a visible mark in the form of a circular clear zone within which all bacteria have been destroyed. Such zones are called plaques. Mutations occurring in the phage genome can be detected by following the phage's effect upon its host, in particular upon the formation of plaques on cultures of different host strains.

In the T_4 bacteriophage three classes of mutation (RI, RII and RIII) have been distinguished according to the types of plaque (normal or r-type) formed in two strains of *E. coli*, as shown in Table 6.3. After different mutations of the three types have been identified, it is possible to map their relative positions on the phage genome by methods essentially similar to those of classical genetics. It is not possible to mate one phage with another, but a bacterial culture can be infected with a mixture of two types of phage in a concentration sufficient to ensure that a high proportion of cells are penetrated by representatives of both types. During the ensuing cycle of intracellular multiplication, events analogous to crossing-over occur. A sprinkling of progeny particles can be detected which apparently derive different parts of their genetic material from one or the other phage type in the original mixture.

TABLE 6.3 Use of two strains of *E. coli* to distinguish four genetic types of bacteriophage.
An r-type plaque is larger than normal and has a sharp instead of a fuzzy margin.

Type of phage used for infection	Types of plaque formed on cultures of two strains of *E. coli*	
	B-strain hosts	K-strain hosts
normal	normal	normal
RI	r-type	r-type
RII	r-type	none
RIII	r-type	normal

In order to exhibit the analogies with the more orthodox Mendelian systems described earlier, we shall consider the consequences of exposing a liquid culture of B-strain bacteria to a mixture of RI and RII phages, so that virtually every cell becomes doubly infected. After the cycle of phage multiplication and bacterial lysis has been allowed to proceed, the culture can be centrifuged and a suspension of progeny particles harvested from the supernatant and seeded on to plates of fresh B-strain hosts. In the absence of crossing-over only r-type plaques should be formed, as can be seen from Table 6.3. In fact, however, a small proportion of normal plaques appear as well, and the phage particles isolated from such plaques breed true for this trait. This can be interpreted as arising by a process of crossing-over reminiscent of the exchange of material between homologous chromosomes in the meiotic division of diploid organisms; the frequency with which recombinant types are recovered can be used as a measure of distance along the phage genome. A consistent map can thus be made in just the same way as has been done for the chromosomes of *Drosophila*, though in the case of T_4 the map is found to be circular rather than linear.

We shall now confine attention to the RII region, in which all mutations result in one and the same phenotype, namely r-type plaques on B-strain hosts and no plaques on K strain. It has been possible by detailed application of the *cis-trans* test to dissect this region into two cistrons. At first sight it is not at all clear how the *cis-trans* test could be applied to this material: using the symbols r_A^- and

r_B^- to denote mutations within the RII region and r_A^+ and r_B^+ to denote the corresponding non-mutant forms, we should like to compare the heterozygote $r_A^- r_B^+ / r_A^+ r_B^-$ with the heterozygote $r_A^+ r_B^+ / r_A^- r_B^-$. But the phage is genetically haploid, and so the possibility of forming a heterozygote by mating two phages does not arise. However mixed infection mimics the heterozygous state not only in allowing crossing-over to occur between the infecting particles, but also in showing the *cis-trans* effect already discussed. Provided that there is one intact version of every cistron on one or other of the infecting particles, normal plaques are formed. Neither r_A^- nor r_B^- phage will form plaques on cultures of strain K bacteria, since as stated earlier all phage with mutations in the RII region lack this ability. But infection with a mixture of the two types results in the appearance of normal plaques, as shown in Table 6.4. The conventional oblique line, earlier used in symbolizing heterozygous genotypes in diploid organisms, is here used by analogy to separate the two components of a mixed infection. The inference from the behaviour of the last class listed is that inactivation of the B segment as on the left-hand side of the oblique line, does not impair the functioning of the A segment adjacent to it on the chromosome, and conversely for the inactivation of the A segment on the right of the line. Thus A and B are indeed separate cistrons and complement each other's activity when present on different particles within the same host cell.

TABLE 6.4 Results of treating a culture of the K strain of *E. coli* with suspensions of bacteriophage particles.
A mutation in the A cistron shows complementation with a mutation in the B cistron.

	Single infection	Single infection	Single infection	Double infection
Types used for infection	$r_A^- r_B^+$	$r_A^+ r_B^-$	$r_A^+ r_B^+$	$r_A^+ r_B^- / r_A^- r_B^+$
Phenotypic expression	no plaques	no plaques	normal plaques	normal plaques

For two mutations within the same cistron a quite different result is obtained. Denoting two such mutant forms by the symbols r_{A1}^- and r_{A2}^- we obtain the picture set out in Table 6.5. The result of mixed infection is italicized in the table

TABLE 6.5 Failure of two mutations located in the same cistron to complement each other.

	Single infection	Single infection	Single infection	Double infection
Types used for infection	$r_{A1}^- r_{A2}^+$	$r_{A1}^+ r_{A2}^-$	$r_{A1}^+ r_{A2}^+$	$r_{A1}^- r_{A2}^+ / r_{A1}^+ r_{A2}^-$
Phenotypic expression	no plaques	no plaques	normal plaques	*no plaques*

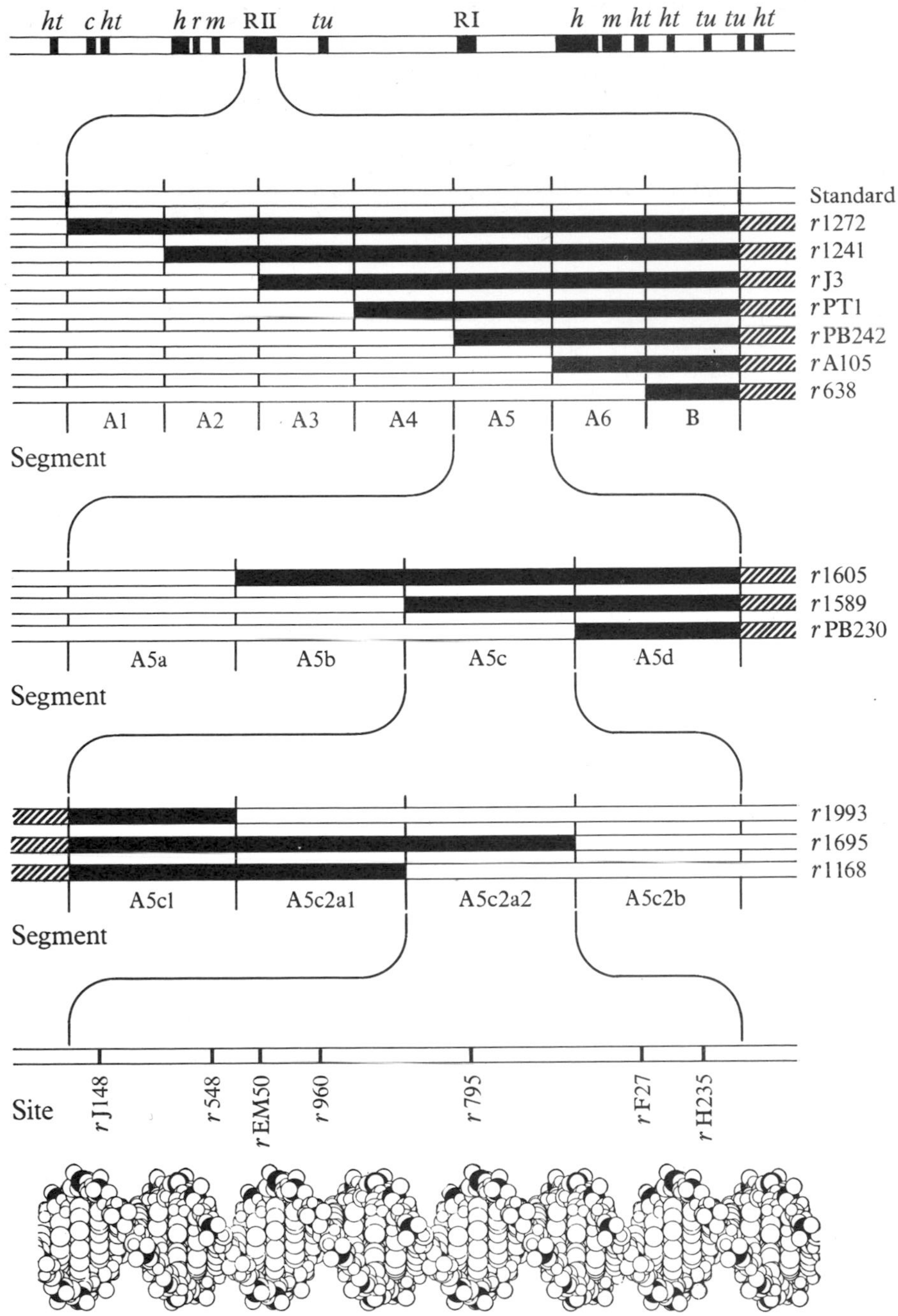

Fig. 6.11 The genetic map of the T_4 phage chromosome with the RII region, and parts of the RII region, shown in successive enlargement. The segment of the genome molecule included in this figure is of a length estimated to correspond to the genetic segment shown in the bottom row of the map. The molecular structure of the genome is discussed in Chapter 7. (*From* Benzer, 1962.)

to emphasize that complementation does not here occur. If we wish to be certain that r^-_{A1} and r^-_{A2} are in fact different mutations, producing defects in different parts of the A cistron, we have to show that the two mutations are separable by recombination. Strain B bacteria are infected with a mixture of the two mutant types, both of which multiply in this host. After extensive lysis has occurred, the supernatant is applied to a strain K culture, on which neither of the components of the original mixture would be expected to form plaques. Normal plaques do, however, appear in low frequency. These are attributable to the formation of recombinant $r^+_{A1}r^+_{A2}$ phages during multiplication in the first host (strain B) by crossing-over *within* the A cistron. The case is similar in essence to that schematized in Fig. 6.9 for *Aspergillus*. The refinement of analysis achieved by Benzer using these methods is illustrated in Fig. 6.11.

The modern definition of a gene

We have seen that the classical view of a gene as a unit of function, mutation and recombination had to be abandoned – there are many potential mutation points and crossover points within the segment of a chromosome or genome controlling one unit of function. In the early 1940's the new view of the gene was summed up by Beadle and Tatum in the slogan 'one gene – one enzyme', the gene carrying, in some form, 'instructions' for making a given enzyme. However this is not quite an adequate definition, since an enzyme is often made up of more than one type of sub-unit (Chapter 3) and genes can carry instructions for making non-enzyme proteins, haemoglobin, for example. The modern definition of a gene becomes 'a segment of a chromosome or genome carrying instructions for making one polypeptide chain'. Exactly how these instructions are coded in the gene, and acted on by the cell, form the subject of the next two chapters. A further complicating factor comes later (Chapter 10) in that some control segments of the genome map as genes, but do not code for polypeptide chains. In this book we keep to the above definition of a gene and refer to these control segments as 'genome regions'.

A cistron is a chromosome segment or genome segment identified by the *cis-trans* test, and could, in principle, include more than one gene. For example, it might sometimes be necessary for synthesis of a given functional enzyme made from two different sub-units, for the genes for these sub-units to be both intact on adjacent positions on a chromosome. The *cis-trans* test would then group the two genes together as one cistron. Genetic experiments, in themselves, can only identify cistrons, though biochemical study of mutants can establish the nature of the proteins synthesized under control of a given cistron.

It has been found that the results of a *cis-trans* test are not always clear-cut, owing to two interesting effects, *intracistronic complementation* and *polarity* which must be discussed in concluding this chapter. We must also briefly describe, as a background to discussion in later chapters, an experimental advance of great importance, the use of *conditional lethal* mutants.

In general, if a wild-type gene codes for a protein sub-unit, then mutation in this gene will lead to formation of a defective protein. Sometimes mutation may lead to modification of the protein without impairment of its function (for example, some of the abnormal haemoglobins discussed in Chapter 9), but amino-acid sequence studies are required to detect such a change, and a mutation of this kind will not show up in genetic experiments. Sometimes mutation will lead to failure of the gene to direct the synthesis of any protein at all. Sometimes a non-functional protein may be made, sufficiently similar in structure to the normal protein to carry some of the same antigen sites. The presence of this modified protein can then be detected by immunological assay. In these two latter situations genetic experiments will simply indicate the absence of functional protein.

Intracistronic complementation is really a contradiction in terms, but allowable because intracistronic complementation is usually only partially effective, and hence can be experimentally distinguished from the full intercistronic complementation discussed above. Intracistronic complementation apparently arises from the fact that some enzymes are made up of a number of identical sub-units. An enzyme of this kind is specified by one gene, which maps as one cistron in the *cis-trans* test, showing normal phenotype in a diploid cell provided one genome or chromosome carries an intact version of the gene (*cis* arrangement) and something like mutant phenotype in the *trans*-arrangement (illustrated at the top of Fig. 6.9) when the allelomorphic genes on the two genomes are making two kinds of polypeptide chains, both with structural defects but with the defects at different points in the chain. Sometimes in this situation the *trans*-arrangement phenotype shows evidence that a small amount of the relevant enzyme is present – it is not completely absent as would be expected. This is intracistronic complementation; the two good parts of the cistron appear to complement one another, though on different genomes. One possible explanation of this effect is as follows. A defective polypeptide chain has an altered tertiary structure and even if the altered sub-units can still aggregate to form a multi-subunit protein similar to the enzyme, the enzyme activity is lost. However in co-aggregation of two types of altered chain, both defective at different points, the sub-unit interactions of the quaternary structure can sometimes push the tertiary structure of the subunits back to their normal form and some enzyme activity is restored. Alternatively, in some examples of intracistronic complementation it seems that a protein sub-unit, which in the wild-type organism is a single chain, can be formed in the mutant by aggregation of two chain fragments, each made on the good bits of the gene on different genomes.

In conditional lethal mutants the effects of mutations are conditionally expressed, depending on whether conditions are permissive or non-permissive. Two classes of conditional lethal mutants have been mainly used in modern phage work – conditional mutants of one type can maintain virulent infection cycles in certain mutant strains of the host bacterium but not in the wild strain, and conditional mutants of the second type can multiply at one temperature but not at another. The former are called amber and ochre mutants and the latter temperature-sensitive mutants. These conditional lethal mutants can be isolated with relative ease. Some temperature-sensitive mutants, for example, propagate normally at 25–30°C but, unlike wild-type phage, do not grow at 40°C. Failure to propagate at 40°C is readily noted experimentally when screening a large batch of samples. The mutation involved in these cases is usually a point mutation leading to an altered protein which is a less stably folded molecule, and unfolds or changes structure to become non-functional at 40°C (temperature-sensitive mutants are prone to show intracistronic complementation).

In the amber and ochre mutants, a point mutation leads to premature termination of polypeptide chain synthesis. In a non-permissive host bacterium inactive peptide fragments are formed. A permissive host, which allows multiplication of the mutant phage, is one in which synthesis of the polypeptide continues in spite of the phage point mutation, to give a protein which is functional and allows phage multiplication, although this protein may carry an altered amino acid, or may be larger than normal owing to inefficient chain termination. Amber and ochre mutants may show polarity, that is, for two adjacent genes, mutation in gene 1 may prevent proper synthesis of the product of gene 2 as well as that of gene 1, while mutations in gene 2 have no effect on synthesis of the gene 1 product. Premature termination of polypeptide chain synthesis in gene 1 can affect read-out of gene 2. Polarity effects complicate the classical *cis-trans* type of experiment since mutants with mutations in different genes can fail to complement or show only weak complementation (two mutants with defects in genes 1 and 2 respectively may both be deficient in gene 2 product). However by good experimental design these effects, and intracistronic complementation, can be recognized and used to give additional information. The easy preparation of large numbers of conditional lethal mutants, with point mutations, allows detailed genetic mapping, and what is more important, by using mutants of this kind it becomes possible for the first time to map genes which are essential to function. Mutations in genes essential to function (in phage, or in any haploid organism) will normally be lethal, and prevent further genetic experimentation, unless permissive conditions can be found which allow the mutant phage, or organism, to be isolated and cultured.

Modern genetics, both as regards the detail of mutation and recombination, and also as regards the effects of these genetic changes, comes right down to the molecular level. We need to consider in the next chapter the chemical composition and molecular structure of the genome; then we can carry this discussion further and consider in Chapter 8 the genetic control of amino-acid assembly, and in Chapter 9 the various ways in which mutations can alter the detailed structure of the proteins under their control. These are the molecular events underlying evolutionary change.

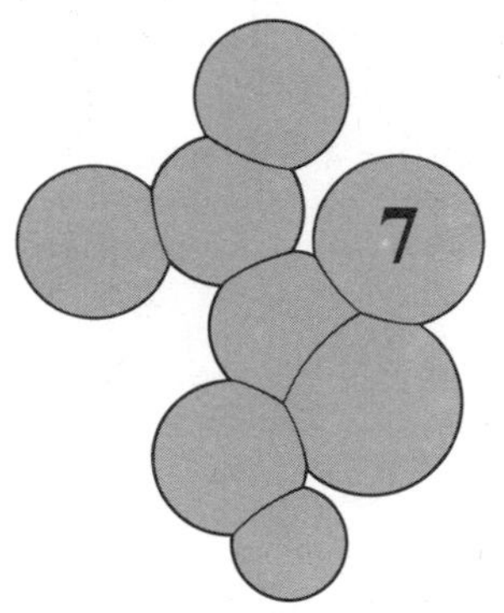

The structure of nucleic acids and nucleoproteins

The chemical nature of the genes

Nucleic acids, as their name implies, were at one time thought to be predominantly present in cell nuclei. Later it was shown that nucleic acids are of two kinds, deoxyribonucleic acid (DNA) found mainly in the nucleus, and ribonucleic acid (RNA) found mainly in the cytoplasm. Specific staining shows that in a dividing cell the nuclear DNA is localized in the chromosomes. Chromosomes are, in fact, nucleoproteins, containing protein and DNA.

Nucleic acids are long-chain molecules whose structure is to be discussed in this chapter. Prior to about 1940 it was thought that the DNA must be a relatively uninteresting component of the chromosome, containing perhaps a simple repeating sequence of groups along the molecular chain and acting as some sort of thread on which the genes were strung. Only proteins, it then seemed, were sufficiently complex in structure to carry the information content of the genes.

The first serious challenge to this view came in 1944, with Avery and MacLeod's experiments on bacterial transformation. It had been shown by Griffith in 1928 that inherited characters could be acquired by pneumococci from admixture with dead organisms of a variant strain. Griffith injected mice with living pneumococci of a pure-breeding strain lacking the polysaccharide capsule characteristic of most members of this species. Since the capsule is an essential prerequisite of virulence, these mice did not succumb. But when dead pneumococci from a normal encapsulated strain were added to the inoculum, the mixture proved lethal, and living bacteria isolated from the infected mice proved to be of the encapsulated variety and bred true to type. This phenomenon, now called bacterial transformation, is included in Fig. 6.10. The heritable transmission of the acquired characteristic suggests that a small segment of bacterial genome has been transferred, as indicated in Fig. 6.10, but to establish this point further evidence is needed, for the effects of bacterial transformation could be due to some transferred factor activating or otherwise modifying an existing segment of the recipient bacterium's genome.

Avery and MacLeod set out to pursue this question by purifying and making chemical study of the factor from encapsulated pneumococci which could effect transformation of the non-encapsulated strain. Their conclusions were outlined by Avery in a letter to his brother written in May, 1943:

I have not published anything about it – indeed have discussed it only with a few because I am not yet convinced that we have (as yet) sufficient evidence.

It is the problem of the transformation of pneumococcal types. You will recall that Griffith in London, some fifteen years ago, described a technique whereby he could change one specific type into another specific type through the intermediate R form. For example: Type $II_R \rightarrow$ Type III. This he accomplished by injecting mice with a large amount of heat killed Type III cells together with a small inoculum of a living R culture derived from Type II. He noted that not infrequently the mice so treated died and from their heart blood he recovered living encapsulated Type III pneumococci. This he could accomplish only by the use of mice. He failed to obtain transformation when the same bacterial mixture was incubated in broth. Griffith's original observations were repeated and confirmed both in our laboratory and abroad by Neufeld and others. Then you remember Dawson with us reproduced the phenomenon *in vitro* by adding a dash of anti-R serum to the broth culture. Later Alloway used filtered extracts prepared from Type III cells in the absence of formed elements and cellular debris and induced the R culture derived from Type II to become a typical encapsulated III pneumococcus. This you may remember involved several and repeated transfers in serum broth, often as many as five or six before the change occurred. But it did occur and once the reaction was induced, thereafter without further addition of the inducing extract, the organisms continued to produce the Type III capsule that is to say the change was hereditary and transmissible in series in plain broth thereafter.

For the past two years, first with MacLeod and now with Dr. McCarty, I have been trying to find out what is the chemical nature of the substance in the bacterial extract which induces this specific change. The crude extract of Type III is full of capsular polysaccharide, C (somatic) carbohydrate, nucleoproteins, free nucleic acids of both the yeast and thymus type, lipids, and other cell constituents. Try to find in the complex mixtures the active principle! Try to isolate and chemically identify the particular substance that will by itself, when brought into contact with the R cell derived from Type II, cause it to elaborate Type III capsular polysaccharide and to acquire all the aristocratic distinctions of the same specific type of cells as that from which the extract was prepared! Some job, full of headaches and heartbreaks. But at last *perhaps* we have it. The active substance is not digested by crystalline trypsin or chymotrypsin, it does not lose activity when treated with crystalline ribonuclease which specifically breaks down yeast nucleic acid. The Type III polysaccharide can be removed by digestion with the specific Type III enzyme without loss of transforming activity of a potent extract. Lipids can be extracted . . . and . . . the extract can be deproteinized . . . until protein free and biuret negative. When extracts, treated and purified to this extent, but still containing traces of protein, lots of C carbohydrate, and nucleic acids of both the yeast and thymus types are further fractionated by the dropwise addition of absolute ethyl alcohol, an interesting thing occurs. When alcohol reaches a concentration of about 9/10 volume there separates out a fibrous substance which on stirring the mixture wraps itself about the glass rod like thread on a spool and the other impurities stay behind as granular precipitate. The fibrous material is redissolved and the process repeated several times. In short, this substance is highly reactive and on elementary analysis conforms *very* closely to the theoretical values of pure desoxyribose nucleic acid (thymus) type (who could have guessed it). This type of nucleic acid has not to my knowledge been recognized in pneumococcus before, though it has been found in other bacteria.

Of a number of crude enzyme preparations from rabbit bone, swine kidney, dog intestinal mucosa and pneumococci and fresh blood serum of human, dog, and rabbit, only those containing active depolymerase capable of breaking down known authentic samples of desoxyribose nucleic acid have been found to destroy the activity of our substance – indirect evidence but suggestive that the transforming principle as isolated may belong to this class of chemical substance. We have isolated a highly purified substance of which as little as 0·02 of a microgram is active in inducing transformation. In the reaction mixture (culture medium) this represents a dilution of one part in a hundred million – potent stuff that – and highly specific. This does not leave much room for impurities but the evidence is not good enough yet.

In dilutions of one to a thousand the substance is highly viscous as are authentic preparations of desoxyribose nucleic acid derived from fish sperm. Preliminary studies with the ultracentrifuge indicate a molecular weight of approximately 500 000 – a highly polymerized substance.

We are now planning to prepare new batches and get further evidence of purity and homogeneity by use of the ultracentrifuge and electrophoresis. This will keep me here for a while longer. . . .

If we are right, and of course that is not yet proven, then it means that nucleic acids are not merely structurally important but functionally active substances in determining the biochemical activities and specific characteristics of cells and that by means of a known chemical substance it is possible to induce predictable and hereditary changes in cells. This is something that has long been the dream of geneticists. The mutations they induced by Xray and ultraviolet are always unpredictable, random, and chance changes; if we prove to be right – and of course that is a big if – then it means that both the chemical nature of the inducing stimulus is known and the chemical structure of the substance produced is also known, the former being thymus nucleic acid, the latter Type III polysaccharide, and both are thereafter reduplicated in the daughter cells and after innumerable transfers without further addition of the inducing agent and the same active and specific transforming substance can be recovered far in excess of the amount originally used to induce the reaction. Sounds like a virus – may be a gene. But with mechanisms I am not now concerned. One step at a time and the first step is what is the chemical nature of the transforming principle? Some one else can work out the rest. Of course the problem bristles with implications. It touches the biochemistry of the thymus type of nucleic acids which are known to constitute the major part of chromosomes but have been thought to be alike regardless of origin and species. It touches genetics, enzyme chemistry, cell metabolism and carbohydrate synthesis. But today it takes a lot of well documented evidence to convince anyone that the sodium salt of desoxyribose nucleic acid, protein free, could possibly be endowed with such biologically active and specific properties and that is the evidence we are now trying to get. It is lots of fun to blow bubbles but it is wiser to prick them yourself before someone else tries to.

Interest in DNA began to grow though still only slowly. The bacterial transformation experiments suggested that DNA must be either a segment of donor genome or must exert some specific effect on a segment of the recipient's genome. When virus experiments began to suggest that the infectious component of a virus was its nucleic acid it became more difficult to think of nucleic acids as having simply a rather specific effect on an existing gene. Small viruses are nucleoproteins, some containing DNA and some RNA. Hershey and Chase showed in 1952 by radioactive labelling of the protein and DNA of T_2 bacteriophage that most or all of the DNA, but very little of the phage protein, passed into the cytoplasm of the infected bacterium. As we now know from electron micrographs the protein of the head and tail structures remains outside, and the main component injected into the infected bacterium is the DNA core. In phage infection DNA was apparently capable of redirecting the synthetic activity of the infected bacterium to make a whole set of new proteins, which had presumably not previously been made by the bacterium, which became assembled into new phage particles. Thus at this time (around 1952) it began to seem quite possible that the information-carrying part of the chromosomes might be the DNA rather than the protein, for under some circumstances DNA molecules could clearly exert considerable control over cellular activity.

Structure of nucleic acids

As noted above, nucleic acids are long-chain molecules. The individual units which link together to form these chains are called *nucleotides* with the general formula

```
                             B
                             |
                             CH      H or OH
                            /   \   /
             O             O     CH
             ||             \   /
       HO—P—O—CH2—CH—CH—OH
             |
             OH
```

where B represents one of a variety of *bases*. Those most commonly found in natural nucleic acids are included in Table 7.1. Figure 7.1 shows the backbone of *ribonucleic acid* (RNA) with its alternating phosphate groups and ribose sugar rings, and illustrates how the nucleotides are linked together in nucleic acid

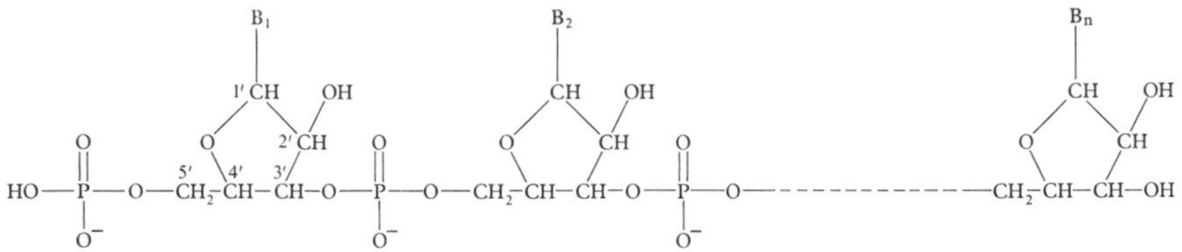

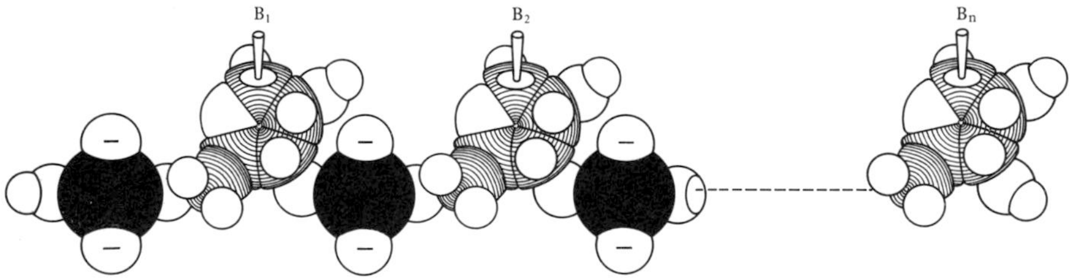

Fig. 7.1 The 'backbone' of a nucleic acid chain. In ribonucleic acid (RNA) shown in this figure, the backbone is made up of alternating phosphate groups and *ribose* sugar rings. (As in earlier drawings, phosphorus atoms are shown black.) In deoxyribonucleic acid (DNA) the backbone is made up of alternating phosphate groups and *deoxyribose* sugar rings, i.e. the —OH group on each sugar ring at position 2′ is replaced by hydrogen (—H). The pegs labelled B_1, B_2, . . ., B_n represent points of attachment for the *bases* shown in Table 7.1. In the complete nucleic acid chain, one or other of these bases is attached to each sugar ring at these points (position 1′ on each sugar ring).

chains. In the complete nucleic acid molecule one or other of the bases of Table 7.1 is attached to the backbone at each of the points marked by the pegs B_1, B_2, ... In *deoxyribonucleic acid* (DNA) the —OH group on each sugar ring at position 2′ (Fig. 7.1) is replaced by hydrogen (—H).

TABLE 7.1 Chemical formulae and drawings of molecular models of nucleotide bases. *The arrows indicate holes in these models fitting the pegs shown on the backbone model of Fig. 7.1. In 5-methyl-uracil (thymine) and 5-methyl-cytosine, —CH_3 groups replace the hydrogens at position 5. In 5-hydroxy-methyl-cytosine, a hydroxy-methyl (—CH_2OH) group replaces the hydrogen at position 5, and in 5-glycosyl-hydroxy-methyl-cytosine a glucose ring is attached to this hydroxy-methyl group. In 5-ribosyl-uracil ('pseudouracil') the uracil ring is attached to the backbone ribose through the carbon at position 5 in the uracil ring, rather than through the nitrogen at position 3.*

	6-keto (=O)	6-amino (−NH_2)
PYRIMIDINES	Uracil 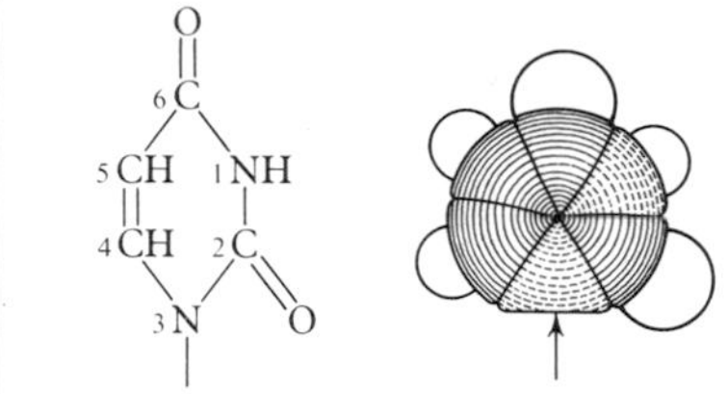	Cytosine 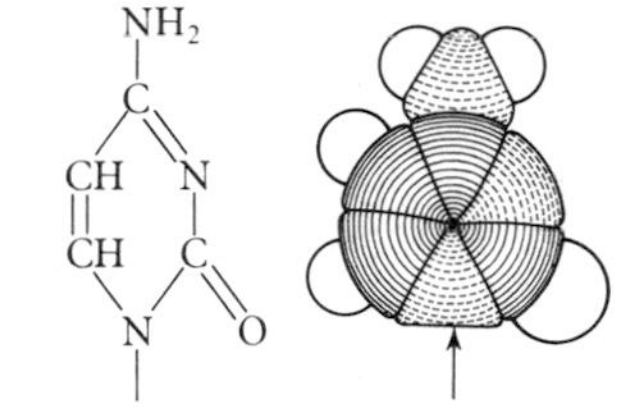
PURINES	Guanine 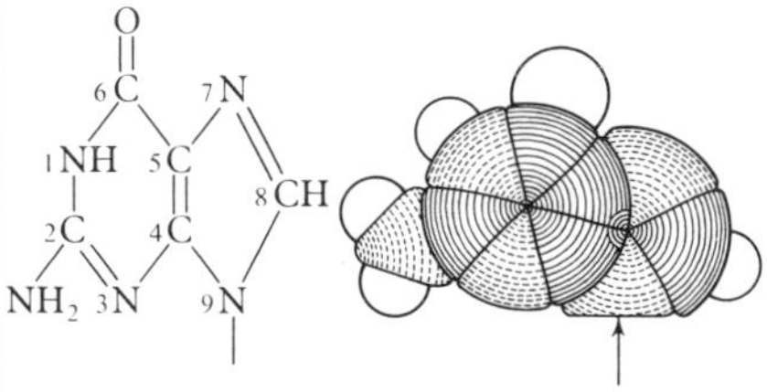	Adenine 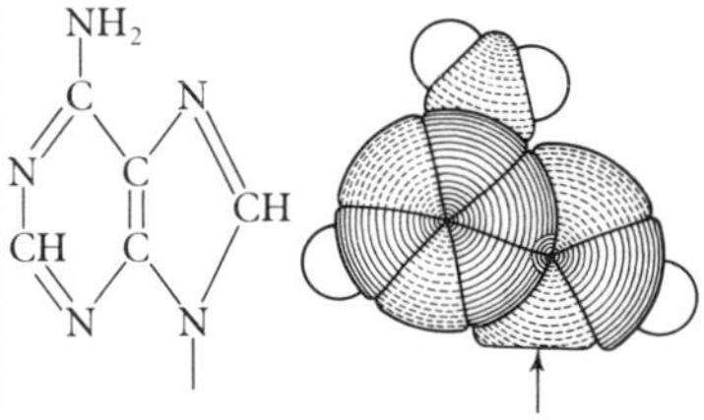

The bases found in naturally occurring nucleic acids are of four types (Table 7.1): 6-keto and 6-amino pyrimidines and 6-keto and 6-amino purines. All four types of base are found in appreciable amount in all naturally occurring nucleic acids (Table 7.2), but in RNA the 6-keto pyrimidine is uracil, while in DNA it is 5-methyl-uracil (thymine). A regularity is apparent in the relative proportions of the different bases. In DNA from nearly all sources, and in RNA from some sources, purines and pyrimidines are found in equal proportions, and also 6-keto and 6-amino nucleotides are present in equal proportions (Table 7.2). It follows that for these nucleic acids the number of adenine type (6-amino purine) nucleotides is equal to the number of thymine type (6-keto pyrimidine), and the number of guanine type (6-keto purine) is equal to the number of cytosine type (6-amino pyrimidine). The significance of these relationships will become clear when we consider the three-dimensional structure of nucleic acid molecules in later sections.

Molecular weights for purified RNA are found in the range 20000 to 2×10^6, corresponding to chains of about 60–6000 nucleotides. Molecular weights for purified DNA are found in the range 100000 to 120×10^6, but very large DNA molecules are rather readily broken up by the treatments used in their extraction, so that their molecular weight *in vivo* may be higher than this.

The amount of DNA in a mammalian cell corresponds to a total of about 10000 million nucleotides.

Structure of DNA double-helices

Studies of the physico-chemical and optical properties of DNA solutions, for DNA from most sources, show the molecules to be long thin, relatively rigid threads. Their diameter, as we shall see, may be estimated from X-ray diffraction studies to be about 2 nm, and their length in electron microscope preparations may range up to 30 μm or more. Treatments such as raising the temperature of the DNA solution, taking the pH beyond certain limits, or lowering the salt concentration well below physiological concentrations, produce changes in the physico-chemical properties of the solution, as well as changes in optical rotation and ultraviolet absorption. It must be supposed that the nucleic acid chains of native DNA are not in a fully flexible, randomly-folded form, but are arranged in the thread-like molecule in a well-defined three-dimensional structure which can be broken down under certain conditions, the nucleic acid undergoing denaturation analogous to protein denaturation. Extending the term used for proteins, we can refer to the type of folding or coiling of the chains in a nucleic acid molecule as the *secondary* structure of the molecule.

From consideration of the flow birefringence of DNA solutions, and from X-ray diffraction studies, Astbury and Bell showed, as long ago as 1938, that the bases in the DNA molecule must lie with their planes perpendicular to the long axis of the molecule, probably stacked one above another 'like a pile of plates'. Little further progress was made in elucidating the secondary structure of the molecule until the dramatic advances of 1952–53. At that time the chemical detail of the backbone structure shown in Fig. 7.1 had been recently established, and X-ray diffraction studies of nucleotide crystals had provided detailed information on bond lengths and angles which made it possible to build accurate scale models of nucleic acid chains and to decide, for example, the most probable orientation of the backbone sugar rings. Furberg suggested in 1952, in a speculative way, two possible structures for the DNA molecule formed by coiling or folding a single nucleic acid chain; in one of these models the bases were stacked to form a long column with the backbone spiralling up the outside. An equally speculative model was proposed by Pauling and Corey, early in 1953, in which three chains twisted together to form a long helix, with the bases projecting outwards from a central core formed by the backbone phosphate groups.

Meanwhile, Wilkins, Rosalind Franklin and others at King's College in

London, by drawing out threads from concentrated DNA solutions with careful control of humidity, had been able to get the molecules into a highly-oriented quasi-crystalline form and to obtain in this way more detailed X-ray diffraction patterns than those of Astbury and Bell. These patterns were similar to, though not at that time so detailed as, the patterns of Fig. 2.17. Their studies were beginning to provide direct evidence for the helical nature of the DNA molecule, and to suggest in contrast to Furberg's and Pauling and Corey's proposals, that it contained two intertwined strands, with the phosphate groups arranged round the outside of a helix about 2 nm in diameter, and with the bases turned inwards towards the axis of the molecule.

TABLE 7.2 Base composition of nucleic acids (%).

	adenine	guanine	cytosine	5-methyl-cytosine	5-hydroxy-methyl-cytosine	5-glucosyl-hydroxy-methyl-cytosine	uracil	5-methyl-uracil (thymine)	5-ribosyl-uracil	purine / pyrimidine	6-amino / 6-keto
Bovine thymus DNA [1]	28	22	21	1	0	0	0	28	0	1·00	1·00
Herring sperm DNA [1]	28	20	21	3	0	0	0	28	0	0·92	1·08
Wheat germ DNA [1]	27	23	17	6	0	0	0	27	0	1·00	1·00
Yeast DNA [1]	31	19	17	0	0	0	0	33	0	1·00	0·92
E. coli DNA [1]	26	25	25	0	0	0	0	24	0	1·04	1·04
T_2 phage DNA [1, 2]	32	18	0	0	3	14	0	33	0	1·00	0·96
Mouse fibroblast mitochondrial DNA [9]	29	21	20	0	0	0	0	30	0	1·00	0·96
T_2 specific messenger RNA [8]	28	21	16	0	0	0	35	0	0	0·96	0·79
Sea urchin egg RNA [3]	23	29	27	0	0	0	21	0	0	1·08	1·00
Yeast RNA [3]	25	25	23	0	0	0	27	0	0	1·00	0·92
TMV RNA [4]	28	26	20	0	0	0	26	0	0	1·17	0·92
TYMV RNA [7]	23	17	38	0	0	0	22	0	0	0·67	1·56
ϕX174 DNA [5]	25	24	18	0	0	0	0	33	0	0·96	0·75
Rat liver ribosomal RNA [6]	19	32	29	0	0	0	20	0	0	1·04	0·92
Rat liver soluble RNA [6]	20	29	29	0	0	0	18	0	4	0·96	0·96
E. coli ribosomal RNA [6]	25	31	23	0	0	0	21	0	0	1·27	0·92
E. coli soluble RNA [6]	20	31	29	0	0	0	19	0	1	1·04	0·96

1. DAVIDSON, J. N. (1960) *The Biochemistry of the Nucleic Acids*, 4th edn. London, Methuen.
2. JESAITIS, M. A. (1956) *Nature, Lond.* **178**, 637.
3. ELSON, D. and CHARGAFF, E. (1955) *Biochim. biophys. Acta* **17**, 367.
4. LORING, H. S. (1955) in *The Nucleic Acids*, ed. E. Chargaff and J. N. Davidson, New York, Academic Press.
5. SINSHEINER, R. L. (1959) *J. mol. Biol.* **1**, 43.
6. OSAWA, S. (1960) *Biochim. biophys. Acta* **42**, 244.
7. MARKHAM, R. and SMITH, J. D. (1951) *Biochem. J.* **49**, 401.
8. NOMURA, M., HALL, B. D. and SPIEGELMAN, S. (1960) *J. mol. Biol.* **2**, 306.
9. NASS, Margit M. K. (1969) *J. mol. Biol.* **42**, 529.

At this point, Watson and Crick in Cambridge, convinced that the structure of DNA would prove of central importance in understanding the molecular action of genes, were exploring with molecular models the possible coiling of nucleotide chains. In his autobiographical account of this period, *The Double Helix*, Watson tells how he had briefly seen one of Rosalind Franklin's X-ray diffraction pictures of DNA, during a discussion with Wilkins, and how he

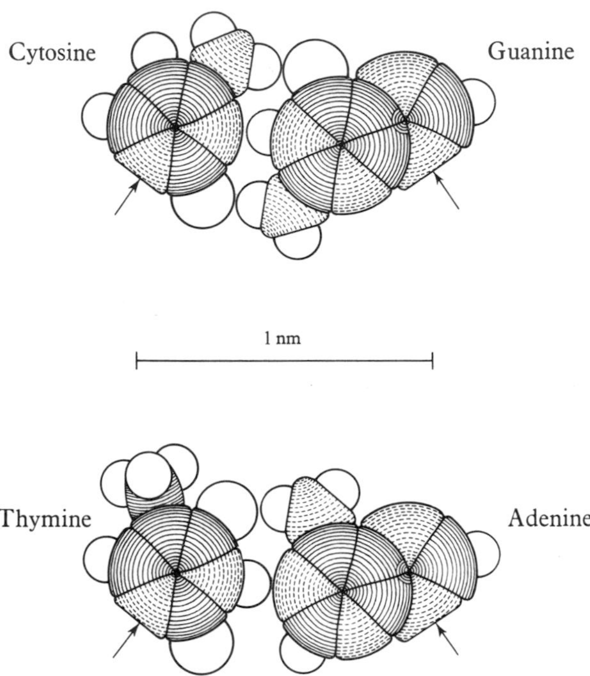

Fig. 7.2 The base pairs which Watson and Crick suggested must be present in the double-helical DNA molecule. The arrows indicate the points of attachment of the bases to the backbones of the two intertwined nucleic acid chains. The distance between points of attachment is the same for both base pairs (1·1 nm). Two hydrogen bonds are formed between thymine and adenine, and three between cytosine and guanine. Uracil can form the same bonds with adenine as thymine does (compare this drawing of thymine with the drawing of uracil in Table 7.1).

learnt also from this discussion that the (unpublished) X-ray work suggested that the phosphate groups were probably on the outside of the structure, with the nucleotide bases pointing inwards.* In considering models to fit the X-ray

* Watson gives his personal memoirs and these make good reading but do not give, as Klug has pointed out (*Nature* **219**, 808, 1968), a fair account of Rosalind Franklin's personality or contribution.

data Watson and Crick were impressed by a chemical feature Chargaff had noted in 1951, that in DNA from all sources then analysed, purines and pyrimidines and 6-keto and 6-amino bases were present in equal proportions (Table 7.2). Further, they felt that if DNA were the genetic material, its structure must somehow provide a key to the duplication of stored genetic information. With these various considerations in mind Watson and Crick put forward their famous and dramatic suggestion for the secondary structure of the DNA molecule. They showed that a regular structure could be formed from two nucleic acid chains twisted into a double helix if adenine bases were always paired with thymine (5-methyl-uracil) by hydrogen bonding, and guanine bases with cytosine (Fig. 7.2).

This double helix is shown in Fig. 7.3. The backbone chains of phosphate groups and sugar rings run in spirals up the outside of the long rod-like structure. The core of the rod is made up of the base pairs of Fig. 7.2, stacked 'like a pile of plates', as Astbury and Bell had suggested. Each base pair in the helix is rotated approximately 36° about the axis, relative to the pair immediately below, so that there are approximately ten base pairs in each complete turn of the double helix.

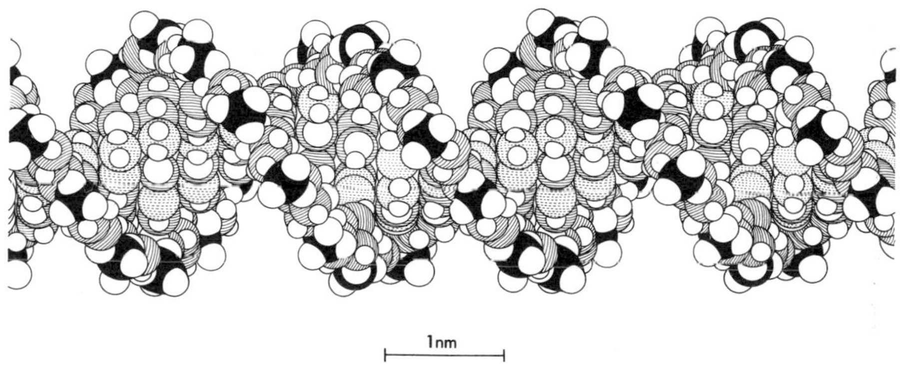

Fig. 7.3 The double-helical DNA molecule formed by intertwining of two DNA chains of complementary base sequence. The base pairs of Fig. 7.2 are stacked like a 'pile of plates' in the centre, and the two backbone chains spiral up round the outside of this central core. Note the alternating shallow and deep spiral grooves so formed. The repeat distance along the axis of the molecule is approximately 0·34 nm per base pair. (*From* Feughelman *et al.*, 1955.)

The specific pairing of the bases in this structure accounts of course, for the observed regularity in the base composition of DNA. Watson and Crick pointed out that the double-helical structure would also provide, in principle, a mechanism for the duplication of genetic material. If we represent a half-molecule of DNA as a chain of some such form as the following:

A—A—C—T—G—T—T—A—G—etc.

where A stands for adenine, C for cytosine and so on, then since adenine can only be partnered by thymine and cytosine only by guanine the half-molecule fully specifies the base-composition of the complete molecule, which must take the form:

```
A—A—C—T—G—T—T—A—G—
|  |  |  |  |  |  |  |  |    etc.
T—T—G—A—C—A—A—T—C—
```

This picture suggests a simple mechanism for the replication of the molecule in such a way as to preserve the entire sequence of base-pairs. We have only to imagine a longitudinal splitting of the molecule into its constituent halves, like this:

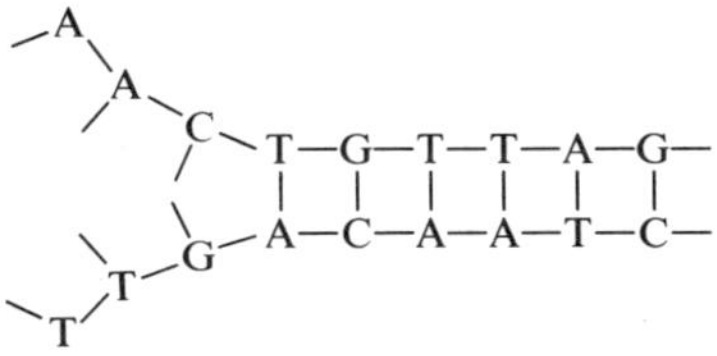

```
 /A
    \
     A
    /  \
        C
       /  \T—G—T—T—A—G—
            |  |  |  |  |  |
       \   /A—C—A—A—T—C—
        G
    \  /
     T
    /
 \T
```

followed by attachment of bases from the medium to the complementary bases of each half-molecule now lacking a partner, as below:

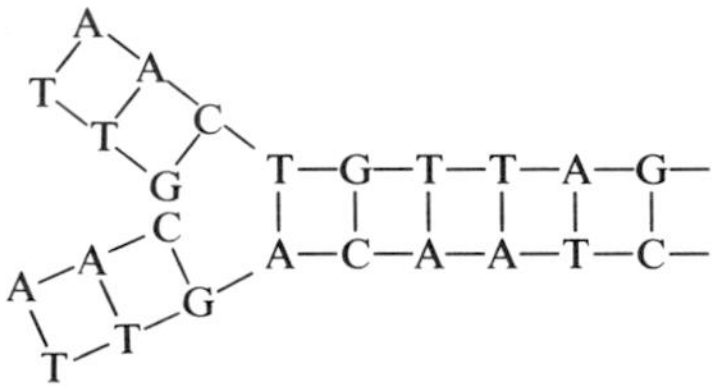

```
     A
    /  \
   T     A
    \  /  \
     T      C
      \  /    \
        G        T—G—T—T—A—G—
                 |  |  |  |  |  |
        C        A—C—A—A—T—C—
     /    \    /
   A—A      G
   \   \  /
    T—T
```

The daughter molecules are thus automatically assured of being replicas of the parent molecule.

This scheme of DNA replication exhibits a key feature which must be possessed by any substance claiming the title of 'genetic material', namely a means of faithfully replicating the finest details of structure. A corollary is that errors in the copying process ('mutations' in genetic terms) should be capable, once they have occurred, of being replicated with similar fidelity. This too is guaranteed by the above scheme.

The mode of replication pictured above can be described as 'semi-conservative', in the sense that each daughter molecule consists of one half-molecule conserved from the parent molecule and one half-molecule assembled from the medium. Any experiment tending to establish a semi-conservative mechanism for DNA synthesis strengthens the argument given above and hence supports the general case.

The simplest and most convincing demonstration of such a mechanism is that of Meselson and Stahl. They grew *E. coli* in a medium in which all the nitrogen had been replaced by its heavy isotope ^{15}N, until a similar substitution had been effected in all the bacterial constituents, including the DNA. DNA labelled with heavy nitrogen was extracted from the organisms and mixed with DNA from the same organism grown in a normal medium. It was found possible to effect a clean separation of the two components of the mixture by strong centrifugation in caesium chloride solution. The caesium chloride solutions in these experiments are made up to a concentration which makes their density comparable with that of the DNA. Strong centrifugation establishes a caesium chloride density gradient in the tube, with the density at the bottom higher than that of the DNA, and the density at the top less than that of the DNA. DNA molecules come to equilibrium during centrifugation at a point part-way down the tube where their density is equal to that of the caesium chloride solution.

In Meselson and Stahl's experiment the heavy DNA settled at a lower point in the tube than the normal DNA, each forming a distinct and clear-cut layer, as shown in the first part of Fig. 7.4. The crucial step now consisted in re-transferring labelled bacteria to normal medium and allowing one generation of cell division to occur. DNA was then extracted and centrifuged as described. Only one layer was formed, of material of a density halfway between 'heavy' DNA and normal. In other words, 'hybrid' molecules had been formed, exactly one-half of each consisting of material from the old, heavy DNA, and one-half of new unlabelled material derived from precursors in the medium. This is precisely what would be predicted from the semi-conservative mechanism illustrated in the foregoing diagrams, the last of which can be redrawn as follows, using asterisks to indicate material labelled with heavy nitrogen.

```
        A*
       /  A*
      T  /  C*
       T  /   T
        G       G*—T*—T*—A*—G*—
                |   |   |   |   |
       C       C*—A*—A*—T*—C*—
   A—A  \    A*
   \  \  G*
    T*—T*
```

New 'hybrid' molecules being formed

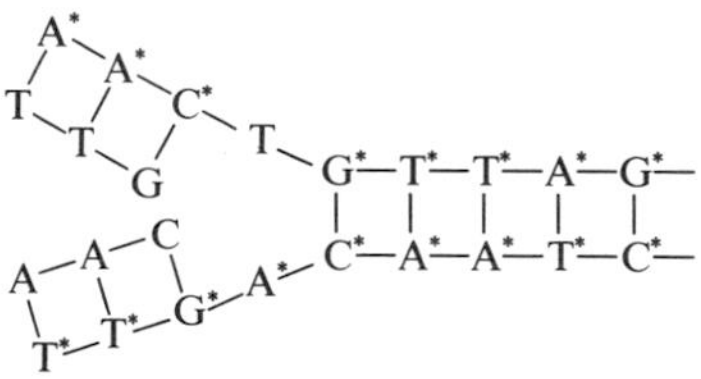

Further confirmation was obtained by allowing *two* generations of cell division to occur in normal medium. This time the extracted DNA formed two separate layers each containing half of the total DNA, one corresponding to 'hybrid' DNA as before, and the other to unlabelled DNA. The reader can verify by studying the last diagram that this is indeed what would be expected from the next cycle of replication. The main features of this elegant experiment are summarized in Fig. 7.4.

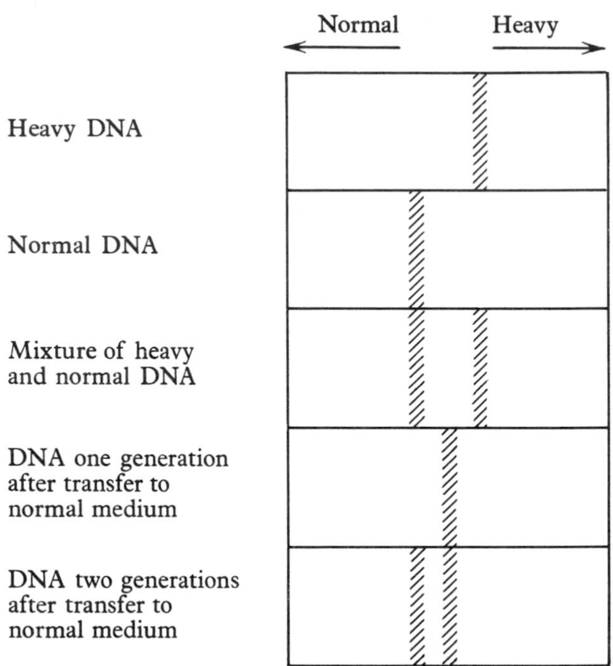

Fig. 7.4 The Meselson–Stahl experiment. For explanation see text.

Fully to appreciate the power of Meselson and Stahl's evidence, it is worth reflecting on the kind of result which would be obtained if DNA replication were *not* semi-conservative. If it were *fully* conservative, equal amounts of heavy and normal DNA would be formed in the first generation, rather than DNA of intermediate density. If the replication were neither fully nor semi-conservative, and the integrity of the original fully labelled base-chains were broken up during replication, then the inescapable consequence would be a mixture of heavy and light bases in the daughter molecules. There would be no possibility of the reconstitution into two distinct types, each present in equal amount, which occurred in the second generation of Meselson and Stahl's experiment.

Similar results have subsequently been obtained with eukaryotic cells, including human cells in tissue culture.

Watson and Crick's proposal for the mode of duplication of the genetic material, so simple in essence, but so profound and far reaching in its implications in genetics and evolution, has naturally stimulated a great deal of other experimental work and theoretical speculation. There is every indication that their ideas are essentially correct. Refinement of the X-ray diffraction analysis for extracted DNA has led to minor modifications of Watson and Crick's original molecular arrangement, but confirmed the essential features of the double-helical structure. The DNA extracted from a wide variety of cells, in varying states of activity, has been shown by X-ray diffraction studies to be in double-helical form. Where DNA is accessible to such studies in a natural or more nearly natural state, in intact sperm, in nucleoproteins extracted from cell nuclei, and in whole nuclei isolated from thymus cells and fowl red cells, it is also found in this form. Synthetic polynucleotides have been shown to form similar double helices (discussed in the next section). Exceptional cases (such as ϕX 174 phage) in which DNA is found in single-stranded form will be discussed in due course.

Further support for the Watson and Crick pairing scheme is provided by Kornberg's experiments on *in vitro* synthesis of DNA, from either the more common bases, or from the substituted bases: 5-bromo-uracil, 5-bromo-cytosine, 5-methyl-cytosine, etc. It will be noted from Fig. 7.2 that substituents in the 5-position do not interfere with hydrogen bonding in Watson and Crick's base pairs, so that if their ideas are correct we should expect that these 5- substituted bases could be built into a double helix just as readily as the corresponding unsubstituted bases. This is in fact what is found for *in vitro* synthesis of DNA. Uracil or 5-bromo-uracil can be incorporated into DNA *in vitro* in place of thymine (5-methyl-uracil), but cannot replace any of the other bases. The substituted cytosines can be incorporated in place of cytosine, but again cannot replace any of the other bases.

The conditions under which *in vitro* synthesis of DNA has been achieved provide in themselves further confirmation of Watson and Crick's ideas. In Kornberg's enzyme system all four component nucleotides must be present in the triphosphate form for synthesis of new DNA, as well as a DNA polymerase, and certain cofactors such as Mg^{++} ions. If a small amount of DNA is added to this system it acts as a template, or *primer*, for the synthesis of new DNA. Under these conditions the new DNA formed has the same base composition as the primer DNA.

The synthesis of DNA can also be studied *in vivo* by autoradiography. Experiments of this kind, in the first place, confirm the Meselson–Stahl evidence for semi-conservative replication. Secondly, for the relatively simple situation of DNA replication in bacteria, they give further information about the mechanics of replication. By a very elegant technique Cairns made the *E. coli* genome visible in the light microscope during duplication (Fig. 7.5). For one and part of a second generation he grew the bacteria in a medium containing radioactive thymidine. He then lysed the bacteria by very gentle means so that the DNA molecules remained unbroken, collected some of these DNA molecules on a support film, coated this with a layer of photographic emulsion and left it for two months in the dark. After development of the emulsion the position of the labelled DNA could be seen from the track of silver grains created by the labelled thymidine (Fig. 7.5). In the first generation one strand of the DNA became labelled, and new DNA formed along the labelled strand during the second generation, which thus becomes doubly labelled, can be distinguished, by the denser grain track, from singly labelled strands. The *E. coli* genome is circular and about 1 mm long (molecular weight about 2×10^9). In order for a closed ring of double-stranded DNA not to become totally tangled, as a duplicating enzyme

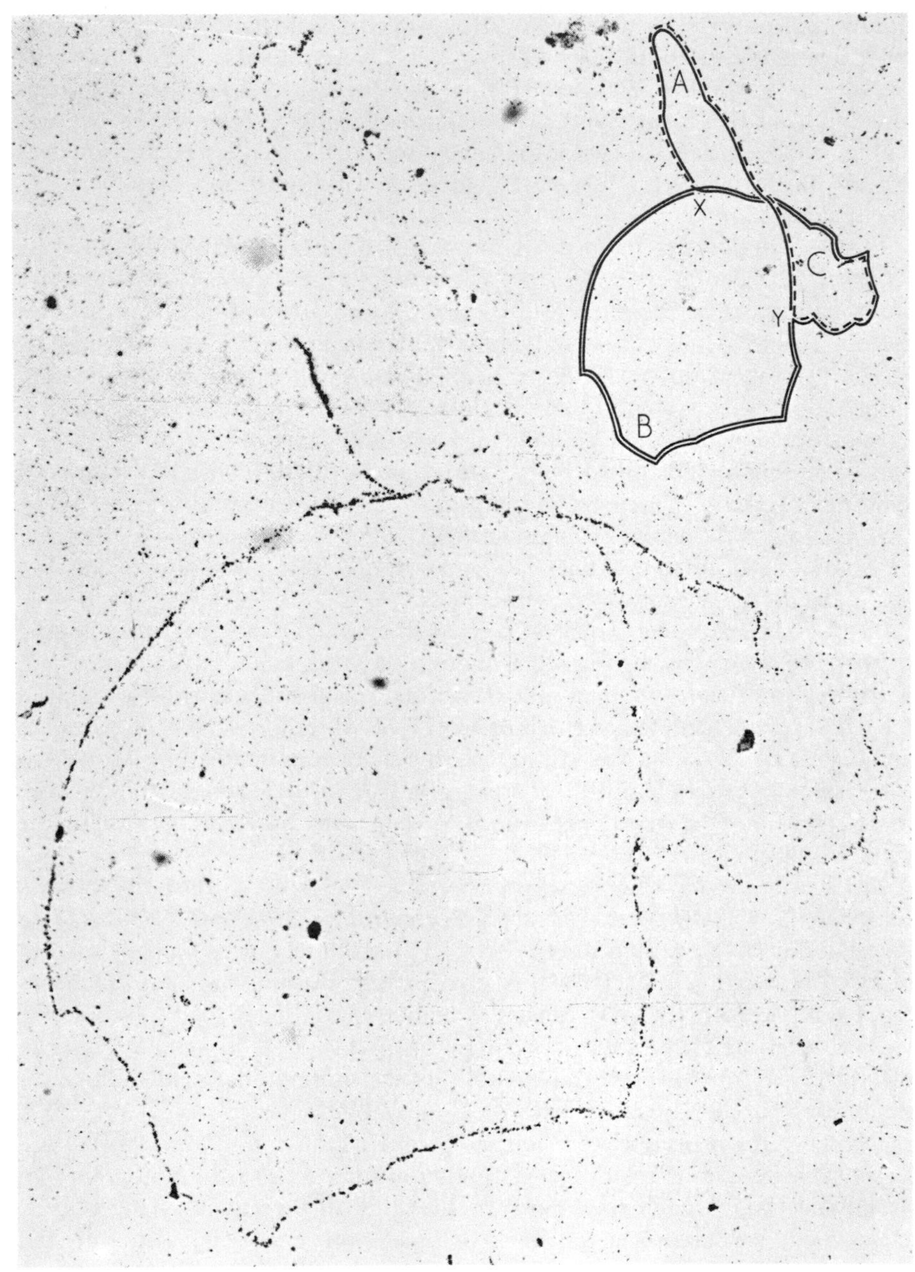

(*See foot of p. 225*)

moves along the DNA, Cairns supposed that there must be some 'swivel' structure incorporated into the ring at some point. It is now known that no special inserted structure is needed, but that this spinning effect can be achieved by a simple nick in one of the strands (a 'nick' being an enzyme-mediated break of one covalent bond linking successive nucleotides). If duplication is to proceed in a simple way nucleotides have to be added to one chain at the 5′ end and to the other chain at the 3′ end. No enzyme, or double-enzyme complex, has yet been isolated which can do this. The Kornberg DNA polymerase and other DNA polymerases more recently isolated can only add nucleotides to the 3′ end of a DNA chain. It may be that the true DNA replicase has not yet been isolated and all the DNA replicases studied to date are repair enzymes (concerned *in vivo* with repairing double-stranded DNA damaged in one strand). Alternatively, perhaps it has not yet been possible to reproduce *in vitro* the exact condition required for addition of nucleotides to the 5′ end. There is the further possibility that the 5′ strand grows by stepwise addition of segments of new DNA which have been synthesized *backwards* along the template strand near the duplication fork. This seems a very odd way for a cell to carry out a process of such central importance as DNA duplication but, suggested first by Okazaki's experimental evidence that labelled nucleotides incorporated into duplicating DNA appear first in short polynucleotide fragments and later in high-molecular-weight DNA, this seems at the present time the most probable mechanism.

There is evidence from thin-section electron micrographs that the *E. coli* genome in the intact cell may be attached at some point to the bacterial cell membrane. Possibly this is the point on the DNA at which duplication can be initiated. Jacob and Brenner have suggested such attachment might provide an important control mechanism. They suppose that after a round of duplication new membrane might be synthesized between the two replicated genome elements so that these are carried further apart during growth. Their separation beyond a certain distance could then affect duplication control mechanisms to allow a further round of duplication.

The molecular mechanism of crossing-over during DNA replication (see Chapters 1 and 6) is still not understood in detail. As noted in Chapter 6 this process takes place more readily when there is close homology between two genomes. In molecular terms close homology means close similarity in base sequence in the two DNA molecules. If the strands of double-helical DNA molecules are sometimes partly unravelled by thermal motion or interaction with proteins, and if these unravelled segments tend to be broken and rejoined (by endonucleases and repair enzymes) one can imagine that if two double-

---

Fig. 7.5 Autoradiograph of *E. coli* DNA taken by light microscopy. Magnification ×340. For details of preparative method see text. In the inset drawing labelled chains are shown by full lines, unlabelled by broken lines. The parent genome, labelled in one strand and part of another, began to duplicate at X. Synthesis of new strands has continued to Y. (*From* Cairns, 1966.)

stranded molecules with closely similar sequences come partly unravelled there might be a tendency for the strand of one molecule to coil up again with the complementary strand of the other (Fig. 7.6). This uncoiling and coiling might

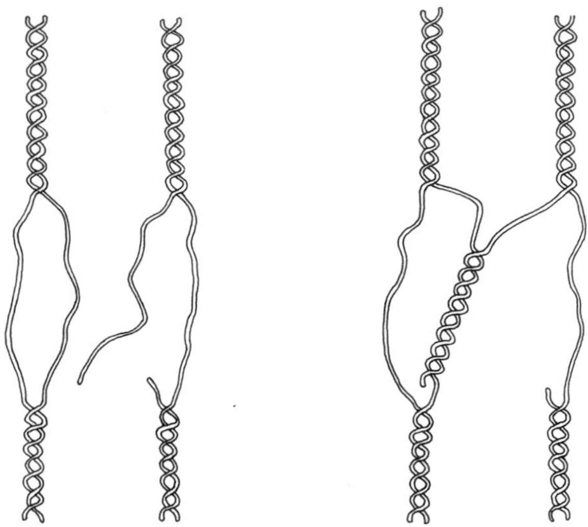

Fig. 7.6 Drawings showing the kind of uncoiling and coiling that might take place between homologous regions of two double-helical DNA molecules in the early stages of 'crossing-over'.

be taking place all the time for DNA as a result of random thermal motion, or might take place only under certain conditions. Figure 7.6 is intended only to illustrate the type of mechanism that can be suggested for crossing-over. Although a lot of evidence is available on various aspects of this process, particularly for replication of viral and bacterial DNA, no simple molecular ideas have yet proved adequate to explain the true nature of the process.

Polynucleotides

Naturally occurring nucleic acids, as we have seen, contain four different types of base in varied sequences. To study the properties of molecules of this kind it is useful to consider first the simpler synthetic nucleic acids that can be made with a single type of base, by incubating a solution containing one type of nucleotide with a polynucleotide phosphorylase. If the enzyme is incubated with a mixture of two or more bases, nucleic acid chains can be made with these bases only, in random sequence. Nucleic acids formed by *in vitro* synthesis with restricted base composition are termed polynucleotides.

In distilled water, or in very low salt concentrations, most polynucleotides exist as single-stranded molecules with the molecular chains in randomly-coiled flexible form. As the salt concentration is increased there is a tendency for these chains to twist together to form helices. For example, in a solution of physiological salt concentration (i.e. of salt concentration comparable with that of cell cytoplasm) containing both polyriboadenylic acid (poly rA) made from adenine ribonucleotides only, and also polyribouridylic acid (poly rU) containing uracil bases only, the polynucleotides form double-stranded helices

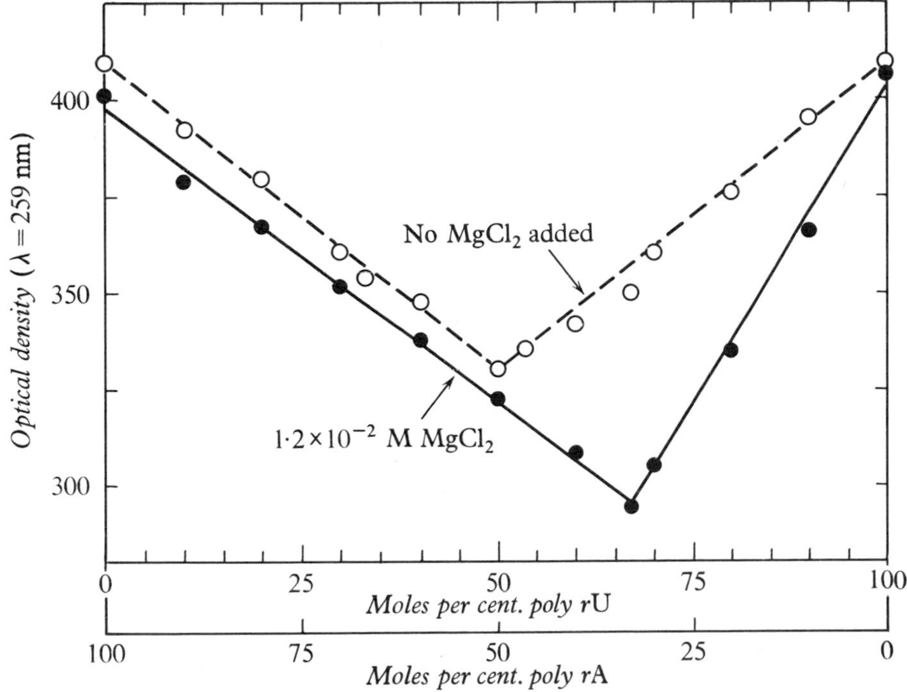

Fig. 7.7 Ultraviolet absorption (at a wavelength of 259 nm) plotted against composition, for mixtures of polyribouridylic acid (poly rU) and polyriboadenylic acid (poly rA) in the absence of Mg^{++} ions, and with $MgCl_2$ present at a concentration of $1{\cdot}2 \times 10^{-2}$ molar. The sharp absorption minima indicate complexing in 1:1 and 2:1 ratios in the two cases. (*From* Felsenfeld and Rich, 1957.)

which have been shown by X-ray diffraction studies to be essentially similar to those of natural DNA. The presence of helical structure is shown up also by studies of the optical rotation and ultraviolet absorption of the solution. (Interaction between the bases in the double helices decreases their ultraviolet absorption.) In the absence of Mg^{++} the aggregation is maximal when the two types of nucleotide are present in equal concentration, thus showing that the double helices formed under these conditions contain one strand of poly rA and one of poly rU (Fig. 7.7). If Mg^{++} is added to the mixture of poly rA and poly rU, at an Mg^{++} concentration of about 10^{-2} molar, triple helices form in which one poly rA chain entwines with two chains of poly rU (Fig. 7.7). The second poly rU chain spirals up the deep groove of the poly rA–poly rU double helix. The physiological significance of this triple helix formation is not known; possibly the triple helices only form under *in vitro* conditions.

Synthetic polynucleotides form a variety of double and triple helices. Polydeoxyribonucleotides can be synthesized *in vitro*, as well as polyribonucleotides, and mixed double helices will form, with one ribonucleotide strand and one deoxyribonucleotide strand, if the bases in the two strands can form suitable hydrogen bonds. The presence or absence of the 2′ ribose —OH group thus has little effect on the physical properties of the two types of nucleic acid.

Single-stranded nucleic acids

We have so far discussed two types of structure for nucleic acid molecules: the randomly-folded, constantly changing, fully-flexible chains found at low salt concentration, and the long, relatively rigid, thread-like molecules formed by nucleic acid and polynucleotide chains twisting together to form double and triple helices. But this is not the whole story. When the polynucleotide phosphorylase acts on a mixture of all four ribonucleotides in equal concentration, it forms a synthetic nucleic acid chain, poly rAGUC, in which the four bases are present in roughly equal proportion, but in a random order. This synthetic RNA shows a type of behaviour different from that of simple, one base, polynucleotides. At low salt concentration the chains are in single-stranded randomly folded form, but as the salt concentration is increased the poly rAGUC strands show no tendency towards aggregation to double or triple helices. The molecules remain single-stranded, but double-helical regions begin to form within each molecule, as a result of hydrogen bonding between bases in different parts of the chain.

This coiling into double-helical regions may be demonstrated, and its extent estimated, from the change in ultraviolet absorption of poly rAGUC solutions, of physiological salt concentration, on heating (Fig. 7.8). The ultraviolet absorption of poly rAGUC shows a gradual increase, with increasing temperature, along a curve which flattens off at 80–90°C. This increase in absorption is a result of the decreased interaction between bases as the coiled regions of the molecules unfold. By comparison, double-helical DNA shows a larger and more sudden transition, at a temperature around 80°C (Fig. 7.8). If the solutions are now cooled again, the behaviour of poly rAGUC is found to be fully reversible, but the behaviour of DNA is reversible only if the solution is cooled slowly, or held for a time at a temperature around 60–70°C. Even then DNA from many sources shows only partial reversion; the DNA curve, on cooling, more readily follows that of the poly rAGUC.

The explanation of these effects, supported by evidence from optical rotation and sedimentation studies, is that the sudden DNA transition corresponds to complete unravelling of the double helix, and that above 90°C thermal agitation is sufficient to bring all nucleic acids to the randomly-folded, single-stranded

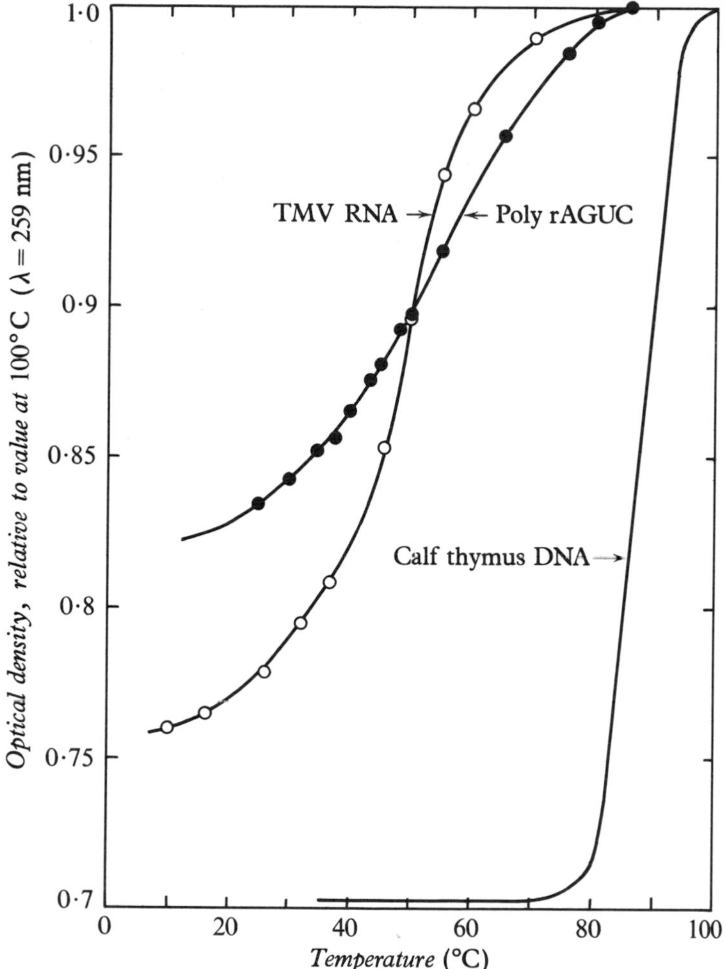

Fig. 7.8 The change in ultraviolet absorption on heating (at a wavelength of 259 nm) for solutions of poly rAGUC, TMV RNA, and calf thymus DNA. (*From* Doty *et al.*, 1959.)

form, even at physiological salt concentrations. Comparison of the magnitude of the total absorption change, on going from 20 to 90°C, for DNA and poly rAGUC, shows that in poly rAGUC, at 20°C, approximately 60 per cent of the bases are in double-helical regions. These double-helical regions of the single-stranded poly rAGUC molecule are relatively short, and also of variable length and stability. As poly rAGUC is heated some double-helical regions begin to 'melt' at quite low temperatures, and the ultraviolet absorption curve shows a gradual rise, rather than the sudden transition characteristic of the melting of the very long double-helix of DNA. If phage or bacterial DNA is heat denatured in this way and then cooled slowly, or held for a time at a temperature around 60–70°C, double-stranded molecules reform to a large extent, but unless these precautions are taken the DNA cools to the single-stranded form. DNA from other sources requires more prolonged treatment for renaturation after heating. This is probably an indication of the greater diversity of base sequence present in material from higher organisms.

DNA can exist naturally in single-stranded form, and RNA normally occurs in this form. The RNA in TMV is a single chain of about 6 500 nucleotides (see Chapter 3). Transfer-RNA and messenger-RNA molecules (whose role will be discussed in Chapter 8) are single strands. RNA extracted from ribosomes and from small spherical viruses, and DNA from the ϕX174 virus, are also single-stranded.

The behaviour of these single-stranded nucleic acids on heating is essentially similar to that of poly rAGUC, although the extent of hydrogen bonding at 20°C is somewhat variable for single-stranded nucleic acids from different sources. RNA extracted from TMV, for example (Fig. 7.8) and transfer-RNA, show a higher helical content than poly rAGUC. This presumably implies that there are more, or longer, segments of complementary sequence in the chains of TMV RNA and transfer-RNA, than are found by chance in the random sequence of poly rAGUC.

From the study of polynucleotide interactions, Fresco, Alberts and Doty proposed a model for the sort of structure that might be adopted by a single-stranded nucleic acid molecule with approximately 60 per cent of its bases in double-helical regions, under physiological conditions (Fig. 7.9). (By analogy with protein terminology we can talk of double-helical secondary structure in part of the molecule.) They suggested that as the secondary structure is formed, on cooling single-stranded nucleic acid chains, there is constant breaking and reforming of hydrogen bonded regions, until a structure is reached in which hydrogen bonding and helix formation is maximal. They suggested that the reason such a high proportion of bases can be involved in this bonding, even if the base sequences in different regions of the chain are not particularly favourable for complementary pairing, is that base sequences which do not fit can be thrown out in loops from the double helix (Fig. 7.9). Alternatively, as we shall see later, non-Watson–Crick base pairs may be incorporated into a double-helix with some local irregularity. The formation of several short helices (as in Fig. 7.9) rather than a single helix, allows greater freedom in reducing the

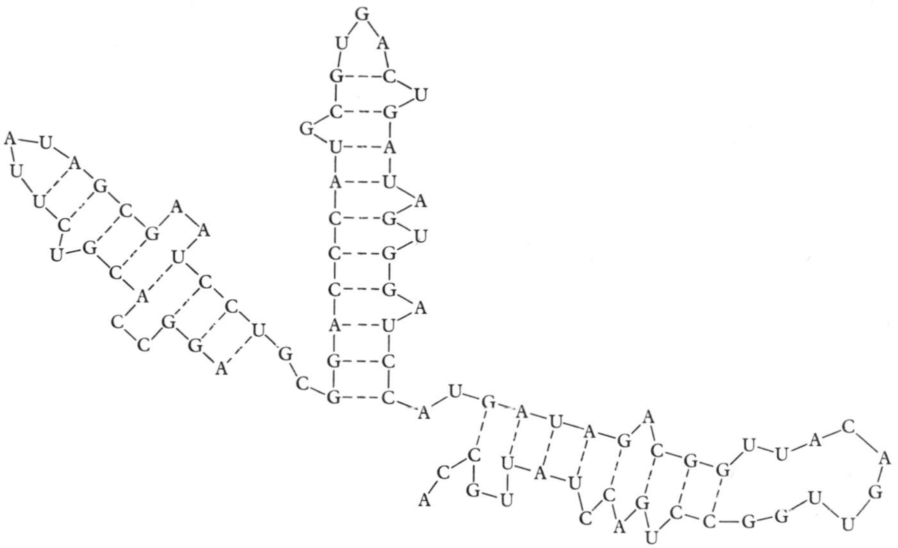

Fig. 7.9 Schematic illustration of the secondary structure of a single-stranded nucleic acid molecule. In deriving this conformation Fresco, Alberts and Doty chose a random RNA sequence of 90 nucleotides in which the four types of base were present in roughly equal proportions. They then tried to maximize the number of adenine-uracil and guanine-cytosine base pairs, by forming the double-helical regions, with loops and short lengths of unpaired bases between these regions. The structure illustrated is one of a number of possible arrangements with 50–70 per cent base pairing (see also Fig. 8.1 for further development of this idea). (*From* Fresco, Alberts and Doty, 1960.)

number of loops to bring the maximum number of bases into the double-helical regions. Where the proportion of bases in helical regions at 20°C is greater than 60 per cent, as it is for the RNA extracted from TMV and for transfer-RNA, we may expect the double-helical regions to be more regular, with fewer loops.

The structure of Fig. 7.9 was determined theoretically as the most probable of a large number of possible configurations for this particular nucleotide sequence. In this sequence, the four bases are present in equal proportions but in a randomly chosen order. It must be supposed that under physiological conditions the helical regions would still continue to break up and reform to some extent even when a condition of maximum hydrogen bonding has been reached, so that if the base sequence allowed other configurations, similar to that of Fig. 7.9, in which the amount of hydrogen bonding was around 60 per cent, the single-stranded molecule might be constantly going over from one configuration to another. If the base sequence allowed only one configuration with favourable base pairing, the nucleic acid chain would tend to form a stable three-dimensional structure characteristic of its nucleotide sequence, in the same way as a polypeptide chain folds up to form a compact structure characteristic of its amino-acid sequence.

The secondary structure of a single-stranded nucleic acid chain can be modified by interaction with protein, or by interaction with other nucleic acids. When TMV RNA becomes incorporated into the virus structure, the secondary structure of the isolated RNA is completely lost; the configuration of the nucleic acid chain in the virus is determined by the protein component (see Chapter 3).

For nucleic acid chains, which contain only four types of base, the problem of determining primary structure is more difficult than in the case of peptide chains which contain twenty types of amino acid. If the nucleic acid chain is fragmented by partial digestion the short segments formed cannot be so readily separated and distinguished as peptide fragments. However primary structure has been determined for a number of transfer-RNA molecules, since these are relatively short (70–80 nucleotides) and contain a number of abnormal bases, which aid identification of chain fragments. For these molecules bonding arrangements can be proposed using the principles illustrated in Fig. 7.9, as will be discussed in Chapter 8.

In discussing the structure of DNA and of single-stranded nucleic acids we have so far assumed that only the Watson–Crick base pairs A–T (or A–U) and G–C can fit into a regular double-helical structure. But in fact the interactions of single-base polynucleotides, and study of molecular models show that a variety of different base pairs can be formed by hydrogen bonding, and some of these, notably one type of adenine–guanine pair, can fit quite well into a regular double helix of A–T and G–C pairs. It seems that bonding interaction is particularly strong for the Watson–Crick pairs as a result perhaps of resonances within the purine and pyrimidine rings when these pairs form, but it is possible that in the double-helical regions of single-stranded nucleic acids other base pairs are present. Double-helical regions might in fact be quite stable without strong hydrogen bonds between base pairs as a result of hydrophobic interactions, as discussed in a later section. If this is the case then fewer small 'loops' would be expected than are shown in the Fig. 7.9 structure.

Mutation

If an abnormal base pair is formed occasionally during DNA replication this will lead to a mutation and this effect may contribute to the natural mutation rate that is found for cells dividing under normal conditions (see Chapter 6). Another contributing factor may be the fact that nucleotide bases spend a small part of their time in alternative tautomeric forms to those of Table 7.1, and this may lead to occasional faulty copying. The molecular replication process is also subject to modification by a variety of physical and chemical agents. These agents increase mutation rates and are termed mutagenic. They include cosmic radiation, X-rays, ultraviolet radiation, nitrous acid, acridine dyes and a variety of other chemicals. These mutagenic agents modify the DNA structure in a variety of ways and we will discuss here the effects of the two chemical agents noted, to illustrate the sort of effects that are involved.

The chemical effect of nitrous acid is to remove amino ($—NH_2$) groups from amino acids and from amino-containing purines and pyrimidines. Its effect on

DNA chains is to remove the 6-amino groups from cytosine and adenine bases. This modifies the hydrogen bonding possibilities for these altered bases (Fig. 7.2) and at the next round of replication bases other than guanine and thymine will be able to pair with the modified cytosines and adenines. The effect of the acridine dyes is rather different. These are flat molecules that tend to insert themselves into the DNA structure between base pairs. Their mode of action as mutagens is not fully understood but the effect is either *insertion* of an additional base into a growing chain (presumably because the template is spaced out at this point by a dye molecule) or an absence (*deletion*) of one of the bases in the growing chain (presumably because a dye molecule is temporarily held at this nucleotide-binding site and growth continues with addition of the nucleotide that pairs with the next base along the template sequence). Once a mutation has taken place, so that the base sequence along the DNA is altered, the altered sequence is subsequently replicated and carried through to succeeding generations.

The study of recovery of cells from exposure to ultraviolet radiation, however, shows that alteration of a base will not always lead to a stable mutation, since cells appear to contain 'repair' enzymes which act to maintain the DNA structure unchanged. Ultraviolet at 260 nm is absorbed directly by nucleotides, which are thereby raised to a high-energy, more reactive, state. Adjacent thymidines on the same strand of a DNA molecule then become covalently linked to form dimers, but this altered segment in one strand can be cut out by an endonuclease, the original sequence resynthesized by a DNA polymerase, using the 'good' strand as a template, and the new segment linked into the cut strand by a ligase, to restore the original DNA (an endonuclease is an enzyme which produces a break, or nick, in a nucleic acid chain, and a ligase is an enzyme which joins broken chain ends). Mechanisms of this kind help to maintain genetic continuity, by reducing the natural mutation rate. Natural mutations may be presumed to arise from the effects of natural radiations, natural chemical mutagens, occasional failures in exact DNA duplication, and occasional failure of enzyme repair mechanisms.

Factors determining the configurations adopted by nucleic acid chains

We have seen in Chapter 2 that a polypeptide backbone carries no net charge, although the electronic distribution of fractional positive and negative charges on the backbone groups is important in causing the polypeptide chain to coil up into a helix. Other factors causing polypeptide chains to adopt a compact configuration are the high proportion of hydrocarbon side chains in proteins, and electrostatic attraction between positive and negative side-chain groups.

By comparison, the nucleic acids at pH ~ 7 carry a negative charge of one unit on each phosphate group along the backbone (indicated in Fig. 7.1 by minus signs, each representing half a unit of negative charge, on the phosphate oxygen atoms). The internal electronic distribution in the nucleotide bases leads to local concentrations of negative and positive charge at the groups which form hydrogen bonds, but the bases carry no net charge at pH $\sim$7. Thus all nucleic acid molecules carry a large net negative charge at pH $\sim$7 due to the backbone groups. A large net charge on a long-chain molecule, as we have noted when discussing synthetic peptides and the denaturation of proteins in acid or alkaline solutions, mitigates against other forces which might lead to compact folding. Nucleic acids in aqueous solution at very low salt concentration are single-stranded, with the molecular chains in randomly-folded, constantly changing, fully-flexible form.

The presence of salts in aqueous solutions reduces the effects of electrostatic forces. Regions of negative charge on a large molecule increase the proportion of positive to negative ions in the immediately surrounding solution. This has a shielding effect, reducing the electrostatic forces between different parts of the large molecule and also the electrostatic forces between molecules. For polynucleotides or nucleic acids in physiological salt solutions, particularly in the presence of Mg^{++} or other divalent cations, electrostatic forces become of less dominant importance and other factors come into play. The nucleotide bases are flat, or disc-like, with the upper and lower faces of the discs hydrophobic, and with hydrogen bonding N—H and =O groups round the edge. In an aqueous environment there is a tendency for these bases to stack together, excluding water from the spaces between them. (Compare discussion of similar hydrophobic effects in Chapters 2 and 4.) This effect is enhanced by specific, cohesive forces which arise from electronic interaction between resonating aromatic molecules.

In the DNA molecule, and in the double and triple helices discussed in the polynucleotide section, two or three chains twist together to bring their bases into planar contact. These structures are stable, under the appropriate conditions, provided those hydrogen bonding groups round the edges of the bases which are removed from contact with water are able to form satisfactory hydrogen bonds in the helices. These bonds then provide an additional factor stabilizing the helical structures; it has been estimated that the stability of DNA double-helices may be attributed about equally to hydrophobic factors and hydrogen bonding effects. At physiological salt concentrations these forces of attraction between bases become dominant over electrostatic repulsions between phosphate groups. (The arrangement of the phosphate groups round the *outside* of the helical structures does separate these groups and reduce the repulsion forces as far as the other considerations allow.)

For the nucleic acids, as for the proteins, the secondary structures which exist under physiological conditions result from a slight imbalance between almost equally opposed forces favouring the random or the folded forms of these flexible-chain molecules. These secondary structures are thus very labile, allowing variation and subtlety of physiological role – allowing, for example, the strands of a DNA double helix to come apart during replication, and (as we shall see) during read-out of genes.

Nucleoprotamines and nucleohistones

In sperm, and in the nuclei of somatic cells, the bulk of the DNA is in structural association with protamines and histones. These are proteins containing an abnormally high proportion of the basic side chains, arginine and lysine, and a high proportion of their glutamic and aspartic acid side chains in the amide form (Table 2.2). At pH $\sim$7 they therefore carry a large net positive charge. Nucleic acids carry a large net negative charge, and the protein and nucleic acid components of nucleoprotamines and nucleohistones are held together largely by electrostatic forces. Protamines are distinguished from histones by their more restricted amino-acid composition (Table 2.2), and are found as the main protein component in the sperm of some species: salmine from salmon sperm, clupeine from herring sperm, etc. In the nuclei of all somatic cells, including those of salmon and herring, and in the sperm of some species, the main deoxyribonucleoprotein proteins are histones.

X-ray diffraction studies have shown that in nucleoprotamines, both in their extracted form and in intact sperm, the DNA is in double-helical form. A structure suggested for nucleoprotamines is shown in Fig. 7.10. The protamine polypeptide chain in this structure is wound round the double helix in the shallow groove, in such a way that the positively-charged basic side-chain groups lie close to the negatively-charged phosphate groups of the double helix. In clupeine, non-basic side chains have been found to occur in pairs, alternating with tetrapeptide sequences of arginine along the polypeptide chain; in the structure of Fig. 7.10 the non-basic pairs are shown forming folds between each tetrapeptide arginine sequence. The structure and position of the histones in the nucleohistones is not so well established as that of the protamines in the nucleoprotamines. It seems that a proportion of the histone polypeptide chains may be in α-helical form, rather than in the more extended form of the nucleoprotamine polypeptide chain.

One role of protamines and of histones in sperm, and perhaps their only role, is to neutralize the net charge on the DNA molecules and facilitate their close packing in the sperm head. Polyamines play this role in the T even (T_2 T_4 . . .) phage heads. The histones of somatic cells however are of greater potential interest. Do they have a purely structural role as bridges linking DNA molecules together, or bringing the DNA into a super-coiled or folded, condensed form, or are they 'control' molecules, switching genes on and off? In the next chapter we shall consider in detail how genes direct protein synthesis and in Chapter 10 how the genes are themselves controlled, but the problem of gene control is essentially this: mitotic division ensures that the somatic cells of the various tissues of an organism contain the same complement of genes, yet in cells of a given type most of these genes are 'switched-off'. The genes which specify the polypeptide chains of haemoglobin are active in the erythroblast cell where haemoglobin is synthesized, and yet inactive in all the other cells of the body, which possess the identical diploid chromosome complement to that of the erythroblast. In the same way, the genes specifying, for example, the amino-acid sequences of

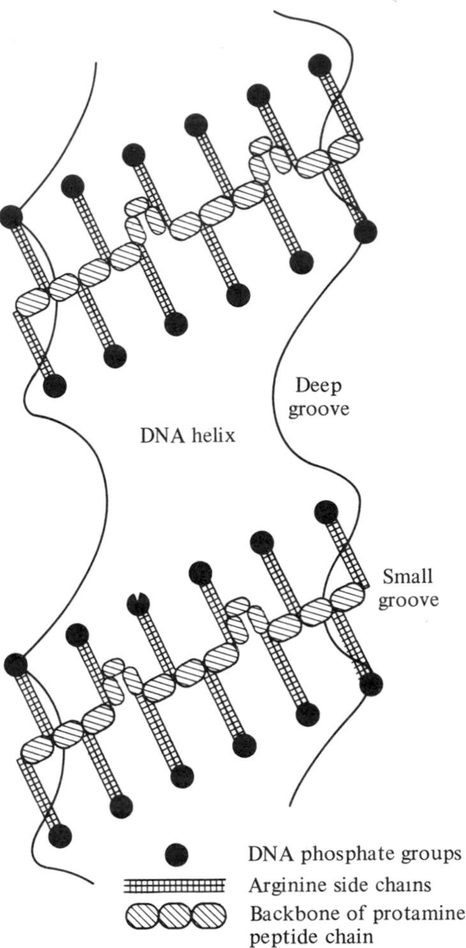

Fig. 7.10 A structure suggested for nucleoprotamines on the basis of X-ray diffraction and protamine sequence studies. The protamine peptide chain is supposed to lie along the shallow groove of the DNA double-helix, with the positively-charged basic ends of the arginine side chains held by electrostatic attraction to the negatively-charged phosphate groups of the DNA. Non-basic side chains are supposed to occur in pairs folded out of the shallow groove (so that only basic side chains lie right in the groove). Each phosphate group binds a positively-charged side chain. (Longer sequences of non-basic side chains could be accommodated in this structure as larger loops, and it is not essential to the structure that the basic side chains should occur in tetrapeptide groups.) (*From* Wilkins, 1956.)

the polypeptide chains of insulin, which is synthesized in the islet cells of the pancreas, must in some sense be lying dormant in the erythroblast and other body cells. The insulin gene must suffer functional modification in order to ensure that it does *not* initiate the synthesis of insulin when it finds itself in any other cell type than the specialized insulin-secreting cells of the pancreas. Mechanisms for keeping a gene 'switched off' must exist, and, further, the switched-off state must be stable in cell heredity. Not only do neurones not secrete detectable amounts of insulin, but neither do the undifferentiated cells from which they are descended. The difference in functional activity between the neurone and the pancreatic cell must originate elsewhere than within the gene for insulin synthesis, since the two contrasted cells must each have derived their gene content by faithful copying from a common ancestral cell. As evidence that whatever modification of gene *function* may occur in course of metazoan differentiation, the nucleus of a specialized cell still retains a full complement of genetic information, some experiments by Gurdon, for example, show that normal adult *Xenopus* frogs may be grown from enucleated eggs, if these eggs are injected with nuclei from differentiated cells taken from the intestinal epithelium of feeding tadpoles.

Evidence that differences in functional activity of somatic cells originate from local environmental agencies is afforded by classical studies of experimental embryology. To take an example, ectodermal tissue destined to develop into skin can be transplanted to a presumptively neural site in the developing embryo. If transplantation is made before the tissue has passed a certain stage in its differentiation, the transplant will be switched along the new developmental pathway characteristic of its adopted environment, and will develop into neural tissue. Such changes, once consolidated, become fixed in cell heredity. Thus if in an adult the cornea of the eye is used to replace a piece of skin removed from the thorax, the transplant preserves its specific features including its transparency. It is just arguable that the phenomena of differentiation might be controlled by a purely cytoplasmic system of determinants, independent of, and in no way affecting, the activities of the genes of the nucleus. The modern view is, however, strongly influenced by the known mechanisms of adaptive enzyme formation in micro-organisms, reviewed in Chapter 10. These studies indicate that a gene can be switched on and off as a result of metabolic events in the cytoplasm, without undergoing structural changes of the type involved in mutation.

The role of the histones in differentiation is not yet clear. Experiments by Allfrey and Mirsky, and by Bonner and co-workers, have shown that removal of part of the histones from isolated thymus cell nuclei, by treatment with trypsin, leads to a large increase in gene expression and corresponding increase in protein synthesis in an *in-vitro* system. But this does not show whether removal of histones is simply a non-specific loosening of the nuclear structure, so that genes become more accessible to enzymes, control molecules, etc., or whether the histones are a group of proteins of many different types, each type specifically blocking a given gene.

Histones can be separated into five fractions: very-lysine-rich (F1), lysine-rich

(F2B), lysine-arginine-rich (F2A2), arginine-glycine-rich (F2A1) and arginine-rich (F3). Some of these fractions, F2A1 for example, appear to contain only a single protein. The same protein is found in this fraction in histones extracted from different tissues of the same animal and there is little variation between species (Fig. 7.11). Other fractions are more heterogeneous. F1, for example,

10
Ac-SER-GLY-ARG-GLY-LYS-GLY-GLY-LYS-GLY-LEU-GLY-LYS-GLY-GLY-ALA-
16 20
LYS(*Ac*)-ARG-HIS-ARG-LYS(*Me*)-VAL-LEU-ARG-ASP-ASN-ILE-GLN-GLY-ILE-
30 40
THR-LYS-PRO-ALA-ILE-ARG-ARG-LEU-ALA-ARG-ARG-GLY-GLY-VAL-LYS-
50 60
ARG-ILE-SER-GLY-LEU-ILE-TYR-GLU-GLU-THR-ARG-GLY-VAL-LEU-LYS-VAL-
70
PHE-LEU-GLU-ASN-VAL-ILE-ARG-ASP-ALA-VAL-THR-TYR-THR-GLU-HIS-ALA-
77 80 90
LYS-ARG-LYS-THR-VAL-THR-ALA-MET-ASP-VAL-VAL-TYR-ALA-LEU-LYS-ARG-
100
GLN-GLY-ARG-THR-LEU-TYR-GLY-PHE-GLY-GLY-OH

Fig. 7.11 Amino acid sequence of calf thymus arginine–glycine-rich histone. In the same histone from pea seedlings VAL in position 60 and LYS in position 77 are replaced by ILE and ARG respectively. Also LYS in position 20 is not methylated. (*From* Wilhelm *et al.*, 1971.)

contains from three to five different proteins, in different species, and these show greater sequence differences between species. On the whole, however, the work on sequence determination for histones seems to be showing that there are a limited number (say, less than 20) different histones in the cells of a given species and that only in a few exceptional cases do histones show tissue specificity. The variation in sequence in different species for the histone of a given fraction or subfraction are usually small. These are not the characteristics expected for proteins playing a direct role in control of protein synthesis in tissue differentiation. The histones are more likely to be playing an essential structural role, in condensing chromatin, or in maintaining a general repression of gene activity, which is specifically lifted when a given gene is activated. Probably control of gene activity is mainly mediated by non-histone nuclear proteins, but the nature of the gene-control molecules in higher organisms is still uncertain.

Also still uncertain is the fascinating question of the detailed molecular structure of chromosomes, and of the chromosomal material (chromatin) in interphase nuclei. The diameter of a DNA double helix is about 2 nm, while the finest chromosomal threads seen in the light microscope are 100–200 nm in diameter. These threads may thus contain a large number of DNA double helices organized into a complex nucleoprotein structure, or a single long DNA double helix coiled in a complex way. In electron micrographs fine chromosomal threads can be seen intermediate in size between the DNA double helix and the threads seen in the light microscope. The constituent threads of lampbrush chromosomes in amphibian oocytes, for example, appear to be about 20 nm in diameter. Almost certainly the DNA threads of chromosomes are coated with proteins, both histones and non-histone proteins. The nucleoprotein threads are then folded

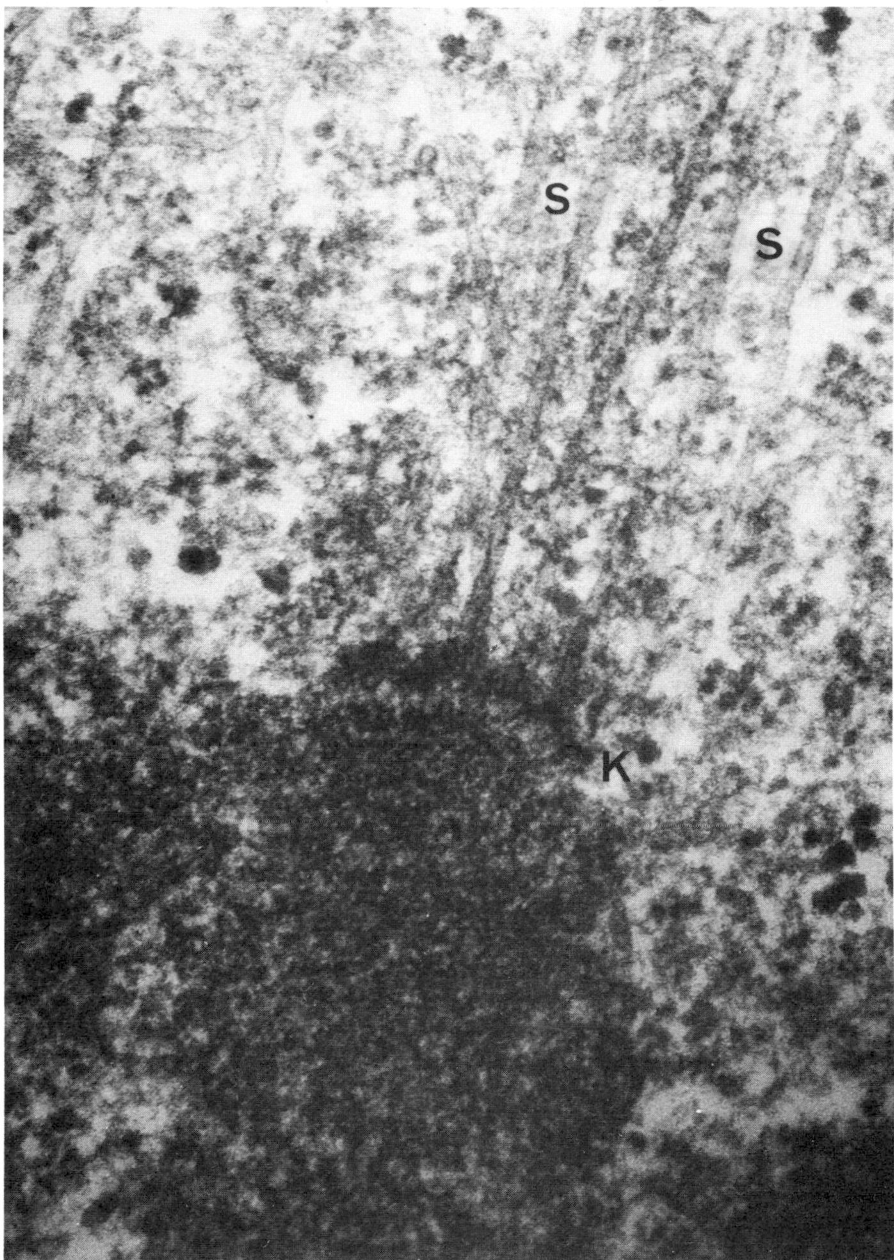

Fig. 7.12 A thin-section micrograph showing the attachment of spindle microtubules (S) to a chromosome in a dividing HeLa cell. K = attachment plate. Magnification × 120 000. (*From* Robbins and Gonatas, 1964.)

TABLE 7.3 Ribosome and virus data.

		Sedimentation coefficient (S)	Mol. wt. ($\times 10^6$)	Diam. (hydrated) (nm)	Diam. (dry) (nm)	Percentage and type of nucleic acid (the RNA is in each case single-stranded)	Mol. wt. of nucleic acid ($\times 10^6$)	Number and types of protein sub-units	Mol. wt. of protein sub-units
RIBOSOMES	*E. coli* ribosomes (2, 6, 7, 8, 9)	70	2·8	30	16	63 (RNA)	1·8	Over 50 different types	3000–15000
RIBOSOMES	Calf liver ribosomes (1, 11, 18)	80	3·7	35	–	50 (RNA)	1·9	–	–
RIBOSOMES	Pea seedling ribosomes (4, 5, 21)	80	4·2	–	–	40 (RNA)	1·7	Heterogeneous	~18000
	Turnip yellow mosaic virus (TYMV) (7, 10, 12, 13, 14)	–	5·7	28	25	37 (RNA)	1·8	180 all identical	20500
	R17 and Qβ viruses (3)	–	3·6	–	20–25	30 (RNA)	1·1	180 all identical and one of a different type	14000
	φX174 virus (16, 17, 19)	–	6·2	–	25	25 (single-stranded DNA)	1·6	192 of four different types	
								60	48000
								60	19000
								60	5000
								12	36000

	Sedimentation coefficient (S)	Mol. wt. ($\times 10^6$)	Diam. (hydrated) (nm)	Diam. (dry) (nm)	Percentage and type of nucleic acid (the RNA is in each case single-stranded)	Mol. wt. of nucleic acid ($\times 10^6$)	Number and types of protein sub-units	Mol. wt. of protein sub-units
Poliomyelitis virus (15)	–	6·7	30	–	30 (RNA)	2·0		–
Polyoma and SV40 viruses (20)	–	17·3	–	40	15 (double-stranded DNA)	2·5	920 (420 each of two types and about 80 of a third type)	all three types around 16000

1. HALL, B. D. and DOTY, P. (1959) *J. mol. Biol.* **1**, 111.
2. MOORE, P. B. *et al.* (1968) *J. mol. Biol.* **31**, 441.
3. LODISH, H. F. (1968) *Prog. Biophys.* **18**, 287.
4. T'SO, P. O. P., BONNER, J. and VINOGRAD, J. (1958) *Biochim. biophys. Acta* **30**, 570.
5. T'SO, P. O. P. (1958) in *Microsomal Particles and Protein Synthesis*, ed. R. B. Roberts. London, Pergamon.
6. TISSIERES, A., WATSON, J. D., SCHLESSINGER, D. and HOLLINGWORTH, B. R. (1959) *J. mol. Biol.* **1**, 221.
7. ZUBAY, G. and WILKINS, M. H. F. (1960) *J. mol. Biol.* **2**, 105.
8. KURLAND, C. G. (1960) *J. mol. Biol.* **2**, 83.
9. HUXLEY, H. E. and ZUBAY, G. (1960) *J. mol. Biol.* **2**, 10.
10. KLUG, A. and FINCH, J. T. (1960) *J. mol. Biol.* **2**, 201.
11. DOTY, P. (1962) *Biochem. Soc. Symposia* **21**, 8.
12. HUXLEY, H. E. and ZUBAY, G. (1960) *J. mol. Biol.* **2**, 189.
13. HARRIS, J. I. and HINDLEY, J. (1961) *J. mol. Biol.* **3**, 117.
14. DE ROSIER, D. J. and HASELKORN, R. (1966) *J. mol. Biol.* **19**, 52.
15. FINCH, J. T. and KLUG, A. (1959) *Nature, Lond.* **183**, 1709.
16. SINSHEIMER, R. L. (1959) *J. mol. Biol.* **1**, 37.
17. KLUG, A. and CASPAR, D. L. D. (1960) *Adv. Virus Research* **7**, 225.
18. CURRY, J. B. and HERSH, R. T. (1961) *Biochem. biophys. Res. Comm.* **6**, 415.
19. CARUSI, E. A. and SINSHEIMER, R. L. (1961) *Fed. Proc.* **20**, 438.
20. KOCK, M. A. and ANDERER, F. A. *et al.* (1967) *Virogology* **32**, 503 and 523 (two papers).
21. SETTERFIELD, G., NEELIN, J. M., NEELIN, E. M. and BAYLEY, S. T. (1960) *J. mol. Biol.* **2**, 416.

or coiled in some way, possibly differently coiled in the parts of the nuclear material where genes are switched on or switched off. The F1 histone fraction appears less tightly bound to DNA than the other fractions, and it is thought that the histones of the other fractions may lie in the narrow groove of DNA, like protamines (Fig. 7.10), forming, with the DNA, a super-coiled nucleoprotein thread 10–20 nm in diameter. The F1 histones might then cross-link these threads at various points to bring about further condensation. The problem of visualizing these structures is that in thin-section electron micrographs, even with serial sections, it is not possible to follow the run of the nucleoprotein threads, and if methods are used to expand the structure, like swelling chromosomes in hypotonic solutions, or treatment with trypsin, to resolve the threads, then the compact chromosomal structure is lost. Some features can be seen in thin sections, like the attachment plate by which the centromere is drawn to the poles of the cell by spindle microtubules (Fig. 7.12). It is possible that the nucleoprotein threads of chromosomal material are condensed into an irregular tangle in which the exact run of the threads is of no significance. The important questions are: exactly how do histones or other proteins interact with the DNA to bring about this condensation, how is the interaction modified in active and inactive genetic segments, how long are the individual strands of DNA, are they circular, are they attached to the inner nuclear membrane, how do they duplicate during interphase, and finally, of course, how are all these features controlled? We return to discussion of the structure of the active part of chromosomal material in Chapter 10.

Ribosomes

We have noted in Chapter 5 that ribosomes are small particles found in the cell cytoplasm, free or often attached to internal membranes, which are the sites of protein synthesis. The details of the protein assembly process are to be discussed in the next chapter, and we set the scene for that discussion by considering in this section the structure of ribosomes. In thin-section micrographs ribosomes appear about 15 nm in diameter, but when ribosomes are isolated and studied by physico-chemical means (for example, rate of sedimentation in an ultra-centrifuge) their hydrated diameter is found to be rather larger than this. There are two main classes of ribosomes: those from the cytoplasmic matrix of the cells of plants and higher animals which have a sedimentation coefficient around 80 *S*, an estimated molecular weight of about 4×10^6 and a hydrated diameter of about 35 nm, and ribosomes from bacteria and from cytoplasmic organelles (mitochondria and chloroplasts) which all have sedimentation coefficients around 70 *S*, molecular weights around 3×10^6, and hydrated diameters of about 30 nm. (The sedimentation coefficient is related to molecular weight, a higher coefficient indicating a higher molecular weight, but sedimentation rate is also affected by shape and hydration.)

Ribosomes contain RNA and protein. Both classes of ribosome contain about the same amount of RNA (total molecular weight $\sim 1{\cdot}8 \times 10^6$, about

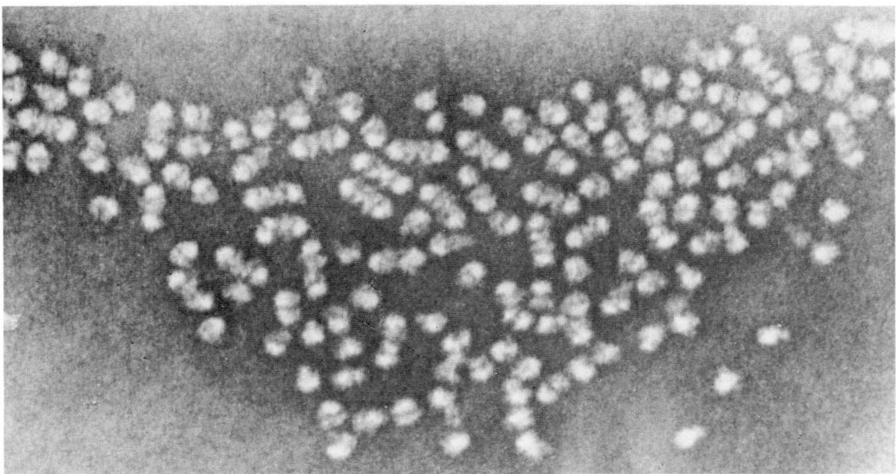

Fig. 7.13 A mixture of 70 *S* and pairs of 70 *S* *E. coli* ribosomes seen in the electron microscope by negative contrast (compare virus pictures of Chapter 3). The 70 *S* ribosomes are composed of one large ($\frac{2}{3}$) and one small ($\frac{1}{3}$) unit. The dimers are seen to be made up of two 70 *S* ribosomes making contact at the centre of the small ($\frac{1}{3}$) unit to form a linear aggregate with the appearance: large–small–small–large. (*From* Huxley and Zubay, 1960.)

5 000 nucleotides). The 80 *S* ribosomes thus contain a higher proportion of protein (60% protein, 40% RNA) than the 70 *S* ribosomes (40% protein, 60% RNA). In size and composition ribosomes are very comparable with small viruses like TYMV (Table 7.3) but their detailed structure seems to be very different. In the case of the small viruses the RNA forms a core and the protein a surrounding shell (Fig. 3.4). In ribosomes the RNA seems to be entwined with the protein in a looser, more hydrated, structure. The protein shells of viruses are made up of one, or a limited number, of types of protein sub-unit. Ribosomes contain over fifty *different* types of protein sub-unit.

In negative-contrast preparations a 70 or 80 *S* ribosome can be seen to be formed of two unequal parts (Fig. 7.13) and this is confirmed by study of ribosomes in solutions of varying Mg^{++} concentrations. At a Mg^{++} concentration of $\sim 10^{-3}$ molar the ribosomes are in 80 or 70 *S* form. At higher Mg^{++} concentration they readily form the paired structures seen in Fig. 7.13, and at lower Mg concentrations readily break up into units 2/3 and 1/3 of the molecular weight of the 80 or 70 *S* particle. The component units each show the protein to nucleic acid ratio of the 80 or 70 *S* particle from which they are derived, so that both protein and nucleic acid split in roughly 2/3 : 1/3 amounts. The component units exist in dynamic equilibrium in the living cell, and this proves important during the process of protein assembly. For *E. coli* this equilibrium may be written:

$$70\ S\ \text{ribosome} \rightleftharpoons 50\ S\ (2/3\ \text{unit}) + 30\ S\ (1/3\ \text{unit}).$$

The main RNA strands in the 50 and 30 *S* units are about 3400 and 1700 nucleotides long, and the 50 *S* sub-unit contains a further short strand 120 nucleotides long, called 5 *S* RNA (the sedimentation coefficients for these three component strands, in the case of *E. coli*, are 23, 16 and 5 *S*). These RNA molecules are all present in the ribosome in single-stranded form. In eukaryocytes the two larger components at least are synthesized first as a single strand, which is then enzymically cut to yield the final ribosomal RNA strands.

In view of the situation for small spherical viruses, where a limited number of different protein sub-units are used to form the protein shell, and where, in fact, the symmetry and nature of the protein arrangement arises from geometrical principles based on sub-unit identity or near-identity, it was a great surprise to all involved in study of ribosome structure to find that the 30 *S* ribosome sub-units contain over 20 different types of protein sub-unit, and the 50 *S* ribosome sub-units at least a further 30 different types, each ribosome probably containing one molecule each of these different proteins. These results are established by fractionation of solubilized ribosomal proteins, and also by some experiments on streptomycin-sensitive and streptomycin-resistant strains of *E. coli* to be described in the next chapter.

It seems that the difference between viruses and ribosomes reflects, on the one hand, factors of genetic economy, allowing a viral genome of only three genes in some cases to code and control the complete virus infectious cycle (Chapter 10) and reflecting for ribosomes, on the other hand, complexities of protein assembly which are not yet understood. It is tantalizing, and challenging, in view of the important and rather central role of ribosomes in molecular biology, that the heterogeneous protein composition of these ribonucleoprotein particles has so far prevented their being crystallized, as small viruses are crystallized, in a form suitable for detail study by the X-ray diffraction technique. It may be however that it is wrong to think of ribosomes as having a formed structure at all. We have seen that they are highly hydrated particles, that is of wet diameter much greater than their dry diameter. Possibly ribosomes consist of enzymes and factors required for different steps of protein assembly, strung along an RNA chain folded in a rather random constantly-changing tangle. The ribosome 'tangle' cannot be too random during protein assembly for at that time the ribosome is interacting with other RNA molecules. These interactions are to be discussed in Chapter 8.

For proteins which accumulate, after synthesis, in the internal sacs of rough endoplasmic reticulum (Chapter 5) the ribosomes are attached to membranes by the 50 *S* sub-unit at the opposite end to the 50–30 *S* attachment area. Presumably the ribosomes penetrate into the membrane, and the growing polypeptide chain comes off the ribosome, in this situation, into the lumen of the reticulum.

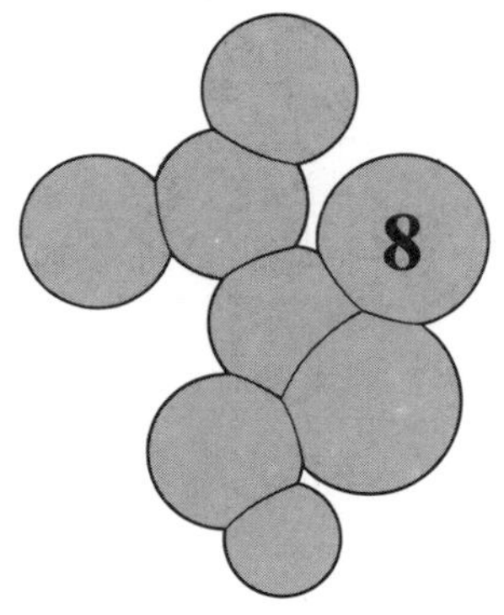

8 Protein synthesis

We have discussed in earlier chapters the important functional and structural roles of proteins in cells, and seen that the properties of a given protein are determined by the variety and sequence of the amino-acid building blocks in its component polypeptide chains. Protein synthesis, then, is one of the crucial activities of living matter, and requires the assembly of amino acids in specific sequences.

Early theories of protein synthesis were of two kinds. One possibility was that, in suitable conditions, the reactions by which enzymes break down proteins might be reversed, and that proteins might be synthesized by these enzymes, either from the individual amino acids, or from small peptides. This view was supported by the work of biochemists who were able to join small numbers of amino acids together to form short polypeptide chains by ordinary chemical condensations. It was thus thought that pieces of protein of increasing complexity could be put together to form longer polypeptide chains, and that their properties might then be changed enzymically by altering or adding amino acids until the required protein was made.

Although these experiments, as it turned out, were not very relevant to protein synthesis *in vivo*, they did indicate that the formation of bonds between amino acids requires a supply of chemical energy for the reaction. About 4 kcal/mole are needed to join two amino acids, and although this figure is reduced if peptides are joined, it still remains a considerable barrier to polypeptide synthesis by direct condensation. As might be expected from this, it was soon discovered that protein synthesis was closely coupled to the energy-producing system in the cell, and that it would not, for instance, occur in the presence of poisons such as cyanide, which attack this system. Later it was found that ATP and other co-factors were essential for synthesis and that indeed ATP is intimately linked with the first stages in protein synthesis.

Theories of a second type were designed to provide some method by which the order, or sequence, of the amino acids in the protein chain might be determined during synthesis. The original 'template' theory of Pauling and Haurowitz postulated that each protein had, possibly in the chromosomes, a template or copy of itself as an extended polypeptide chain. This template was supposed to provide a series of sites which, by a process like crystallization, would allow amino acids to condense out of the surrounding medium on to the corresponding amino acid in the template. Each amino acid would then be situated correctly adjacent to its neighbours ready to be joined up by an enzyme to give a polypeptide chain identical in every way to the template. This would then be peeled off, possibly by another enzyme, and allowed to fold to give a specific protein.

This template hypothesis was conceived at a time when most biologists held that only proteins among cellular constituents had sufficient versatility in their properties to act as genes and therefore to control the extreme complexity of a living cell. At the time, no evidence for the existence of such a template or of the necessary enzymes was found, and it has become increasingly clear that the shape of the side-chain groups of proteins cannot provide the necessary sites to which identical amino acids could be bound. This objection could in principle be overcome by a further refinement of the theory. Nucleic acids or other molecules might take up a configuration complementary to that of the template, like a mould. Then the mould could leave the template and so provide a string of sites for new amino acids to bind on to, the new polypeptide chain being identical with the original template.

All these speculations have eventually had to be abandoned because new methods and new techniques have provided the facts on which a much more satisfactory picture of the processes of protein synthesis has now been constructed.

Modern techniques for the study of protein synthesis

Some of the techniques which have led to these recent advances have been described in general terms in Chapter 1, but it is worth while considering some of them in more detail in this particular context. The most important of these has been the use of radioactive isotopes. Amino acids labelled with S^{35}, C^{14} or H^3 may be used as precursors in protein synthesis and can, with care, be made to show where and when the labelled amino acid is incorporated into protein, and to show whether it is associated with other substances, or structures, before and during its final incorporation. It is necessary to make sure that the labelled amino acid is not changed in the cell into some other compound which may then attach itself to proteins. To establish that protein synthesis has occurred in any experiment, it must be shown that the amino acid is incorporated in the protein with a proper peptide linkage not only at an end, but also in the middle of the polypeptide chains.

Briefly, the method is either to feed an animal with a solution of radioactive precursors or to place the cells (or their components) in a solution containing the radioactive amino acid for a known length of time. The cells are next removed from the animal, or the solution, and washed clean of surplus radioactivity. It is then necessary to isolate the different cellular components and locate the labelled amino acids by measuring the radioactivity of each fraction with a suitable meter.

Parallel to advances in the use of radioactive tracers, there have been great refinements in the techniques of separating cellular components and in their chemical characterization (see Chapter 1). The most important development has been in techniques of centrifugation; not only has it been possible to make simple machines which are able to exert a very large centrifugal force (up to 250000 *g* is common, and twice that possible), but the design of a special instrument called the analytical centrifuge has enabled the biochemist to study and make measurements on large molecules and biological structures as they are being centrifuged. This is achieved by an arrangement of lenses and mirrors which illuminates the centrifuge vessel as it rotates and throws an enlarged image of the solution on to a screen, or photographic plate, once every revolution. This image can be arranged to show either the concentration of ultraviolet absorbing substances like nucleic acids, or the concentration of all material. In this way the movement of fractions with different sedimentation coefficients can be studied as they are centrifuged. Particles moving at different rates can be identified and characterized.

In protein synthesis, the most important fractions have proved to be two which are left behind after all the nuclei, mitochondria and outer cell membranes have been spun down from the solution at 10000–20000 *g*. If the supernatant solution from this first centrifugation is further centrifuged at 100000 *g*, a small gelatinous pellet forms at the botton of the tube, which contains the ribosomes and the remains of the endoplasmic reticulum. The clear supernatant after this second centrifugation still contains, as we shall see, many other substances important in protein synthesis.

In addition to these two main techniques, there has been a more subtle advance which stems from our increasing knowledge of the chemical environment within cells. It has proved useful to the biochemist to make the reactions he studies increasingly independent of any uncertainties due to their biological environment. From this point of view, a single tissue like a liver is better than a whole animal, a clone of cells is better than a liver, and a pure preparation of nuclei or mitochondria better than either.

In early biochemical work it was discovered that slices of, for example, liver, could be made to carry out, for a period, many of the metabolic functions of a whole liver. More recently, incubation solutions have been made which will allow cellular components to retain part, at least, of their metabolic activity after the cells have been broken up and the various fractions separated. As the biological system was simplified, it was found necessary to be more and more particular about the various factors which had to be added to the incubation medium before the biochemical reactions would occur. Broadly, our greatly increased knowledge of protein synthesis has been brought about by the isolation of those cellular components essential to protein synthesis and the identification of the ions, ATP, enzymes and so forth which are needed before amino acid incorporation will occur *in vitro*.

The cell components essential for protein synthesis

In the period from 1935 to 1941, it became possible to investigate the distribution of nucleic acids in cells of various kinds. This was primarily due to two groups of workers. Caspersson and his colleagues in Stockholm were able to adapt the microscope to the measurement of absorption at various wavelengths in ultraviolet light, and to detect in this way the distribution of nucleic acids, in the chromosomes, nucleoli and cytoplasm of cells. At about the same time Brachet in Brussels was using staining techniques, in conjunction with specific enzymes, to detect and identify nucleic acids in his preparations. One of the most interesting and significant conclusions of this early work was that cells which were actively making protein, either during rapid growth or in secreting enzymes, all had a greatly increased amount of cytoplasmic nucleic acid (RNA) compared to less active cells. It was not, however, until much later that evidence began to appear which indicated that RNA was not only associated with protein synthesis, but was essential to it.

This was first shown by the work of Gale in Cambridge, and Zamecnik and his colleagues in Boston. The latter workers, in a notable series of papers, began to reveal the several steps in protein synthesis in which RNA is involved. Their first important experiment was one in which rats were injected with radioactive leucine, and the lobes of the liver were removed one by one after 5, 12 and 20 minutes. The cells from each lobe were homogenized, and the nuclei, outer cell membranes, and mitochondria centrifuged down at 18 000 g. The supernatant was spun down at 100 000 g to give a microsome fraction, which contained most of the RNA of the cell. This pellet was further fractionated by treatment with deoxycholate, which solubilizes the microsomal membranes, to give a ribosome fraction and a deoxycholate-soluble fraction which contained newly formed protein. They found that in the lobe excised five minutes after the labelled leucine was injected, most of the radioactivity was in the ribosomes, but later the soluble protein became increasingly labelled. More important, the amount of label in the ribosomes decreased in the later excisions. This meant that amino acids must first be incorporated into protein on the ribosomes before they could appear in the soluble proteins – so that the ribosomes must be implicated in protein synthesis.

This work was then markedly extended by altering the method of the experiment. The liver cells were first broken up and separated into their various fractions, and then some or all of these fractions were incubated in the presence of the labelled precursor. This is a much more precise method since it is possible to study under controlled conditions the effects of adding or leaving out a single fraction. It was found that the essential cellular components needed for incorporation of labelled amino acids into proteins were:

(i) the microsomal fraction;
(ii) the supernatant fraction;
(iii) ATP;
(iv) a biochemical system for making more ATP.

If any one of these were omitted no incorporation occurred, but the addition of mitochondria or other substances did not improve the rate of uptake. It was possible to simplify this list even further when certain tumour cells were used, which, unlike cells from liver, do not have an internal membranous system, and contain smaller amounts of the enzymes which break down ATP. In extracts from these cells, the only components that were essential were:

(i) the ribosomes;
(ii) the supernatant fraction;
(iii) ATP.

The attachment of amino acids to transfer-RNA molecules

Following the experiments described in the previous section, Zamecnik and his colleagues made further studies on the supernatant fraction and soon found that it contained two major components: the first was an enzyme fraction containing a number of different enzymes, each able to form a specific complex with one type of amino acid. The second was an RNA of low molecular weight, differing from the RNA of the ribosomes in base composition (listed as soluble RNA in Table 7.2).

These enzymes have several interesting properties: they are able to catalyse a reaction between ATP and an amino acid, in which these molecules are joined together to form an adenyl–amino-acid compound, and in the process part of the chemical energy stored in ATP is transferred to this new compound. These enzymes are therefore termed 'activating enzymes'. Many of the activating enzymes have been obtained in a highly purified state. There are twenty different enzymes of this type, each specific to a particular amino acid.

The next stage in protein synthesis is shown by the following experiment. If a mixture containing:

(i) activating enzymes;
(ii) ATP;
(iii) the low-molecular-weight RNA fraction described above;
(iv) amino acids

are incubated together, the amino acids become attached to the RNA by a covalent bond. This attachment is dependent on the presence of the appropriate activating enzyme, which thus takes part in two steps: first, the formation of the activated amino acid and then its subsequent transfer to the low molecular weight RNA.

A further experiment can now be done: this RNA of low molecular weight, with labelled amino acids attached, can be isolated from the incubation mixture with phenol, and if these purified RNA amino-acid compounds are incubated with

(i) ribosomes;
(ii) ATP;
(iii) guanosine triphosphate (GTP);
(iv) certain other factors found in the soluble fraction

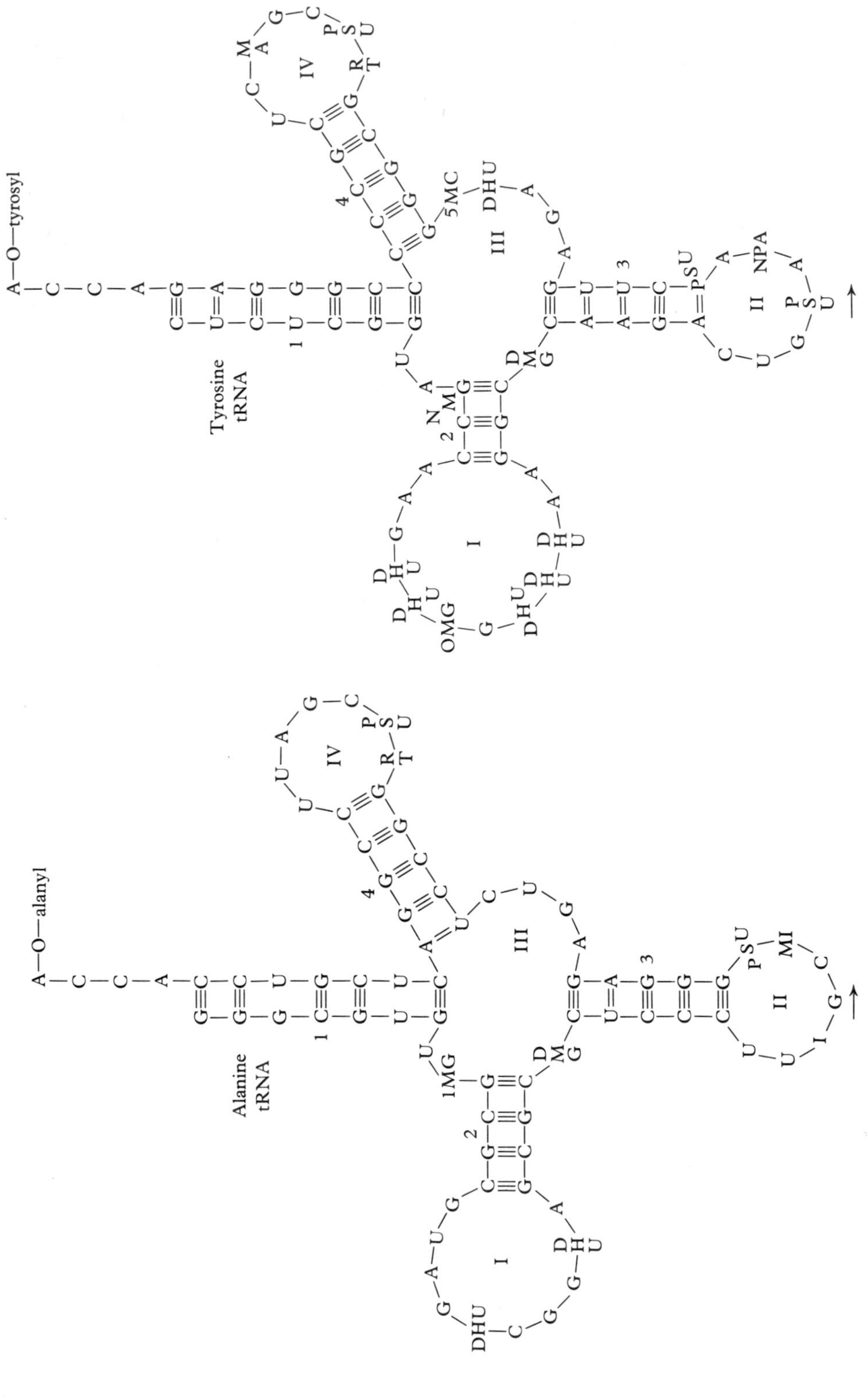

Tyrosine tRNA

A—O—tyrosyl

Alanine tRNA

A—O—alanyl

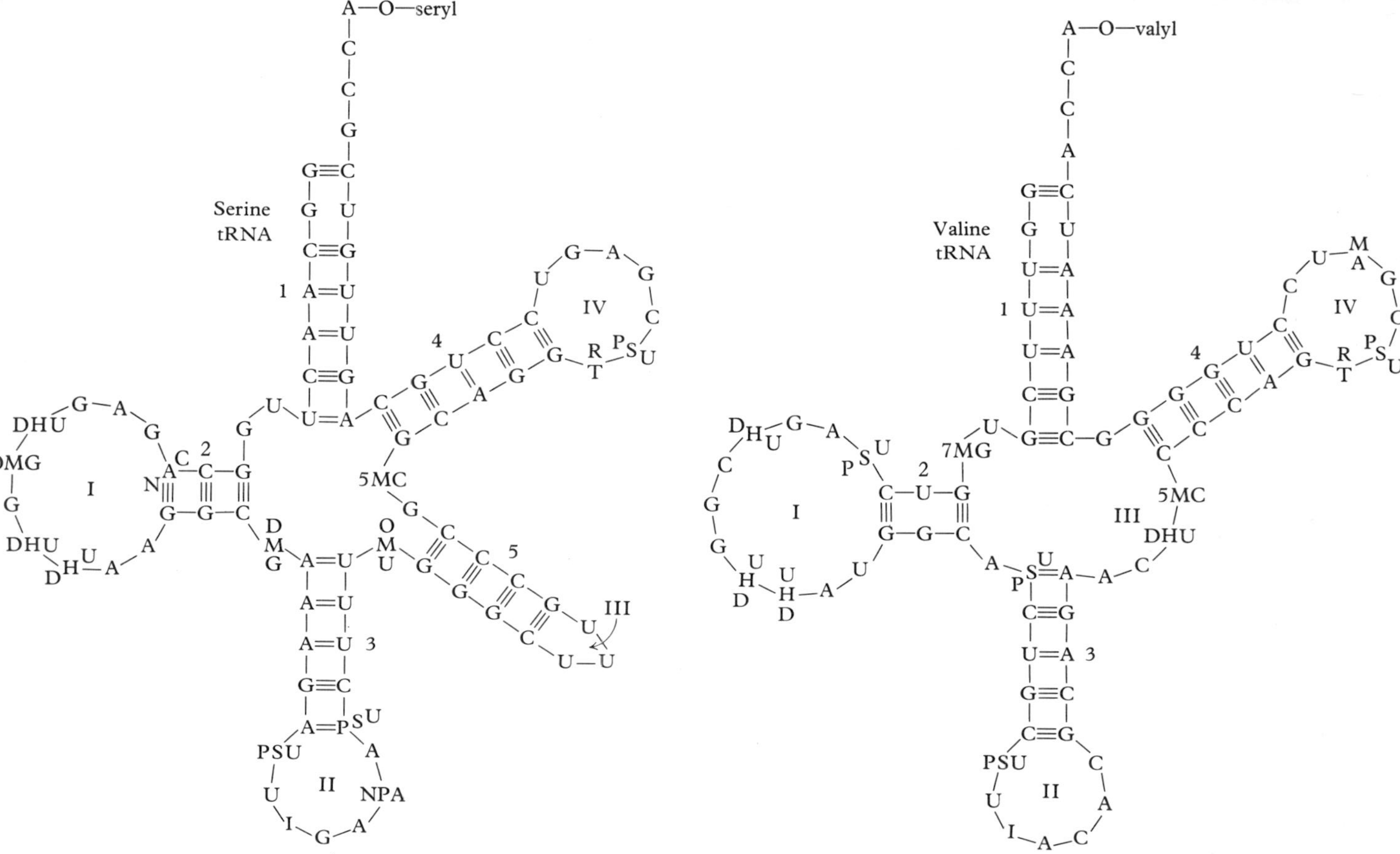

Fig. 8.1 Yeast transfer RNA sequences arranged in 'clover-leaf' pattern. DHU = 4,5-dihydrouridine, PSU = pseudouridine, I = inosine, RT = ribothymidine, 1MG = 1-methylguanosine, DMG = N2-dimethylguanosine, MI = methylinosine, MA = 1-methyladenosine, 5MC = 5-methylcytosine, NPA = N6-isopentenyladenosine, OMG = 2′-O-methylguanosine, NMG = N2-methylguanosine, NAC = N4-acetylcytidine, 7MG = N7-methylguanosine. (*From* Dayhoff, 1969.)

it is found that the amino acids are transferred very rapidly to the ribosomes and there become incorporated irreversibly into protein. Because of its special function this RNA is now generally called *transfer*-RNA, although it is also sometimes referred to as soluble-RNA, or as acceptor-RNA.

Like the activating enzymes, the main class of transfer RNAs (tRNAs) are of twenty different kinds, each specifically accepting one amino acid. They are single-chain molecules of about 80 bases, which in addition to the bases of Table 7.1 include also a number of methylated derivatives and other substituted bases. (In synthesis of the tRNAs, precursor chains are made first with the 'normal' bases of Table 7.1 and the bases are later modified by specific enzymes.) The presence of unusual bases in the tRNAs has facilitated sequence analysis (the methods used are similar in principle to those used for polypeptide chains, involving sequence studies of fragments produced by partial digestion with ribonucleases). The first tRNA for which a complete sequence was determined was that of yeast alanine tRNA by Holley and his co-workers. Other yeast tRNA have now been studied, and some of these are shown in Fig. 8.1. They are shown in a 'clover leaf' pattern but actually this arrangement is still speculative. When Holley *et al.* determined the first tRNA sequence, for alanine tRNA, they suggested from study of the base-pairing possibilities, maximized according to the principles illustrated in Fig. 7.9, that this chain might take up such a clover-leaf, or alternatively they suggested a hair-pin structure, or perhaps a double hair-pin, all of these allowing about the same amount of base pairing. As more sequences have been determined the clover-leaf arrangement has gained in popularity since sequences for other tRNAs can be represented in a similar way with some variation in the size of the loops, and in some cases (as for serine tRNA in Fig. 8.1) the insertion of an additional arm. We probably have to wait, however, for full X-ray diffraction analysis of tRNA crystals to see whether this clover-leaf model is in fact correct, and also to learn how these double-helical regions and loops are arranged in three-dimensions. In comparing Fig. 8.1 and Fig. 7.9 note that there has been a change in thinking since Fresco *et al.* first tackled this problem around 1960. It is no longer considered probable that small loops of one or two bases will protrude out from a double-helical region. The tendency for bases to stack on top of one another (which, as we noted in Chapter 7, contributes as much as hydrogen bonding between bases to the stability of the double helix) is considered now to be strong enough to hold an occasional G–U or U–U pair within a double helix made up predominantly of G–C and A–U pairs.

The end of the tRNA molecule to which the amino acid is attached always has a terminal triplet –C–C–A. This triplet is essential to tRNA function, but carries no specificity. In some other part of the tRNA molecule specific groups must be recognized by the enzymes which carry the activated amino acids, so that these enzymes bind to, and transfer the amino acid to the correct tRNA. (The activating enzymes must thus carry sites specific for their amino acid and for their tRNA.) The part of the tRNA which is involved in this binding to the activating enzyme is not known. In later stages of protein synthesis, as the amino acid is incorporated into a growing polypeptide chain, another part of the tRNA carries specific recognition groups which ensure that the correct amino acid is inserted at each point. The details of this process we shall come to discuss in later sections. We shall see further that in addition to the main class of tRNA molecules, there are separate tRNAs involved in the initiation of polypeptide synthesis.

The incorporation of amino acids into polypeptide chains

We now consider what happens to the transfer-RNA with its attached amino acid at the ribosome. We have seen that there is good evidence that in the ribosomes the amino acids become joined together to form polypeptide chains, which are then released into the cytoplasm, or to the intrareticular spaces of the endoplasmic reticulum. This assembly process was first studied in detail by Dintzis in a series of elegant experiments published in 1961.

Dintzis treated rabbits with phenylhydrazine, which causes an acute anaemia due to the destruction of red cells. The animal reacts to this by discharging immature red cells (reticulocytes) into the circulation while they are still actively making haemoglobin, but very little other protein. Since the peptide fragments of haemoglobin can be easily identified by the finger-printing technique (see Chapter 9), it is possible to study the way in which these are made. In Dintzis' experiment, reticulocytes from a treated animal were suspended in a solution containing labelled leucine. Protein synthesis in the reticulocytes was stopped at different times from 3 to 60 minutes later by cooling the suspension in ice-cold water. The ribosomes were next separated from the soluble proteins by centrifugation, and the haemoglobin was removed from each of these fractions, digested with trypsin as described in Chapter 9, and the various peptides separated and identified by finger-printing. The distribution of labelled leucine was then measured in the various peptides.

The following hypothesis was suggested to fit the results which were revealed: the polypeptide chains of haemoglobin are made serially from one end by laying down in the ribosomes the amino acids one after another. As soon as a chain is complete the whole of it is shed into the cytoplasm. If this hypothesis is correct and if we consider only the completed protein, labelling for short times followed by immediately stopping protein synthesis should result in only those parts being labelled which are near the end of the chain and which were synthesized just before the molecule was completed (Fig. 8.2*a*). Longer times of labelling will cause all fragments of the completed polypeptide chain to be more or less equally labelled (Fig. 8.2*b*).

Let us now consider the partially formed protein still attached to the ribosomes at the time synthesis was arrested. Since in a population of ribosomes there will be bits of haemoglobin at all stages of synthesis, short periods of labelling will only give enough time for one or two fragments to be made; each peptide, whatever its position in the molecule, will have an equal chance of being labelled (Fig. 8.2*a*). For long times of labelling, many whole proteins will have

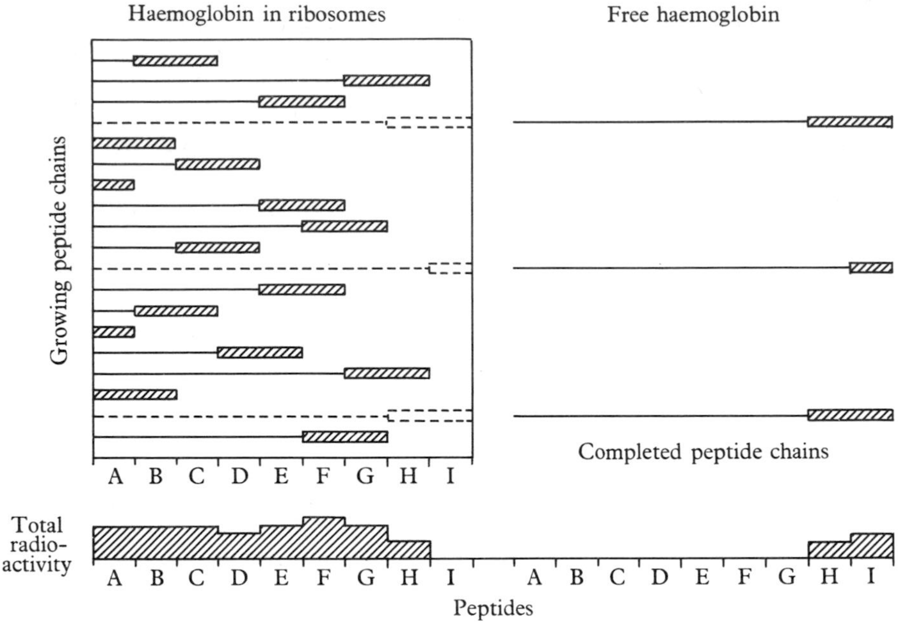

(a) Time of labelling less than time of synthesis

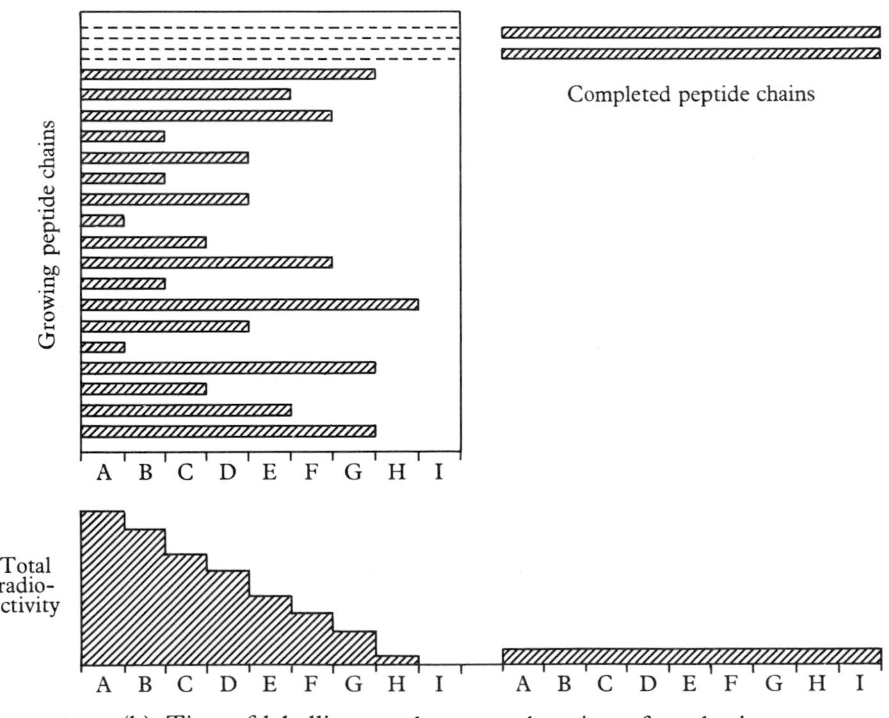

(b) Time of labelling much greater than time of synthesis

been completed and have left the ribosomes, but there will also be, at all stages of synthesis, partially formed molecules still left on the ribosomes which will all be fully labelled. As can be seen from Fig. 8.2*b*, this will result in a sample containing many more labelled peptides formed early, rather than late, in synthesis, and we should find a gradient of labelling, with its peak at the first peptide and falling to a low value for the last. This pattern is exactly the opposite to that found for complete haemoglobin in solution after *short* times of labelling. The experimental result for the α-chain (illustrated in Fig. 8.3*a* and *b*) gives a clearer

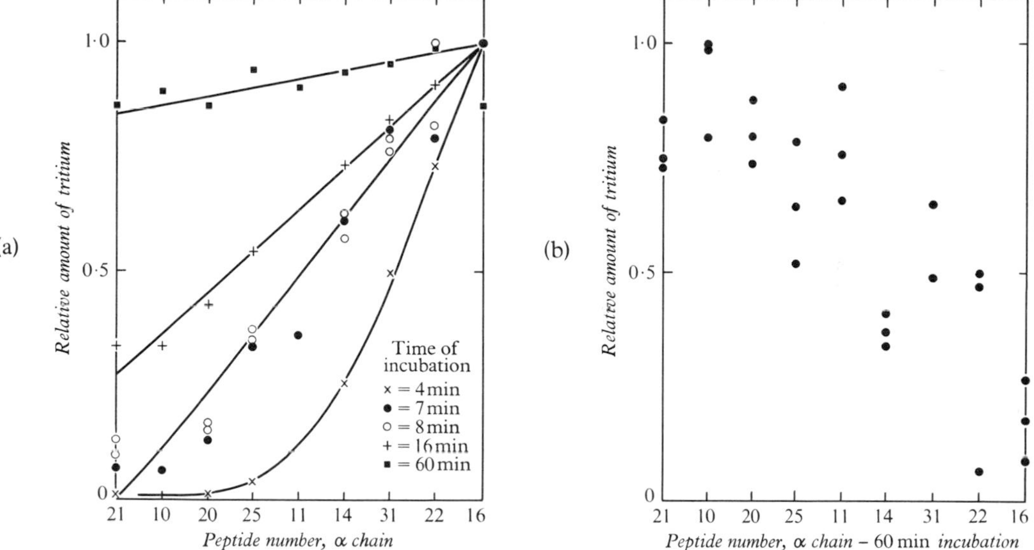

Fig. 8.3 Dintzis' experimental results for the α-chain of haemoglobin: (*a*) completed haemoglobin molecules in solution, and (*b*) incomplete haemoglobin on the ribosomes. (*From* Dintzis, 1961.)

picture than that for the β-chain. It will be seen (Fig. 8.3*a*) that for the completed protein at short times of labelling a fairly clear order of increasing incorporation can be made for the fragments, starting at 21 and 10 and ending with 16. This is the opposite of that found at long labelling times for incomplete protein in the ribosomes (Fig. 8.3*b*).

Fig. 8.2 Diagrams illustrating Dintzis' experiments: (*a*) time of labelling less than time of synthesis, and (*b*) time of labelling much greater than time of synthesis. The two rectangles on the left include a number of growing polypeptide chains. The thick and thin segments of each line represent labelled and unlabelled parts of the peptides respectively. All completed chains are found on the right. The histograms below each section show the total radioactivity found in the peptides A to I for the uncompleted chains on the left and the completed chains on the right. In the lower half of the figure it is assumed that very many more totally labelled chains have been made, compared to any partially labelled but complete chains which appear initially.

From these experiments Dintzis was able to arrange the fragments in the order in which he thought they were made, although at the time, the amino-acid sequence of haemoglobin had not yet been determined and the order of the peptides in the primary structure was unknown. He did not, therefore, have this independent check on the validity of his conclusions. It inspires confidence in these important results that the haemoglobin sequence is now established (Table 9.1), and the order of the peptides agrees with that assigned by Dintzis.

Further experiments with labelled amino acids for a variety of proteins synthesized in cells, or *in vitro*, make it possible now to generalize the results of Dintzis' early work on haemoglobin: proteins are assembled on ribosomes, one amino acid at a time, starting from the N-terminal end. The next major question that we have to consider is this: what is the connection between this mechanism for protein synthesis and the genetic material, in other words, how does the DNA in the chromosomes control the *order* in which the amino acids are laid down in the ribosomes?

Evidence for a 'messenger' carrying genetic information to the sites of protein synthesis

Since in higher cells protein synthesis occurs mainly in the cytoplasm, and the genetic material is in the nucleus, there are general grounds for believing in some intermediary, which can pass information from the genes to the protein-making parts of the cell. Further, there is evidence, from studies on cells incubated with radioactive precursors of RNA, that these are first incorporated into the RNA of the nucleus and then later appear in the cytoplasm. These observations, although also open to other interpretation, suggest that RNA may be synthesized in the nucleus and later become incorporated into the ribosomes. Indeed, the most favoured candidate for the intermediary between nucleus and cytoplasm was at one time the ribosome itself.

It was supposed, as an attractively simple hypothesis, that ribosomal RNA was synthesized in the nucleus with the same base sequence as one, or both, strands of the DNA, and that in the ribosome this RNA then acted as a template for the assembly of transfer-RNA molecules with their attached amino acids. This hypothesis required that a given ribosome should only make one type of protein, or one set of proteins. For a cell to later stop making a particular protein, ribosomes would have to be either destroyed or inactivated. It was found, however, that ribosomes were long-lived in bacteria although these cells can show rapid and large changes in the kinds of protein they make. Perhaps more important, the base ratio of ribosomal RNA was found to be different from that of the bulk of the DNA, and did not change very much in different species of bacteria in which the DNA base ratios varied considerably. If it was assumed that DNA passed its genetic information to the protein by controlling the kind of RNA made, then ribosomal RNA did not appear to fulfil the necessary requirements. Further, a correlation might have been expected between the size of the protein molecule made, and the amount of RNA in the ribosome, yet the analytical centrifuge shows that ribosomes have a very uniform RNA content.

None of these criticisms of the simple ribosome hypothesis was decisive, but they did cause several research workers to investigate an alternative scheme which came to be called the 'messenger-RNA' hypothesis. In order to understand it better, we must now consider several new facts.

The first comes from the genetic work of Monod, Jacob and Wollman in Paris (of which some aspects have already been discussed in Chapter 6). This provided the first evidence for the existence of a short-lived intermediate between the genes and the protein-synthesis mechanism. The experimental material was again the bacterium *E. coli* and the particular genetic locus studied was that concerned with the production of β-galactosidase, an enzyme which splits lactose into glucose and galactose (Chapter 10). This locus can be conveniently studied by mating experiments.

Certain strains of bacteria will mate, when one strain acting as the male will transfer part only of its genetic material to the female (Chapter 6). It is possible to arrange the experiment so that the female cell is unable to make the β-galactosidase because of a mutation in the corresponding gene, while the male cell, although genetically able to make the enzyme, is prevented from doing so by a genetic control segment situated a long way from the galactosidase gene, on a part of the chromosome which is not transferred to the female. While the cells remain unmated, no enzyme can be made, but as soon as the normal galactosidase gene gets into the female cell, the zygote is able to make the enzyme. Not only is this found to occur, but the enzyme is made almost immediately after mating, and at its full normal rate. There are two possible explanations for this. In one, the gene itself could make the protein directly, in the other a gene product, or messenger, is made very quickly and this is able to control the synthesis of new protein. The latter is very much the more likely explanation, since, as we have seen, protein synthesis occurs mainly in the ribosomes and not the chromosomes. Further genetic experiments show that in bacteria, at least, the messenger can only control the ribosome for a very short time after it has left the gene, because destroying or withdrawing the gene results in the immediate cessation of enzyme production.

Further evidence for the messenger hypothesis came from a study of the switch in the synthetic mechanism of a bacterium which occurs after it is infected by bacteriophage, causing it to stop making bacterial protein and start making the enzymes and coat proteins peculiar to the phage. For some time it had been known that, after infection, a new kind of RNA appeared in the bacterium and that this was very rapidly synthesized. In 1961 Hall and Spiegelman showed, by a most elegant technique termed hybridization, which will now be described, that this RNA had a base sequence complementary to one or both of the strands of the phage DNA and not complementary to that of the bacterial DNA.

If a solution of DNA is heated, the hydrogen bonds responsible for base pairing between the two chains break (see Chapter 7); if such a solution is now cooled slowly these bonds may re-form. Such a renatured DNA regains its biological activity and can be used to transform bacteria. Spiegelman heated bacteriophage DNA and the new RNA formed after infection and allowed them to cool together, when they formed a specific DNA-RNA complex, presumably a double helix with one DNA and one RNA strand. If the experiment was repeated using bacterial DNA, or DNA from other sources, these complexes did not form. It seems likely from other experiments on DNA, that the sequence of bases in the two chains must be complementary over a substantial part of their length, before these complexes will form. Naturally occurring DNA-RNA complexes have also been found in infected bacteria. It is therefore likely that the RNA formed after infection has a base sequence complementary to that of the phage DNA.

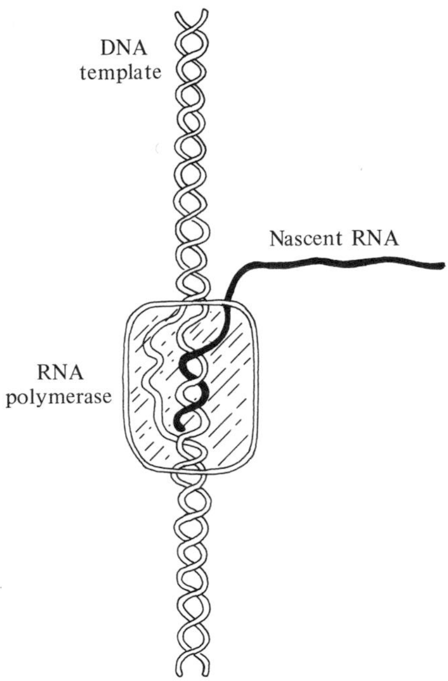

Fig. 8.4 Schematic representation of synthesis of a messenger RNA with sequence complementary to one strand of a double-helical DNA molecule. (*From* Stent, 1966.)

At the same time as these experiments were being done, another group, including workers from Cambridge, Paris and California, studied what happened to the various RNA particles and fractions immediately after phage infection. They found first that no new ribosomes were formed in the cytoplasm of the bacterium, but that newly made protein destined for the phage was made on bacterial ribosomes present before infection. Further, the new RNA which was synthesized after infection became attached to those ribosomes which were actively engaged in protein synthesis.

Although these first experiments were made on phage-infected bacteria, evidence was also found, in normal bacteria, for an RNA fraction with rapid turnover, which became attached to active ribosomes. Later an RNA fraction was identified in a wide range of cell types other than bacteria, with base ratios more like that of their DNAs than of the RNA of their ribosomes. It thus began to appear that the bulk of the ribosomal RNA was 'structural' rather than 'template' RNA, and that other RNA molecules, called *messenger-RNAs*, with a base sequence determined by a given length of DNA, could be incorporated into the ribosome as a template, and that these messenger-RNAs direct the assembly of new polypeptide chains. The synthesis of a messenger-RNA complementary to one strand of a double-helical DNA is indicated in Fig. 8.4. It is supposed that the messenger-RNA nucleotide sequence is determined by Watson–Crick pairing between bases, as in synthesis of complementary DNA strands. The enzyme; RNA polymerase, responsible for messenger-RNA synthesis in *E. coli* has now been characterized. It has the sub-unit structure $\alpha_2\beta\beta'\sigma$ with sub-unit molecular weights of 40000, 150000, 160000 and 90000 respectively.

We must now discuss what has come to be called the 'coding problem'.

The code

If DNA or messenger-RNA controls the synthesis of specific proteins, we must ask how a sequence of nucleotides, made up from the four different kinds of base present in DNA or messenger-RNA, can determine the order in which the twenty or so different amino acids are joined together to form a polypeptide chain. A problem of this type involving the transition between a message written in a four letter alphabet to the same message written in a twenty letter alphabet is a coding problem. As soon as the double-helical structure of DNA was proposed, and with it the implication of an exact mechanism for the replication of the two complementary strands, consideration was given as to how such a code might work. Clearly one base, either A, T, C or G (where A stands for Adenine, T for Thymine, C for Cytidine and G for Guanine) cannot by itself uniquely determine more than one amino acid, that is four in all. Equally it is easy to see that combinations of pairs of bases:

AA	AT	AC	AG
TA	TT	TC	TG
CA	CT	CC	CG
GA	GT	GC	GG

will not be enough to code for twenty amino acids even if we allow that the order of the bases is important, i.e. that AT codes differently from TA, etc. If, however, there are three bases coding for each amino acid, the number of possible triplets is 64 ($4 \times 4 \times 4$), which would give enough combinations to code each amino acid with plenty to spare. It seems probable, therefore, that each 'word' in the DNA code, representing one amino acid, must contain at least three letters.

Since DNA contains no structural feature which would interrupt the sequence of bases to provide the spaces between words, the question arises as to how the reading mechanism distinguishes the end of one word from the beginning of the next. If we write part of a sentence in our code

ATTCGCATAGCA...

and assume that each word has three letters, then it is clear that we can get different words depending on where we start

ATT CGC ATA...

or, for example, A TTC GCA TAG...

To overcome this difficulty, Crick and his colleagues suggested an ingenious theory in a paper called 'Codes without commas'. They suggested, using our example above, that if ATT, CGC and ATA are meaningful words, then all those triplets which *overlap* these, like TTC, TCG, GCA and CAT and so on, are meaningless or nonsense. They then calculated that the sixty-four triplets could be divided, in a number of ways, into two classes, with twenty meaningful triplets in one class and forty-four nonsense triplets in the other. Twenty is the number of amino acids found in nature, if we exclude a few such as cystine and hydroxyproline, which might be considered to be derived from one or other of the magic twenty. Was this just coincidence? It now looks as if it was, but at the time it looked plausible, and scientists, like other people, like neat and tidy solutions.

An alternative way of overcoming the problem of where a word begins and ends is to assume that all words are triplets and that there is some marker which determines the beginning of each sentence. Crick, Barnett, Brenner and Watts-Tobin provided the first evidence that the code might be of this kind. They used Benzer's genetic system, which we have discussed in Chapter 6, and studied the two cistrons governing the host range of the T_4 phage of *E. coli* strain K12. They concentrated, however, on the unusual mutations caused by acridine orange, which involve the deletion or addition of only one base in DNA.

It will be simplest to explain these important results by setting up a model and studying its behaviour. Let us assume that a string of bases can only be read from left to right, that there is nothing to distinguish the end of one word from the beginning of the next, and that the code is read in groups of three starting at some marker (M) at the beginning of each sentence.

For example

M ATTCGCATAGCAGCT...

is normally read as

ATT CGC ATA GCA GCT...

Now let us assume that, by chance, the mutagen removes the first C from the left. The sentence will now read

$$\text{ATT}\quad \overset{-}{\text{G}}\text{CA}\quad \text{TAG}\quad \text{CAG}\ldots$$

where the $^{-}$ sign indicates the position of the deletion. This results, as can be seen, in the wrong words appearing everywhere to the right of the minus sign, and the strain will be recognizably defective. Supposing we can now produce a further mutation which *inserts* another base, indicated by $\overset{+}{\text{X}}$, to the right of our previous deletion. We now have

$$\text{ATT}\quad \overset{-}{\text{G}}\text{CA}\quad \text{TAG}\quad \text{C}\overset{+}{\text{X}}\text{A}\quad \text{GCT}\ldots$$

where X can be any of the four bases. It can be seen that the fifth, sixth and all following triplets will now read correctly, and only the second, third and fourth wrongly. We might, then, expect that if the three-'wrong' amino acids do not affect the structure of the protein significantly, this strain of phage would revert to the wild type. This is what was found in these experiments; phages with a plus and minus mutation close together can grow normally, but the further the plus and minus are separated the more likely it is that their effects will fail to cancel out.

Quite a number of strains were accumulated with plus and minus mutations at different sites on the T_4 B cistron, and the recombinants between them studied. It was found that two plus mutations gave a mutant phage that was recognizably defective, as expected from the sentence

$$\text{A}\overset{+}{\text{X}}\text{T}\quad \text{TCG}\quad \text{CA}\overset{+}{\text{X}}\quad \text{TAG}\quad \text{CAG}\ldots$$

and similar results were obtained for two minus ones.

But, dramatically, three plus or three minus mutations gave a wild type phenotype as in

$$\text{AT}\overset{+}{\text{X}}\quad \text{TC}\overset{+}{\text{X}}\quad \text{GCA}\quad \overset{+}{\text{X}}\text{TA}\quad \text{GCA}\quad \text{GCT}\ldots$$

It can be seen that in these triple mutants, the original fourth (now fifth) and subsequent triplets revert to the normal kind. Presumably in such cases the extra triplet does not prevent protein synthesis, and if an extra amino acid is added at this point, it does not affect the function of the protein. Unless the mutagen removes or adds two or more bases at a time, which is less probable, this last result is very strong evidence that the triplet code which we assumed is correct. It seems that the difficulty of deciding which letter begins each word is not overcome by means of a 'code without commas', but by providing some sort of marker which ensures that each sentence is begun at the right place.

Crick and his colleagues describe an experiment which strongly supports this idea of a marker. All the acridine mutations thus far described are near that end of the B cistron which is next to the A cistron of the phage. It was possible in this further experiment to use a different type of mutant discovered by Benzer in which a part of the DNA including the adjacent ends of the A and B cistrons has been lost, without impairing the function of the B cistron. Normally, without this deletion, any mutation in the A cistron has no effect on the B function, and vice versa. With the deletion, however, the effect of an acridine mutation, which removes or adds one base in the A cistron, spreads over into the B cistron and destroys its function (an example of polarity – see Chapter 6). This implies the loss in the mutant with the deletion of some full stop or marker which normally prevents an acridine mutation in one cistron affecting neighbouring cistrons. Such a full stop may well be a special sequence of bases different from those coding for amino acids.

Breaking the code

The determination of the actual words of the code, that is which triplets determine each amino acid, can be theoretically achieved in a number of ways. One method is to use specific mutagens, which alter one kind of base in the DNA, and to see which amino acids have been altered in the protein. This technique, as we shall see, has been used with considerable success in TMV studies, where a number of single amino-acid replacements have been found.

A much simpler approach was discovered in the early 1960s by Nirenberg and Matthaei in a brilliant series of experiments. These workers were investigating the incorporation of amino acids into the protein formed on ribosomes extracted from *E. coli* in the presence of transfer-RNA and the necessary co-factors. They found that they could separate from the ribosomes an RNA of high molecular weight which could markedly stimulate amino-acid incorporation. More important, an artificial polyribonucleotide containing only the one base, uridine, synthesized with polynucleotide phosphorylase (Chapter 7), caused the ribosomes to incorporate only one amino acid, phenylalanine, into polypeptide chains of relatively high molecular weight. No other amino acid was incorporated into the polypeptide chains, and uridine in unpolymerized form was ineffective.

It seems that the polyuridylic acid (poly rU) here acts as a messenger-RNA and UUU is therefore the first word known in the code: it 'means' phenylalanine. Immediately this discovery was announced, many workers, including Ochoa, tried the effect of other polyribonucleotides on the ribosomes. They found that poly rA codes for lysine and poly rC for proline (poly rG cannot be polymerized with this polynucleotide phosphorylase). In addition, mixed polymers containing two or more bases will also stimulate the incorporation of many amino acids.

The technique used in these experiments is to make polyribonucleotides with varying proportions of, say, U and A, and to study the different amino acids whose incorporation is stimulated by such a polymer. Since the polymerizing enzyme puts the U and A together at random, equal proportions of A and U will result in an equal number of words either all A or all U, and a larger number containing two U and one A or two A and one U. Since these mixed triplets will include all possible triplets of these compositions, with the bases in different orders, we may expect that the incorporation of a number of amino acids will be stimulated, and in this example it was found that in addition to phenylalanine, isoleucine, lysine and tyrosine were incorporated. Further experiments in which the proportion of U and A was varied showed that isoleucine and tyrosine were stimulated to a greater extent by a mixture containing more U than A, and lysine by more A than U. If we assume a triplet code, then presumably 2U + A codes isoleucine and tyrosine, with the actual order distinguishing between the two, and 2A + U or AAA codes lysine.

These experiments give no information about the order of the bases in the triplets. This information was derived from experiments using short-chain polynucleotides of known base sequence. These were used in two ways: Nirenberg incubated ribosomes with known trinucleotides and showed that one transfer-RNA with its attached amino acid was then preferentially bound to the ribosome. The trinucleotide of known sequence was assumed to be the *codon* (i.e. code triplet) for that amino acid. Secondly, Khorana and his colleagues in a most elegant series of experiments not only confirmed the codons for some amino acids but also provided further proof of the triplet nature of the code. They first made two decadeoxyribonucleotides TCTCTCTCTC and AGAGAGAGAG. These were then used as primers with Kornberg's enzyme to produce a long artificial DNA with a repeating sequence TCTCTC . . . in one chain and a sequence AGAGAG . . . in the other. If this DNA was then used as a primer for RNA synthesis, with RNA polymerase, in the presence of only UTP and CTP, it gave a long polyribonucleotide UCUCUC . . . . This could be used as the artificial messenger for protein synthesis and gave a repeating polypeptide with alternating serine and leucine residues.

From these and related experiments the assignments given in Table 8.1 have been deduced. The most important feature of these results is that many amino acids are coded by more than one triplet. Indeed it seems that for most amino acids only the first two bases of the codon need be specified. A code of this type in which a number of triplets code for the same amino acid is termed *degenerate*.

The results obtained from these *in vitro* systems are supported by another independent technique. This is the study of the effects on the RNA of TMV of nitrous acid, which specifically alters the bases C to U and A to G. These results are interesting also as a study of amino-acid replacements arising from mutations. The RNA of the virus not only controls the kind of lesion caused by the virus, which enables mutants to be selected, but also controls the synthesis of the protein of the virus particle whose complete amino-acid sequence in the wild strain is now known (Fig. 8.5). It has been found in the quite large number of nitrous acid mutants so far investigated, that for most of those in which the primary structure of the protein is affected, there has been replacement of only one amino acid. Of the 158 amino acids in this protein only about twenty have been found to be altered by this treatment. Also when an amino acid at a particular site is changed, it is found that the choice of its replacement is very limited. For example, leucine can only be replaced by phenylalanine (Table 8.2).

TABLE 8.1 The code derived from *in vitro* polynucleotide experiments.

Codon	Amino acid	Codon	Amino acid	Codon	Amino acid	Codon	Amino acid
UUU	phe	UCU	ser	UAU	tyr	UGU	cys
UUC		UCC		UAC		UGC	
UUA	leu	UCA		UAA	chain terminating	UGA	chain terminating
UUG		UCG		UAG			
						UGG	try
CUU	leu	CCU	pro	CAU	his	CGU	arg
CUC		CCC		CAC		CGC	
CUA		CCA		CAA	$gluNH_2$	CGA	
CUG		CCG		CAG		CGG	
AUU	ileu	ACU	thr	AAU	$aspNH_2$	AGU	ser
AUC		ACC		AAC		AGC	
AUA		ACA		AAA	lys	AGA	arg
AUG	met	ACG		AAG		AGG	
GUU	val	GCU	ala	GAU	asp	GGU	gly
GUC		GCC		GAC		GGC	
GUA		GCA		GAA	glu	GGA	
GUG		GCG		GAG		GGG	

(Data from NIRENBERG, M. W. *et al.* (1965), *Proc. nat. Acad. Sci. Wash.*, **53**, 1161.)

Other amino acids can be replaced by either of two alternatives: proline can be replaced by serine or leucine, but not by any other amino acid.

If a list of these limited replacements is compared with the base substitutions induced by nitrous acid, the base composition of the triplets coding certain amino acids can be determined, although this does not give the order of bases in each triplet. These amino-acid replacements are in good agreement with the results of the *in vitro* polynucleotide experiments. Similar results can be obtained from a study of abnormal haemoglobins, in which a natural mutation has caused the substitution of one amino acid by another (Table 9.1). All these substitutions are compatible with the code shown in Table 8.1, and therefore support the view that the code is universal, i.e. that the same triplets code for the same amino acids in all viruses, bacteria and other cells.

A priori we might expect the code to be the same for all living organisms, because we think that evolution has occurred as the result of many discrete mutations at different parts of the DNA. It is very hard to conceive of a way in which the code could change, by altering its reading mechanism and turn, say, every leucine in an organism into a tyrosine without dire results. On the other hand, it is equally unlikely that a universal code appeared complete and immaculate with the first living organism. The code must have evolved, and a number of different codes may have evolved independently, so that, for example, vertebrates have slightly different codes from bacteria.

This question was forcefully raised some years ago when it was discovered that DNA base ratios of different species of bacteria were enormously different.

	1	2	3	4	5	6	7	8	9	10	11	12	13	14	15	16	17	18	19	20	21	22
Acetyl	*Ser* –	*Tyr* –	*Ser* –	*Ileu* –	*Thr* –	*Thr* –	*Pro* –	*Ser* –	*GluN* –	*Phe* –	*Val* –	*Phe* –	*Leu* –	*Ser* –	*Ser* –	*Ala* –	*Try* –	*Ala* –	*Asp* –	*Pro* –	*Ileu* –	*Glu*

23	24	25	26	27	28	29	30	31	32	33	34	35	36	37	38	39	40	41	42	43	44	45
Leu –	*Ileu* –	*AspN* –	*Leu* –	*Cys* –	*Thr* –	*AspN* –	*Ala* –	*Leu* –	*Gly* –	*AspN* –	*GluN* –	*Phe* –	*GluN* –	*Thr* –	*GluN* –	*GluN* –	*Ala* –	*Arg* –	*Thr* –	*Val* –	*Val* –	*GluN*

46	47	48	49	50	51	52	53	54	55	56	57	58	59	60	61	62	63	64	65	66	67	68
Arg –	*GluN* –	*Phe* –	*Ser* –	*GluN* –	*Val* –	*Try* –	*Lys* –	*Pro* –	*Ser* –	*Pro* –	*GluN* –	*Val* –	*Thr* –	*Val* –	*Arg* –	*Phe* –	*Pro* –	*Asp* –	*Ser* –	*Asp* –	*Phe* –	*Lys*

69	70	71	72	73	74	75	76	77	78	79	80	81	82	83	84	85	86	87	88	89	90	91
Val –	*Tyr* –	*Arg* –	*Tyr* –	*AspN* –	*Ala* –	*Val* –	*Leu* –	*Asp* –	*Pro* –	*Leu* –	*Val* –	*Thr* –	*Ala* –	*Leu* –	*Leu* –	*Gly* –	*Ala* –	*Phe* –	*Asp* –	*Thr* –	*Arg* –	*AspN*

92	93	94	95	96	97	98	99	100	101	102	103	104	105	106	107	108	109	110	111	112	113	114
Arg –	*Ileu* –	*Ileu* –	*Glu* –	*Val* –	*Glu* –	*AspN* –	*GluN* –	*Ala* –	*AspN* –	*Pro* –	*Thr* –	*Thr* –	*Ala* –	*Glu* –	*Thr* –	*Leu* –	*Asp* –	*Ala* –	*Thr* –	*Arg* –	*Arg* –	*Val*

115	116	117	118	119	120	121	122	123	124	125	126	127	128	129	130	131	132	133	134	135	136	137
Asp –	*Asp* –	*Ala* –	*Thr* –	*Val* –	*Ala* –	*Ileu* –	*Arg* –	*Ser* –	*Ala* –	*Ileu* –	*AspN* –	*AspN* –	*Leu* –	*Ileu* –	*Val* –	*Glu* –	*Leu* –	*Ileu* –	*Arg* –	*Gly* –	*Thr* –	*Gly*

138	139	140	141	142	143	144	145	146	147	148	149	150	151	152	153	154	155	156	157	158
Ser –	*Tyr* –	*AspN* –	*Arg* –	*Ser* –	*Ser* –	*Phe* –	*Glu* –	*Ser* –	*Ser* –	*Ser* –	*Gly* –	*Leu* –	*Val* –	*Try* –	*Thr* –	*Ser* –	*Gly* –	*Pro* –	*Ala* –	*Thr*

Fig. 8.5 The primary structure of wild-strain TMV protein. (*From* Dayhoff, 1969.)

TABLE 8.2 Amino acid replacements observed in the coat protein of tobacco mosaic virus.

Mutant	Replacement	Position
Spontaneous		
A 14	Ilu → Thr	129
	A*U*U A*C*U	
B 13	Asn → Lys	33
	AA*U* AA*A*	
CP 415	Asn → Lys	140
E 66	AA*U* AA*A*	
GK 1	Asn → Ser	73
	A*A*U A*G*U	
flavum	Asp → Ala	19
	G*A*U G*C*U	
necans	Phe → Leu	10
	UU*U* UU*A*	
	Ala → Val	19
	G*C*U G*U*U	
	Ser → Phe	138
	U*C*U U*U*U	
reflavescens	Leu → Phe	10
	UU*C* UU*U*	
revirescens	Leu → Phe	10
	UU*C* UU*U*	
	Val → Asp	19
	G*U*U G*A*U	
Fluorouracil-induced		
FU 27	Val → Ala	58
	G*U*U G*C*U	
FU 41	Thr → Ala	28
	*A*CU *G*CU	
FU 243	Ser → Gly	65
	*A*GU *G*GU	
Nitrite-induced		
Ni 102	Asp → Gly	66
	G*A*U G*G*U	
Ni 109	Glu → Gly	97
	G*A*A G*G*A	
Ni 118	Pro → Leu	20
	C*C*C C*U*C	
Ni 445	Ser → Phe	138
	U*C*U U*U*U	
Ni 458	Thr → Ilu	59
Ni 470	A*C*U A*U*U	
Ni 462	Thr → Ilu	5
	A*C*U A*U*U	
	Ser → Leu	55
	U*C*G U*U*G	
Ni 568	Thr → Ilu	5
	A*C*U A*U*U	
	Thr → Met	107
	A*C*G A*U*G	
Ni 630	Thr → Met	107
Ni 725	A*C*G A*U*G	
Ni 1055	Ilu → Met	21
	AU*A* AU*G*	
Ni 1103	Asn → Ser	73
	A*A*U A*G*U	
	Ilu → Val	125
	*A*UU *G*UU	
Ni 1045	Pro → Ser	63
Ni 1196	*C*CC *U*CC	
Ni 1234	— —	
Ni 1118	Ilu → Val	24
	*A*UU *G*UU	
	Ilu → Val	125
	*A*UU *G*UU	
Ni 1688	Pro → Ser	63
	*C*CC *U*CC	
	Pro → Leu	156
	C*C*C C*U*C	
Ni 1927	Pro → Leu	156
Ni 2029	C*C*C C*U*C	
Ni 2032	Thr → Ilu	136
	A*C*U A*U*U	
Ni 2068	Tyr → Cys	139
	U*A*C U*G*C	
Ni 2204	Ser → Leu	15
	U*C*G U*U*G	
	Thr → Ilu	153
	A*C*U A*U*U	
Ni 2239	Ser → Leu	15
	U*C*G U*U*G	
PM 2	Thr → Ilu	28
	A*C*U A*U*U	
	Glu → Asp	95
	GA*A* GA*U*	

(*From* YCAS, M. (1969) *The Biological Code*, Amsterdam, Elsevier.)

Indeed, the ratio $\frac{A+T}{G+C}$ varies from 0·35 to 2·2. Could it be that bacteria had such very different proteins that they needed such variable DNAs, or was the code different for some or all of them? We would expect that many enzymes and other proteins would be similar in all bacteria, but it has been found that proteins with the same function can have different overall amino-acid composition, and for some amino acids there is a correlation between the proportion of the amino acid in the bacterium and its $\frac{A+T}{G+C}$ ratio. Since there is also considerable degeneracy, and many amino acids can be coded by several triplets with different base compositions, the range of $\frac{A+T}{G+C}$ ratios in bacteria is compatible with a universal code.

We shall discuss further the evolution of the code, and other aspects of molecular evolution in Chapter 9, but we may note here that spontaneous mutations (Tables 8.1 and 9.1), like nitrous acid mutations, can lead to an altered triplet and hence an altered amino acid. This is one of the fundamental molecular processes of biological evolution.

The interaction between messenger-RNA and transfer-RNA at the ribosome

We can now look more closely at the details of protein assembly on the ribosomes. If a sequence of triplets (codons) along the messenger-RNA (mRNA) is to determine the order in which amino acids are assembled into a polypeptide chain it seems that the transfer-RNA (tRNA) which carries a given amino acid must somehow recognize the codon for this amino acid. Even before tRNAs had been discovered Crick suggested that a low molecular weight RNA might play this role. He noted that there is no very obvious way by which an amino acid can recognize (that is, bind specifically to) a short sequence of nucleotides. The shapes of the molecules do not suggest any obvious type of complementary fit. If however the amino acid is linked to a low molecular weight RNA then a triplet of bases in this RNA could be complementary to the codon and bind to it specifically, by the Watson–Crick pairing found in the DNA double helix. This is in fact the role played by the tRNAs.

The special role of tRNA in protein synthesis is clearly shown by the following ingenious experiment. The amino acid cysteine was first attached to its tRNA by means of its specific activating enzyme. The cysteine was then converted to alanine with the aid of a catalyst called Raney nickel, without, however, alteration in either the amino-acid bond to tRNA or the structure of the tRNA. It was then possible to exploit the finding in the Nirenberg kind of experiments (Table 8.1) that poly rUG normally stimulates the incorporation of cysteine, but not alanine, into proteins, and to ask whether the poly rUG 'recognizes' the amino acid or the RNA. The results showed that poly rUG will stimulate the incorporation of alanine into protein if it is attached to cysteine-specific tRNA. This strikingly confirms the hypothesis that tRNA acts as the 'adapter' between the amino acid and the sequence-determining RNA.

In the tRNA sequences of Fig. 8.1 it is found that certain triplets which always come in loop II, IGC for yeast alanine tRNA, G(PSU)A in the tyrosine tRNA, IGA and IAC for the serine and valine tRNAs, are complementary to the codons for these amino acids (the loops in Fig. 8.1 are labelled I, II, III and IV and the double-helical regions 1, 2, 3 and 4 for convenience of reference). These triplets in the tRNAs are called *anticodons*, and the fit between codon and anticodon actually allows a little 'wobble' which accounts for the degeneracy of the code. Thus the alanine codon, which is GCX (where X can be any of the four bases – see Table 8.1) is complementary to the anticodon IGC as a result of bonding

```
....G   C   X.....
    |||  |||  ⋮
....C   G   I .....
```

Inosine, which occurs frequently at this position in the anticodon, is similar to guanine but without the amino group at position 2 (see Table 7.1). It can form two hydrogen bonds with cytosine in a regular double helix (Fig. 7.2), but when codon and anticodon come together during assembly of amino acids in protein synthesis only a short segment of double helix forms and the folding of the tRNA, and its freedom of movement, allow I to fit to any of the four bases at this point in the codon (Fig. 8.6). Note that the G–U abnormal pair found in tRNA molecules (Fig. 8.1) requires only as much distortion of the double helix as is allowed in the codon-anticodon fit. The U–U pair of alanine tRNA which is a pyrimidine–pyrimidine pair requires rather more distortion.

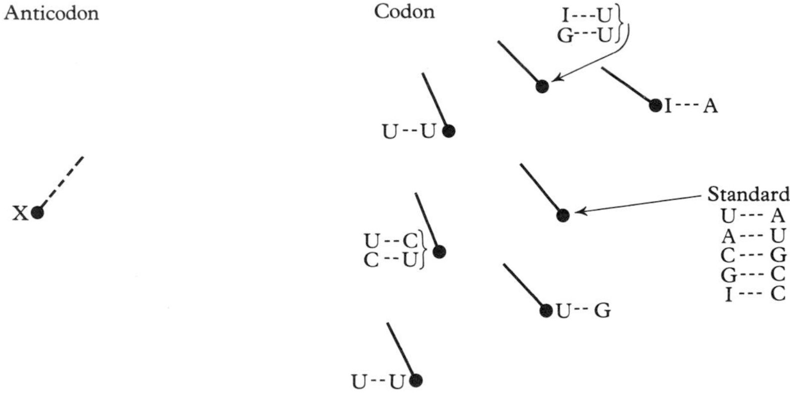

Fig. 8.6 The point X represents the position of the C_1' atom of the glycosidic bond (shown dotted) in the anticodon. The other points show where the C_1' atom and the glycosidic bond fall for the various base pairs. The amount of wobble allowed in interaction between codons and anticodons during normal protein synthesis appears to allow the four positions to the right of the diagram and also I—G, but not the three close positions. (*From* Crick, 1966.)

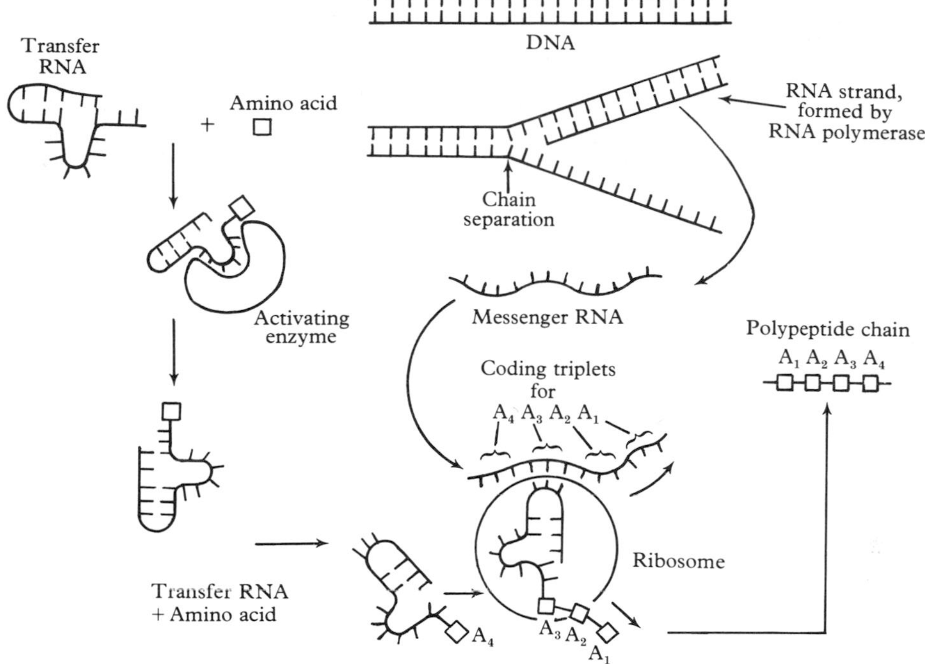

Fig. 8.7 A schematic outline of protein synthesis. The ladder-like drawings represent two-stranded nucleic acids, with the bases represented by the short cross-lines. The secondary structure of transfer-RNA is indicated in a purely schematic way. No attempt is made to illustrate the detail of ribosome structure. (Modified *from* Rich, 1962, and Watson, 1963.)

In broad outline then protein synthesis may be visualized as in Fig. 8.7. We know that synthesis takes place on ribosomes, we know that amino acids are brought to the ribosome attached to specific tRNAs, we know that the sequence of codons along mRNA determines the sequence in which amino acids are joined together. It may be supposed that different tRNAs bring up different amino acids in turn, but the amino acid which is accepted and incorporated into the growing polypeptide chain at any point is the one whose anticodon fits the codon at the ribosome acceptor site at that time. During protein synthesis the mRNA must move across the ribosome, or the ribosome move along the mRNA, to bring each codon successively to the acceptor site.

We can now consider the varied experimental evidence in support of this scheme and the light that this throws on the details of the process. First there is evidence from centrifugation rates that ribosomes active in protein synthesis are joined together in clusters (sometimes called polysomes). In thin-section micrographs one can see that ribosomes tend to associate in rows, and in micrographs of positively-stained preparations of ribosomes active in protein synthesis one can see that they are linked by thin strands. These strands are mRNA.

In protein synthesis ribosomes attach to mRNA molecules, sometimes at the 5′-end, sometimes at a point further down the strand (see Chapter 10). They travel varying distances along the mRNAs (towards the 3′ end – see Chapter 10) synthesizing protein sequences dictated by the mRNA codon sequence, and the ribosomes then fall away.

The molecular weights of mRNAs range from 120000 to $2{\cdot}4 \times 10^6$ or more (single strands of from 400 to 8000 or more nucleotides, ranging in length from 200 nm to 4 μm or longer). Ribosomes can travel down these strands about 50 nm apart. Thus polysomes can contain from a few to eighty or more individual 70 *S* ribosomes, depending on the size of protein molecule being synthesized and the number of proteins that are being coded by a single polycistronic messenger-RNA. Messenger-RNAs have a long life in some cells, such as mammalian reticulocytes, a short life in others, the rate of degradation of mRNA being somehow linked to the needs of the cell for different proteins and forming part of the control mechanisms for protein synthesis which are to be discussed in Chapter 10. Bacterial mRNA is short-lived, but bacterial ribosomes as we have noted earlier, have a relatively long life, recircling from one mRNA to another, synthesizing the proteins needed at different times. Each time a 70 *S* ribosome falls off the mRNA, however, it splits into its 50 *S* and 30 *S* components. A 50 *S* and a 30 *S* component come together again at the start of a new mRNA run, first the 30 *S* sub-unit binding to the mRNA and then the 50 *S* sub-unit joining the complex. Possibly the mRNA lies in a groove between the sub-units in the active 70 *S* ribosome, so that the ribosome cannot fall off the mRNA part way along without splitting into its 50 *S* and 30 *S* sub-units.

Transfer-RNA molecules are bound to the ribosomal surface, and the ribosome carries in fact two tRNA binding sites which have been termed the A (acceptor) and the P (peptidyl) sites. As protein synthesis proceeds the amino acid most recently linked to the growing polypeptide chain, let us say amino acid *n*, remains covalently linked to $tRNA_n$ which is bound at the P-site. $tRNA_{n+1}$ carrying the next amino acid $(n+1)$ now binds at the A-site, the polypeptide chain breaks its bond to $tRNA_n$ and forms a peptide bond to amino acid $n+1$, while $tRNA_{n+1}$ moves across from the A-site to the P-site to make way for the next incoming amino acid.

A number of enzymes, protein factors and also GTP are involved in this process but before considering these steps in more detail we must first note that the initiation of protein synthesis, that is the binding to the ribosome of the first tRNA and amino acid of the chain is a different process from chain elongation, since the first incoming tRNA has either to bind directly to the P-site, or bind to the A-site and move across to an unoccupied P-site. It is found in fact that a separate class of tRNAs are required for this initiation step. The first amino acid bound in *E. coli* is always formylmethionine (in higher organisms methionine) though this may not be apparent in the final protein sequence because the formyl group or the formylmethionine may be later enzymically split off the chain. In *E. coli*, only a specific tRNA, formylmethionine tRNA, can bind to the ribosomes to initiate protein synthesis. It initiates synthesis at the methionine codon of mRNA (AUG – Table 8.1) and, probably because of the special sequences which precede it, can recognize this triplet at the beginning of a protein-specifying sequence and initiate protein synthesis there, and *not* at AUG codons further down the sequence, where methionine will be in due course incorporated during chain elongation. It seems that there must be base sequences on mRNA, preceding the AUG codon at the start of a protein-specifying sequence, which are somehow recognized in the initiation process.

We can now consider the role of the various factors that have been found to be required for protein synthesis in addition to the ribosomes, tRNAs, amino acids, activating enzymes and ATP discussed so far. These factors have been discovered by more detailed analysis of the components of the supernatant fraction of homogenized cells required for protein synthesis to proceed, or be initiated, or terminated on ribosomes *in vitro*. For chain initiation three protein factors are required in *E. coli*, termed F_1, F_2 and F_3, and for chain elongation three further factors T_s, T_u and G as well as GTP. A further factor or factors (R factors) are required for release of the completed polypeptide chain. Possibly, as we shall see, the 5 *S* RNA of the ribosomes (see Chapter 7) is involved also in chain elongation.

Considering first the steps of chain elongation, starting from a point at which $tRNA_n$ with its partially completed polypeptide chain is bound to the P-site, and $tRNA_{n+1}$ with the next amino acid is heading towards the A-site, the first step is the binding of $tRNA_{n+1}$ at the A-site. The $mRNA_{n+1}$ codon must be involved here in allowing only the correct tRNA to be bound. It may be that other tRNAs become attached to the site in random thermal motion but only the correct tRNA can make a good fit to the mRNA codon and become firmly bound. The protein factor T_u is bound also near the A-site at this stage, together with a molecule of GTP. The terminal high-energy phosphate is now split off the GTP, and P_i (inorganic phosphate) and a T_u-GDP complex are released (T_s plays a role in splitting this complex to free T_u for a further round of activity). The growing polypeptide chain is at this stage transferred from $tRNA_n$ to form a peptide bond with the new amino acid. It is here that 5 *S* ribosomal RNA may play a part, with the growing polypeptide chain transiently forming a covalent bond to the 5 *S* RNA prior to formation of the peptide bond to the new amino acid.

At some point in the chain elongation process $tRNA_n$ has to leave the P-site and $tRNA_{n+1}$ has to move to the P-site from the A-site, to make room for the incoming $tRNA_{n+2}$. This translocation, which is accompanied by a movement of the ribosome one codon along the mRNA, is mediated by the G factor, and at this point a further molecule of GTP is split. It would seem therefore that the first GTP (at the stage mediated by T_u) is essentially providing the energy for peptide bond formation, and this second GTP (at the stage mediated by the G factor) is providing the energy for translocation. It is difficult, as yet, to exactly visualize the mechanics of the whole process since, as we have seen in Chapter 7, too little is known of the details of ribosome structure. It seems improbable that the growing polypeptide chain is bodily moved from one part of the ribosome surface to another when it is transferred from $tRNA_n$ at the P-site to the amino acid of $tRNA_{n+1}$ at the A-site, and bodily moved back again at the translocation stage. It seems more probable that the growing chain stays bound to one part of the ribosomal surface while the flexibility of ribosomal structure allows relative movement of the groups of the A- and P-sites. During translocation $tRNA_{n+1}$ presumably stays bound to the $mRNA_{n+1}$ codon as the ribosome moves three nucleotides along the mRNA, with the A-site groups detaching from $tRNA_{n+1}$, and the P-site groups taking their place.

Some of the experimental methods which have been used to elucidate this sequence of events are described in the review by Lipmann and the paper by Gupta *et al.*, given as references for this section, and these authors, in turn, cite other papers for accounts of some of the decisive experiments. The interactions of tRNAs and mRNAs with different parts of the ribosome can be inferred from study of the binding of these RNAs to separate fractions of 50 *S* and 30 *S* subunits. The protein factors essential for chain elongation are isolated, and their role elucidated, by study of protein synthesis in *in-vitro* systems. Subtle tricks are used: in the work of Gupta *et al.*, for example, formylmethionyl tRNA was first bound to ribosomes attached to the RNA of f2 bacteriophages at the initiation point of the sequence coding for f2 coat protein (this phage RNA acts as a messenger-RNA in an infected bacterium and in *in-vitro* protein synthesis). The next amino acid in the coat protein sequence is known to be alanine, hence the formylmethionine tRNA–ribosome–phage RNA complex was next allowed to react with T_s, T_u, GTP and *only* alanine tRNA. Later the G factor and more GTP were added to bring about translocation. Gupta *et al.* use this sytem not only to study steps in chain elongation but also, by digesting away with ribonuclease all RNA not protected by attachment to the ribosome, to study the nucleotide sequence of mRNA near the initiator codon.

We can now consider together the chain termination and chain initiation events. The chain terminating codons are UAA, UAG and UGA (Table 8.1). When the ribosome reaches one of these codons no tRNA can be bound, and a protein factor R mediates the splitting of the bond between the polypeptide chain and the final tRNA. The completed polypeptide chain falls away from the ribosome. The ribosome may now continue on down the mRNA to synthesize another protein, if the mRNA is a polycistronic message. Alternatively one of the initiation factors, F_3, may displace the 50 *S* ribosomal sub-unit and bind to the 30 *S* sub-unit. In this case both sub-units fall away from the mRNA. F_3 now plays its initiation role in facilitating binding of the 30 *S* sub-unit to another mRNA, and perhaps carries some specificity, in binding to one mRNA rather than another at the sequence preceding the AUG initiation codon, so exerting some control over which proteins are synthesized. The 50 *S* sub-unit is now joined to the 30 *S* sub-unit on the mRNA, the binding of formylmethionine tRNA is promoted by two further factors F_1 and F_2, the initiation factors fall away, and synthesis of a new protein can proceed. Thus protein synthesis seems to be quite a complex process. Much remains to be learnt about the details and the roles of all these components. mRNAs, for example, carry a poly A tail 50–200 nucleotides long of as yet unknown significance

There now remains a final question to be considered. Is the sequence of amino acids the necessary and sufficient condition for the specific folding of the peptide chain to form a protein? The classical argument against such a hypothesis invoked the irreversibility of protein denaturation. Proteins, once unwound and their three-dimensional shape destroyed, do not readily regain their original shape even though the linear order of their amino acids is preserved. This argument may be countered in two ways. First, as we have seen, proteins are made serially from one end. Folding from the free end, even if it is not controlled by the structure of the ribosome, is unlikely to be the same as the folding which would occur if the whole extended polypeptide chain were thrown off into the medium. In other words, specific folding in the first part of the chain may well govern how the remainder folds. Second, there is now direct evidence that the polypeptide chain of ribonuclease and many other proteins can be unfolded and refolded to their native conformations.

If the four S–S bridges present in ribonuclease are all broken (by chemical reduction) the molecule loses its biological activity and the chain unfolds. If conditions are restored which will allow the bridges to reform (i.e. a return from reducing to oxidizing conditions) they do so at random, and the wrong bits of the chain are then joined together. But the tertiary structure of the molecule with abnormal bridges is found to be less stable than the native form, and if suitable oxidation-reduction conditions are maintained, and free —SH groups are present to allow constant breaking and reforming of S–S bonds, the molecule is able, so to speak, to have another try at making the right bridges. It then regains its specific shape and biological activity, by a trial and error process. Proteins without S–S bonds more readily come to the equilibrium state.

It seems that the order of amino acids rather precisely determines the most stable three-dimensional structure of a protein, and that conditions during protein synthesis allow the molecule to take up this form. We do not have to postulate any special mechanism in the cell determining the shape of a protein, apart from the sequence determination controlled by messenger-RNA.

Of course, not all proteins spring complete and active from the ribosomes. In many cases enzymes and hormones are synthesized in an inactive form (e.g. trypsinogen and insulin – see Chapter 3) and require splitting of peptide bonds before they become active. Additionally, polysaccharides, lipids, metal ions and other substances are often found in proteins. These are attached during synthesis on the ribosome or after the protein has left the ribosome, either by enzymes, or by the spontaneous formation of specific complexes.

Protein synthesis in mitochondria and chloroplasts

Mitochondria and chloroplasts contain their own machinery for protein synthesis, quite independent of that of the cytoplasmic matrix, including DNA, transfer RNAs, amino-acid-activating enzymes, and 70 *S* ribosomes.

Mitochondrial DNA is about 5 μm long (molecular weight 11×10^6, 10–30 genes) and forms a double-stranded ring. It is not known for certain how mitochondria are formed (see Chapter 5) but most probably they can divide like bacteria, by elongation and division, with DNA duplication keeping one, or a few, steps ahead of mitochondrial division, so that each mitochondrion carries several copies of its genome. The similarity in size between bacterial and mitochondrial ribosomes is paralleled by a similar sensitivity to antibiotics. A number of very useful antibiotics (streptomycin, tetracycline, chloramphenicol) inactivate the 70 *S* ribosomes of bacteria but not 80 *S* ribosomes of plant and animal cell cytoplasm; these antibiotics also inactivate the 70 *S* ribosomes of animal and plant mitochondria. In streptomycin-sensitive strains of bacteria the streptomycin binds to the bacterial ribosomes and causes misreading of codons during translation. For *E. coli* it is possible to compare streptomycin-sensitive and streptomycin-resistant strains and show that the difference lies in just one of the twenty or so proteins of the 30 *S* ribosomal unit, which must therefore be placed in the ribosome structure so that it affects the fit of codon and anticodon of the bound mRNA and tRNAs.

The similarities between bacteria and mitochondria have led to the idea that mitochondria perhaps originated as free-living organisms which became symbiotic in primitive cells. If this is the case, the symbiosis has now evolved to a point of considerable interdependence. The functions of mitochondrial genes can be distinguished from those of chromosomal genes by their departure from the rules of Mendelian inheritance and by behaviour characteristic of maternal cytoplasmic inheritance. Studies of mutants showing effects of this kind have been combined in some cases, notably in yeasts, with biochemical work, to determine which of the proteins of oxidative phosphorylation, for example, might be coded by mitochondrial DNA. It seems that most, or all, of these enzymes are coded by chromosomal DNA. The mitochondrial DNA seems to carry genes which specify the mitochondrial ribosomal–RNA sequences (shown by hybridization) some at least of the mitochondrial ribosomal proteins, and some protein components of the inner mitochondrial membrane. The evidence that some of the mitochondrial ribosomal proteins are coded by the mitochondrial DNA comes from study of some antibiotic-resistant strains of yeast which show non-Mendelian maternal inheritance of this trait. In the mutants a ribosomal protein is presumed to be altered, so that the mutants are no longer sensitive to the antibiotics that prevent mitochondrial protein synthesis in normal yeasts.

The evidence that other mitochondrial proteins, such as cytochrome c, are coded by nuclear genes comes from study of yeast mutants (some of the so-called petite strains) with defective respiration and morphologically abnormal mitochondria, which show Mendelian inheritance of these defects. The mitochondrial proteins which are coded by mRNA from chromosomal genes seem to be made on cytoplasmic rather than on mitochondrial ribosomes. Cytochrome c is located in the intact mitochondrion on the *outer* surface of the inner membrane and hence has to pass, after synthesis, only across the outer membrane. Alternatively, cytochrome c could be synthesized, as for export, into the lumen of the endoplasmic reticulum, and then by temporary fusion of the reticulum membrane and the outer mitochondrial membrane it could pass to the space between the two mitochondrial membranes.

As regards the main role of the mitochondrial DNA, we are left with the hypothesis that the synthesis of a mitochondrial membrane protein, or perhaps a limited number of specific membrane components *within* the mitochondrion, confers on the inner mitochondrial membrane specific binding properties which lead to the assembly of the enzymes of oxidative phosphorylation in functional form in and on this membrane. If this is the case it would be the local synthesis of these specific membrane proteins which would lead to the assembly of oxidative-phosphorylation multi-enzyme-carrier complexes in mitochondria, and not on other membranes of the cell. This might be one reason for having DNA in mitochondria, and a separate protein assembly line. Another possible advantage in this way of doing things is that it may allow control of mitochondrial division and DNA duplication, independent of cell division and chromosomal DNA duplication, thus allowing increase in mitochondrial numbers where energy is needed in the form of ATP.

Chloroplasts also contain their own DNA, and in much larger amounts than mitochondria. The total amount of DNA in a chloroplast (mol. wt. $2\text{–}6 \times 10^9$) is comparable with the amount in bacteria, though whether the chloroplast DNA is a single molecule of this size (as in *E. coli*) or made up of a number of smaller molecules, or a number of copies of a smaller genome, is not yet known. Chloroplasts, like mitochondria, contain their own apparatus for protein synthesis with 70 *S* ribosomes sensitive to those antibiotics which inactivate bacterial ribosomes. Hybridization experiments show that chloroplast DNA codes for the chloroplast ribosomal RNA. Experiments in which uptake of labelled amino acids into different proteins is studied, under conditions where chloroplast protein synthesis is inhibited by chloramphenicol, suggest that chloroplast DNA may carry the genes for some chloroplast cytochromes and ribosomal proteins, but this is still an area of considerable uncertainty. For chloroplasts, as for mitochondria, there may be an advantage in having DNA, ribosomes and enzymes, etc., for protein synthesis within these organelles, so that their division and multiplication is independent of cell multiplication and division, and can be 'turned on' when photosynthetic activity is needed.

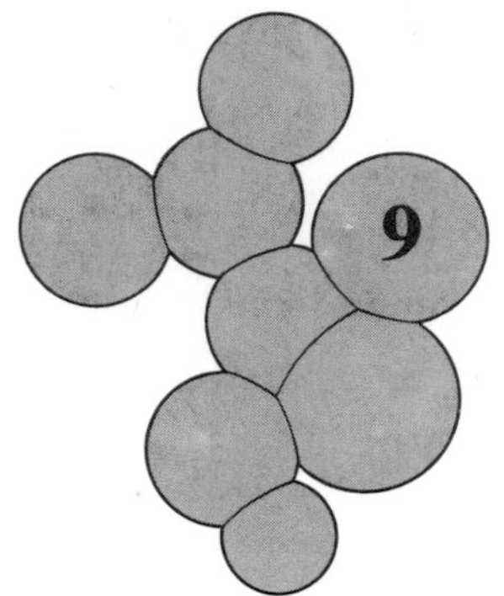

Haemoglobin and cytochrome c: molecular disease and molecular evolution

We have seen in previous chapters, that significant advances in modern genetics have come from work on one of the simplest of all organisms – bacteriophage; it is surprising that equally dramatic advances have been made in a field which we might expect to be much more complex: human genetics. The same principle, the selection of rare mutants from very large populations, can be exploited in the two types of study. In the case of human abnormal haemoglobins, world-wide investigations offer the possibility of picking up rare variants. Recent progress in the study of human haemoglobins derives from such studies, and from our detailed knowledge of the structure of this protein.

Haemoglobin is a medium-sized protein, having a molecular weight of about 67000; each molecule contains four haem groups (Fig. 9.1). Each haem group

```
                  CH3    CH=CH2
                   |      |
                   C======C
                   |      |
             HC—C         C——CH
              ||   \\   /      |
    CH3—C—C         N         C=C—CH3
        ||    \     :      /     |
        ||      N··Fe—N          |
        ||    //    |      \     |
    CH2—C—C         N         C=C—CH=CH2
    |      |      /   \        |
    CH2   HC=C         C===CH
    |          |       |
    COOH       C=======C
               |       |
               CH2     CH3
               |
               CH2
               |
               COOH
```

Fig. 9.1 The chemical formula of the haem group (ferrous protoporphyrin)

is attached to a molecular sub-unit, of which there are two pairs; each sub-unit is formed by the complex folding of a single polypeptide chain (see Chapter 2). The sub-units of the haemoglobin found in normal adult men and women

(HbA) are called α and β, so that the molecule may be represented by a figure:

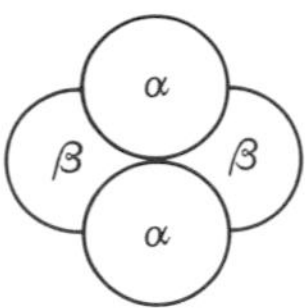

or by the formula $\alpha_2\beta_2$. There are 141 amino acids in the α-chain and 146 in the β-chain. The arrangement of the sub-units is illustrated in Fig. 2.15.

Sickle-cell anaemia

More than sixty years ago, Herrick, in Chicago, investigated the blood of an anaemic West Indian student and noticed that his red cells had a peculiar elongated, sickle shape. They took up this form in a low partial pressure of oxygen, when for instance they had been left for some time under a sealed cover-glass (Fig. 9.2). In well-oxygenated blood they were a more normal shape.

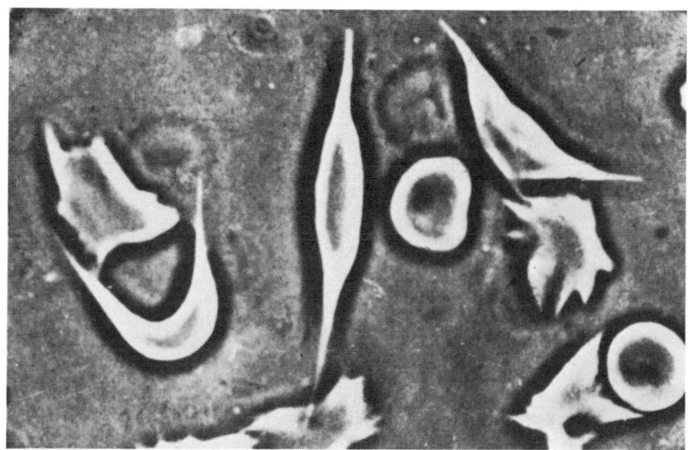

Fig. 9.2 Photomicrograph of sickle cells by phase-contrast microscopy. Magnification approximately ×1 500. (*From* Bessis, 1956.)

It soon became apparent that this anaemia, associated with sickling, was to be found in other patients and that Herrick had described a new disease entity. There are two forms of the condition; in one there is severe anaemia, and in the other the patients have no symptons except under severely anoxic conditions as, for example, in unpressurized aircraft. In both forms of the condition the red cells sickle when they are kept in low enough concentrations of oxygen. The severe anaemia, and the mild form – which came to be called the sickle-cell *trait* – are widespread for instance in Central and West Africa and amongst men and women whose ancestors came from these areas. The two forms of the disease are found to occur in one and the same family. In many parts of Africa the child with sickle-cell anaemia often dies before he is two years old, but with higher living standards the children tend to live longer. It is estimated that there are about 50 000 blacks with sickle-cell anaemia in the United States, and about 2 million with sickle-cell trait.

In 1949, Pauling and his colleagues at the California Institute of Technology were investigating the haemoglobin of red cells from sickle-cell anaemic patients, using the technique of electrophoresis. They found that the haemoglobin of these patients behaved differently from the haemoglobin of normal red cells:

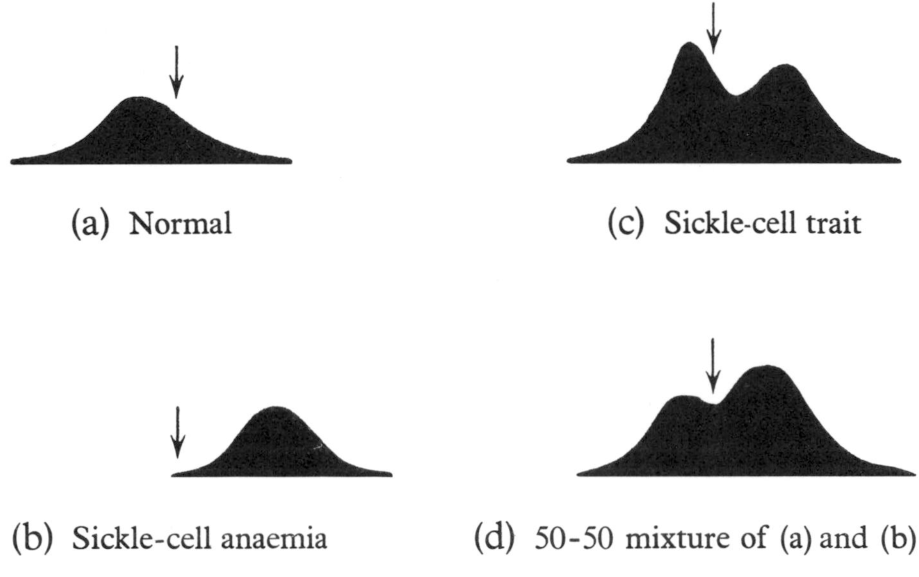

Fig. 9.3 Moving boundary electrophoresis patterns for normal and sickle-cell carbon-monoxy-haemoglobins in phosphate buffer at pH 6·9. At this pH normal haemoglobin carries net negative charge, and the sickle-cell haemoglobin net positive charge. The arrows indicate the starting positions. (*From* Pauling *et al.*, 1949.)

if the two haemoglobins, as carbon-monoxy-haemoglobin, were run at pH 6·9 in the Tiselius apparatus (see Chapter 1), the sickle-cell haemoglobin moved towards the cathode and the normal adult haemoglobin moved towards the anode (Fig. 9.3). The two haemoglobins differed in the net charge on the molecules and there was therefore a chemical difference between them. It was soon shown that the prosthetic groups, in this case the carbon monoxide derivative of haem, were similar in the two compounds, and it was therefore in the protein moiety that they differed.

In the same year as Pauling published this work, Neel and Beet made independent investigations into the pedigrees of families in which there was a high incidence of sickling. A typical family tree is shown in Fig. 9.4. We can see that a child who inherited the sickle-cell character from both parents had the severe

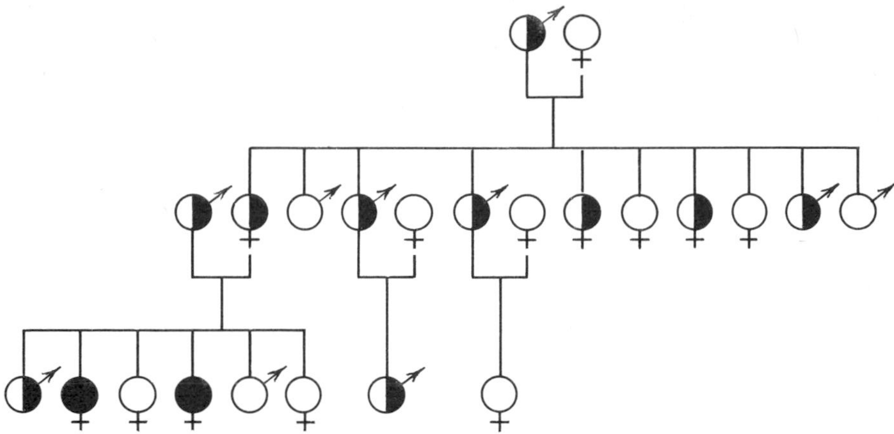

Fig. 9.4 A family tree showing an example of the inheritance of sickle-cell anaemia (●) and sickle-cell trait (◑). Members of the family whose red cells do not sickle are indicated by open circles (○). (*From* Harris, 1959, after Neel.)

form of the disease (sickle-cell anaemia) while the child who inherited the character from only one parent had the mild form of the disease (sickle-cell trait). In other words, the homozygous form had a severe anaemia; the heterozygous form was symptomless or only had a mild anaemia. The electrophoretic picture showed that only the abnormal haemoglobin, called HbS, was present in the homozygous disease, while it was present together with normal haemoglobin (HbA) in the heterozygous form (Fig. 9.3). In the subject with sickle-cell trait all the red cells have an equal tendency to sickle at low oxygen partial pressures, which means that HbA and HbS are produced by the same erythropoietic cell and are present together in each red cell.

By 1950 it was thus known that the sickling disorder was inherited and was characterized by the production of a slightly different haemoglobin in the red cells, the difference being in the protein part of the molecule. In the following year, Perutz and Mitchison at Cambridge University showed that HbS crystallized out of solution more readily than HbA when the two proteins were tested at a low oxygen partial pressure, that is in the reduced form. At high oxygen tensions, that is as oxyhaemoglobin, the solubilities were similar (Fig. 9.5). Moreover, their investigations showed that the haemoglobin in cells which had taken up the sickle shape was in para-crystalline form. It is this that distorts the red cell into a sickle shape; haemoglobin is of course in high concentration in red cells. Blood containing sickle cells is more viscous than normal; the abnormal cells lock together. This happens *in vivo* in the peripheral parts of the circulation when the oxygen tension falls; the blood vessels may become occluded with masses of impacted clumps. Further anoxia and increased sickling result. Some of the impacted cells may haemolyse. Such a sequence is seen in patients with sickle-cell anaemia, giving rise to sudden

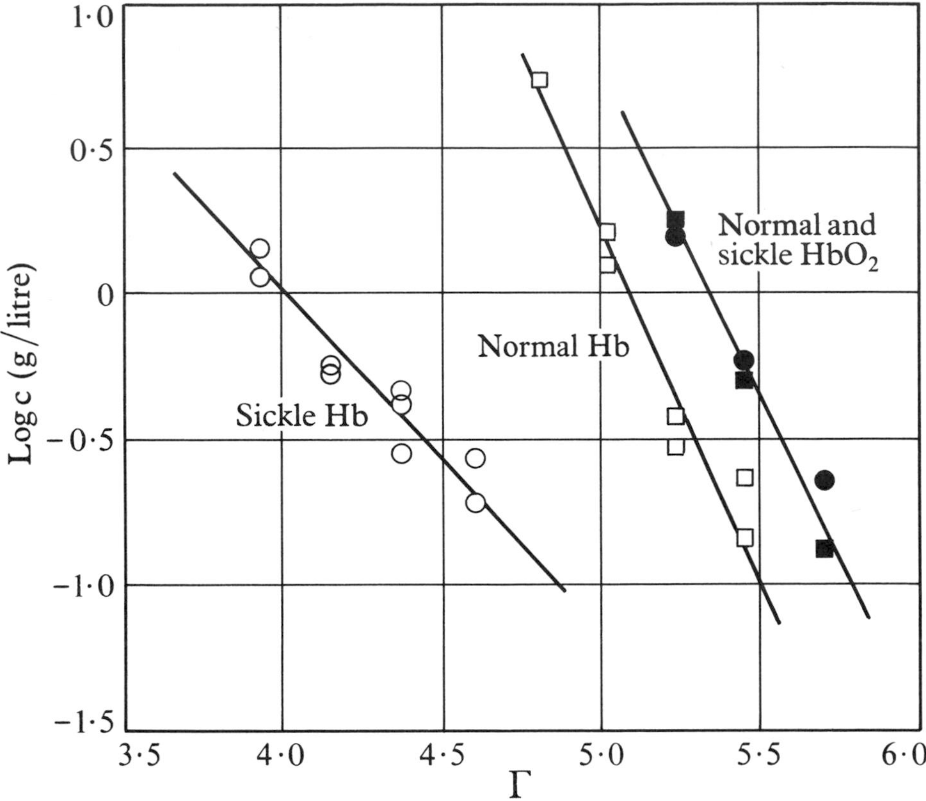

Fig. 9.5 Solubilities of normal and sickle-cell haemoglobins. ○ sickle haemoglobin (reduced); ● sickle oxyhaemoglobin; □ normal haemoglobin (reduced); ■ normal oxyhaemoglobin. Solubility ($\log_{10}$ of saturated concentration in g/litre) is plotted against ionic strength of phosphate buffer ($\Gamma = \frac{1}{2}\Sigma\, c_i z_i$, where c_i is the concentration of each ion in moles/litre and z_i its valency). (Modified *from* Perutz and Mitchison, 1950.)

painful 'crises' in the joints, bones and muscles, and vascular accidents to the brain, kidney and spleen.

In 1955–57, Ingram, also of Cambridge, discovered the chemical differences between the proteins HbA and HbS using a very precise method. The technique was that called 'finger-printing' and had been devised by Sanger to elucidate the amino-acid sequences in the insulin molecule. The protein is first hydrolysed by alkali, or by proteolytic enzymes. During an alkaline hydrolysis peptide linkages are broken at random; if proteolytic enzymes are used a precise splitting of the molecule is achieved, for each of these enzymes breaks the polypeptide chains only at specific peptide linkages. (Trypsin splits polypeptides at linkages adjacent to lysine and arginine side chains, leaving all other linkages untouched.) In the next stage of the technique, the resultant mixtures of peptides are placed on filter paper and separated by electrophoresis. The paper is then turned at right angles and a further separation achieved by subjecting the spread-out polypeptides to chromatography. The sort of pattern obtained using tryptic digestion of normal adult haemoglobin (HbA) is shown in Fig. 9.6. A total of 26 fragments have separated, each one representing a small peptide. When HbS was treated in this way, the pattern obtained was seen to be similar, except in

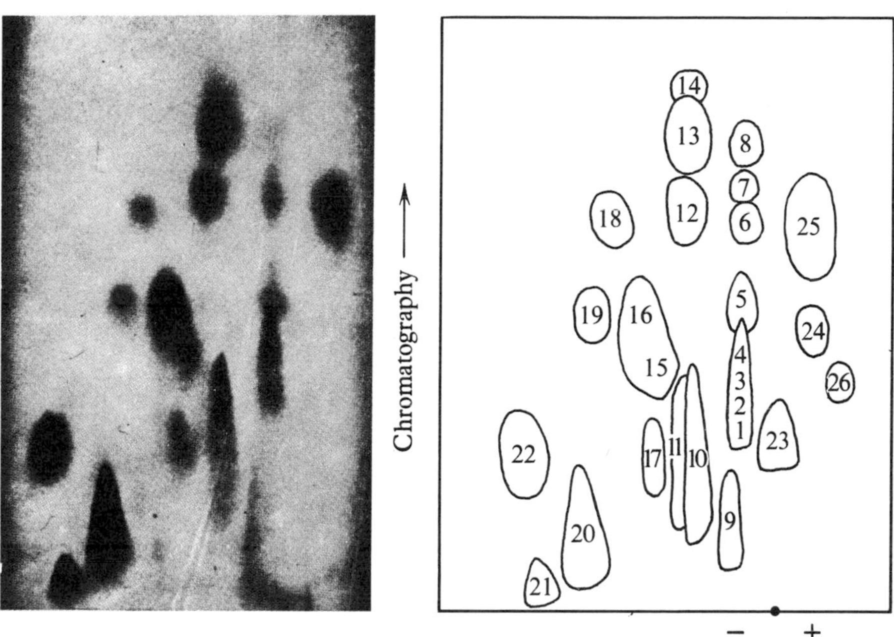

Fig. 9.6 'Finger-print' pattern of a tryptic digest of HbA. The filter paper on the left shows a typical pattern, each spot corresponding to a peptide or group of peptides. The sketch on the right shows the different peptides that can be resolved in HbA patterns of this kind, by varying conditions, and shows also the number allocated to each peptide. (*From* Ingram, 1961.)

one respect – the peptide numbered 4 had been shifted in position during the electrophoresis, towards the negative pole (Fig. 9.7*a*). This part of the paper was cut out and analysed; peptide 4 was found to be a peptide of 9 amino acids. Ingram and Hunt then determined the sequence of these amino acids for HbS and HbA. There was only one difference: valine had replaced glutamic acid in the sixth position along the chain (position A3 in Table 9.1).

We have seen already that a haemoglobin molecule consists of two pairs of sub-units. When split in two, like this:

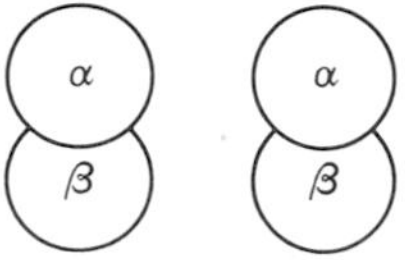

the halves are identical and each contain about 300 amino acids. Of these, only one was changed in HbS. The sequence was altered at this one point only. The replacement of charged glutamic acid by uncharged valine is responsible for the electrophoretic differences first observed by Pauling and later used by Ingram

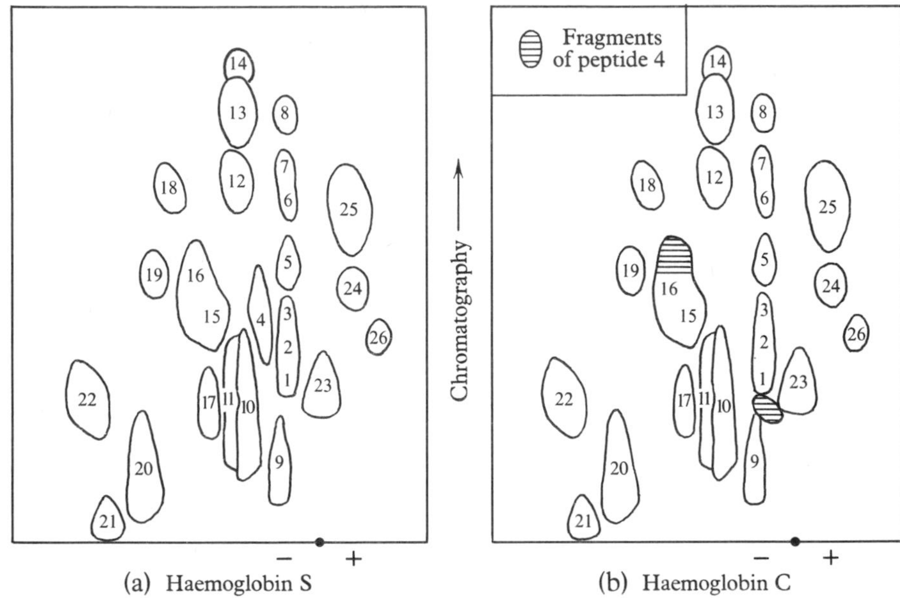

Fig. 9.7 'Finger-print' patterns for HbS and HbC. The pattern of HbS is identical with that of HbA (Fig. 9.6) except that peptide 4 is found a little to the left of its original position. In the pattern for HbC part of peptide 4 (now split into two peptides – see text) is found above peptide 16 and part between peptides 9 and 23. (Modified *from* Ingram, 1961.)

in his separation. A probable explanation of the relative insolubility of deoxygenated HbS is that the amino acid replacement allows intermolecular hydrophobic aggregation in the paracrystalline form.

Pauling was therefore justified in his far-sighted prediction when, in 1949, he called sickle-cell anaemia a molecular disease: an inherited alteration of only one amino acid in each half of a haemoglobin molecule leads to a haemoglobin whose solubility is low at low oxygen pressures; this haemoglobin will form crystals within the red cells, leading to a distortion of the cell and to obstructive vascular lesions characteristic of the disease.

The elucidation of the defect, or inborn error, of haemoglobin synthesis in sickle-cell anaemia and trait has a bearing on our ideas of protein synthesis in general.

First, we can see how accurately proteins are made. The polypeptide chain is a precise sequence of particular amino acids; alteration in certain cases of even one of these may lead to profound changes. In sickle-cell anaemia the changes can represent the difference between life and death. However, as we shall see later, it seems that only some amino acids in the sequence are as important as this. Some can be changed without any effect on the protein's function.

Secondly, we can see that the precise amino-acid sequence of a protein can be determined genetically. In the homozygous sickle-cell anaemia only the abnormal protein is produced, in the heterozygous trait the normal and the abnormal proteins are both produced; in normal individuals there is no sickle-cell haemoglobin present. In the heterozygous condition, haemoglobin of one sequence is being produced under the control of the maternal gene, and another of slightly different sequence under control of the paternal gene.

These facts can be accounted for by assuming that the sequence of bases in the DNA of the gene responsible for sickle-cell anaemia has been altered at one point, so that the corresponding amino-acid sequence is altered also. In the heterozygous condition, sickle-cell trait, neither gene is dominant and both HbA and HbS are synthesized, although the former is made at a slightly faster rate, the two being present in the proportion 60/40.

Geographical distribution of the sickle-cell gene

We have already seen that the incidence of sickling in Central and West Africa is high. The gene is also to be found elsewhere in the world (Fig. 9.8). In certain parts of Africa as many as 40 per cent of people in a local area may show the sickle-cell trait; it is not uncommon to find an incidence of 15 per cent. This is an astonishing situation for it means that a gene is being perpetuated which, in the homozygous form, is so disadvantageous that a person with this form rarely survives to reproductive maturity. We can rule out continuing mutation as the factor responsible for this situation, for the rate needed to keep an incidence of 15 per cent in a population would have to be impossibly high. We have to ask therefore what selective advantage there is to the heterozygous person which outweighs the greatest possible disadvantage to the homozygous. Allison made the very interesting suggestion that people with the sickle-cell trait had a lower incidence of malaria than those with normal haemoglobin. This suggestion would explain the geographical distribution of the disease, since the sickle-cell areas are in the malarious areas of the world, or those areas that have until recently been malarious. It seems well demonstrated that the severe forms of malaria, such as cerebral malaria, do not develop as commonly in people with the sickle-cell trait as in the normal members of the population. Moreover, for *P. falciparium* malaria, infants with the trait are relatively resistant to the initial infection. (The situation is less clear-cut in adults who have developed some degree of immunity to the disease.) We may then expect that when malaria is abolished from an area, the sickling gene will be at a selective disadvantage and will tend to die out. This may be happening, for example, in the West Indies and the U.S.A. It is informative to note that during the years of the slave trade, Africans were taken from what is now Ghana, where there is a high incidence of

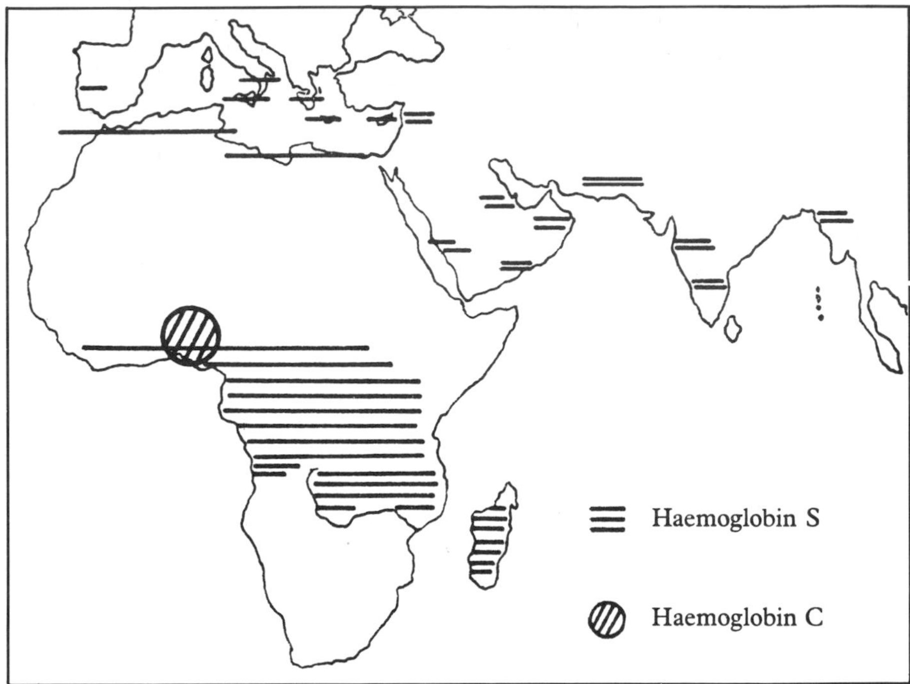

Fig. 9.8 A map showing the areas of major incidence of HbS and HbC. (*From* Lehmann, 1959.)

the HbS gene and endemic sickle-cell anaemia, to various areas of America and the Caribbean. Among their descendants in Surinam there is a high incidence of malaria, and the HbS gene is commonly found; in Curaçao, however, malaria has not been common, and the population has a low frequency of HbS – the gene is apparently dying out. The fact that the gene is found somewhat more frequently in the poor, in various parts of the world, may also fit in with these ideas.

Why the malaria parasite finds the blood of a person with the trait less satisfactory than normal blood is not known. It may be that the parasite does not thrive so well in a cell in which some of the haemoglobin is HbS, or possibly the parasite causes a low oxygen tension in the cells, producing sickling and destruction of both cell and parasites; or some secondary effect of the sickle-cell condition may confer increased resistance to infection by the parasite.

Other abnormal haemoglobins

Pauling's discoveries led to an intensive investigation, using electrophoretic techniques, into the blood of people with anaemias of unknown cause. Patients were discovered who apparently had a sickle-cell anaemia (the homozygous condition) but, surprisingly, the red cells of only *one* parent showed the sickling phenomenon. The haemoglobin of these apparently homozygous cases showed, however, an unusual electrophoresis pattern: about half migrated as HbS while the rest moved towards the cathode faster than HbS. There were therefore two haemoglobins present, HbS which caused the cells to show sickling, and HbC, as it is now called, the fast migrating one. The homozygous form of the HbC disorder has also been found. But it is important to note that no case has been discovered in which all three haemoglobins, HbS, HbC and HbA, coexist in the blood of the same individual. The genes governing the production of these proteins are apparently alleles and in any one individual can exist in only two forms, one derived from each parent.

It is obviously of considerable interest to find out where the abnormality of HbC lies in the molecule. This question was investigated by Ingram, and the finger-printing pattern obtained with HbC is shown in Fig. 9.7*b*. Instead of peptide 4, there are now two peptides; the position in the primary structure of haemoglobin occupied by glutamic acid in HbA and valine in HbS is, in HbC, occupied by lysine. Since trypsin breaks the lysine linkages, two peptides will appear in the pattern obtained after finger-printing. Moreover, the replacement of the glutamic acid by the uncharged valine in HbS, and by the positively charged lysine in HbC, will explain the greater rate of migration of HbC towards the cathode during electrophoresis. In HbC disease, the DNA is apparently altered at the same small region as in HbS but in a different way, for the resulting proteins are altered at the same point.

The changes of amino acids in HbS and HbC affect the β sub-units of the haemoglobin molecule. We may therefore write HbA, HbS and HbC as:

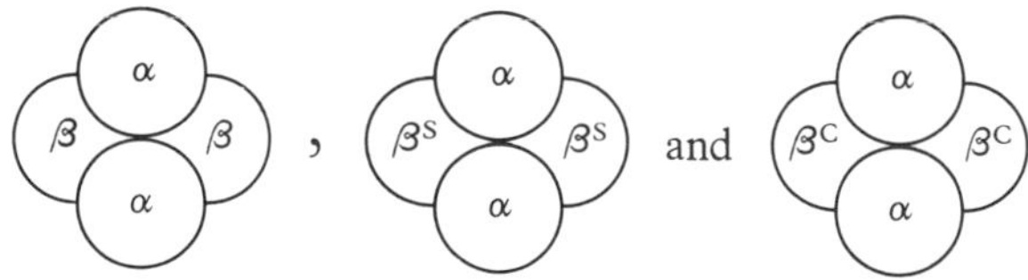

or $\alpha_2\beta_2$, $\alpha_2\beta_2^S$ and $\alpha_2\beta_2^C$. It is possible that the HbC gene arose as a mutation, not from HbA, but from HbS.

Not all abnormalities of the haemoglobin molecule are associated with disease. In one symptomless haemoglobin abnormality there is again a change in the β sub-unit and again it is in peptide 4; but in this case the seventh, and not the sixth, amino acid is affected. The seventh amino acid in this peptide in normal haemoglobin is also glutamic acid, and it is replaced in this abnormal haemoglobin by glycine.

More than 100 abnormal haemoglobins have now been described in which the particular 'lesion' is at other sites of the β-chain, or in the α-chain (Table 9.1). (Since each sub-unit contains a single polypeptide chain, the same labels can be used for the chains as for the sub-units.) In a case described by Atwater, and by Baglioni and Ingram, there were four haemoglobins present in the blood: $\alpha_2\beta_2$, $\alpha_2^G\beta_2$, $\alpha_2\beta_2^C$ and $\alpha_2^G\beta_2^C$. Two different α-chains were present, and two

TABLE 9.1 Sequences of myoglobin and haemoglobin chains. Substitutions characteristic of a number of abnormal human haemoglobins are as follows: in the α-chain, HbI, asp at A14; Norfolk, asp at EG; M (Boston) tyr at E7; B (Philadelphia) lys at E17. In the β-chain, HbS, val at A3; HbC, lys at A3; G (San Jose) gly at A4; E, lys at B8; M (Emory) tyr at E7; M (Milwaukee 1) glu at E11; D (Punjab) glu NH_2 at GH4.

	NA1	2	3	A1	2	3	4	5	6	7	8	9	10	11	12	13	14	15	A16	AB1	B1	2	3	4	5	6
MYOGLOBIN	val	…	leu	ser	glu	gly	glu	trp	gln	leu	val	leu	his	val	trp	ala	lys	val	glu	ala	asp	val	ala	gly	his	gly
HAEMOGLOBIN Horse α	val	…	leu	ser	ala	ala	asp	lys	thr	asn	val	lys	ala	ala	trp	ser	lys	val	gly	gly	his	ala	gly	glu	tyr	gly
β	val	gln	leu	ser	gly	glu	glu	lys	ala	ala	val	leu	ala	leu	trp	asp	lys	val	asn	…	…	glu	glu	glu	val	gly
Human α	val	…	leu	ser	pro	ala	asp	lys	thr	asn	val	lys	ala	ala	trp	gly	lys	val	gly	ala	his	ala	gly	glu	tyr	gly
β	val	his	leu	thr	pro	glu	glu	lys	ser	ala	val	thr	ala	leu	trp	gly	lys	val	asn	…	…	val	asp	glu	val	gly
γ	gly	his	phe	thr	glu	glu	asp	lys	ala	thr	ilu	thr	ser	leu	trp	gly	lys	val	asn	…	…	val	glu	asp	ala	gly
δ	val	his	leu	thr	pro	glu	glu	lys	thr	ala	val	asn	ala	leu	trp	gly	lys	val	asn	…	…	val	asp	ala	val	gly

	B7	8	9	10	11	12	13	14	15	16	C1	2	3	4	5	6	7	CD1	2	3	4	5	6	7	8	D1
MYOGLOBIN	gln	asp	ilu	leu	ilu	arg	leu	phe	lys	ser	his	pro	glu	thr	leu	glu	lys	phe	asp	arg	phe	lys	his	leu	lys	thr
HAEMOGLOBIN Horse α	ala	glu	ala	leu	glu	arg	met	phe	leu	gly	phe	pro	thr	thr	lys	thr	tyr	phe	pro	his	phe	…	asp	leu	ser	his
β	gly	glu	ala	leu	gly	arg	leu	leu	val	val	tyr	pro	trp	thr	gln	arg	phe	phe	asp	ser	phe	gly	asp	leu	ser	gly
Human α	ala	glu	ala	leu	glu	arg	met	phe	leu	ser	phe	pro	thr	thr	lys	thr	tyr	phe	pro	his	phe	…	asp	leu	ser	his
β	gly	glu	ala	leu	gly	arg	leu	leu	val	val	tyr	pro	trp	thr	gln	arg	phe	phe	glu	ser	phe	gly	asp	leu	ser	thr
γ	gly	glu	thr	leu	gly	arg	leu	leu	val	val	tyr	pro	trp	thr	gln	arg	phe	phe	asp	ser	phe	gly	asn	leu	ser	ser
δ	gly	glu	ala	leu	gly	arg	leu	leu	val	val	tyr	pro	trp	thr	gln	arg	phe	phe	glu	ser	phe	gly	asp	leu	ser	ser

	D2	3	4	5	6	7	E1	2	3	4	5	6	7	8	9	10	11	12	13	14	E15	16	17	18	19	20
MYOGLOBIN	glu	ala	glu	met	lys	ala	ser	glu	asp	leu	lys	lys	his	gly	val	thr	val	leu	thr	ala	leu	gly	ala	ilu	leu	lys
HAEMOGLOBIN Horse α	…	…	…	…	…	gly	ser	ala	gln	val	lys	ala	his	gly	lys	lys	val	ala	asp	gly	leu	thr	leu	ala	val	gly
β	pro	asp	ala	val	met	gly	asn	pro	lys	val	lys	ala	his	gly	lys	lys	val	leu	his	ser	phe	gly	glu	gly	val	his
Human α	…	…	…	…	…	gly	ser	ala	gln	val	lys	gly	his	gly	lys	lys	val	ala	asp	ala	leu	thr	asn	ala	val	ala
β	pro	asp	ala	val	met	gly	asn	pro	lys	val	lys	ala	his	gly	lys	lys	val	leu	gly	ala	phe	ser	asp	gly	leu	ala
γ	ala	ser	ala	ilu	met	gly	asn	pro	lys	val	lys	ala	his	gly	lys	lys	val	leu	thr	ser	leu	gly	asp	ala	ilu	lys
δ	pro	asp	ala	val	met	gly	asn	pro	lys	val	lys	ala	his	gly	lys	lys	val	leu	gly	ala	phe	ser	asp	gly	leu	ala

		EF1	2	3	4	5	6	7	8	F1	2	3	4	F5	6	7	8	9	FG1	2	3	4	5	G1	2	3	4
MYOGLOBIN		lys	lys	gly	his	his	glu	ala	glu	leu	lys	pro	leu	ala	gln	ser	his	ala	thr	lys	his	lys	ilu	pro	ilu	lys	tyr
HAEMOGLOBIN	Horse α	his	leu	asp	asp	leu	pro	gly	ala	leu	ser	asp	leu	ser	asn	leu	his	ala	his	lys	leu	arg	val	asp	pro	val	asn
	β	his	leu	asp	asn	leu	lys	gly	thr	phe	ala	ala	leu	ser	glu	leu	his	cys	asp	lys	leu	his	val	asp	pro	glu	asn
	Human α	his	val	asp	asp	met	pro	asn	ala	leu	ser	ala	leu	ser	asp	leu	his	ala	his	lys	leu	arg	val	asp	pro	val	asn
	β	his	leu	asp	asn	leu	lys	gly	thr	phe	ala	thr	leu	ser	glu	leu	his	cys	asp	lys	leu	his	val	asp	pro	glu	asn
	γ	his	leu	asp	asp	leu	lys	gly	thr	phe	ala	gln	leu	ser	glu	leu	his	cys	asp	lys	leu	his	val	asp	pro	glu	asn
	δ	his	leu	asp	asn	leu	lys	gly	thr	phe	ser	gln	leu	ser	glu	leu	his	cys	asp	lys	leu	his	val	asp	pro	glu	asn

		G5	6	7	8	G9	10	11	12	13	14	15	16	17	18	19	GH1	2	3	4	5	6	H1	2	3	4	5
MYOGLOBIN		leu	glu	phe	ilu	ser	glu	ala	ilu	ilu	his	val	leu	his	ser	arg	his	pro	gly	asn	phe	gly	ala	asp	ala	gln	gly
HAEMOGLOBIN	Horse α	phe	lys	leu	leu	ser	his	cys	leu	leu	ser	thr	leu	ala	val	his	leu	pro	asn	asp	phe	thr	pro	ala	val	his	ala
	β	phe	arg	leu	leu	gly	asn	val	leu	ala	leu	val	val	ala	arg	his	phe	gly	lys	asp	phe	thr	pro	glu	leu	gln	ala
	Human α	phe	lys	leu	leu	ser	his	cys	leu	leu	val	thr	leu	ala	ala	his	leu	pro	ala	glu	phe	thr	pro	ala	val	his	ala
	β	phe	arg	leu	leu	gly	asn	val	leu	val	cys	val	leu	ala	his	his	phe	gly	lys	glu	phe	thr	pro	pro	val	gln	ala
	γ	phe	lys	leu	leu	gly	asn	val	leu	val	thr	val	leu	ala	ilu	his	phe	gly	lys	glu	phe	thr	pro	glu	val	gln	ala
	δ	phe	arg	leu	leu	gly	asn	val	leu	val	cys	val	leu	ala	arg	asn	phe	gly	lys	glu	phe	thr	pro	gln	met	gln	ala

		H6	7	8	9	10	11	12	13	14	15	16	17	18	19	20	H21	22	23	24	HC1	2	3	4	5
MYOGLOBIN		ala	met	asn	lys	ala	leu	glu	leu	phe	arg	lys	asp	ilu	ala	ala	lys	tyr	lys	glu	leu	gly	tyr	gln	gly
HAEMOGLOBIN	Horse α	ser	leu	asp	lys	phe	leu	ser	ser	val	ser	thr	val	leu	thr	ser	lys	tyr	arg						
	β	ser	tyr	gln	lys	val	val	ala	gly	val	ala	asn	ala	leu	ala	his	lys	tyr	his						
	Human α	ser	leu	asp	lys	phe	leu	ala	ser	val	ser	thr	val	leu	thr	ser	lys	tyr	arg						
	β	ala	tyr	gln	lys	val	val	ala	gly	val	ala	asn	ala	leu	ala	his	lys	tyr	his						
	γ	ser	trp	gln	lys	met	val	thr	gly	val	ala	ser	ala	leu	ser	ser	arg	tyr	his						
	δ	ala	tyr	gln	lys	val	val	ala	gly	val	ala	asn	ala	leu	ala	his	lys	tyr	his						

gln = glutamine, aln = asparagine. The myoglobin sequence is that of whale myoglobin.

(*From* WATSON, H. C. *and* KENDREW, J. C. (1961) *Nature Lond.*, **190**, 670 and DAYHOFF, Margaret O. (1969) *Atlas of Protein Sequence and Structure* **4**. National Biochemical Research Foundation, Silver Spring, Maryland, U.S.A.)

different β-chains. This patient's father has therefore transmitted different α genes, and different β genes, from those of the mother. All four genes have full expression, since all four haemoglobins are present in about the same concentration. The genes determining the sequences in the α- and β-chains were found in this investigation to segregate independently. It is very improbable therefore that these two genes occupy adjacent positions on a chromosome, and they probably lie on different chromosomes.

In the abnormal haemoglobin disease, HbH, there seems to be a complete failure to make α-chains. The β-chains are, however, made quite normally. These, in absence of any α-chains, associate to make a haemoglobin whose formula may be written β_4. It is only when α-chains are absent, or made in relatively small quantities, that β-chains associate in this way.

Figure 9.9 shows the electrophoretic separation of a few of the abnormal haemoglobins that have been described; many more are known (Table 9.1). There are also, most probably, abnormal haemoglobins that remain undiscovered. Where a mutation has produced a change in the amino-acid sequence

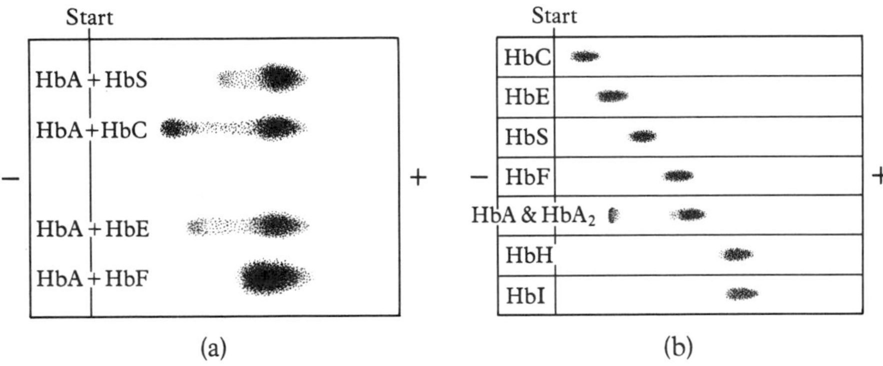

Fig. 9.9 Filter-paper electrophoresis of haemoglobins. The various haemoglobins included in this figure (HbA, HbC, HbE, etc.) are described in the text of this chapter or in Table 9.1. In these electrophoresis runs, at pH 8·6, all the haemoglobins carry a net negative charge. Haemoglobins with greater net charge move further to the right during the run. The starting line is indicated on the left of each pattern. The upper picture (*a*) shows the actual results obtained for some of the different haemoglobins. (Modified *from* Huisman, 1959.) The sketch (*b*) indicates relative mobilities more clearly.

in which one amino acid is replaced by another having a similar charge the molecule is electrophoretically unchanged. There may be no clinical abnormality, and other techniques in common use may not separate this abnormal haemoglobin from HbA. The mutation therefore remains 'silent' and undiscovered. There are 438 bases in the DNA coding for the 146 amino acids of the β-chain. In a mutation it is most likely that there will be one base change only. Theoretically 25% of such base changes would not alter the amino acid sequence at all, 3% would result in premature chain termination and 72% would produce a change in a single amino acid, a third of these being electrophoretically silent.

If the substitution of an amino acid affects regions of the chain at the surface of the molecule, there will probably be no change in the functional characteristics of the haemoglobin. However, in the region where sub-units interact, or adjacent to the haem, or at an internal site affecting the stability of the molecule, then either the mutation will be lethal or give rise to physiological abnormality.

The oxygen uptake curve of haemoglobin and the Bohr effect

The interaction between sub-units in the four sub-unit structure of the haemoglobin molecule has two effects on oxygen binding. First, as O_2 molecules become bound to one or more of the haemoglobin sub-units, this increases the affinity for oxygen at the binding sites on the other sub-units. This effect produces an S-shaped curve of oxygen uptake against partial pressure of oxygen (Fig. 9.10) and is physiologically important, since the blood can change from

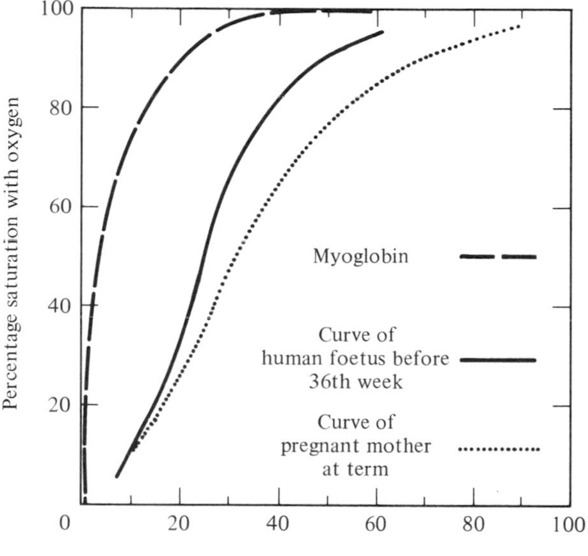

Fig. 9.10 Oxygen dissociation curves for human foetal and maternal blood and for myoglobin. (Modified *from* Walker and Turnbull, 1955.)

being fully saturated to being largely deoxygenated in moving from the lungs to the tissues, when the partial pressure of oxygen changes from 100 to 20 mm of mercury. The second feature of haemoglobin oxygen binding which becomes modified by interaction between sub-units is the dependence of oxygen uptake on pH. This is called the Bohr effect. Under acid conditions the oxygen uptake curve is shifted to the right, if plotted as in Fig. 9.10, the haemoglobin molecule binding less oxygen at the same partial pressure. This effect is also of physiological importance for it leads to greater release of oxygen from the blood in the acid conditions of a fatigued muscle.

X-ray diffraction analysis of the haemoglobin structure in the oxygenated and deoxygenated states has now become precise enough for Perutz to suggest how the sub-units may interact in the haemoglobin molecule to produce these effects.

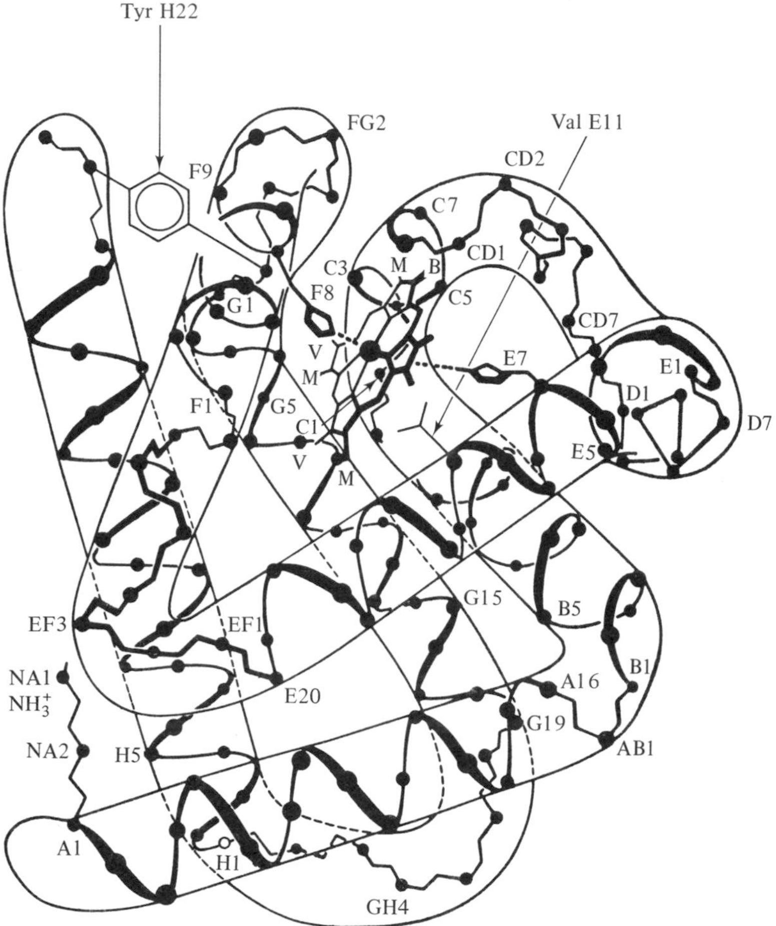

Fig. 9.11 The tertiary structure of the haemoglobin β-chain in the deoxygenated state. The primary structure of the chain is given in Table 9.1. (*From* Perutz, 1970.)

The tertiary structure of the deoxygenated haemoglobin β-chain is shown in Fig. 9.11. The folding of the α-chain is very similar. When the haem group binds an oxygen molecule, this O_2 moves in in place of a water molecule between the histidine residue E7 and the haem iron. This alters the electron distribution within the iron atom, and its interaction with the nitrogens of the haem group (Fig. 9.1) in such a way that the iron atom moves about 0·08 nm into the plane of the haem group. The histidine F8, which is complexed to the iron atom, and with it the whole α-helical segment F are drawn in closer to the α-helical region H, and the tyrosine residue H22 near the carboxyl end of the chain (Table 9.1) is expelled from its pocket between the two helices (Fig. 9.12). In deoxygenated haemoglobin, with the tyrosines tucked between the helices, the carboxyl ends

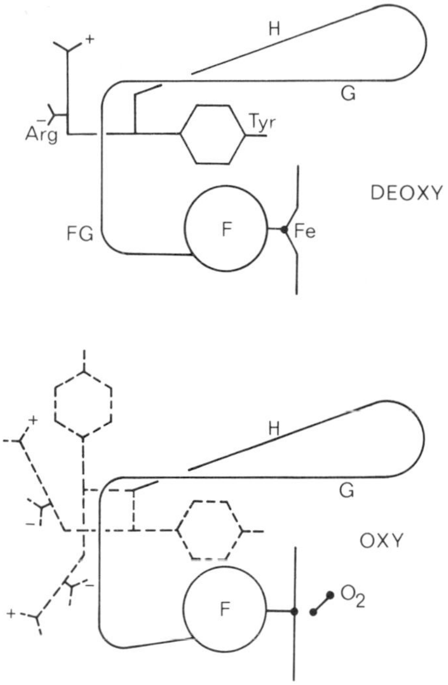

Fig. 9.12 A diagram illustrating the expulsion of tyrosine (H22 in the sequence of Table 9.1) and the carboxy-terminal arginine of the haemoglobin α-chain (H23) when oxygen is bound to the haem iron atom. (*From* Perutz, 1970.)

of the chains of all four sub-units lie in the surface of the molecule. They are electrostatically bound to other charged groups in the surface through the terminal $—COO^-$ groups and the positive side chains of the terminal residues, which are arginine for the α-chain and histidine for the β-chain (Table 9.1). In oxygenated haemoglobin, when the tyrosine is expelled from between the helices, the terminal residues are no longer held to the molecular surface, but protrude out, with freedom of movement, into the surrounding solution. This effect is included schematically in Fig. 9.12.

In the haemoglobin molecule the contact region between α- and β-chains is dovetailed so that the CD region of one chain fits into the FG region of the other (Fig. 2.15). If, starting from the deoxygenated state one sub-unit takes up an oxygen and goes over into the conformation with the tyrosine expelled, the 'fit' at the FG region is energetically less favourable. The haemoglobin molecule is stressed. But if more oxygen molecules are now taken up, by the other sub-units, the stress can be relieved as the molecules move into a new symmetrical fully-oxygenated state. This leads to increased oxygen affinity at the unoccupied sites of a partially oxygenated molecule. The oxygen affinity of haemoglobin is, in the red cell, modified by the presence of 2,3-diphosphoglycerate (DPG). This compound, present in normal red cells in approximately equal molar concentration to that of haemoglobin, binds strongly to the deoxygenated form, providing a further factor contributing to the uptake curve of Fig. 9.10.

The Bohr effect arises from the movement of the terminal carboxyl residues with their basic side-chains out from the haemoglobin surface. This modifies the pKs of the side-chain group, the terminal $—COO^-$ group, and also of the groups to which they were previously bound in the protein surface. Consider the terminal histidine of the β-chain, protruding into solution with its p$K \sim 7$. Under slightly acid conditions this group spends a higher proportion of its time in the positively charged state, and it is in this positively charged state that it binds into the surface in the deoxygenated conformation. The deoxygenated conformation is favoured by acid conditions. A number of groups are involved, on both the α and β chains, which act in this way, cooperatively contributing to the Bohr effect.

The fine changes in tertiary and quaternary structure of haemoglobin on oxygen binding have been taken as a model for the kind of *allosteric* effect that may take place in feedback inhibition or modification of enzyme activity (Chapter 3) where a conformation change in the inhibitor-binding sub-units of the multi-subunit enzyme leads to a conformation change in the catalytic sub-units, modifying their enzymic activity. The haemoglobin sub-unit may be considered in a sense as a mechanical amplifier in which atomic movement of the iron atom of less than 0·1 nm leads to atomic rearrangements of about 0·7 nm at the boundary between sub-units.

The haemoglobins during development

The blood of newborn children has a higher affinity for oxygen than adult blood. As may be seen from the oxygen dissociation curve for human foetal and adult bloods (Fig. 9.10), at any given partial pressure of oxygen, foetal blood picks up more oxygen than that of the adult. This assists the transfer of oxygen from the maternal to the foetal haemoglobin. It has long been recognized that foetal or newborn blood contains a different haemoglobin, more resistant than that of the adult to denaturation with either acid or alkali. The foetal haemoglobin molecule (HbF) actually has much the same affinity for oxygen as adult haemoglobin, but foetal blood has higher oxygen affinity than adult blood because HbF is less sensitive than HbA to the presence of DPG.

The replacement of foetal haemoglobin by adult haemoglobin has been studied mostly in mammals and particularly in man. When the child is born the level of foetal haemoglobin is about 80 per cent; during the next few weeks it falls quite rapidly (Fig. 9.13) and the level of HbA rises accordingly. It has been suggested that the stimulus to the replacement may be the different oxygen partial pressures of the foetus and the infant. But the situation must be more complicated than this. HbA has been detected at an early stage of intra-uterine life: one observer reports the appearance of HbA as early as thirteen weeks, while at birth about 20 per cent is present. The oxygen supply to the foetal haematopoietic tissue is low in late pregnancy, yet HbA production is rising during this period.

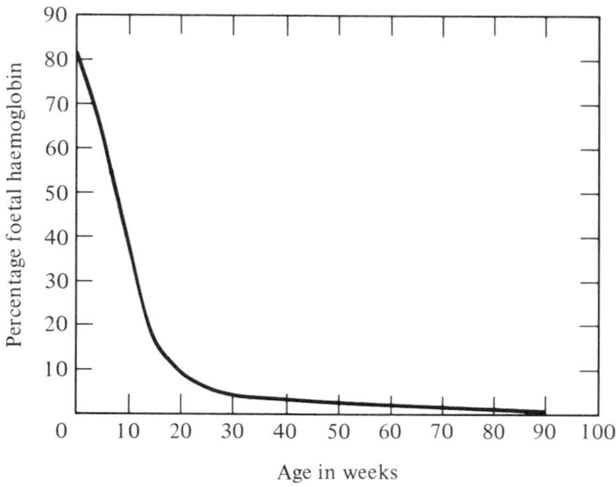

Fig. 9.13 The disappearance of foetal haemoglobin from the blood of the normal child in the weeks following birth. (Modified *from* Jonxis, 1959.)

HbF production normally continues into early infancy but by about the fourth month almost all the haemoglobin is of the adult form; in adult life there is no HbF in the blood. Moreover, once lost, the ability to produce any (or all but minimal quantities) of HbF cannot be recovered. It is of interest to note that people brought up since infancy at low oxygen partial pressures, for example Sherpas or Peruvian Indians, make the change-over from HbF to HbA in a normal manner.

There is evidence that in the early stages of human embryonic life (at about thirteen weeks) another haemoglobin is present in addition to HbF. Less is known of this primitive or embryonic haemoglobin, but other species of mammals have a similar succession of haemoglobins during development.

The haemoglobin produced by foetuses in a homozygous sickle-cell inheritance is a normal HbF. Studies of HbF show why this is so. HbF contains the same α-chains as adult haemoglobin but another dimer replaces β_2, different in many respects from either β_2 or β_2^S. This has been called γ_2. It is only when the production of γ-chains gives place to the production of β-chains that the genetic defect of sickle-cell disease shows, for this defect only affects the production of β-chains,

β_2^S being formed rather than β_2. The β-chains of normal adult human haemoglobin and those of abnormal haemoglobins such as HbS or HbC differ, as we have seen, only in one amino acid. The γ-chains of foetal haemoglobin differ from the β-chains of HbA to a much greater extent (Table 9.1).

Just as there are mutations in the β-chains of HbA producing abnormal haemoglobins such as HbS, so there are mutations in the γ-chain producing abnormal foetal haemoglobins. There is also a situation in which the production of α- and β-chains is completely suppressed giving rise to an association of two dimers of γ: Hb Bart's has a formula which may be written γ_4. More recently it has been shown that there are two, non-allelic, types of γ-chain produced, which differ in only one amino acid; glycine or alanine may be present at position H13. These chains are called γ^G and γ^A respectively.

In 1955, Kunkel and his colleagues noticed that there was a small amount of haemoglobin in normal adult blood which, on electrophoresis at an alkaline pH, moved more slowly than HbA (Fig. 9.9*b*). About 2·5 per cent of this minor component is present in normal blood. Analysis of this haemoglobin, usually called HbA_2, has shown that the α-chain is the same as that present in the major normal component, but that a modified form of the β-chain is present in this haemoglobin. These modified chains have been called δ-chains (see Table 9.1).

It is interesting to note that in HbG disease with a major component $\alpha_2^G\beta_2$, the minor component is $\alpha_2^G\delta_2$, which is electrophoretically different from HbA_2 ($\alpha_2\delta_2$). There are also variants of HbA_2 in which the δ-chain has undergone a mutation, similar to the mutations we have seen in the α-, β- and γ-chains.

A description may now be given of haemoglobin synthesis which draws together many of the discoveries described in this and other chapters.

A precise sequence of bases in a particular DNA molecule determines the precise amino-acid sequence found in the α-chains of human haemoglobin. The chains are produced by stepwise addition of amino acids beginning at the N-terminal end. In people with abnormalities, a mutation has occurred which changes the DNA base sequence; the chains are made, therefore, with the amino-acid sequence altered at one point. α-chains, normal or abnormal, are produced in foetal life and continue to be produced in the adult.

In foetal life, a different DNA sequence determines a different haemoglobin chain, the γ-chain. In the foetus, α- and γ-chains are normally made at the same time and combine to form the four-unit haemoglobin $\alpha_2\gamma_2$. This is foetal haemoglobin. Soon after birth, there is suppression of the activity of the DNA base sequence determining γ-chains, and another DNA base sequence becomes active resulting in the production of β-chains. The β-chains join with α-chains to form $\alpha_2\beta_2$ – adult haemoglobin. Another DNA base sequence determines the amino-acid sequence of the δ-chains, which likewise associate with α-chains to form $\alpha_2\delta_2$. This haemoglobin is produced in a smaller amount and is therefore present only as a minor component in normal adult blood. Each of these chains, α, β, γ or δ, is thus controlled by a different DNA base sequence (i.e. a different gene) and each is liable to an alteration, or mutation, leading to the formation of abnormal sequences in these chains and hence to abnormal haemoglobins.

We have discussed in this section conditions in which haemoglobins are structurally abnormal. There are a variety of other clinical conditions, the thalassaemias, in which there are abnormalities in the *rates* of synthesis of different haemoglobin chains.

In a later section of this chapter we shall discuss the evolution of these different haemoglobin genes, but we must consider first, in the next section, the early stages in the evolution of biological molecules.

The origin of life

At the beginning of the present century, the spontaneous formation of a unit as complex as a living cell, from small molecules, seemed so improbable an event as to suggest, if not divine intervention, at least some curious special circumstances, initiating living processes in primeval times. Ideas current since medieval and earlier times of the spontaneous generation of living organisms under present-day terrestrial conditions, from mud or slime, had apparently been refuted by scientific investigation.

However, as more became known about the molecular structure of cell constituents and of organisms such as viruses, which lie on the borderline between living and non-living, it became apparent that the problem was not being properly conceived. Consider, for example, the comparable problem in biological evolution: the sudden emergence of man in a world of unicellular organisms would be highly improbable, but his gradual evolution by natural means through a series of organisms of increasing complexity is quite conceivable. In a similar way the immediate development of a cell from small molecules is highly improbable, but if the cell is the culmination of a long process of biochemical and biomolecular evolution preceding biological evolution, a process in which small molecules aggregate to form large molecules, and large molecules aggregate into cell precursors, each step in this process could have been a natural and quite probable event.

The evolutionary side of molecular biology includes both the study of the period preceding the appearance of units as complex as the cells of present-day organisms, and also the study of the molecular events underlying the specialization and interaction of cells during biological evolution. We are concerned in this section with the earlier period. Most of the experiments in this field have been designed to show that certain types of molecule *could* have been formed by natural means, under primeval conditions. It is another matter, of course, to prove that the reactions demonstrated in this way were important reactions in bimolecular evolution, particularly as we cannot be certain of the exact composition of the atmosphere, and other conditions, prevailing in primeval times.

In view of the central role played by proteins in all biochemical processes, some experiments carried out by Miller seem of particular significance. He showed that if a mixture of the gases supposed to be present in the earth's primeval atmosphere – hydrogen, methane, ammonia vapour, water vapour, carbon monoxide, hydrogen sulphide, etc. – are enclosed in a glass vessel above a water surface, and an electric discharge passed through the gas mixture for a time, then a variety of organic molecules, including amino acids, appear in solution in the water. Electric storms could have produced a similar effect in primeval times, as could ultraviolet sunlight, at that time much more intense in the lower atmosphere than it is now, because of the absence of oxygen, and hence of the ultraviolet-absorbing ozone layer in the upper atmosphere.

Before enzymes or other proteins could have been formed, two apparently rather improbable events would have had to take place: the amino acids, present presumably at very dilute concentration in the early seas and lakes, would have had to be concentrated in some way, and they would have had to be linked together into polypeptide chains. The formation of peptide bonds requires energy, and at the present time protein synthesis in cells requires the presence of enzymes, nucleic acids, and ATP. 'High-energy' compounds of various sorts (metaphosphates, phosphocyanate, etc.) might have been present in the primeval seas, but there could, of course, have been no protein enzymes available to catalyse the formation of the first proteins.

Extending Miller's work, with these problems in mind, Wilson has shown that in the reactive conditions of a discharge passing through a mixture of gases simulating the primeval atmosphere, methane molecules become linked together to form larger hydrocarbon molecules. Analysis of the gas mixture at the end of his experiments showed hydrocarbons containing up to five carbon atoms. He found also that at an early stage of these experiments, a surface layer of polar lipids formed at the air-water interface (polar lipids are introduced and discussed in Chapter 4), and as the experiment continued these molecules became cross-linked to form an insoluble film. The fact that no molecules containing more than five carbon atoms were found in the gas mixture does not show that no larger hydrocarbon molecules were formed, for larger molecules would be expected to condense as a liquid lipid layer on the water surface. Wilson concluded therefore, that molecules synthesized mainly from methane, containing six carbon atoms or more (both pure hydrocarbons and other molecules with additional —SH and $—NH_2$ groups contributed by reaction with the hydrogen sulphide and ammonia vapour) were condensing onto the surface in his experiments and were subsequently being cross-linked to form an insoluble film, either by direct effect of the discharge, or by reactive atoms and molecules produced by the discharge. As in the case of Miller's experiments we may suppose that similar effects might be produced by ultraviolet radiation.

The possibility of the formation, under primeval conditions, of surface films predominantly hydrocarbon in nature, but with an array of —SH, $—NH_2$ and =O groups on the aqueous side, offers promising ground for further speculation and experiments on biomolecular evolution. In the first place a film of this type provides a surface onto which small molecules such as the amino acids, present in the primeval seas and lakes, could be adsorbed and concentrated. In the second place, the polar groups of the film might sometimes be fortuitously so disposed that they catalysed chemical reactions between the molecules adsorbed to this surface, including possibly the joining together of amino acids to form polypeptide chains. The first step in this process might have been the covalent linking of amino acids to the surface film by 'high-energy' bonds (similar to the linking of amino acids to transfer-RNA in present-day protein synthesis). The energy for the formation of such bonds could have been derived, as the energy for cross-linking molecules in the surface film is supposed to be derived, from the energy of electric storms, ultraviolet radiation, etc., either directly, or via reactive 'high-energy' compounds.

There is a further most interesting possible result of the formation of surface films. Suppose, as illustrated in Fig. 9.14, that a wind blows over a lake covered with a surface film. Droplets torn off the surface by the wind, as they fall back into the lake, could be carried below the surface, covered by a biomolecular layer of polar lipids (cf. discussion of membrane structure in Chapter 4). Provided the surface layers are stable enough (as a result of cross-linking between their constituent molecules), this droplet will be stable and enclose a small volume of fluid cut off from its environment by a membrane. This membrane, moreover, could have catalytic groups on both its inner and outer surfaces so that we have the possibility of specific and active transport of small molecules or ions across the membrane. A primitive cell of this type could grow by accretion of further material to its surface. (As its size increased it would be broken up by currents, particularly in turbulent waters, and the hydrophobic edges of broken membrane would tend to rejoin to form smaller 'daughter' cells.)

The accretion of material, either to a membrane or to a molecular aggregate would have been at first an undirected process leading to 'growth' by addition of similar but not necessarily identical material. Accurate self-reproduction requires material which will either catalyse its own synthesis, or bind to itself only the molecules of which it is composed. A crystal does this in a simple way, and it is possible that some small molecules or polymers may have acted as condensation sites for identical small molecules or polymers in primitive membranes (or molecular aggregates) to initiate self-reproduction of at least a part of the growing material. It seems possible also that nucleotides and nucleic acid chains may have been formed at a relatively early stage, and controlled the further synthesis of nucleic acids to form chains of complementary sequence, as DNA does today.

It is much more difficult to visualize how nucleic acids could have come to control the assembly of amino acids in specific sequences. Early proteins were presumably formed by the linking together of amino acids in a random way, and the complex interaction between proteins and nucleic acids, which is at present found to control protein synthesis (see Chapter 8) must be the result of long ages of biomolecular evolution. (Factors such as ability to survive and reproduce, potential for further development, etc., are supposed to prove the mechanism for the evolution of primitive cells and molecular aggregates just as they do for later biological evolution.) The detailed structure of ribosomes, when this is elucidated, may yield important clues about the evolution of present-day

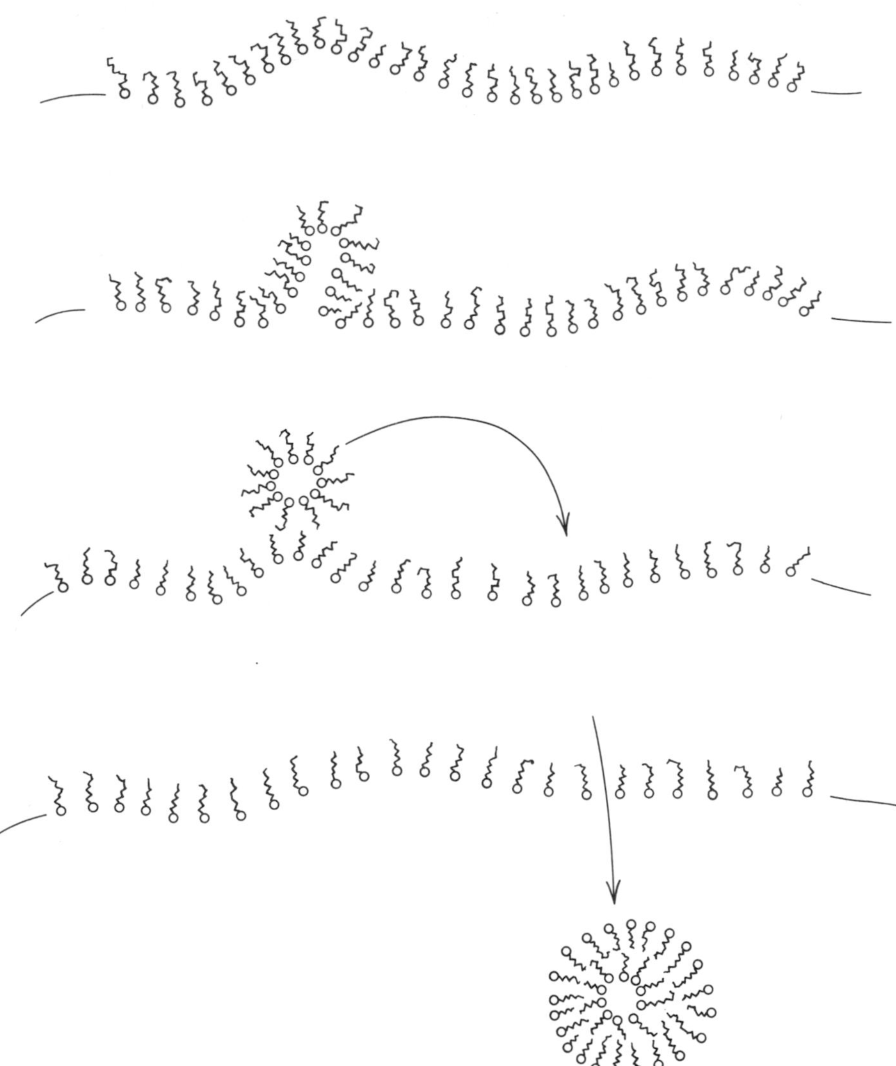

Fig. 9.14 Schematic picture of the way in which primitive cells might have been formed. A wind blowing across a lake covered with polar lipids is supposed to tear off droplets, which then fall back into the lake. Polar lipids are represented as in Fig. 4.13.

processes of protein synthesis. Ribosomes can be regarded as multi-protein complexes with the assembly of the component proteins controlled by the nucleotide sequence of ribosomal RNA. Possibly the nucleic-acid chains of primitive cells did not control *synthesis* of proteins but only the *assembly* of primitive enzymes into multi-enzyme complexes, and in this way self-replicating molecules became linked to 'useful' function (useful function in this sense meaning a function which enabled the primitive cell to survive, grow, divide and evolve).

It is possible to study experimentally the synthesis of polypeptides under primitive (non-enzymic) conditions and the effects of nucleotides on such synthesis. For example, Paecht-Horowitz and Fox and co-workers have studied the formation of polyalanine by condensation of alanine adenylate. Alanine adenylate contains alanine linked to adenine by a high-energy phosphate bond. It is found that condensation to polyalanine is favoured by adsorption to montmorillonite. Here we have the beginnings perhaps of the study of simple mechanisms by which nucleotides could influence amino acid polymerization and some experimental support for the idea originally put forward by Bernal, that adsorption of polymers to clays might have been important in prebiotic evolution. However, we cannot be certain that the first self-duplicating polymers were nucleic acids or the first code a nucleotide code. It is possible that primitive chains of say, amino sugars might have played a coding role in earlier times. Cairns-Smith has suggested even that mineral sheets of micas or clays could have acted as early self-replicating genes. Possibly so, but these clay genes could not have evolved into DNA genes, and at some point the present-day protein assembly mechanism must have got started in a simple way with genes influencing amino acid assembly made of either nucleic acid or some more primitive polymer, carrying $—NH_2$ and $=O$ groups to control self-replication through hydrogen bonding.

An interesting question in study of prebiotic evolution is that of molecular asymmetry. At the present time only L-amino acids are found in proteins, and D-amino acids are found only in some bacterial peptides (which might represent survival of some molecular mechanism from very early times). Most biologically important sugars found in nature are D-sugars. In the early formation of amino acids, nucleotides, polypeptides or polynucleotides there is no reason to expect predominance of L- or D-forms. Possibly individual polypeptide chains might have tended to polymerize all L- or all D- rather than as L-D mixed chains, but in this case one would have expected both L- and D-polypeptides to be present in the prebiotic environment. At a later stage as polynucleotides and polypeptides evolved some interaction, as precursor to the present genetic code, it might have happened that L-polypeptides spiralled into helices more readily with polynucleotides containing D-sugars, and conversely. Thus two types of early organism could have been competing for survival, some with L-polypeptides and D-sugar-polynucleotides, some with D-polypeptides and L-sugar-polynucleotides. There is no reason *a priori* why man should drive his horse-carriages, or modern cars, on the left or the right of the road. In fact at the present stage of man's development some countries drive on the right, some on the left. But it seems probable that in fifty years time or less (partly for economic reasons connected with the cost of manufacturing parts for cars) we shall all drive on the right. In some such way, in the competition between primitive organisms, the D-amino-acid organisms could have been starved out of existence in competition with the L-class for scarce nutrients they both required. The D-class, of course, would have had a 'common market' of a kind, for they could prey on one another, but could not use as nutrients the amino acids of the L-class. Similarly the L-class would have preyed mainly on one another. Once the D-class began to lose out, finding fewer bits around to incorporate into themselves of proteoids (protein aggregates) or primitive cells of their own kind, the organisms of the D-amino-acid class would have all gone down together. Admittedly these ideas are highly speculative, but they illustrate the kind of thinking that follows, as we begin to think of primitive cells and molecular aggregates as 'living', and suggest the kinds of experiment that are needed to trace out more certainly the most probable course of biomolecular evolution.

The transition from the prebiotic to the biological period of evolution, that is, the actual origin of life, is difficult to define with any precision. It may be taken perhaps as the point at which macromolecular complexes, proteoids or primitive cells, first came to incorporate a polymer which catalysed directly or indirectly its own replication. Without this property of self-duplication the evolutionary continuity and development that characterizes 'life' cannot be considered to have started. The time of these events may be estimated as somewhere around 3000–4000 million years ago. The age of the earth is estimated to be about 5000 million years and the fossil record of living organisms goes back about 2000 million years. This allows a 1000 million years or so for prebiotic evolution, and a further 1000 million years or so for development of the forms found in the earliest fossils.

The evolution of proteins

Although the early stages in the formation of proteins are still little understood, later stages of protein evolution are much more accessible to study. Since the formation of cells and the establishment of the present genetic code and protein assembly mechanisms, random base changes, chance deletions, and occasional abnormal duplications have constantly led to formation of new DNA sequences and new polypeptide chains. Proteins that were useful in primitive algae 1000 million years ago have survived in modified form in present day organisms. Protein amino-acid sequences provide a fascinating library of detailed information about molecular events over this whole period of biological evolution.

In Table 9.1 which shows the amino-acid sequences for human haemoglobin α-, β-, γ- and δ-chains, for horse α- and β-chains, and for whale myoglobin, it will be seen that at certain positions, or for short lengths of the sequence, the amino acids are identical in all seven chains, or sometimes differ only in the myoglobin chain. Where amino-acid sequences show this kind of correspondence in different proteins, or different chains, these sequences are said to be *homologous*. Often the amino acids in corresponding positions of two homologous sequences are not identical but one amino acid is replaced by another of similar type as, for example, isoleucine in place of leucine, or glutamic acid in place of aspartic acid. In earlier sections of this chapter we have seen how a single amino-acid substitution can sometimes have a profound effect on function, as in HbS. We now have to balance the picture by noting that in spite of the extensive differences in amino-acid sequence, as between haemoglobin α- and β-chains, both types of chain have similar tertiary structure (Chapter 2) and similar function. It is only amino acids in key positions which cannot be altered without loss of function.

Where two sequences show significant homology (that is, significantly more similarity than would be expected by picking two sequences at random from the twenty or so natural amino acids) there are in principle two possible explanations. Either both sequences have arisen by independent evolution from a common ancestral gene, or if both proteins have a similar function, as in the case of the α- and β-subunits of haemoglobin, certain amino acids may be required in defined positions in the chain for this function; then similarities of sequence could have arisen in evolution from two quite different initial sequences. In the evolution of haemoglobin and myoglobin, for example, two quite different sequences could have had the property of wrapping around a haem group, both proteins acting in a primitive way as oxygen carriers. Through evolution the sequences could have become modified to improve this function, and efficient function might have dictated in each case that certain types of amino acid be found at certain positions in the chain. It seems however highly improbable that this is what actually happened, and it is now generally accepted for a number of proteins, that homologous sequences have evolved from common ancestral sequences. The reason for this view is that comparison has now been made for some proteins, notably cytochrome c, for a wide variety of species ranging from yeast to man. It is found that if we make the assumption that the sequences found for cytochrome c arise from a common ancestral sequence and that random mutations have led to viable stable amino-acid substitutions at a roughly constant rate, then a phylogenetic tree can be constructed on sequence data alone which closely correspond to the phylogenetic tree constructed on other evidence. This seems to be true for cytochrome c, haemoglobin and myoglobin, and other proteins also. The greater the number of amino-acid substitutions the further a protein is in evolution from its ancestral form. Moreover the amino-acid substitutions that are found, now that the genetic code is known (Table 8.1) can usually be seen to be consistent with a single-base change in a coding triplet such as might be brought about by random mutation or, less commonly, changes involving two successive single-base changes. It is also necessary to allow that random mutation can cause occasional small insertions or deletions, so that when comparison is made between two sequences for homology, gaps are left as necessary in one or other chain to produce maximum correspondence of identical or similar amino acids. Finally it is also necessary to accept the possibility of genetic duplication. The underlying mechanism for this is not yet understood, but it is as though a DNA replicating enzyme can copy a genetic segment then go back and recopy it and continue on, so that the new DNA contains a repeat of this genetic segment. Thus originally, far enough back, the mammalian germ line carried only one haemoglobin-myoglobin type of gene,

TABLE 9.2 Amino-acid sequences for cytochrome c. (*From* DAYHOFF, Margaret O. (1969) *Atlas of Protein Sequence and Structure* **4**.)

		-8	-7	-6	-5	-4	-3	-2	-1	1				5					10					15					20					25				
Man, chimpanzee									a	G	D	V	E	K	G	K	K	I	F	I	M	K	C	S	Q	C	H	T	V	E	K	G	G	K	H	K	T	G
Rhesus monkey									a	G	D	V	E	K	G	K	K	I	F	I	M	K	C	S	Q	C	H	T	V	E	K	G	G	K	H	K	T	G
Horse									a	G	D	V	E	K	G	K	K	I	F	V	Q	K	C	A	Q	C	H	T	V	E	K	G	G	K	H	K	T	G
Donkey									a	G	D	V	E	K	G	K	K	I	F	V	Q	K	C	A	Q	C	H	T	V	E	K	G	G	K	H	K	T	G
Cow, pig, sheep									a	G	D	V	E	K	G	K	K	I	F	V	Q	K	C	A	Q	C	H	T	V	E	K	G	G	K	H	K	T	G
Dog									a	G	D	V	E	K	G	K	K	I	F	V	Q	K	C	A	Q	C	H	T	V	E	K	G	G	K	H	K	T	G
Rabbit									a	G	D	V	E	K	G	K	K	I	F	V	Q	K	C	A	Q	C	H	T	V	E	K	G	G	K	H	K	T	G
California grey whale									a	G	D	V	E	K	G	K	K	I	F	V	Q	K	C	A	Q	C	H	T	V	E	K	G	G	K	H	K	T	G
Great grey kangaroo									a	G	D	V	E	K	G	K	K	I	F	V	Q	K	C	A	Q	C	H	T	V	E	K	G	G	K	H	K	T	G
Chicken, turkey									a	G	D	I	E	K	G	K	K	I	F	V	Q	K	C	S	Q	C	H	T	V	E	K	G	G	K	H	K	T	G
Pigeon									a	G	D	I	E	K	G	K	K	I	F	V	Q	K	C	S	Q	C	H	T	V	E	K	G	G	K	H	K	T	G
Pekin duck									a	G	D	V	E	K	G	K	K	I	F	V	Q	K	C	S	Q	C	H	T	V	E	K	G	G	K	H	K	T	G
Snapping turtle									a	G	D	V	E	K	G	K	K	I	F	V	Q	K	C	A	Q	C	H	T	V	E	K	G	G	K	H	K	T	G
Rattlesnake									a	G	D	V	E	K	G	K	K	I	F	T	M	K	C	S	Q	C	H	T	V	E	K	G	G	K	H	K	T	G
Bullfrog									a	G	D	V	E	K	G	K	K	I	F	V	Q	K	C	A	Q	C	H	T	C	E	K	G	G	K	H	K	V	G
Tuna									a	G	D	V	A	K	G	K	K	T	F	V	Q	K	C	A	Q	C	H	T	V	E	N	G	G	K	H	K	V	G
Dogfish									a	G	D	V	E	K	G	K	K	V	F	V	Q	K	C	A	Q	C	H	T	V	E	N	G	G	K	H	K	T	G
Samia cynthia					h	G	V	P	A	G	N	A	E	N	G	K	K	I	F	V	Q	R	C	A	Q	C	H	T	V	E	A	G	G	K	H	K	V	G
Tobacco horn worm moth					h	G	V	P	A	G	N	A	D	N	G	K	K	I	F	V	Q	R	C	A	Q	C	H	T	V	E	A	G	G	K	H	K	V	G
Screw worm fly					h	G	V	P	A	G	D	V	E	K	G	K	K	I	F	V	Q	R	C	A	Q	C	H	T	V	E	A	G	G	K	H	K	V	G
Fruit fly (*Drosophila*)					h	G	V	P	A	G	D	V	E	K	G	K	K	L	F	V	Q	R	C	A	Q	C	H	T	V	E	A	G	G	K	H	K	V	G
Baker's yeast (iso-l)				h	T	E	F	K	A	G	S	A	K	K	G	A	T	L	F	K	T	R	C	E	L	C	H	T	V	E	K	G	G	P	H	K	V	G
Candida krusei			h	P	A	P	F	E	Q	G	S	A	K	K	G	A	T	L	F	K	T	R	C	A	E	C	H	T	I	E	A	G	G	P	H	K	V	G
Neurospora crassa					h	G	F	S	A	G	D	S	K	K	G	A	N	L	F	K	T	R	C	A	E	C	H	G	E	G	G	N	L	T	Q	K	I	G
Wheat germ	a	A	S	F	S	E	A	P	P	G	N	P	D	A	G	A	K	I	F	K	T	K	C	A	Q	C	H	T	V	D	A	G	A	G	H	K	Q	G

	30					35					40					45					50					55					60				
Man, chimpanzee	P	N	L	H	G	L	F	G	R	K	T	G	Q	A	P	G	Y	S	Y	T	A	A	N	K	N	K	G	I	I	W	G	E	D	T	L
Rhesus monkey	P	N	L	H	G	L	F	G	R	K	T	G	Q	A	P	G	Y	S	Y	T	A	A	N	K	N	K	G	I	I	W	G	E	D	T	L
Horse	P	N	L	H	G	L	F	G	R	K	T	G	Q	A	P	G	F	T	Y	T	D	A	N	K	N	K	G	I	T	W	K	E	E	T	L
Donkey	P	N	L	H	G	L	F	G	R	K	T	G	Q	A	P	G	F	S	Y	T	D	A	N	K	N	K	G	I	T	W	K	E	E	T	L
Cow, pig, sheep	P	N	L	H	G	L	F	G	Q	K	T	G	Q	A	P	G	F	S	Y	T	D	A	N	K	N	K	G	I	T	W	G	E	E	T	L
Dog	P	N	L	H	G	L	F	G	R	K	T	G	Q	A	P	G	F	S	Y	T	D	A	N	K	N	K	G	I	T	W	G	E	E	T	L
Rabbit	P	N	L	H	G	L	F	G	R	K	T	G	Q	A	V	G	F	S	Y	T	D	A	N	K	N	K	G	I	T	W	G	E	D	T	L
California grey whale	P	N	L	H	G	L	F	G	R	K	T	G	Q	A	V	G	F	S	Y	T	D	A	N	K	N	K	G	I	T	W	G	E	E	T	L
Great grey kangaroo	P	N	L	N	G	I	F	G	R	K	T	G	Q	A	P	G	F	T	Y	T	D	A	N	K	N	K	G	I	I	W	G	E	D	T	L
Chicken, Turkey	P	N	L	H	G	L	F	G	R	K	T	G	Q	A	E	G	F	S	Y	T	D	A	N	K	N	K	G	I	T	W	G	E	D	T	L
Pigeon	P	N	L	H	G	L	F	G	R	K	T	G	Q	A	E	G	F	S	Y	T	D	A	N	K	N	K	G	I	T	W	G	E	D	T	L
Pekin duck	P	N	L	H	G	L	F	G	R	K	T	G	Q	A	E	G	F	S	Y	T	D	A	N	K	N	K	G	I	T	W	G	E	D	T	L
Snapping turtle	P	N	L	N	G	L	I	G	R	K	T	G	Q	A	E	G	F	S	Y	T	E	A	N	K	N	K	G	I	T	W	G	E	E	T	L
Rattlesnake	P	N	L	H	G	L	F	G	R	K	T	G	Q	A	V	G	Y	S	Y	T	A	A	N	K	N	K	G	I	I	W	G	D	D	T	L
Bullfrog	P	N	L	Y	G	L	I	G	R	K	T	G	Q	A	A	G	F	S	Y	T	D	A	N	K	N	K	G	I	T	W	G	E	D	T	L
Tuna	P	N	L	W	G	L	F	G	R	K	T	G	Q	A	E	G	Y	S	Y	T	D	A	N	K	S	K	G	I	V	W	N	N	D	T	L
Dogfish	P	N	L	S	G	L	F	G	R	K	T	G	Q	A	Q	G	F	S	Y	T	D	A	N	K	S	K	G	I	T	W	Q	Q	E	T	L
Samia cynthia	P	N	L	H	G	F	Y	G	R	K	T	G	Q	A	P	G	F	S	Y	S	N	A	N	K	A	K	G	I	T	W	G	D	D	T	L
Tobacco horn worm moth	P	N	L	H	G	F	F	G	R	K	T	G	Q	A	P	G	F	S	Y	S	N	A	N	K	A	K	G	I	T	W	Q	D	D	T	L
Screw worm fly	P	N	L	H	G	L	F	G	R	K	T	G	Q	A	A	G	F	A	Y	T	N	A	N	K	A	K	G	I	T	W	Q	D	D	T	L
Fruit fly (*Drosophila*)	P	N	L	H	G	L	I	G	R	K	T	G	Q	A	A	G	F	A	Y	T	N	A	N	K	A	K	G	I	T	W	Q	D	D	T	L
Baker's yeast (iso-l)	P	N	L	H	G	I	F	G	R	H	S	G	Q	A	Q	G	Y	S	Y	T	D	A	N	I	K	K	N	V	L	W	D	E	N	N	M
Candida krusei	P	N	L	H	G	I	F	S	R	H	S	G	Q	A	Q	G	Y	S	Y	T	D	A	N	K	R	A	G	V	E	W	A	E	P	T	M
Neurospora crassa	P	A	L	H	G	L	F	G	R	K	T	G	S	V	D	G	Y	A	Y	T	D	A	N	K	Q	K	G	I	T	W	D	E	N	T	L
Wheat germ	P	N	L	H	G	L	F	G	R	Q	S	G	T	T	A	G	Y	S	Y	S	A	A	N	K	N	K	A	V	E	W	E	E	N	T	L

	65					70					75					80					85					90					95					100				104
Man, chimpanzee	M	E	Y	L	E	N	P	K	K	Y	I	P	G	T	K	M	I	F	V	G	I	K	K	K	E	E	R	A	D	L	I	A	Y	L	K	K	A	T	N	E
Rhesus monkey	M	E	Y	L	E	N	P	K	K	Y	I	P	G	T	K	M	I	F	V	G	I	K	K	K	E	E	R	A	D	L	I	A	Y	L	K	K	A	T	N	E
Horse	M	E	Y	L	E	N	P	K	K	Y	I	P	G	T	K	M	I	F	A	G	I	K	K	K	T	E	R	E	D	L	I	A	Y	L	K	K	A	T	N	E
Donkey	M	E	Y	L	E	N	P	K	K	Y	I	P	G	T	K	M	I	F	A	G	I	K	K	K	T	E	R	E	D	L	I	A	Y	L	K	K	A	T	N	E
Cow, pig, sheep	M	E	Y	L	E	N	P	K	K	Y	I	P	G	T	K	M	I	F	A	G	I	K	K	K	G	E	R	E	D	L	I	A	Y	L	K	K	A	T	N	E
Dog	M	E	Y	L	E	N	P	K	K	Y	I	P	G	T	K	M	I	F	A	G	I	K	K	T	G	E	R	A	D	L	I	A	Y	L	K	K	A	T	K	E
Rabbit	M	E	Y	L	E	N	P	K	K	Y	I	P	G	T	K	M	I	F	A	G	I	K	K	K	D	E	R	A	D	L	I	A	Y	L	K	K	A	T	N	E
California grey whale	M	E	Y	L	E	N	P	K	K	Y	I	P	G	T	K	M	I	F	A	G	I	K	K	K	G	E	R	A	D	L	I	A	Y	L	K	K	A	T	N	E
Great grey kangaroo	M	E	Y	L	E	N	P	K	K	Y	I	P	G	T	K	M	I	F	A	G	I	K	K	K	G	E	R	A	D	L	I	A	Y	L	K	K	A	T	N	E
Chicken, turkey	M	E	Y	L	E	N	P	K	K	Y	I	P	G	T	K	M	I	F	A	G	I	K	K	K	S	E	R	V	D	L	I	A	Y	L	K	D	A	T	S	K
Pigeon	M	E	Y	L	E	N	P	K	K	Y	I	P	G	T	K	M	I	F	A	G	I	K	K	K	A	E	R	A	D	L	I	A	Y	L	K	Q	A	T	A	K
Pekin duck	M	E	Y	L	E	N	P	K	K	Y	I	P	G	T	K	M	I	F	A	G	I	K	K	K	S	E	R	A	D	L	I	A	Y	L	K	D	A	T	A	K
Snapping turtle	M	E	Y	L	E	N	P	K	K	Y	I	P	G	T	K	M	I	F	A	G	I	K	K	K	A	E	R	A	D	L	I	A	Y	L	K	D	A	T	S	K
Rattlesnake	M	E	Y	L	E	N	P	K	K	Y	I	P	G	T	K	M	V	F	T	G	L	S	K	K	K	E	R	T	N	L	I	A	Y	L	K	E	K	T	A	A
Bullfrog	M	E	Y	L	E	N	P	K	K	Y	I	P	G	T	K	M	I	F	A	G	I	K	K	K	G	E	R	Q	D	L	I	A	Y	L	K	S	A	C	S	K
Tuna	M	E	Y	L	E	N	P	K	K	Y	I	P	G	T	K	M	I	F	A	G	I	K	K	K	G	E	R	Q	D	L	V	A	Y	L	K	S	A	T	S	–
Dogfish	R	I	Y	L	E	N	P	K	K	Y	I	P	G	T	K	M	I	F	A	G	L	K	K	K	S	E	R	Q	D	L	I	A	Y	L	K	K	T	A	A	S
Samia cynthia	F	E	Y	L	E	N	P	K	K	Y	I	P	G	T	K	M	V	F	A	G	L	K	K	A	N	E	R	A	D	L	I	A	Y	L	K	E	S	T	K	–
Tobacco horn worm moth	F	E	Y	L	E	N	P	K	K	Y	I	P	G	T	K	M	V	F	A	G	L	K	K	A	N	E	R	A	D	L	I	A	Y	L	K	Q	A	T	K	–
Screw wormfly	F	E	Y	L	E	N	P	K	K	Y	I	P	G	T	K	M	I	F	A	G	L	K	K	P	N	E	R	G	D	L	I	A	Y	L	K	S	A	T	K	–
Fruit fly (*Drosophila*)	F	E	Y	L	E	N	P	K	K	Y	I	P	G	T	K	M	I	F	A	G	L	K	K	P	N	E	R	G	D	L	I	A	Y	L	K	S	A	T	K	–
Baker's yeast (iso-l)	S	E	Y	L	T	N	P	X	K	Y	I	P	G	T	K	M	A	F	G	G	L	K	K	E	K	D	R	N	D	L	I	T	Y	L	K	K	A	C	E	–
Candida krusei	S	D	Y	L	E	N	P	X	K	Y	I	P	G	T	K	M	A	F	G	G	L	K	K	A	K	D	R	N	D	L	V	T	Y	M	L	E	A	S	K	–
Neurospora crassa	F	E	Y	L	E	N	P	X	K	Y	I	P	G	T	K	M	A	F	G	G	L	K	K	D	K	D	R	N	D	I	I	T	F	M	K	E	A	T	A	–
Wheat germ	Y	D	Y	L	L	N	P	X	K	Y	I	P	G	T	K	M	V	F	P	G	L	X	K	P	Q	D	R	A	D	L	I	A	Y	L	K	K	A	T	S	S

A, Ala; C, Cys; D, Asp; E, Glu; F, Phe; G, Gly; H, His; I, Ile; K, Lys; L, Leu; M, Met; N Asn; P, Pro; Q, Gln; R, Arg; S, Ser; T, Thr; V, Val; W, Trp; Y, Tyr; X, ε-*N*-trimethyllysine (coded as Lys); a, acetylated end, CH_3CO—NH—; h, free amino end, H_2N—.

later this duplicated and each gene continued to change and evolve independently to give myoglobin and a primitive single-chain haemoglobin. The haemoglobin in the primitive vertebrate, the lamprey, is a single chain with one haem. Later still the haemoglobin chain duplicated further to give the α- and β-chains, and later still duplicated again for γ-chains, etc. This process is illustrated in Fig. 9.15. Interaction between the haems in the four sub-unit

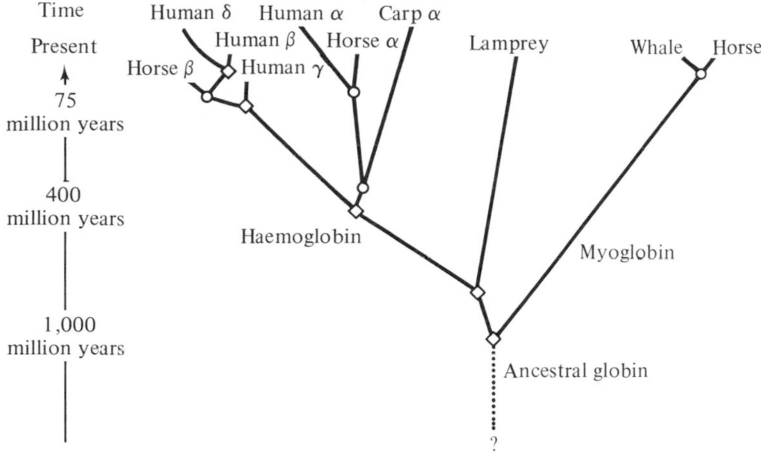

Fig. 9.15 Evolution of myoglobin and haemoglobin chains. (*From* Dayhoff, 1969.)

haemoglobin molecule confers physiological advantage, as a result of its effect on the oxygen uptake curve. Thus fortuitous mutations and amino-acid changes which caused the primitive protein to form a four sub-unit aggregate have been preserved, as have other features with survival value, like a separate type of foetal haemoglobin chain.

Cytochrome c

Cytochrome c (which also incorporates a haem group) because of its widespread occurrence in different organisms allows extensive study of species differences and of the many steps in the evolution of a protein chain (Figs. 9.16, 9.17). The cytochrome c sequence for a number of species is shown in Table 9.2. The haem group in this protein, unlike those of haemoglobin and myoglobin is covalently linked to the polypeptide chain, at cysteine residues 14 and 17.

As an example of how Fig. 9.16 is derived from Table 9.2 we can compare some mammalian cytochrome c sequences. Comparing dog and whale we see two changes: proline instead of valine at position 44 and threonine instead of lysine at position 88. Now we look at the pig sequence: it is similar to dog at position 44 (proline) similar to whale at 88 (lysine) but differs from both at

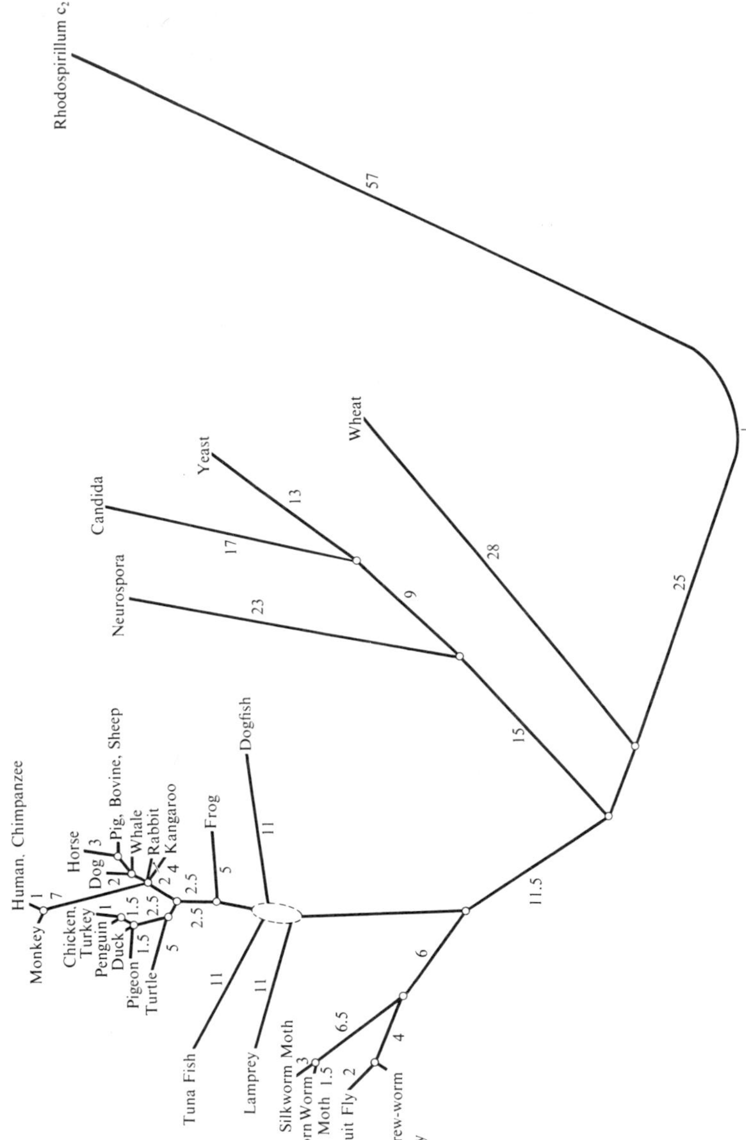

Fig. 9.16 Phylogenetic tree of cytochrome c. The numbers of inferred amino acid changes are shown on the tree. The point of earliest time cannot be determined directly from the sequences; it has been placed by assuming that, on the average, species change at the same rate. (*From* Dayhoff, 1969.)

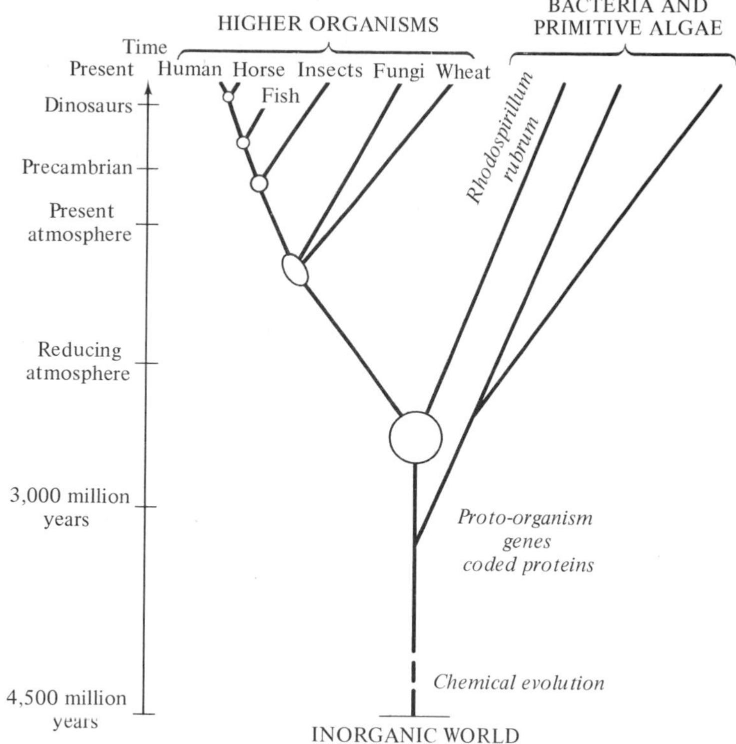

Fig. 9.17 Time scale of cytochrome c evolution. (*From* Dayhoff, 1969.)

position 92 where glutamic acid replaces alanine. Let us postulate now that a sequence existed at some stage of mammalian evolution with proline at position 44, lysine at 88 and alanine at position 92. From this postulated sequence we can get the whale sequence by putting valine instead of proline at 44, the dog sequence by putting threonine instead of lysine at 88, and the pig sequence by putting glutamic acid instead of alanine at 92 – a single amino-acid change in each case. If we now turn to Table 8.1 we see that the necessary recoding for these last two changes can be achieved in each case by a one-step mutation (a single base change). Substitution of valine for proline requires two base changes (hence the '2' on the dog line of Fig. 9.16). In Fig. 9.16, whale, dog and pig are shown branching out from a circle which represents the postulated precursor sequence.

A sequence from residue number 70 through to 80 is remarkedly invariant, and if this sequence is conserved unchanged through evolution it must play a vital role – any change in one of these amino acids is a lethal mutation. By contrast, certain parts of the sequence show wide variations in the amino acids present. It would be expected that these variable parts of the sequence would tend to lie on the outside of the molecule, away from the haem, and in lengths

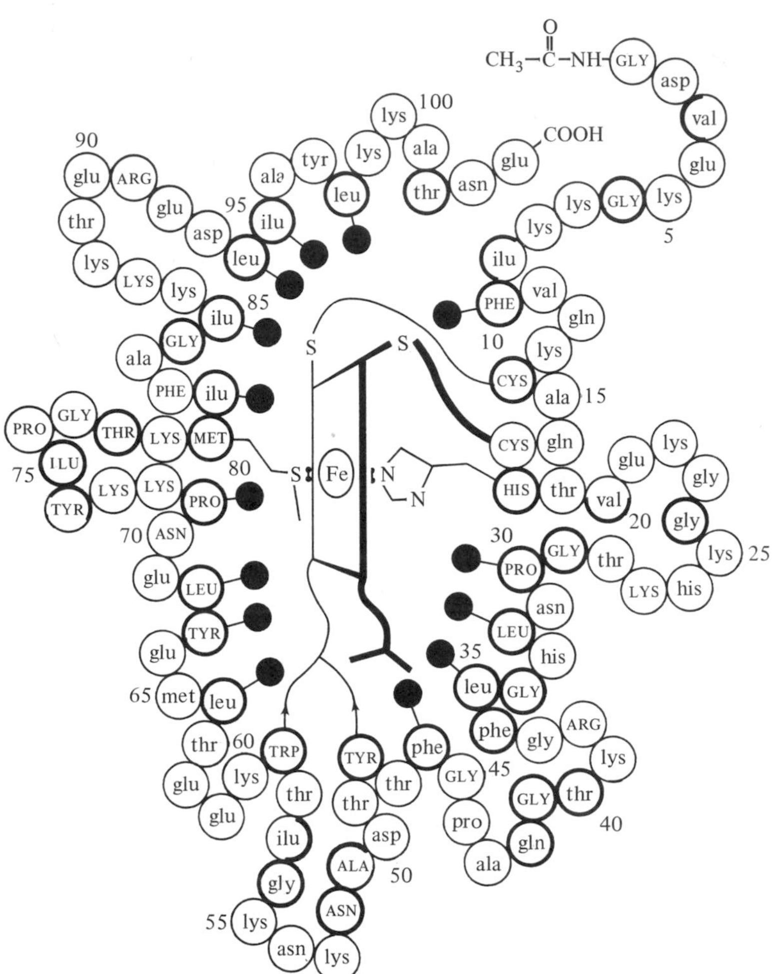

Fig. 9.18 A representation of the tertiary structure of horse cytochrome c. Heavy circles indicate side chains that are buried in the interior of the molecule, and attached black balls mark residues whose side chains pack against the haem. Light circles indicate side chains on the outside of the molecule, and dark half-circles show groups that are half-buried at the surface. Arrows from tryptophan 59 and tyrosine 48 to the buried propionic acid group represent hydrogen bonds. Residues designated by capital letters are totally invariant among the proteins of the species listed in Table 9.2. Interior hydrophobic residues tend to be invariant, as do glycines where these are situated at a bend in the tertiary structure which does not allow room for a larger residue. Both these features keep the chain tightly wrapped around the haem in all species. Other invariant residues are discussed in the text. (*From* Dickerson *et al.*, 1971.)

of the chain which had no crucial effect on the tertiary structure. In such positions the substitution of one amino acid by another does not affect function and does not affect the survival of the organism. These ideas can now be checked in detail, since X-ray diffraction analysis of the cytochrome c structure has been carried by Dickerson and his group to a resolution sufficient to show details of chain folding and the positions of the residues (Fig. 9.18). There is some α-helical folding in the cytochrome c structure (but only residues 92–102), and six points at which *part* of a 3_{10}-helix forms. These 3_{10} 'bends' avoid some of the strain of a full 3_{10}-helix and become energetically favourable in the cytochrome c structure, which is dominated by the tendency for hydrophobic side-chains to orient towards the centre of the molecule and wrap the chain about the hydrophobic haem group. The haem sits in a hydrophobic crevice between one half of the molecule (residues 1–47) and the other half. The invariant residues 67, 68 and 70–80 form one wall of the crevice with the methionine at position 80 complexing with the haem iron atom on this side. At the other side of the haem the two cystines (residues 14 and 17) are invariant and also the histidine at position 18 which complexes with the iron atom on that side. The lysine residues 72, 73, 79, 86, 87, 88 (four of which are invariant) form a positively charged area at the surface of the molecule on one side, and a further surface concentration of positive charge is found at the other side (in mammals these are lysine residues 99, 100, 5, 7, 8, 13, 22, 25, 27). Between these two areas of the surface there is a cluster of negatively-charged acid groups (including 90, 92, 93, 2, 4) and although there is more residue variation in this area, it retains its negative character in all species. For correct function cytochrome c must fit into the mitochondrial membrane at the right point in the electron-transfer chain (Fig. 5.2) and possibly these charged regions are involved in electrostatic binding to the membrane or to other electron carriers. This would represent a vital part of the molecule, in addition to the areas around the haem, which must maintain a constant character through evolution.

It may be noted at this point that the code relations between nucleotide triplets and amino acids (Table 8.1) are not random, but have themselves evolved, so that the present code is subtly suited to its purpose. For example, leucine and isoleucine are found in adjacent blocks of Table 8.1, as are aspartic and glutamic acid, so that single-base changes will relatively frequently lead to substitution of one of these similar amino acids, to give a protein that is probably still functional, perhaps a little more efficient, perhaps a little less, with the more efficient form surviving. A distinction can be made between single-base changes which involve only purine–purine or pyrimidine–pyrimidine changes, called transitions, and purine–pyrimidine or pyrimidine–purine changes, called transversions. Some mutagenic agents tend to produce transitions more readily than transversions. The type of base change involved when one amino acid replaces another gives some clue about the type of mutation mechanism that caused this replacement.

Since the code is common now to all living organisms it must have taken its final form very early in biological evolution. But at that stage there can have been no direct selective pressures for survival of the kind of code that would be useful later. If the code that evolved is one that is well suited to later evolution of higher organisms this must be because of molecular factors inherent in protein and polypeptide structure and in prebiotic conditions and mutation mechanisms. The code must contain clues, as yet undeciphered, about this early period.

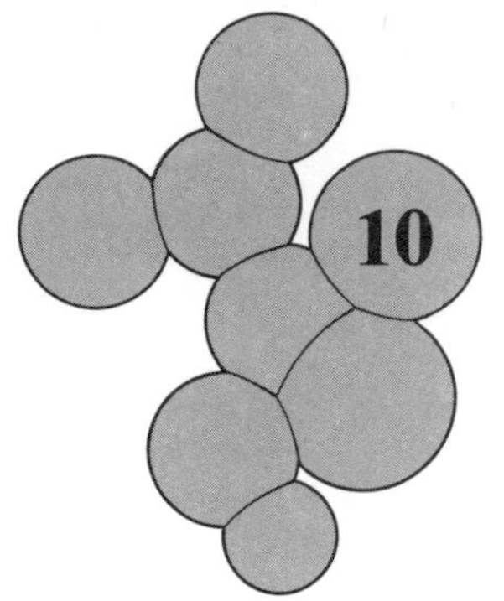

Control of protein and nucleic acid synthesis

(Including a section on tumour viruses by Rose Sheinin)

We have noted in Chapter 7 that the differentiation of cells – the whole process by which cells achieve specialized form and function during the development of an organism – is in large part a question of control of gene expression. The genetic complement of each cell contains the information required to make all the proteins this organism can synthesize, but in any one cell, at any one time, many of these genes are repressed. The control of protein synthesis, of RNA synthesis, and of DNA duplication are thus central themes of modern molecular biology. Some progress is being made in understanding control mechanisms in higher organisms (we shall make brief reference later to the mode of action of peptide and steroid hormones) but much more is known about control mechanisms in certain bacterial and viral systems which have been under close scrutiny in recent years. Another system accessible to detailed investigation, the transformation of cells by tumour-producing viruses, is particularly exciting at the present time. The study of the interaction of these viruses with host cells relates directly to study of normal control of cell division and its breakdown in malignancy.

In this chapter we first consider a relatively simple case of control of protein synthesis, regulation of the formation, in *E. coli*, of the enzymes of lactose metabolism. We then turn to infection of *E. coli* by the phages R17, Qβ, T_4, and λ and explore in outline the control mechanisms which operate here. R17 and Qβ are small RNA phages with very simple genomes. λ is a DNA phage which can either cause virulent infection or become integrated into the host genome, with virus replication repressed. In later sections we consider tumour viruses, antibody production, and some aspects of oocyte development.

The *lac* operon of *E. coli*

In bacteria the synthesis of a given enzyme requires, of course, that the DNA sequence coding for this enzyme be carried by the bacterial genome. However the parts of the genome that are active at any time, and hence the enzymes that are synthesized, are often controlled by small molecules present in the environment, or culture medium. Thus, for wild-type *E. coli*, if lactose is present in the medium, enzymes will be made which metabolize lactose, so that this sugar can be used as an energy source. In the absence of lactose these enzymes will be made in only very small amount. Lactose, and a number of related sugars carrying the same chemical group, act as *inducers* for synthesis of these enzymes, and the phenomenon is termed *enzyme induction*. Alternatively, if an essential molecule is provided in the culture medium, an amino acid for example, bacteria may cease to make the enzymes required for synthesis of this amino

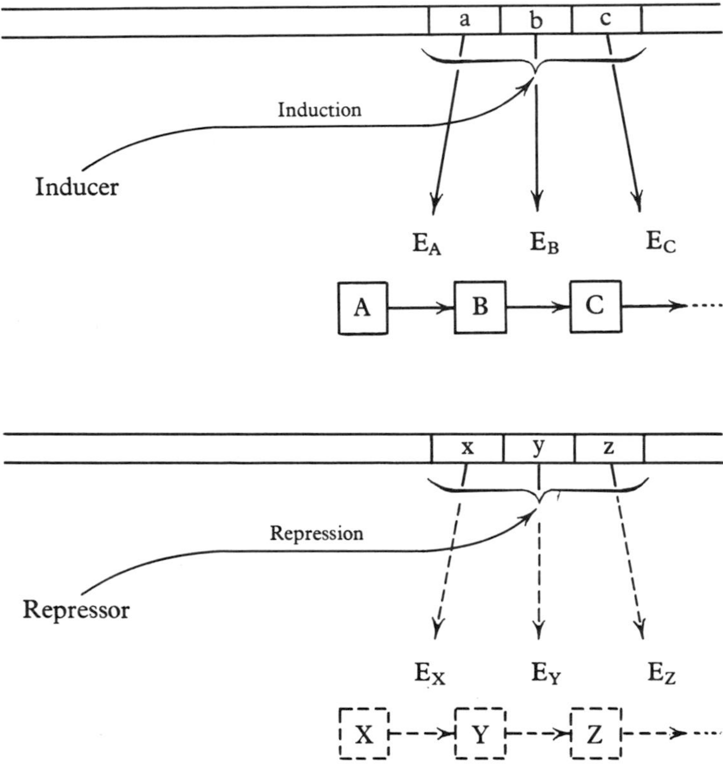

Fig. 10.1 The upper figure shows enzyme induction and the lower figure enzyme repression. Capital letters denote substances formed at successive stages in a metabolic sequence of reactions. E_A, E_B, . . . are enzymes with specificity for substances A, B, . . . ; *a*, *b*, . . . are genes specifying the structure of E_A, E_B, . . .

acid. This effect is termed *enzyme repression*. (Repression must be distinguished from another phenomenon already discussed in Chapter 3, known as 'feed-back inhibition'. This latter term describes the inhibition of the activity – not the synthesis – of the enzymes of a metabolic sequence.) The two phenomena, induction and repression, are shown diagrammatically in Fig. 10.1.

Around 1960, Jacob and Monod made a careful analysis of a variety of experiments on the induction of the enzymes of lactose metabolism by *E. coli*. Wild-type *E. coli* grown in a medium containing glucose as sole source of carbohydrate cannot instantly metabolize added lactose. Synthesis of specific enzymes is required before this sugar can be utilized. *E. coli* mutants can be obtained which are 'constitutive' for the metabolism of lactose. Constitutive mutants synthesize the lactose-metabolizing enzymes at high rates even in the absence of lactose.

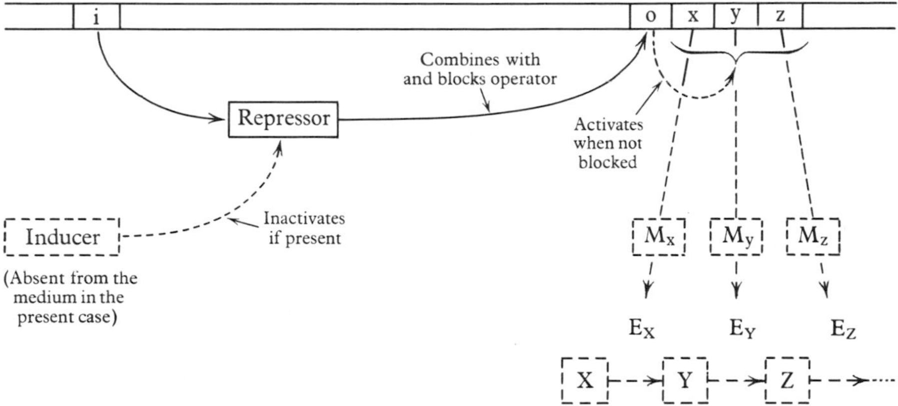

Fig. 10.2 The lower diagram of Fig. 10.1 has been redrawn to accommodate the main features of Jacob and Monod's theory of genetic regulation. i = regulator gene. The genes *x*, *y* and *z* together with the operator o, collectively constitute an operon, and their products, M_x, M_y and M_z are the messenger-RNA molecules coding for the amino-acid sequences of the enzymes E_X, E_Y and E_Z. Broken lines denote parts of the system which are 'switched off' and remain latent in the absence of inducer.

To draw their results together in a unified scheme Jacob and Monod introduced the concept of a genetic unit, which they termed an *operon*, made up of *structural genes* (coding for a number of functionally related enzymes) and *regulator genes* concerned with switching-off or switching-on synthesis of these enzymes. The main features of Jacob and Monod's proposals for *lac* (lactose enzymes) operon control are shown in Fig. 10.2, and a modern map of the *lac* operon in Fig. 10.3. The product of the regulator gene *i* is an intracellular repressor which, in the absence of inducer, combines with an *operator* region *o*, blocking synthesis of messenger-RNA by the structural genes (Fig. 10.2). If inducer is added to the medium the broken lines of Fig. 10.2 transform into continuous lines, while the arrow leading into the operator region *o* is changed from a continuous line to a broken line. Inducer is supposed to combine with the intracellular repressor, preventing it now from combining with the *o* region. Messenger-RNA synthesis at the structural genes of the operon is no longer blocked.

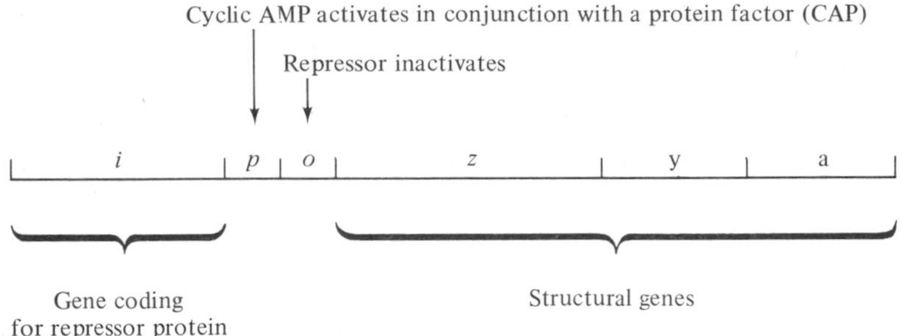

Fig. 10.3 The *lac* region of *E. coli*. The *z*, *y* and *a* genes code respectively for β-galactosidase, β-galactoside permease and galactoside acetylase. The *i* gene codes for repressor protein which binds to the *o* segment in the absence of lactose or other inducer. When the *lac* operon is transcribed into mRNA, the RNA polymerase binds initially to the promoter (*p*) segment. Initiation of transcription requires cyclic AMP and a protein factor (termed CAP), so that transcription is blocked when the concentration of cyclic AMP is low (this is termed catabolite repression). Lactose is a poor energy source and the *lac* operon is only turned on when good sources such as glucose are lacking. Under these conditions the cyclic AMP level is high.

Before discussing the experiments which led Jacob and Monod to put forward this scheme we may note that more recent work has amply confirmed their main hypothesis and added some points of detail (Fig. 10.3). *z*, *y* and *a* are structural genes coding for the proteins of lactose utilization. These are respectively: β-galactosidase (*z*), which splits lactose into glucose and galactose; β-galactoside permease (*y*) a membrane protein required for transport of β-galactosides from the surrounding medium into the bacterial cytoplasm; and a galactoside acetylase (*a*), of unknown function.

The *i* gene product is now known to be a protein (*lac* repressor protein). The regulator gene is thus not fundamentally different from the structural genes. Both types of gene code for amino-acid sequences. Near the *o* region is a *promoter* (*p*) region where messenger-RNA synthesis is initiated if no repressor protein is bound at the *o* region. The *o* and *p* regions of the genome are smaller than the structural genes and the *i* gene, and the DNA sequences of these segments are *not* translated into amino-acid sequences. These control segments of the genome are usually referred to as 'regions' rather than genes though operationally, in genetic experiments, these regions map as genes and can be considered a separate type of control gene.

In the *lac* system of *E. coli* the regulator gene *i* lies adjacent to the *p* region of the genome, but this is not an essential feature of operon control. In systems other than the *lac* system of *E. coli*, regulator genes are often remote from the genetic regions they control, and this does not affect regulation, since the product of the regulator gene is a soluble protein that can diffuse through the bacterial cytoplasm. Monod, Changeux and Jacob have suggested that repressors must be proteins with two different active sites: one specific for the inducer and one specific for the operator region of the genome. They suppose that binding of the inducer may modify the tertiary or quaternary structure of the repressor (allosteric effect) in such a way that the repressor's operator-binding site is modified, so that it can no longer attach to the operator. With the operator region free of repressor, messenger-RNA synthesis initiated at the *p* region continues on to the end of the *a* gene, so that a polycistronic messenger-RNA is produced, coding for all three of these lactose proteins. Their synthesis is switched on and off in a coordinated way.

We can now consider the experimental basis of Jacob and Monod's original scheme and discuss the consequences of mutations affecting the structural genes, the *i* gene, or the *o* and *p* regions which have led to their identification and characterization. Table 10.1 illustrates, for example, the effects of mutation,

TABLE 10.1 Predicted consequences of mutations affecting structural genes, and constitutive mutations affecting the regulator gene or the operator region, in the inducible lactose system of *E. coli.*

Gene or region affected by mutation		Enzymes produced: In presence of inducer		Enzymes produced: In absence of inducer	
		galactosidase	acetylase	galactosidase	acetylase
None		+	+	–	–
Structural genes	*z*	–	+	–	–
	a	+	–	–	–
Regulator gene	*i*	+	+	+	+
Operator	*o*	+	+	+	+

affecting the *z* and *a* structural genes, the *i* gene and the *o* region. Mutations in either the regulator gene (*i*) or the operator region may be expected to modify the repressor protein or the *o* region in such a way that either repressor now permanently binds to the *o* region under all conditions, or repressor can now no longer bind to the *o* region. Mutants of the latter type are detected as constitutive mutants; they synthesize both enzymes at maximum rate irrespective of the presence or absence of inducer. Constitutive mutations in the *i* gene and the *o* region can be distinguished genetically. *E. coli* is genetically haploid, but it is possible by exploiting a special phenomenon, known as sexduction (Fig. 6.10) to bring about a stable state of diploidy for a short length of the chromosome containing all the genes involved in the lactose system. The full genetic analysis which thus becomes possible splits the constitutive mutants neatly into two classes. The first class is recessive to the normal form and the second is dominant. This result can be interpreted by supposing that the two classes of mutation affect the regulator gene and the operator region respectively. A heterozygote possessing one defective regulator gene will continue to produce repressor through the activity of the normal regulator gene present in single dose. The organism will therefore show the normal (inducible) phenotype. But an organism heterozygous for the operator segment will behave phenotypically as a constitutive mutant; the defective operator present in single dose will be insensitive to the repressor and hence will permit the continued function of the *z* and *a* structural genes associated with it on the same chromosome.

The two classes of mutation can be subjected to a second genetic test, based on the *cis–trans* effect. That is, we can extend our predictive scheme to the case of organisms which are heterozygous not only for the regulator gene or the operator region as the case may be, but also for the *z* and *a* structural genes. It should be immaterial whether the regulator gene stands in the *cis* or *trans* relation to *z* and *a* since it affects their activity through a non-chromosomal intermediary (the repressor) and not directly through contiguity on the chromosome. This is found to be the case. On the other hand, the operator is pictured as directly controlling the group of structural genes adjacent to it on the chromosome. Writing o^c for the defective operator characteristic of the second class of constitutive mutants we can see that the *cis* heterozygote o^+z^+/o^cz^- should behave differently from the *trans* heterozygote o^+z^-/o^cz^+. In the former case β-galactosidase will be inducible, but there will be constitutive synthesis of whatever protein is produced by the mutated structural gene z^-. (Such a protein is usually identified immunologically, possessing some of the antigenic features of β-galactosidase, without its enzyme activity.) In the *trans* case, on the other hand, we expect the abnormal protein to be inducible, and galactosidase to be synthesized constitutively. The full scheme is given in Table 10.2, and is verified in practice. The operator acts directly on neighbouring structural genes on the same chromosome, without non-chromosomal intermediary.

TABLE 10.2 The phenotypic difference between the *cis* and *trans* types of double heterozygotes involving the operator and the *z* structural gene within the same operon

CRM = *cross-reacting material, that is, material resembling galactosidase in its antigenic characteristics but lacking the enzyme activity.*

Genotypes		Types of protein synthesized: Inducer present	Types of protein synthesized: Inducer absent
haploids	o^+z^+	galactosidase	nil
	o^+z^-	CRM	nil
	o^cz^+	galactosidase	galactosidase
	o^cz^-	CRM	CRM
diploid heterozygotes	o^+z^+/o^cz^-	galactosidase + CRM	CRM only
	o^+z^-/o^cz^+	galactosidase + CRM	galactosidase only

Further refinement of genetic and biochemical experiments led to discovery of the *p* region and elucidated the role of the *y* gene which lies between the *z* and *a* genes. The *y* gene product does not catalyse a chemical step in lactose utilization, but facilitates entry of lactose into the bacterium across the plasma membrane. As indicated in Fig. 10.3 (and noted in the legend to that figure) a further control acts at the *p* region. Initiation of messenger-RNA synthesis at this point is facilitated by cyclic AMP (cAMP, an important control molecule whose role in other situations will be discussed in later sections of this chapter) acting in conjunction with a protein factor (CAP). The CAP protein is a dimer, with sub-unit molecular weight 22 000.

In addition to genetic experiments of the type so far described ingenious methods have been devised to isolate *lac* repressor protein, and also to isolate the *lac* operon DNA segment itself. The isolation of the repressor protein by Gilbert and Muller–Hill in 1966 was an experimental tour-de-force, for at that time the chemical nature of the repressor was still uncertain (some experiments had suggested it might be RNA) and it could be expected to be present in only minute amounts (they found less than 50 molecules per bacterial cell). They first selected a super-inducible mutant, that is, a mutant in which activity of the *lac* operon could be induced by very low levels of a chosen inducer, isopropyl-β-thio-galactoside. In this mutant, a mutation at some point in the *i* gene had modified the repressor, as compared with that present in wild-type *E. coli* in such a way that it bound this particular inducer relatively tightly. Using ammonium sulphate precipitation to fractionate bacterial extracts, they tracked down the repressor by dialysing the fractions against solutions of radioactive isopropyl-β-thio-galactoside, looking for specific binding of the inducer. They were able to confirm the specificity of inducer binding by using, as control material, extracts of a mutant which could only be induced by relatively high levels of isopropyl-β-thio-galactoside. In this way they were able to show the protein nature of the repressor, estimate its molecular weight from its sedimentation coefficient, and provide direct support for Jacob and Monod's scheme for *lac* operon control. The repressor has a molecular weight of about 152 000 containing four identical sub-units of molecular weight 38 000 and binding four inducer molecules.

To isolate the *lac* operon DNA, Shapiro *et al.* used two transducing phages (transduction is discussed in Chapter 6, see Fig. 6.10), one carrying the *lac* segment *ipozy* and the other the inverted segment *yzopi*. These phages were obtained by working with two strains of *E. coli* in which the *lac* operon had been inserted in opposite orientation by sexduction. From each phage they prepared double-stranded DNA and separated the two strands (H- and L-strands). For one phage the H-strand carried the *lac* 'sense' message, and for the other the 'sense' message was on the L-strand, the H-strand carrying the complementary sequence. They then brought together the H-strands from the two phages, under temperature conditions which would allow double-stranded DNA to reform where the sequences of the two strands were complementary (molecular hybridization). The resulting molecule contained a double-helical segment, where the *lac* operon segments of the two H-strands were of complementary sequence, with single-stranded non-complementary lengths of DNA protruding at either end. All that was then required to clean up the isolated *lac* segment was to digest off the single strands with a single-strand-specific deoxyribonuclease. They backed up this elegant work with electron micrographs showing the hybridized DNA before and after nuclease digestion, from which the length of the double-helical *lac* region could be estimated as 1·4 μm. Their *lac* operon segment included *p* and *o* regions, *z* gene and part of the *i* and *y* genes. From the known molecular weight of β-galactosidase (the *z* gene product) the length of the *z* gene may be estimated to be about 1·26 μm (3700 base pairs). The combined length of the *p* and *o* regions is thus at most about 0·14 μm, consisting of only a few hundred base pairs.

We need to define at this point two terms used in later discussion: *transcription* and *translation*. Transcription is the term used to describe mRNA synthesis, in which genetic information encoded in one nucleic acid chain sequence is transcribed to another nucleic acid chain. Translation refers to protein synthesis, in which the message in nucleotide code is translated into an amino-acid sequence.

We have concentrated attention on the *lac* system of *E. coli* because the genetic control mechanisms have been worked out in this case in some detail. But it seems probable that this general scheme has very wide application, and that enzyme induction and repression in bacteria, and control processes in development and differentiation of higher organisms often involve *interaction of a small control molecule with a genetic control segment indirectly via a specific allosteric protein mediator*. The allosteric protein is not always a repressor: CAP protein acts at the *lac p* region as an allosteric intracellular protein which facilitates rather than blocks transcription. Also, binding of the small control molecule to the allosteric protein may either cause it to change conformation and *bind to* the DNA (as in the case of CAP protein + cAMP at the *lac p* region) or the opposite (*lac* repressor takes up a conformation that binds to DNA when no inducer is present). Other possible alternatives are: a protein facilitating transcription that binds to the DNA when the small control molecule is *not* present, or a repressor protein that binds to DNA when the control molecule *is* present. An example of the latter is found, for example, in control of tryptophan synthesis in *E. coli*. When tryptophan is present in the culture medium, in adequate concentration, a repressor switches off an operon of five genes coding for enzymes required for tryptophan synthesis (which makes sense, since under these conditions the bacterium does not need to make its own tryptophan).

It is interesting that the *lac* repressor is a four-sub-unit protein, and CAP protein a two-sub-unit protein. This suggests that the allosteric effects involved, when genetic control proteins bind small control molecules, may be similar to those involved in oxygen binding by haemoglobin (Chapter 9). It is not difficult to visualize that a minor change in the tertiary structure of the *lac* repressor sub-units, when inducer is bound, might be sufficient to produce slight rearrangement of quaternary structure and a crucial alteration in the part of the protein surface that interacts with the *lac* operator region of the *E. coli* DNA. If haemoglobin provides a valid model for allosteric effects, two reasons can be suggested as to why these proteins might often, or always, contain more than one sub-unit: first the amplification of molecular movement produced by sub-unit interaction (see discussion of haemoglobin in Chapter 9), second the increased sensitivity to concentration of the small control molecule over a limited concentration range (cf. myoglobin and haemoglobin oxygen uptake curves of Fig. 9.10).

Infection of *E. coli* by the small viruses R17, Qβ, MS2 and f2

In the previous section we have been considering *E. coli* adjusting happily to different environmental levels of lactose. We now turn to a different type of event, catastrophic for the *E. coli* involved, invasion by a virulent phage (bacterial virus). The essential features of phage infection have been discussed in Chapter 6. We are going to be concerned now with expression of the phage genes in the bacterial cytoplasm.

Phages inject their genome into bacteria by various means. As discussed later, the T-even phages inject it through the tail structure (described in Chapter 3). Small phages of the type to be discussed in this section apparently adsorb to pili – fine tubules which are found on the surface of some bacteria – and inject their nuclei acid down the hollow pili into the bacterial cytoplasm. Subsequently, in virulent infection, the phage nucleic acid replicates, phage proteins are made, new phages are assembled from their component nucleic acid and protein, and finally the infected bacterium lyses to release hundreds or thousands of new mature phages to infect further bacteria.

For R17, Qβ, MS2 and f2 the genome is in each case a single strand of RNA, coding for three proteins: an RNA replicase (to replicate the phage RNA) a coat protein (to form the outer coat of the phage particle) and a maturation protein (which is essential for correct phage morphogenesis and is itself incorporated into the mature phage particle). Reading from the 5′ to the 3′ end of the phage RNA (which is the direction ribosomes move during translation) these genes are arranged in the order: maturation protein, coat protein, replicase. The RNA strands in these small viruses are about 3 500 nucleotides long (molecular weight $1{\cdot}1 \times 10^6$) contained in a spherical protein shell made of 180 identical coat protein sub-units and one molecule of maturation protein (Table 7.3). R17, Qβ, MS2 and f2 all attack *E. coli*, and R17, MS2 and f2 are closely related.

Initially, after entry of the phage RNA into the *E. coli* cytoplasm, ribosomes of the host bacterium bind to this RNA (which acts as a messenger-RNA) at a translation initiation point at the 'start' side of the coat protein gene. Coat proteins and replicase units are made, as ribosomes travel down to the 3′ end of the phage RNA. The product of the replicase gene is not actually a complete enzyme, but a protein sub-unit which joins with three sub-units made by the host bacterium to form the complete RNA replicase.

The ensuing sequence of events is not known with certainty, but infection appears to proceed along the following lines: one of the newly formed replicase molecules now binds to the translation initiation point at the beginning of the coat protein gene and, binding at this point, prevents further protein synthesis by blocking attachment of ribosomes. When all the ribosomes engaged in protein synthesis have travelled to the 3′ end and dropped off, a replicase now binds to the

3′ end of the phage RNA (this might be a different replicase molecule, or might be the same replicase if, when there are no ribosomes on the RNA, this folds to a tertiary structure which brings the 3′ end near the coat protein gene initiation point). The replicase now travels from the 3′ to the 5′ end of the RNA to synthesize a complementary strand of RNA. The complementary strand is termed the 'minus' strand and the infectious strand the 'plus' strand. The new minus strand can now act as a template for synthesis of further plus strands. After enough plus strands have been formed to mop up all the replicase molecules made originally, new unblocked coat protein gene initiation points become available for a more massive burst of synthesis of coat protein and, as a further refinement, the coat protein, when present in sufficient amount, blocks movement of ribosomes down the plus strand to form replicase. At this stage, then, mainly coat protein is made, and soon the necessary components, plus strands and coat protein sub-units have been made in about the right proportion (180 sub-units for each plus strand) to make the thousands of mature particles that will be released on lysis. One maturation protein molecule is incorporated into each mature phage particle, and the maturation protein gene must be regulated to 'switch on' at some point towards the end of the infection cycle.

In addition to the coding sequences for the three proteins the phage RNA carries at both ends, and also at the coat protein gene initiation point and at the replicase gene initiation point, short sequences which act as control segments and are not translated. At the 5′ end of R17 RNA there is a sequence of 129 bases before the first codon of the maturation protein, and a 36-base untranslated region between the coat protein and replicase genes. For Qβ the untranslated sequence at the 5′ end is 61 bases long and the untranslated sequence between coat protein and replicase genes appears to be 600 bases long. The untranslated sequences presumably play some part in determining the specificity of binding of ribosomes at certain points along the RNA, and in binding replicase and coat protein, to block attachment and movement of ribosomes.

It is important to note, at this point, that we have now considered two quite different control systems: in the previous section we discussed *transcription* controls (in *E. coli*), mediated by repressors and transcription facilitating proteins, interacting with small untranscribed stretches of DNA to modulate the binding of RNA polymerase and its movement along the DNA, thus controlling the rates of synthesis of different mRNAs. In this section we have been discussing *translation* controls (operative during phage infection) in which specific proteins bind to small untranslated stretches of phage RNA (acting as mRNA) to modulate the binding of ribosomes and their movement along the mRNA, thus controlling the rates of synthesis of different proteins.

φX174

We move on now to consider a more complex phage, φX174, with nine or ten genes rather than three, and containing DNA rather than RNA. The mature φX174 particle contains a single-stranded covalently-closed ring of DNA of molecular weight $1{\cdot}6 \times 10^6$ (about 5000 nucleotides). The structure of ϕX174, with its corner 'spikes' has been described in Chapter 3. In the ϕX174 genome (Fig. 10.4) the three genes coding for the capsid protein E and the spike proteins F and A occupy about half the genome length. These proteins have molecular weights ~48 000, ~19 000 and ~36 000 respectively, and the length of their genes are thus about 1 300, 500 and 1 000 nucleotides long, together making up 2800 of the 5000 or so bases of the genome.

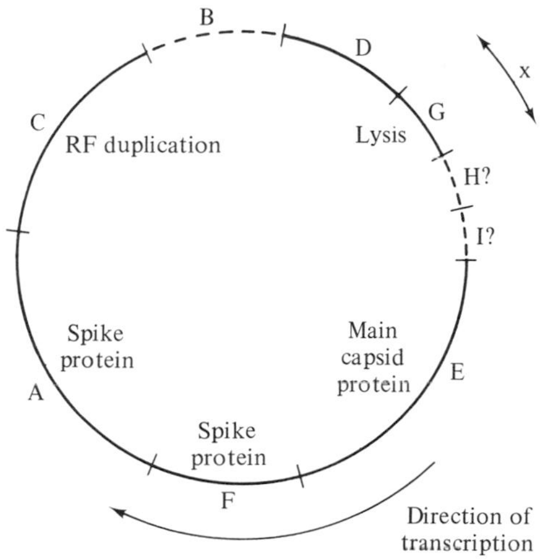

Fig. 10.4 Genetic map of ϕX174. Cistron size is judged from the molecular weight of the corresponding protein. The molecular weights of B, H, and I proteins have not yet been determined. The positions on the map of the H, I, and X genes is not known. The X gene product is a phage structural protein. The product of the G gene plays some part in the final lysis of the infected bacterium. (*From* Hayashi and Hayashi, 1970.)

After infection (ϕX174 is again a phage which infects *E. coli*) a complementary (minus) DNA strand is synthesized and its ends are covalently joined, to give a circle of double-stranded DNA which is the replicative form (RF) of ϕX174 DNA. During the first 10 minutes or so after infection, 10–20 new double-stranded RF DNA's are made. Later, progeny plus strands are synthesized from the minus strand templates of these RF DNAs.

In studies of ϕX174 mRNA synthesized at various stages after infection a similar molecular weight distribution is found at early and late stages. There appears for this phage to be no temporal control of transcription (in contrast to the case of larger phages like T_4 and λ discussed in later sections). ϕX174 mRNA is transcribed from the minus strand of the RF form, and DNA replication either alternates with transcription, or takes place simultaneously on the same DNA minus strand. The size distribution of mRNA molecules found indicates that there are a number of points on the genome at which RNA polymerase can initiate transcription. The phage makes use of the host (*E. coli*) RNA polymerase.

In an *in vitro* system RNA polymerase can be used with RF ϕX174 as primer. In this system the mRNA formed is as long or longer than the complete genome, unless a soluble factor from *E. coli* called transcription termination factor ρ (rho) is added. With ρ factor present a distribution of mRNAs is found more comparable with the distribution found *in vivo*. It seems then that short DNA sequences can act as transcription *termination* controls, in conjunction with the ρ factor. The ϕX174 genome appears to carry a number of these termination sites as well as transcription initiation sites.

The three genes for the E, F and A proteins are transcribed as a polycistronic message (a single long mRNA molecule), but the three proteins are found in infected *E. coli* in ratio 1:1:0·1. There must therefore be factors at the translational stage which control this ratio. These effects are not understood in detail, but there might be, for example, a tendency for E or F protein to bind to the mRNA at the end of the F cistron and prevent ribosomes moving on down the mRNA, as coat protein does at the replicase initiation point in Qβ. The time relations for synthesis of the other proteins (C, B, D, G, H, I, X) and their roles in the infections process have not been studied yet for ϕX174 in the same detail as has been possible in the simpler cases of R17 and Qβ.

To summarize new points brought out in this section: we have seen that in infection of *E. coli* by ϕX174, DNA is single-stranded in the mature particle but replicates from a circular double-stranded form. We have noted that, in transcription, some segments of phage genome DNA act as initiation and others as termination sites, controlling attachment of RNA polymerase to the DNA, and termination of mRNA synthesis. The ϕX174 genome carries a number of such initiation and termination points. In bacterial enzyme induction and repression, control by small molecules is exerted at such initiation sites. In ϕX174 infection there appear to be no protein repressors or transcription facilitating proteins switching off or switching on phage genes during the infectious process, since the same mRNAs are made in early and late infection. In translation control from ϕX174 mRNA proteins probably bind to untranslated sequences between cistrons blocking attachment or movement of ribosomes, and so control translation of the adjacent cistrons.

T_4

T-even phages feature at a number of points in this book (structure, Chapter 3; Benzer's work, Chapter 6; study of the DNA code, Chapter 8) and are introduced again here to note the pioneer work of Kellenberger and co-workers in use of electron microscopy to study the genetic control of assembly of these larger phages.

The T_4 genome is about forty times the size of the ϕX174 genome (mol. wt. of T_4 DNA 120×10^6, a sequence of about 200000 base pairs). The genetic map of T_4 is shown in Fig. 10.5. It will be seen that Benzer's rIIA and rIIB genes are included in this map and about 80 other genes. The circular genetic map implies that the T_4 genome exists at least transitorily in circular double-

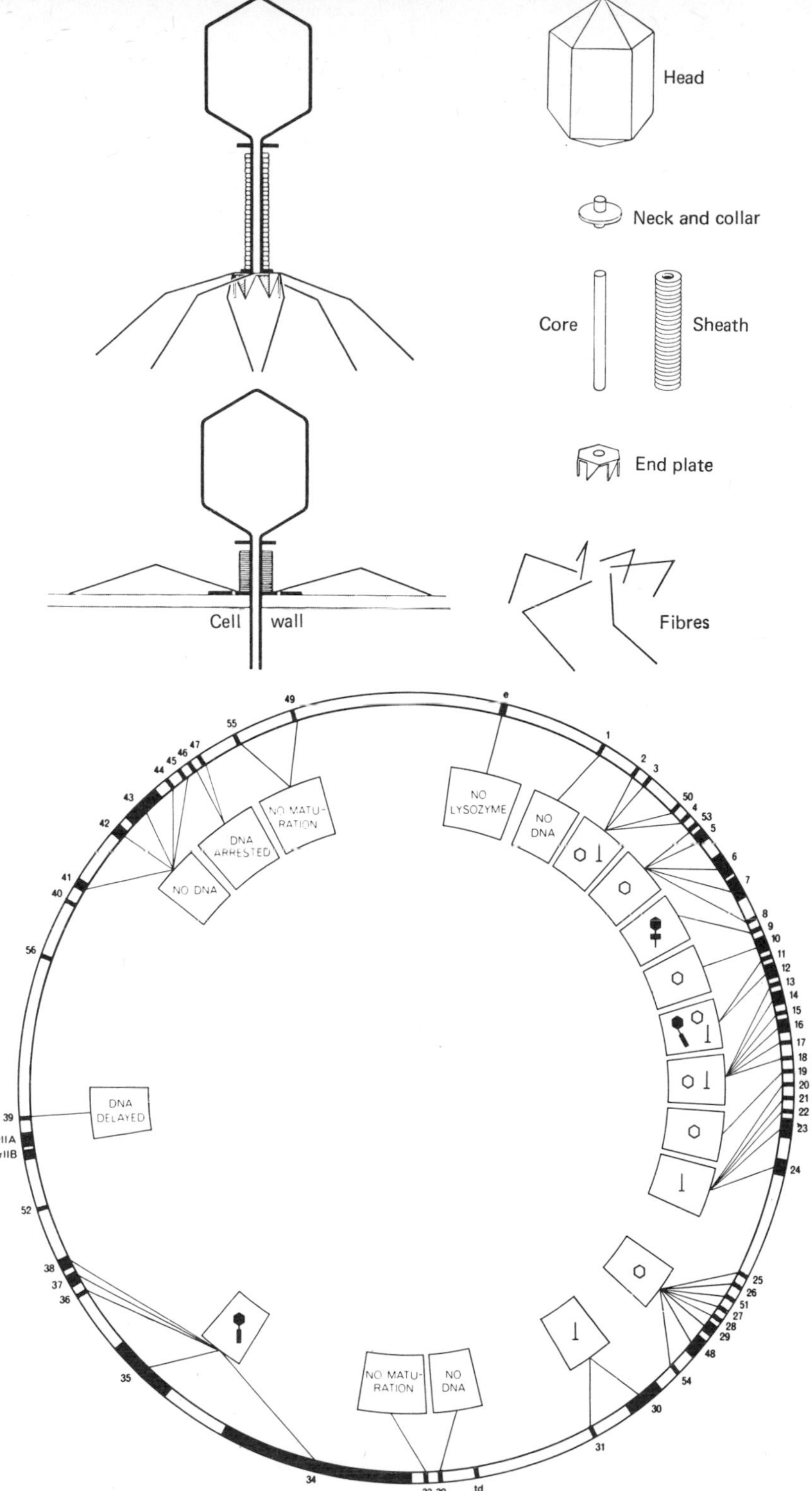

Fig. 10.5 The structure and genetic map of T_4 bacteriophage. Some of the effects of mutations in different genes are indicated in the map drawings. (*From* Edgar and Epstein, 1965.)

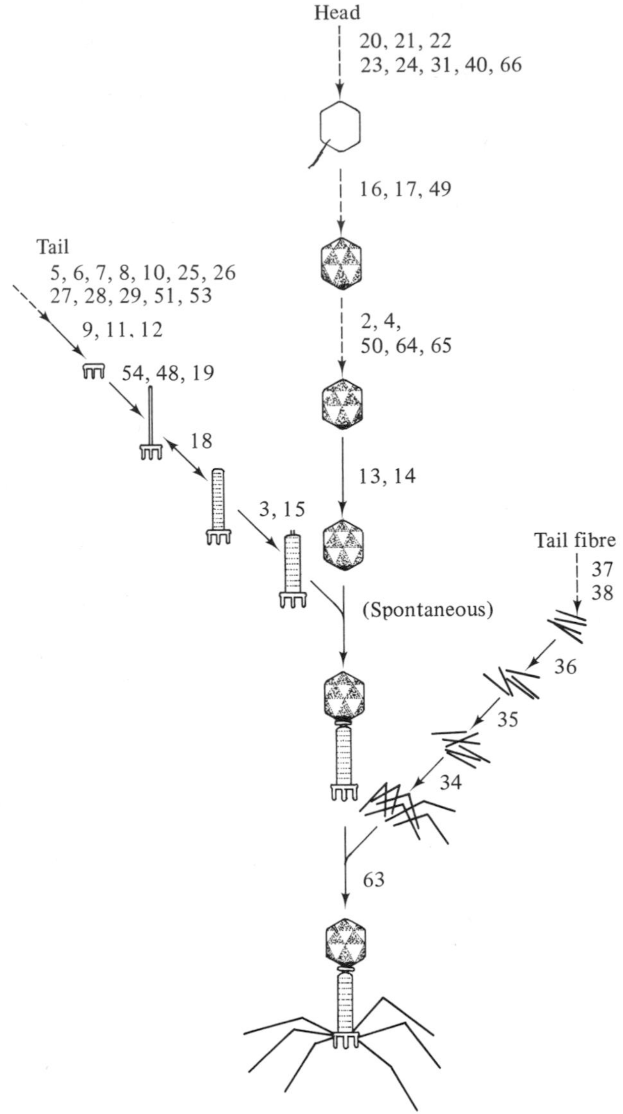

Fig. 10.6 The genes involved in the assembly of T_4 phage. (*From* Edgar, 1968.) In more recent work some stages of this process have been studied in greater detail – see paper by J. King included in the bibliography for this section.

stranded form in the *E. coli* cytoplasm, but it is packaged in the mature phage as a linear double-stranded DNA molecule, folded, and condensed with basic proteins and polyamines in the phage head. During packaging of T_4 the cut in the genome may be made at a variety of points, so that different phages contain the gene sequence in different permutations (that is, starting from different genes

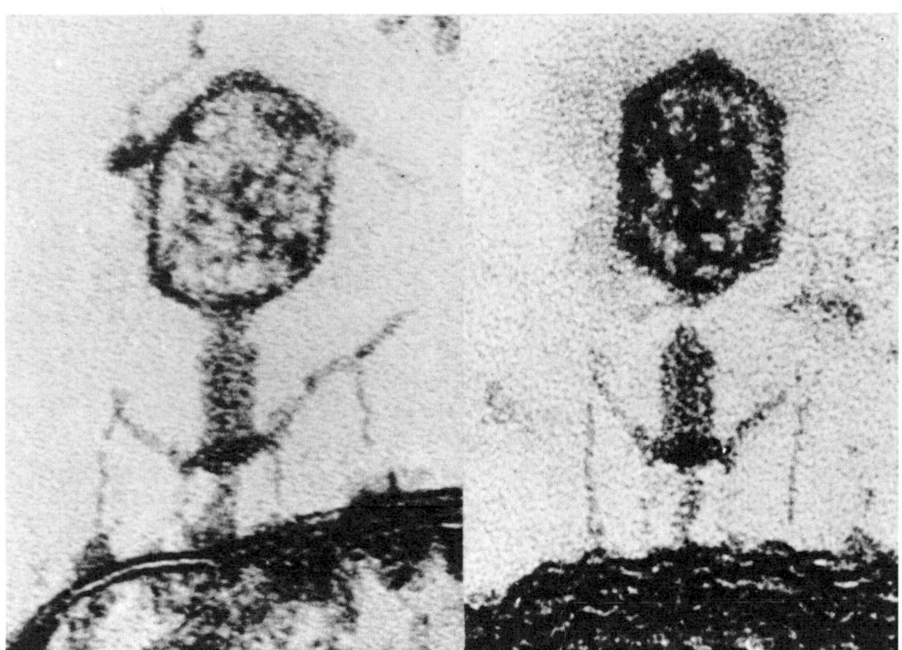

Fig. 10.7 Micrographs showing the attachment of T-even phages to the *E. coli* wall. Contact with the wall is first made by the tips of the long tail fibres. These fibres appear to flex at their 'kink' to bring the tail plate towards the bacterial wall, and a set of shorter fibres then anchor the plate more firmly. Magnification ×320 000. (*From* Simon and Anderson, 1967.)

in the circular sequence). Once the circles reform the identical genome is regenerated.

We are going to discuss in the next section the control processes which operate in infection of *E. coli* by λ phage. Suffice to say here that T phage infection is comparable in complexity with virulent infection by λ phage. (T-even phages and also T_7 have been studied in some detail). We focus attention in this section on assembly of infectious particles from the component capsid and tail proteins.

In general, in phage infection, the genes active early after infection are concerned with replication of the phage nucleic acid and with synthesis of the phage proteins. For viruses of simple structure, like Qβ or even perhaps φX174, assembly may then be simply a spontaneous process. But for T_4, assembly is under the control of a number of genes. The roles of these genes are shown schematically in Fig. 10.6. Their action is not understood yet in molecular detail, but one can speculate that, for example, the product of gene 63 is a protein which either enzymically modifies, or else becomes attached to the end of the tail fibres. The altered tail fibres now attach spontaneously to the six sites around the base plate. The accumulation of intermediate forms, or of structurally defective phages, in mutants with defective assembly genes is studied directly by electron microscopy, the bacteria being lysed and their contents made visible by the negative contrast method (see Chapter 3). As well as incomplete particles other aberrant forms are sometimes found such as long tubular aggregates of head protein (called polyheads) and bizarre phages with two or more tails attached to different corners of the head. T_4 provides a marvellous system for study of assembly of different protein sub-units into complex forms.

The roles of the tail and tail fibres in T-even phage infection are beautifully illustrated in Anderson's micrographs reproduced in Fig. 10.7. The phage particles attach themselves to *E. coli* bacteria by specific interaction between the protein of the tail fibres and certain sites on the *E. coli* wall. Attachment is followed by contraction of the outer sheath of the tubular tail, with release of ADP, previously bound (as ATP) to the tail protein. This splitting of ATP presumably provides the energy for the contraction process, which is followed by injection of the genome from the head, through the hollow tail, and through the bacterial wall and plasma membrane, into the bacterial cytoplasm (Fig. 10.5).

λ (lambda)

The phage lambda is representative of a group of viruses which can interact with host cells in two major patterns. They may initiate virulent infection which results in the production of progeny virus; or they may lysogenize cells. The latter process gives rise to viable cells which do not normally replicate virus, but do carry the viral genome as part of their cellular DNA. In the integrated state the viral genome is termed prophage.

As extracted from infectious phages, λ DNA is a double-stranded, linear molecule with a molecular weight of about 30×10^6. This genome, which comprises about 46 000 base pairs, is packaged into an icosahedral head 50 nm in diameter, to which is attached a tail 150 nm in length. The genetic map of λ DNA is shown in Fig. 10.8.

During productive infection λ DNA undergoes a process of circularization which results from base-pairing of homologous terminal nucleotide sequences. Replication of viral DNA proceeds on such circular templates. Maturation to form the linear DNA which is packaged into phage heads occurs by specific endonucleolytic cleavage of the newly replicated DNA. The two strands are broken about 12 base pairs apart, giving rise to a double-stranded, linear molecule with terminal single-stranded regions which can participate in ring closure as noted above.

In the process of lysogenization, λ DNA does not replicate. Instead the viral DNA becomes inserted at a specific locus into the *E. coli* genome. Establishment of this prophage state is not fully understood, but undoubtedly occurs by crossing over and recombination between viral and bacterial genome, as illustrated in Fig. 6.10 (see Chapter 6). Much has now been learned about the very intricate pattern of control of genome expression of λ phage. In cells which permit virulent infection the phage genome is almost completely transcribed and

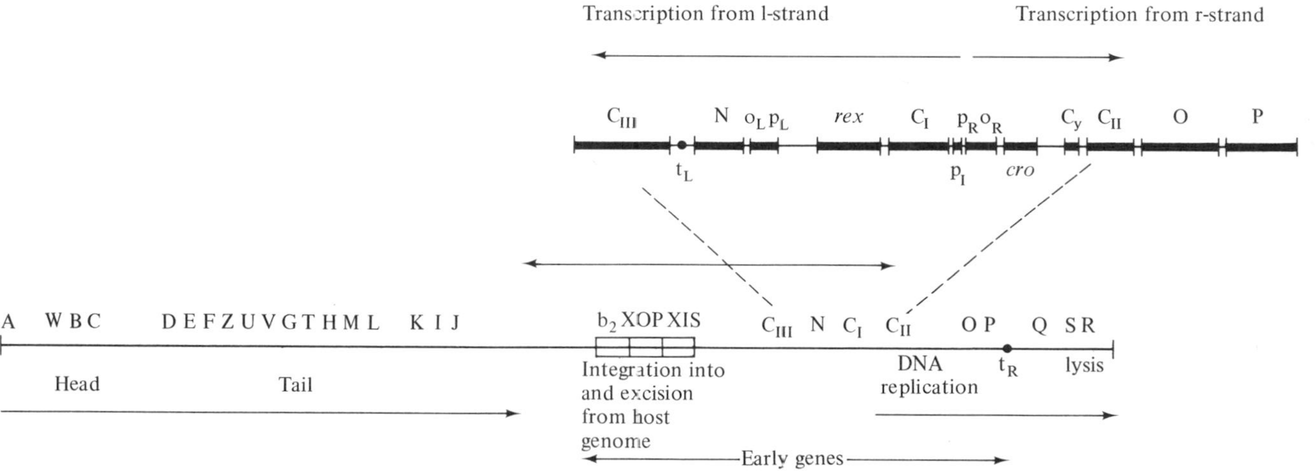

Fig. 10.8 λ genome in the mature phage particle. After infection the two ends of the genome join to form a circle. Only part of the genome (near gene CI) is active in the prophage state. Other genes in this region are active in the early stages of virulent infection. In the later stages of virulent infection genes are activated for synthesis and assembly of head and tail proteins and for cell lysis (these different genetic segments are indicated in the figure). The arrows running to the left indicate transcription from the *l* strand and arrows to the right transcription from the *r* strand. o_L-operator for leftward transcription; p_I and p_L-promoters for leftward transcription; o_R-operator for rightward transcription; p_R-promoter for rightward transcription; $t_L t_R$-terminator sites. (*From* Reichardt and Kaiser, 1971.)

translated in an organized pattern. In lysogeny, expression of a large part of the prophage genome is suppressed. The decision between lysogeny and virulent infection is mediated by the level of cyclic AMP in the bacterial cytoplasm at the time of infection, which is in turn tied to carbohydrate metabolism – the decision for virulent infection being made when the cyclic AMP level is low in a well nourished bacterium, and for lysogeny in lean times.

The initial establishment of the lysogenic state is proving to be a highly complex process not yet fully understood, involving the products of the CI, CII, CIII and *cro* genes and promoter regions $p_I p_L$ and p_R (Fig. 10.8). Once lysogeny is established there is transcription only from the p_I promoter region, in a leftward direction on the map of Fig. 10.8 along one of the DNA strands (*l*-strand). The product of the CI gene, immediately to the left of p_I, is a repressor which binds to the operator regions o_L and o_R (Fig. 10.8). In the lysogenic state this repressor prevents initiation of transcription at the promoter regions p_L and p_R, shutting off the activity of the remainder of the prophage genome. The CI gene product also acts to *facilitate* its own synthesis at the p_I region.

In virulent infection, on the other hand, there is transcription leftward from the p_L promoter region along the *l*-strand and transcription to the right from the p_R promoter along the other DNA strand (*r*-strand). The product of the *cro* gene, immediately to the right of p_R, is a repressor for leftwards transcription from p_I, shutting off expression of the CI gene. The genes involved in the decision between virulent infection and lysogeny thus form what, in electronics, would be called a 'flip-flop' circuit – they interact to drive the system decisively one way or the other. The genes active in the later stages of virulent infection are controlled by a promoter segment located between genes Q and S. This promoter region is the site of action of a positive regulator produced by gene Q, and in the presence of this positive regulator, since the DNA is in circular form at this state, the genes S, R, A through to J can be transcribed.

In addition to promoter regions there are also terminator regions on the λ genome. Extending the original Jacob–Monod operon concept, Szybalski has introduced the term *scripton* to describe a segment controlled by one promoter. If the promoter is *repressed* the whole segment is repressed, but if the promoter is *active* then controls within the scripton become operative, and parts of the corresponding genetic segment may be active or repressed by these subsidiary control mechanisms. Thus early in virulent infection transcription is started at p_L, the product of the N gene is synthesized, and only in the presence of this N-gene product can transcription proceed on beyond the terminator t_L adjacent to the N gene (Fig. 10.8). In the absence of the N gene product, transcription is blocked at t_L by the host termination factor ρ. There are two other terminator sites similarly affected by ρ and N, one near p_R and the second between genes P and Q (t_R in Fig. 10.8). The N-gene product is required for transcription to proceed beyond t_R to Q and thereby activate the late genes. The significance, and survival value for the phage, of these interlocking λ control systems, which at first sight seem unnecessarily complex for the job in hand, will no doubt become clear when the details of λ infection are more fully understood.

The experiments by which λ control mechanisms are studied involve search and selection of a large variety of mutants. For example, mutants defective in gene Q show normal DNA replication but produce no head and tail proteins, and a class of mutants that are defective in the p_L and p_R promoter regions can grow in the presence of the CI gene product, that is, grow on λ-lysogenic *E. coli*. Genetic experiments can also be combined with a more direct biochemical approach to provide detailed information. The CI gene product (λ repressor) has been isolated by Ptashne and found to be a protein of molecular weight around 30000 (under the isolation conditions used). It is possible to extract mRNA from infected bacteria at various times after infection, study hybridization between this mRNA and the phage DNA strands, and also use mRNA extracts to programme protein synthesis in a cell-free extract from uninfected bacteria, to determine the time sequence of synthesis of different enzymes during the infectious process. Electron microscopy can also contribute to study of the genes involved in control of phage assembly, and λ is being studied in the same way as T_4 in the work described in the previous section.

We do not know how representative the λ system may be of controls operative in higher organisms and in embryogenesis. Lysogenic phages may be unique unusual systems. Interaction of the λ and *E. coli* genomes is, however, the most complex system yet studied in detail and it illustrates, at least, the kind of control interactions that can be built into a biological system. The discovery that cyclic AMP plays a controlling role in λ infection is of considerable interest. We have seen that cAMP is involved in *lac* operon control also. Many of the peptide hormones, including adrenocorticotrophic hormone (ACTH), vasopressin and insulin (Figs. 2.3 and 2.4), exert their effects through membrane-bound adenylcyclases, which in turn control cytoplasmic levels of cAMP. Thus cAMP also acts as an important control molecule in higher cells.

As will be described below the phenomenon of lysogeny has obvious relevance for our understanding of the mechanism of action of tumour viruses, for the latter, like the lysogenizing bacteriophages, change the genetic composition of the host cells, and bring to them information which can result in a modification of cellular phenotype.

The tumour viruses

Introduction

The era of the tumour viruses began in 1908 with the demonstration by Ellerman and Bang that a cell-free extract of leukaemic blood cells of chickens, upon injection into young birds, induced the formation of leukaemia. In 1911 came the first indication that a solid tumour, the Rous sarcoma of chickens, could be produced by injection of cell-free tumour material into disease-free birds. Similar stimulation of tumour formation in mammals was carried out in 1931–1933 by Shope, working first with a rabbit fibroma and then with a rabbit papilloma. Since then it has been established without doubt that these neoplastic diseases, as well as many others, are caused by tumour viruses. In the last decade, in particular, tumour viruses have provided exciting tools for the study of many aspects of molecular and cell biology. They are now considered seriously as aetiological agents of some forms of human cancer.

The tremendous recent developments in this area derived basically from three kinds of experiments. The first, by Gross, led to the development of the now classical *in vivo* test for a tumour virus. Gross worked with AKR and C58 mice, strains genetically destined to succumb to leukaemia with essentially 100 per cent incidence late in life. He prepared a cell-free extract of leukaemic cells, which was injected into newborn animals of the C3H strain, normally resistant to leukaemic disease. Gross found that these injected animals did develop leukaemia later in life. The agent responsible, a murine leukaemia virus (MuLV), has since been isolated, purified and characterized.

With the simple test of preparing a cell-free extract of tumour cells, injecting it into *newborn* hosts and watching for the development of a tumour, it has been possible to set in motion the experiments which have already resulted in

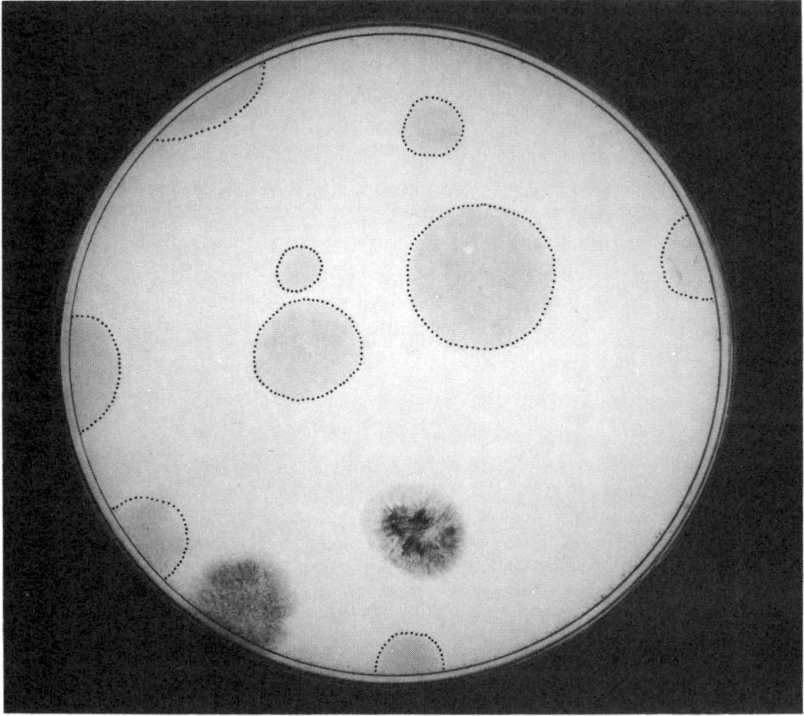

Fig. 10.9 Colonial growth pattern of normal 3T3 fibroblasts and those transformed by SV40 virus. A suspension of 3T3 mouse fibroblasts, infected at high multiplicity with SV40 virus, was plated after suitable dilution. After 14 days, the colonies which had grown up were stained with methylene blue and photographed. Magnification ×1·7. Four fairly large (1–2 cm) and four small (5–7 mm) lightly stained colonies of normal 3T3 cells are visible (within the areas surrounded by dotted lines). Two darkly stained colonies of SV40 transformed 3T3 cells are also present.

the recognition of nearly 200 tumour viruses. Earlier attempts to demonstrate tumour viruses by using adult animals as test systems failed, because of their immuno-resistance to virus.

The second important experimental advance in the study of tumour viruses was the demonstration that they can be grown *in vitro*, using simple cell systems. This provided facilities for the cultivation of large quantities of the various tumour viruses, so that they could be purified and characterized. In addition, it permitted study of their infectious cycle.

Finally came the demonstration that tumour viruses can produce neoplastic transformation *in vitro*. Host cells, infected *in vitro* with tumour virus, were seen to give rise to clones of cells whose growth properties were very different from those of the normal cells, but which greatly resembled those of tumour cells grown *in vitro* (see Fig. 10.9). These cells with altered growth characteristics and with altered cellular and colonial morphology were called transformed, and the phenomenon *virus transformation*. To demonstrate that the virus-transformed cells obtained *in vitro* were indeed tumour cells, they were transplanted into adult hosts, which are resistant to induction of tumours by injected free virus. Such animals supported the growth of transformed cells. The latter, like transplanted cancer cells, grew progressively to form tumours, and ultimately to kill the animals.

These latter experiments established the classical *in vitro* test for a tumour virus, i.e. its ability to transform normal cells; its capacity to alter, in an heritable fashion, the characteristic growth properties of susceptible cells. This very important property of tumour viruses, i.e. the ability to alter the genotype of a target cell, will be discussed below.

The *in vitro* transformation assay offers many experimental advantages for the study of tumour viruses and has been used extensively to gather information concerning the mode of action of tumour viruses, the mode of action of chemical and physical carcinogenic agents, and the factors which control the growth of normal and neoplastic cells.

The families of tumour viruses

With the availability of reproducible *in vivo* and *in vitro* test systems, the search for tumour viruses began in earnest about 1960. The great wealth of information obtained since then, summarized in Table 10.3, has revealed that the many different forms of cancer, observed in nature and induced in the laboratory, are not all caused by one single virus. On the contrary, a very large number of viruses can act as oncogenic, or cancer-producing agents, in plants, reptiles, amphibians, birds, mammals and other living things.

The tumour viruses fall into two major classes, the DNA-containing oncogenic viruses, and the RNA-containing oncogenic viruses, also called oncornaviruses or leukoviruses. With respect to their general structure, chemistry and physiology, the oncogenic viruses of any group greatly resemble the non-oncogenic viruses of that same group. All of the known groups of DNA viruses have members which are oncogenic. Even amongst the most recently discovered

TABLE 10.3 The oncogenic viruses

Group	Nucleic acid		Structural features			Site of maturation	Oncogenic members
	Type	Molecular weight ($\times 10^6$)	Shape	Size (nm)	Envelope		
Poxvirus	Linear, DS*-DNA	160–200	Brick-like	300 × 230	+	Cytoplasm	Shope fibroma (rabbit), Yaba (monkey), Molluscum contagiosum (human)
Herpesvirus	Linear, DS-DNA; fragmented	60–80	Nearly spherical	100–150	+	Nucleus	Lucké (frog), Marek's disease virus (chickens), Herpes *saimiri* (monkey, marmoset), Herpes *ateles* (marmoset), Epstein–Barr (EB; human**, primate)
Adenovirus	Linear, DS-DNA	20–25	Spherical	70–80	–	Nucleus	Avian, hamster, rodent, simian, human adenoviruses
Papilloma	DS-CC*-DNA	5	Spherical	55	–	Nucleus	Canine, bovine, equine papilloma, Shope papilloma (rabbit), human wart
Papova	DS-CC-DNA	2–3	Spherical	45	–	Nucleus	Polyoma (mice), SV40 (monkey), hamster papova virus
Parvovirus	Linear, SS*-DNA	2	Spherical	20	–	?	Minute virus of mice (?)
C-type oncornavirus	SS-RNA, fragmented	10–12	Spherical	100–150	+	Cytoplasm	Avian, murine, cavian, bovine, canine, feline, simian, primate leukaemia-sarcoma complexes
B-type oncornavirus	SS-RNA fragmented	10–12	Spherical	100–150	+	Cytoplasm	Mammary tumour virus (MTV) (mice), Mason–Pfizer virus (monkey)

* DS, double-stranded; SS, single-stranded; CC, covalently-closed. ** See page 344.

parvoviruses, there is a report of one tumour-inducing agent, the minute virus of mice.

Amongst the 9 or 10 major groups of RNA viruses, only one, the oncornavirus group, has so far been shown to have oncogenic members. These fall into two subgroups; the B- and C-type viruses shown in the electron micrographs of Fig. 10.10. Amongst the B-type oncornaviruses, only the mammary tumour virus (MTV) of mice is well characterized. Similar, but not identical, particles have recently been isolated from a mammary tumour of monkeys. Virus-like particles from human milk are at present a possible entry in this virus subgroup.

The C-type RNA viruses comprise the largest number of tumour viruses known to be aetiologic agents of naturally occurring disease (leukaemias and lymphosarcomas in birds and mammals). C-type viruses are ubiquitous, having been detected in a wide spectrum of vertebrates, including man. In addition (as discussed below), amongst certain vertebrate species all individual members can be shown to carry, endogenously, a given C-type oncornavirus.

Recent work which bears directly on the problem of the mechanism of oncogenic action of the viruses, i.e. viral genome replication, the effect of virus infection on cell DNA synthesis and the effect of virus on the surface of transformed cells, will be considered throughout the following discussion.

Mechanism of oncogenesis and cell transformation by tumour viruses

One can discuss the phenomenon of cell transformation in terms of two events: (i) the initiation of transformation; (ii) the expression of transformation in the transformed cell phenotype. The evidence to be discussed in later sections now shows quite clearly that tumour viruses initiate transformation by adding to the genetic machinery of the cell one or more copies of a complete or incomplete viral genome. Expression of the transformed cell phenotype is reflected in alterations, both structural and functional, of the biochemical components of the cell surface

A remarkable feature of the oncogenic papovaviruses (polyoma, SV40), adenoviruses and herpesviruses is that, unlike most non-oncogenic DNA viruses, they can induce the synthesis of cell DNA. In the case of polyoma and SV40 viruses, induction of such synthesis is known to be a virus-encoded function. The available evidence indicates that the entire genetic locus involved in cell DNA synthesis is de-repressed thereby making available to the virus-infected cells those nucleases, polymerases, synthetases and ligating enzymes which function to make cellular DNA. From the interaction of cell and viral processes for DNA synthesis comes the essential molecular event of virus transformation, that is, covalent integration of viral DNA into the chromosomal DNA, perhaps at several specific sites.

Turning to the oncornaviruses, we find a relatively long history implicating DNA synthesis as an essential process both for virus replication and for neo-

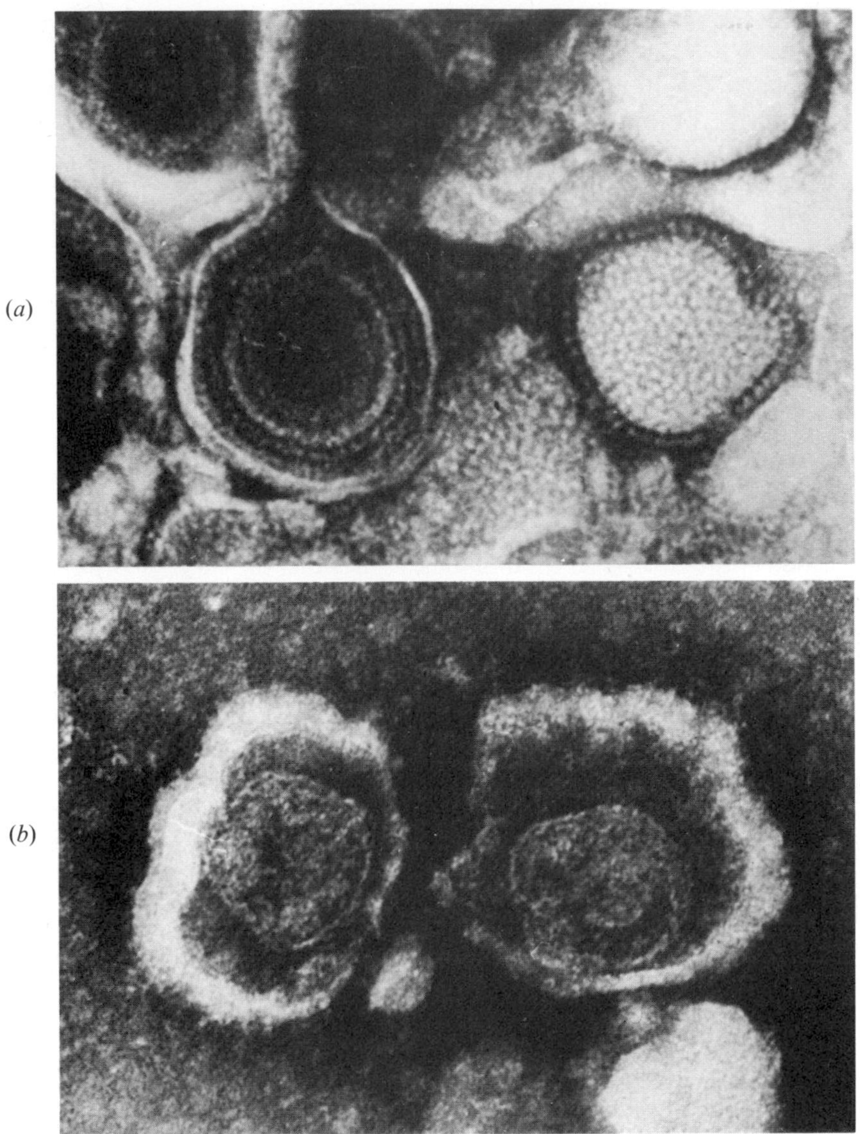

Fig. 10.10 (*a*) B-type oncornavirus (murine mammary tumour virus). Magnification ×280000. (*b*) C-type oncornavirus (murine leukaemia virus). Magnification ×350000. (Electron micrographs by A. F. Howatson.)

plastic transformation of cells. From the study of the inter-relationship between these various phenomena has come evidence that the oncornaviruses also initiate cell transformation by adding virus genetic material to the chromosomal DNA in the form of a DNA copy of the RNA tumour virus genome. In addition these investigations uncovered 'reverse transcriptase', the enzyme whose function may be to make such DNA copies.

Once the DNA of a tumour virus becomes a part of the cellular genome, it is replicated under the control of the cell. It transmits information for cellular transformation, and sometimes for virus replication, in a latent form. Virus-transformed cells are thus analogous to bacteria which have been lysogenized by temperate (non-virulent) bacteriophages.

'Reverse transcriptase'

From attempts to solve the riddle of why DNA synthesis was required both for replication of, and transformation by, an RNA virus came what is perhaps the most exciting biological discovery since that of the structure and function of DNA. In 1970 Baltimore, and Mizutani and Temin, reported that MuLV and Rous sarcoma virus (RSV) had an enzymatic activity which could reverse the normal flow of genetic information by producing a DNA copy of the RNA genome. This enzyme, variously called 'reverse transcriptase' or RNA-dependent DNA polymerase, has been found in all B-type and C-type oncornaviruses examined to date.

Molecular biologists and biochemical geneticists have been provided with an extremely useful tool to study genes and the gene-product relationship, as indicated, for example, by the recent demonstrations that this enzyme can be used *in vitro* to synthesize DNA copies of selected mammalian mRNA molecules. For cell biologists, embryologists and those who study cellular differentiation an intriguing period has arrived, for reverse transcriptase-like activity has been observed in a number of tissues and cell types, apparently free of virus. The significance of this observation will become clear as the enzymes of interest are purified, characterized and tested for their true mechanism of action and their relatedness to virus enzyme. An example of a process in which a cellular RNA-dependent DNA polymerase might function is one in which the genes for a specific RNA molecule are amplified, for example in that stage in the development of frog oocytes in which the DNA which codes for ribosomal RNA is greatly increased (described later in this chapter).

Virologists are clarifying the manner in which reverse transcriptase functions in the replication of the oncornaviruses and in the neoplastic transformation effected by them. That this enzyme is essential for these functions is suggested by the fact that some strains of oncornaviruses, defective for replication and/or transformation, have been shown to lack reverse transcriptase. It seems likely, as was originally suggested by Temin, that the reverse transcriptase transcribes the oncornavirus genome into a DNA copy. This may serve directly as a template for viral RNA synthesis, or it may become integrated into the chromosomal

DNA thereby producing a virus-transformed cell carrying a viral DNA segment.

Imprints of tumour virus on transformed cells

In some instances tumour cells continue to release virus particles, indicating that they carry the full virus genome. However, the majority of virus-transformed cells examined to date do not make virus particles, even though they do carry at least part of the genetic information of the causative agent. There are two major classes of diagnostic parameters which can be used to identify tumour virus genetic information in virus-transformed cells (Table 10.4): those which

TABLE 10.4 Tumour virus imprints in transformed cells

Virus groups	Virion component	Non-virion, virus-specified component	Test
DNA, non-enveloped	Genome DNA	mRNA	Molecular hybridization
DNA, enveloped	Genome DNA Core, envelope antigens	mRNA T-antigen, TSTA Membrane, surface antigens	Molecular hybridization Immunological, physicochemical
Oncorna-virus	60-70S RNA genome	Genome DNA, mRNA	Molecular hybridization
	'Reverse transcriptase'		Immunological, enzymological
	Core (*gs*), envelope antigens	TSTA Membrane, surface antigens	Immunological, physicochemical

define virion components (components, that is, of the mature infectious particle) and those which reveal non-virion, but virus-determined products. We consider first those which define virion components.

In transformed cells which produce virus particles these can be detected by electron microscopy. Where virus particles are absent, other virion components (noted in Table 10.4) can be measured. Virion genome is assessed by molecular hybridization techniques, virion protein usually by serological methods.

In many instances (polyoma, SV40, Shope papilloma, Shope fibroma, adeno and herpesviruses) it has been possible to show directly that cells transformed by the oncogenic DNA viruses contain viral DNA. Three approaches have been used. The DNA of transformed cells has been radioactively labelled and tested for specific hybridization with viral DNA. Reverse labelling in a similar test has also been utilized. In a third system, radioactive viral mRNA has been synthesized *in vitro* (using the *E. coli* RNA polymerase), and used as a specific probe for unlabelled viral DNA as a component of chromosomal DNA.

In those oncornavirus-transformed cells which continue to make virus particles (complete or incomplete), a number of virion components can be detected; in particular the viral RNA, core proteins and at least one envelope glycoprotein. Although the RNA genome of each oncornavirus is unique, as a group these molecules share certain general properties. The RNA of both B- and C-type particles is recovered from virus particles as two major moieties: one of low molecular weight, sedimenting at 5–12 *S*, the other having a molecular weight of about $10–12 \times 10^6$ and sedimenting at about 60–70 *S*. It seems likely that the smaller RNA is of cellular origin. The 60–70 *S* RNA, the actual virus genome, is not a single polyribonucleotide, but can, by dissociation of H-bonds, be separated into at least two subunit classes; each of molecular weight about $3–4 \times 10^6$. Some part of this RNA is associated with sequences of polyadenylic acid of about 100–200 residues in length. The 60–70 *S* RNA is unique amongst naturally occurring RNA molecules, and for this reason oncornavirus genome is readily identifiable.

Recently molecular hybridization has been used to detect virus-specific 60–70 *S* RNA in mouse mammary tumours and in virus-free lymphoid cells transformed by MuLV. Radioactively labelled virus-specific DNA, made *in vitro* from the RNA of the MTV of mice and from the RNA of MuLV (using the virus-specific reverse transcriptase), was used as the specific marker for RNA virus genome. Similar technology has also been applied in an effort to detect oncornavirus genome in human mammary tumour cells and in cells obtained from patients with leukaemia, lymphosarcomas, and from some solid tumours, particularly of the myeloproliferative cell system. Because human tumour viruses have not yet been isolated, the molecular probes used were again DNA copies made *in vitro* from the murine MTV and MuLV RNA. Large molecular weight (70 *S*) RNA molecules were detected in the human tumour cells, in some cases in association with an enzyme with reverse transcriptase activity. The significance of these findings with respect to human tumour viruses have yet to be unequivocally established.

Proteins of the oncornaviruses are particularly important for virus identification. Of specific interest are the reverse transcriptases, the group-specific (*gs*) antigens of the nucleoprotein core and the species-specific antigens of the viral envelope. The former enzyme can be detected on the basis of its immunochemical and enzymological specificity (in terms of template preference). The *gs* antigens, measured by serum reactivity, serve to classify oncornavirus as to genus of origin (e.g. avian, murine, primate, etc.). The envelope antigens (in particular the glycoprotein of the surface projections of the virus) reflect the particular species in which any virus stock was grown, hence the term species-specific antigen. The avian C-type oncornavirus has the same *gs* antigen whether it is grown in cells of the chicken, the turkey, the duck or the quail. A particular murine C-type virus has the same *gs* antigen whether it is grown in cells of the mouse, the rat, the hamster or the human. The six strains of murine MTV isolated to date have the same core *gs* antigens.

Cells transformed by the oncornaviruses all carry the genetic information for synthesis of the *gs* antigen. This may not always be expressed. However, in those cells which do make *gs* antigen, it is always found in the nucleus of the transformed cells. In those oncornavirus-transformed cells which continue to release virus particles, species-specific virion envelope antigens can be detected at the surface of the cells.

We consider now the second group of non-virion, but virus-determined molecules, which can also serve as imprints of the presence of tumour virus genome in virus-free transformed cells. These, listed in Table 10.4, are in turn of two classes: nucleic acids, detectable by molecular hybridization tests; and proteins detected by immunochemical and physicochemical procedures. As in the case of cells transformed by DNA tumour viruses, those transformed by RNA viruses carry viral DNA as a part of the chromosomal DNA. DNA copies of viral RNA genome can be detected in cells transformed by RSV, AMV, MuLV, the feline leukaemia virus and MTV.

One of the most important non-virion, but virus-determined proteins present in cells transformed by the DNA tumour viruses is the nuclear antigen, called the T- (for tumour) antigen (recognized by its interaction with specific antibodies or antisera). All cells transformed by polyoma virus have nuclear T-antigen determined by the polyoma genome; cells transformed by SV40 virus carry SV40-specified nuclear T-antigen; those transformed by adenovirus exhibit adenovirus T-antigen. Herpesvirus-transformed cells also make a distinctive, non-virion, nuclear antigen. It is, however, not yet clear whether this has the physiological and biochemical properties which characterize the known T-antigens.

In the case of polyoma and SV40 viruses it is now clear from studies of virus mutants that the virus genome codes for the formation of the T-antigen. However, it remains to be established whether the T-antigen is itself a gene product of the virus; or whether the pertinent gene codes for a regulatory molecule which de-represses a cellular gene for T-antigen formation. The T-antigen is the expression of an early viral function; it appears transiently during productive infection, during abortive infection and during abortive transformation, a process which yields phenotypically, but not genotypically, transformed cells. Partial purification of the T-antigens of polyoma, SV40 and adenoviruses has been achieved. These molecules, the function of which has not yet been clarified, have the properties of a protein and are of variable molecular weight (60000–250000).

A second non-virion nuclear antigen, the U-antigen, has recently been observed at the nuclear membrane of cells transformed by SV40 virus. The usefulness of this antigen as an expression of virus genome in transformed cells has not yet been adequately tested.

Perhaps the most important non-virion, but virus-specified molecules in transformed cells are those present at the cell surface. These in all likelihood determine the phenotypic expression of the cancer cell. One sub-set of these molecules is operationally defined as tumour-specific transplantation antigen (TSTA), on the basis of *in vivo* tests. Adult animals are immunized with the tumour virus under study and are then challenged with the virus-transformed cells. The immunized animals are markedly resistant to transplanted tumour cells, whereas unimmunized animals succumb to neoplastic disease when given such cells. Other virus-specified 'membrane' and 'surface' antigens can be detected by *in vitro* immunological procedures.

At this point it is useful to comment upon the presence of certain cellular antigens at the surface of virus-transformed, and most (if not all) tumour cells. These are foetal antigens normally synthesized by embryonic or germ cells. As these cells differentiate during neo-natal life, their capacity to make foetal antigens appears to be totally repressed. However, when the adult, differentiated cells are transformed, the neoplastic cells once again begin to synthesize these surface macromolecules. The possible significance of this observation is commented on below. The cellular antigens characteristic of embryonic and of cancer cells have been variously referred to as foetal antigens, retrogenic antigens and, in the case of human cancer cells, as carcino-embryonic, or CEA, antigens.

Biochemistry of the surface of normal and transformed cells

As indicated earlier, there is considerable evidence which suggests that macromolecules at the cell surface play a very important part in the neoplastic expression of virus-transformed cells. A clear understanding of the biochemical lesion which is expressed at the surface of virus-transformed (and other) tumour cells remains to be achieved. Some progress is now being made in examining the chemical structure of the surface macromolecules which mediate the critical growth responses of normal and of virus-transformed cells.

It has been possible to isolate two components of the cell surface which are operationally defined as the plasma membrane and a surface component at its extreme periphery. Comparative analyses have revealed extensive differences between these surface components, as they are recovered from normal and virus-transformed cells. The available evidence implicates the complex carbohydrate-containing glycoproteins and glycolipids of the cell surface as very important targets affected by neoplastic transformation.

Of special interest are the many findings which derived from those initially made by Hakomori and Murakami. They suggest that the pathways of biosynthesis of the complex carbohydrate chains of membrane gangliosides and

TABLE 10.5 Cell surface functions modified as a result of virus transformation

Contact inhibition of movement and growth.
Adhesion, locomotion.
Intercellular communication.
Antigenic surface molecules (e.g. transplantation antigens, foetal antigens, neoantigens).
Receptors for growth factors, serum factors, plant lectins, basic polypeptides.
Permeability of plasma membrane (e.g. uptake of glucose and other sugars).
Activity of membrane enzymes (e.g. of cAMP metabolism).

glycoproteins are regulated in the course of normal growth and differentiation; and that the regulatory processes here involved are genetically altered in the course of virus transformation. Such a model is very attractive in that it provides for a simple, testable mechanism by which many apparently different viruses can bring about the myriad of physiological changes which characterize tumour cells (Table 10.5). The model provides for the generation of new antigens, new surface molecules of many functions, and also for the presence of foetal antigens, the synthesis of which may be de-repressed as a consequence of disturbing the genetic control of formation of surface macromolecules.

Physiology of surface of normal and virus-transformed cells

A whole new area of cell biology, embraced by the term cell–cell interaction, is now becoming more amenable to precise investigation. This designation can be applied to the growth of single cells in culture, the growth of mixed cell populations *in vivo* and *in vitro*; cellular differentiation and morphogenesis; locomotion, adhesion and communication between cells. In examining the physiology of cancer cells and virus-transformed cells, it has become clear that all of these aspects of cellular physiology are altered in the course of virus transformation.

The most dramatic and momentous results are seen in altered patterns of cell growth. This is evident both *in vivo* and *in vitro*, but has been most carefully analysed in the latter situation. Under defined cultural conditions, normal cells grow with a characteristic doubling time to achieve a characteristic saturation density. This saturation density is dependent upon a number of factors, including cell–cell contact, the requirement for specific growth factors (present in the serum of the growth medium) and the efficacy of specific transport mechanisms and receptor molecules at the cell surface. All of these factors undergo modification as a result of virus transformation.

When normal cells make contact, movement in the forward direction ceases. This phenomenon has been termed contact inhibition (see Chapter 1). If space permits, the contact-inhibited cell can move and grow in another direction. Clearly contact inhibition will determine the capacity for cells to grow one over the other, and therefore the colonial morphology and the tissue and organ organization. It will in part contribute to the ultimate concentration of cells present within a confined space, i.e. to the saturation density. Normal cells are very sensitive to contact inhibition. In general, when grown *in vitro* on a solid surface they form a single, well organized sheet of cells, termed by some a 'monolayer'. Virus-transformed cells are usually not very sensitive to contact inhibition. They grow *in vitro* in a very disorganized pattern to form multilayered sheets or clones of cells (see Fig. 10.9).

Recent studies have suggested that these phenomena, in normal cells, may be regulated by local concentrations of cyclic AMP. Thus cells with high intracellular levels of cAMP appear to be more sensitive to contact inhibition than are those with lower levels. In this connection, it is of interest that most tumour and virus-transformed cells examined to date have little intracellular cAMP. In addition their growth properties can be shifted, at least temporarily, in the direction of the normal state by the addition of this nucleotide.

Genetic studies with the tumour viruses

It has now been firmly established that tumour viruses initiate cell transformation by adding viral DNA to the genetic complex of the cell. One might ask whether the whole virus genome is necessary for transformation and/or replication, and one would hope eventually to identify the requisite genes.

In some of the earliest studies directed to this problem, the tumour virus in question was subjected to X-ray and ultraviolet light irradiation at doses known to inactivate DNA. It was found in the case of polyoma, SV40 and adenoviruses that the virus was more resistant to high energy radiation in its transforming function than in its replication. It was concluded that the whole virus genome is not necessary for transformation. This has since been confirmed by the fact that virus particles can be isolated which are defective for replication, but have unimpaired transforming activity. In the case of the oncornaviruses it has been shown that the whole virus genome is not required either for transformation or for replication.

One aspect of the genetic studies with oncogenic DNA viruses was the isolation of infectious nucleic acid. Infectivity of polyoma, SV40 and adenovirus DNA has been demonstrated in *in vitro* systems. DNA from Shope papilloma virus has produced papillomata upon injection into rabbits, while *in vitro* transformation has been achieved with DNA isolated from polyoma and SV40 viruses.

Perhaps the most interesting stage in the genetic study of oncogenic viruses has been the isolation and characterization of mutants. The classes of mutant viruses obtained, some of which have already been analysed by complementation and recombination, are listed in Table 10.6. Of special interest is the characterization of those genes which are essential for transformation. From studies with polyoma virus and RSV it is already clear that there are at least two pieces of genetic information required for stable neoplastic transformation by the tumour viruses; one for initiation of the genetic transformation, the other for the expression of the transformed cell phenotype. In polyoma and SV40, that genetic locus which codes for initiation of viral DNA synthesis is also essential for transformation, perhaps because it mediates integration of the viral DNA into host genome. In the case of the avian leukaemia-sarcoma virus complex the evidence suggests that one segment of the RNA genome carries the information for transformation. It is present in all of the transforming sarcoma viruses, but is absent in the leukosis viruses which are competent for replication. Reverse transcriptase has been implicated both in replication and in transformation, and, as expected on the basis of such functions, defective and non-transforming mutants of MSV and RSV have been isolated which lack this enzyme.

TABLE 10.6 Classes of mutant tumour viruses

Mutant class	Virus	Nature of defect	Replication	Transformation
Non-conditional:				
Defective	Polyoma, SV40, Adenovirus	Deletion of DNA	−	+
	RSV, MSV, HaSV, SSV	Some lacking 'reverse transcriptase' activity	−	+
	RSV (α)	No 'reverse transcriptase' protein	−	+
Non-transforming (NT)	AML, ALV	Absence of RNA subunit	+	−
Defective, NT	RSV, MSV	No 'reverse transcriptase' activity	−	−
Conditional-lethal:				
Host range	MSV	No 'reverse transcriptase' activity	−	+
	Polyoma	Virion protein synthesis	−	+
	Polyoma	No expression of surface changes	−	−
	Human adeno-virus-type 12	No formation of a virion protein (hexon)	−	−
Temperature-sensitive	Polyoma, SV40	Virion protein synthesis	−	+
	Herpesvirus	Virion protein synthesis	−	not known
	Polyoma, SV40	Initiation of viral DNA synthesis	−	−
	Human adeno-virus-5	Not known	−	−
	MSV	Not known	−	−
	MuLV	Not known	−	−
	SR-RSV, B77-RSV, Polyoma,	Cell surface modification	−	−

ALV, avian leukosis virus; AML, avian myeloblastosis virus; HaSV, hamster sarcoma virus; MSV, murine sarcoma virus; MuLV, murine leukaemia virus; RSV, Rous sarcoma virus (SR, Schmidt–Ruppin strain; B, Bryan strain); SSV, simian sarcoma virus.

There are now a number of virus isolates which are mutant in that function required to maintain the phenotype of the transformed cell. The genetic loci involved determine whether changes in the surface structure of cells transformed by these mutant viruses will occur. It is of interest that in the case of B77-RSV-ts75, a member of this virus group, the mutation affects the synthesis of a glycoprotein and a polypeptide of the virion membrane, which in turn may interfere with the glycoprotein metabolism of the cell membrane.

Virogenes, oncogenes and protoviruses

Figure 10.11 presents a simplified model which embraces some of the current concepts about the organization and function of genetic information in a virus-transformed cell. At the core of the model is the virogene, or the virus genome present as part of the cell DNA. It may represent the whole virus genome, or only a part. Where the viral DNA is a copy of an RNA tumour virus it is referred to as protovirus. Genetic material which codes for expression of the

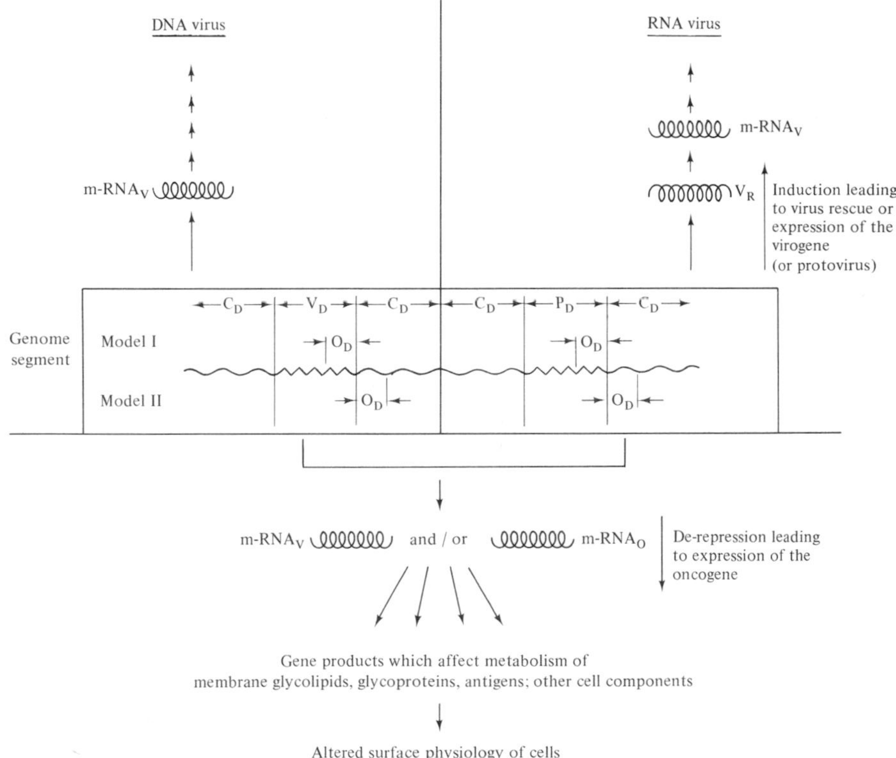

Cell DNA (C_D); Virogene (V_D) or protovirus (P_D); O_D oncogene Viral RNA (V_R); m-RNA of virus (V) or oncogene (O)

Fig. 10.11 Models of genome segment of a cell transformed by an oncogenic virus. Models I and II indicate that virus DNA as virogene (for the DNA virus) or as protovirus (for the RNA viruses) becomes an integral part of the cellular chromosomal DNA. It is not entirely clear whether the oncogene is itself of viral (Model I) or cellular (Model II) origin.

transformed or tumour cell phenotype is termed the oncogene. The oncogene may be a part of the virus genome; it may be a part of the cellular genome; it may be mainly cellular in origin, but include also a segment of the tumour virus genome. Whatever its origin, its expression appears to be under the control of virus genetic information.

Virogene and/or oncogene may be repressed or de-repressed to varying degrees in different virus-transformed cells and under different conditions. Even when fully repressed in the natural state, these genes may be induced to express their genetic information by treatment with irradiation, by treatment with a variety of chemicals (including hormones and known carcinogens), by superinfection with viruses, and by altering the cultural conditions of the cells. Depending upon the particular cell, induction may result in the formation of viral gene products, defective virus or infectious virus.

As the techniques for induction and detection of expression of virus genome in cells have progressed, so has our knowledge about the distribution of tumour viruses in nature; particularly the herpesviruses and the C-type oncornaviruses. It seems likely that many, perhaps all, normal cells of chickens and mice (and of many other vertebrate species) carry genetic determinants of latent tumour viruses. This genetic information, present in all cells, is partially expressed early during embryonic and neonatal life and is then totally repressed as adult life proceeds. Expression of the genome may be activated later in life, depending upon the genetic constitution of the host. Such expression can be induced *in vivo* (as *in vitro*) by physical and chemical carcinogens, giving rise to tumour formation and/or virus replication. There is now good evidence that in some instances of *in vivo* induction by chemicals or radiation, the tumours which result originate from virus-transformed cells.

Any discussion of tumour viruses would not be complete without some reference to a possible viral aetiology for human cancer. There is little doubt that benign human warts are caused either by the human papilloma virus or by the molluscum contagiosum virus. As one moves into the area of malignant disease, absolutely complete identication of a human tumour virus has yet to be made. However, the case for the involvement of the Epstein–Barr herpesvirus in Burkitt's lymphoma is very strong. It is less complete for nasopharyngeal carcinoma. The biological evidence, which reveals that C-type oncornaviruses are the causative agents for a very large number of leukaemic and lymphomatous diseases amongst all kinds of vertebrates, forces one to seek out the analogous virus of human disease. With the conceptual and technical tools which are developing from the study of the experimental tumour viruses, the expectations are very good for the rapid identification of human tumour viruses.

Antibody production

We consider in this section a special type of control of protein synthesis – synthesis of antibodies in response to stimulation by foreign antigens. The wide variety of antigen molecules (100 000 or more) to which specific antibodies can be formed made it reasonable to suppose, as ideas on protein synthesis and conformation were developing in the 1950s and early '60s, that although other proteins might have a tertiary structure determined entirely by the amino-acid sequence of their polypeptide chains, antibodies might be rather different, perhaps more malleable at a certain stage of synthesis and shaped at that stage by interaction with antigen. This was the 'template' theory of antibody formation

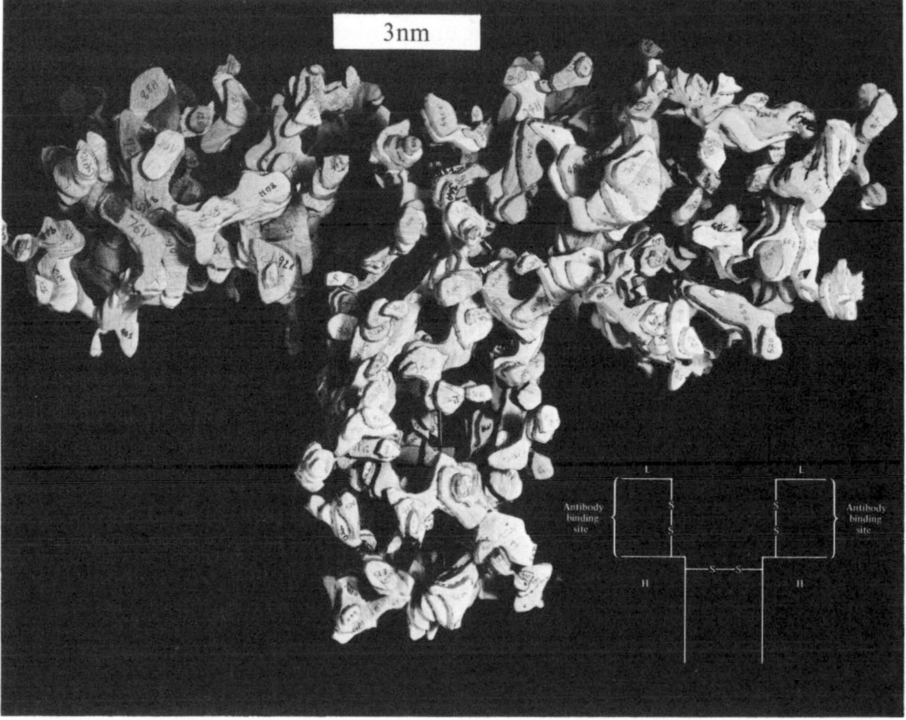

Fig. 10.12 Photograph of a balsa wood model of a γG1 immunoglobulin molecule, from X-ray diffraction analysis at 0·6 nm resolution. White scale mark = 3 nm (*From* Sarma *et al*, 1971.) The insert drawing shows S—S bonding between H- and L-chains and the presumed positions of the antigen-binding sites (the chains themselves being represented purely schematically in this drawing).

put forward originally by Breinl and Haurowitz. Any template mechanism of this kind now seems highly improbable, partly because antibodies of different specificity appear to show differences in amino-acid sequence and partly because antibody molecules can be unfolded in 8 M urea solution and refolded in the absence of antigen with recovery of specific antigen-binding properties. It seems that antibodies, like other proteins, take up a unique tertiary structure determined by their amino-acid sequence. The variety of different antibodies arises from a corresponding variety of coding sequences, though exactly how this variety is, or has been, derived is still uncertain, and it is not yet known how much of this genetic information is carried in the germ line, or how much might arise by selection of sequences in a genetic segment undergoing some special process of rapid somatic mutation or interallelic recombination.

Antibodies, or *immunoglobins*, form the γ-globulin (gamma globulin) fraction of blood serum (see Fig. 1.7). The simplest class, IgG, have molecular weights around 150000 and are made up of two identical heavy (H) and two identical light (L) polypeptide chains. In the IgA class similar 150000 molecular-weight units form dimers or larger aggregates, and the IgM immunoglobulins are pentamers of five 150000 molecular-weight units linked by S—S bonds. Within the basic 150000 molecular-weight units of the different immunoglobulins each L-chain is joined to one of the H-chains by an S—S bond and the two H-chains are linked by one or more S—S bonds (Fig. 10.12). The lengths of the L- and H-chains are about 220 and 440 amino-acid residues respectively (molecular weights around 25000 and 50000). L-chains contain two, and H-chains four, intrachain S—S bonds forming loops, in each case, of about 60 amino-acid residues (Fig. 10.13). The amino-acid sequences in the carboxyl-terminal part of both H- and L-chains are relatively invariable, within a given class of antibodies, but the sequences at the N-terminal end are more variable. These variable parts of the H- and L-chains form the antigen-binding sites.

The amount of genetic information needed to code for a given number of antibodies is reduced somewhat by the fact that the antigen-binding sites contain parts of two different polypeptide chains. The combining of two chains at the binding sites means that, say, 300 different H-chains, with sequence differences in the variable position, and 300 different L-chains could be combined in different ways to form 90000 different antibodies. Only 600 rather than 90000 genes would be needed to code for this variety.

The H-chains of antibodies mainly fall into three classes, the γ, α and μ chains of IgG, IgA and IgM immunoglobulins. The L-chains fall into two classes: κ (kappa) and λ (lambda). (The fact that the IgG heavy chains are called γ-chains is a bit confusing. All these immunoglobulins, with their γ, α or μ chains are found in the γ-globulin fraction of serum and hence are all sometimes called γ-globulins, though in fact only IgG contains γ H-chains.) L-chains of either class may combine with H-chains of any class, and *in vitro* unsymmetrical molecules can be formed. Natural antibodies however are always symmetrical and a molecule of IgG type contains two identical L-chains and two identical H-chains. Within the main H-chain classes there are further sub-classes, there being four types of γ H-chain, two of α H-chain, and two of μ H-chain. Other types of H-chain are also found, more rarely, such as human δ and ε H-chains. Covalently linked to the H-chains there is a carbohydrate component of the immunoglobulin molecule.

The clinical importance of the immune reaction has led to extensive sequence studies on antibody polypeptide chains, starting in the early 1960s, with the Bence Jones proteins which are found in the urine of patients with plasma cell myeloma. Plasma cells (derived probably from lymphocytes) are blood cells

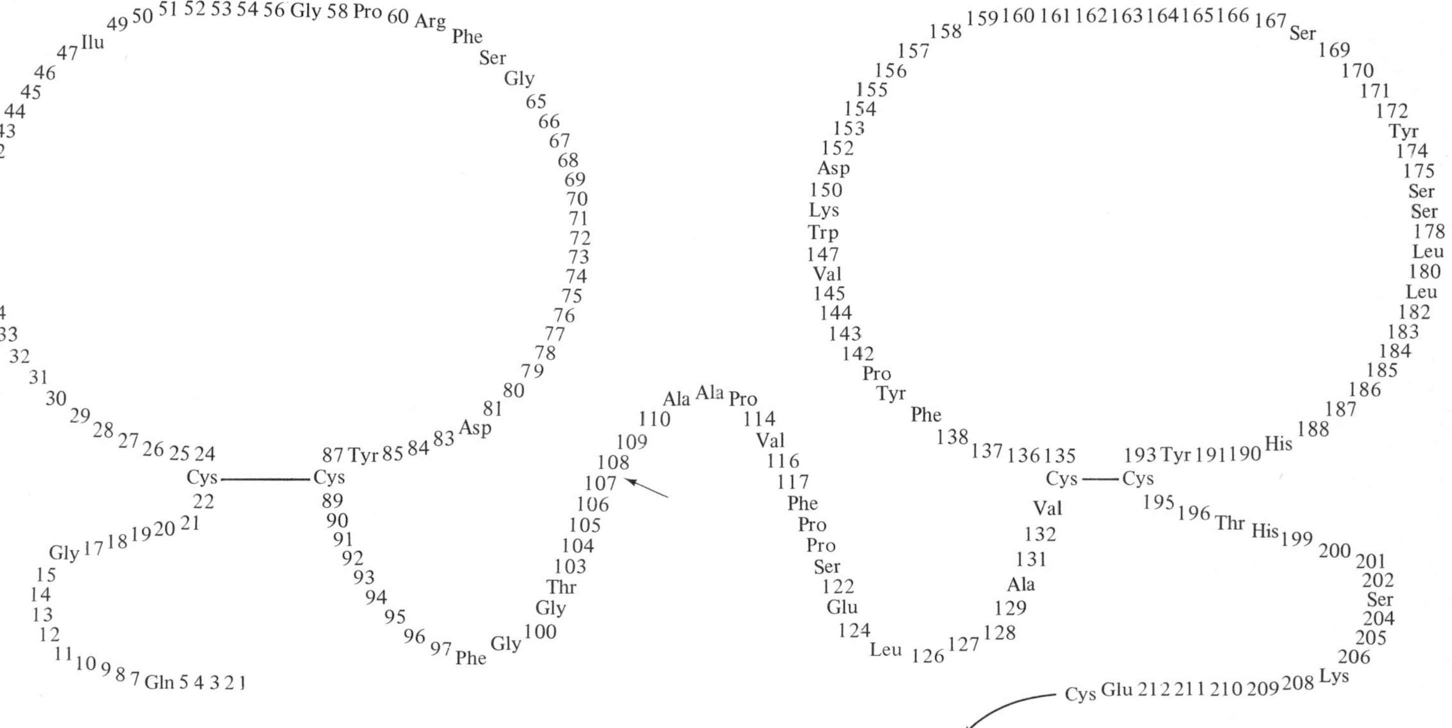

Fig. 10.13 The light chain of immunoglobulins. Positions in the sequence are given for amino acids which are invariant in human κ and λ and mouse κ-chains (see Table 10.7). The arrow indicates the transition from the variable to the constant region. (*From* Kabat, 1968.) This figure differs slightly from a similar figure that could be drawn from the data of Table 10.7. The total number of invariant amino acids is about the same, but many come in different positions, due to differences in the way parts of the κ sequences have been positioned under the λ sequence in establishing maximum homology. Different options are possible in parts of the sequence.

TABLE 10.7 The sequences of human and mouse λ and κ chains in the Bence Jones proteins. (From R. L. Hill *et al.* (1967) *Third Nobel Symposium*, Ed. J. Killander, Interscience, New York.)

	Chain				
Human Bence Jones protein	λ ⟶	H_2N –	^{1}Ser –		
	λ ⟶	^{113}Ala –	Pro –	Ser –	
Human and mouse Bence Jones proteins	κ {mouse human} ⟶	H_2N –	^{1}Asp –	Ile –	
	κ {mouse human} ⟶		^{113}Pro –	Thr Ser –	

λ	^{2}Glu –	Leu –	Thr –	Gln –	Asp –	Pro –	Ala –		Val –	^{10}Ser –	Val –	Ala –
λ	^{116}Val –	Thr –	Leu –	Phe –	^{120}Pro –	Pro –	Ser –	Ser –	Glu –	Glu –		Leu –
κ	^{3}Val Gln –	Leu Met –	Thr –	Gln –	Ser Pro –	Pro Ser –	Ala Ser –	^{10}Ser –	Leu –	Ala Ser –	Val Ala –	Ser –
κ	^{115}Val –	Ser Phe –	Ile –	Phe –	Pro –	^{120}Pro –	Ser –	Asn –	Glu –	Glu –		Leu
λ	^{13}Leu –	Gly –	Gln –	Thr –		Val –	Arg –	Ile –	^{20}Thr –	Cys –	Gln –	Gly –
λ	^{127}Gln –	Ala –	Asn –	^{130}Lys –		Ala –	Thr –	Leu –	Val –	Cys –	Leu –	Ile –
κ	^{15}Leu Val –	Gly –	Glu Asp –	Arg –		Ala Val –	^{20}Ser Thr –	Leu Ile –	Ser Thr –	Cys –	Arg Gin –	Ala –
κ	^{126}Thr Lys –	Gly Ser –	Gly –	Thr –		^{130}Ala –	Ser –	Val –	Val –	Cys –	Phe Leu –	Leu –
λ	^{24}Asp –	Ser –	Leu –	Arg –	Gly –	Tyr –	^{30}Asp –	Ala –	Ala –	Trp –		
λ	^{138}Ser –	Asp –	^{140}Phe –	Tyr –	Pro –	Gly –	Ala –	Val –	Thr –	Val –		
κ	^{26}Ser –	Gln –	(Asx,	Ile,	^{30}Asx,	Ser,	Phe) –	Ser, Met Leu –	Asn –	Trp –		
κ	^{137}Asn –	Asn –	Phe –	^{140}Tyr Pro –	Pro Tyr –	Lys Arg –	Asp Glu –	Ile Ala –	Asn Lys –	Val –		
λ	^{34}Tyr –	Gln –	Gln –	Lys –	Pro –	Lys –	^{40}Gln –	Ala –	Pro –	Leu –	Leu –	Val –
λ	^{148}Ala –	Trp –	^{150}Lys –	Ala –	Asp –		Ser –	Ser –	Pro –	Val –	Lys –	Ala –
κ	^{36}Leu Tyr –	Gln –	Gln –	Lys Gly –	^{40}Pro –	Gly Lys –	Glx Lys –	Pro Ala –	Ile Pro –	Lys –	Arg, Leu Ile –	Leu –
κ	^{147}Lys Gln –	Trp –	Lys –	^{150}Ile Val –	Asp –		Gly Asn –	Ser Ala –			Glu Leu –	Arg Gln –

46 Ile –	Tyr –	Gly –	Arg –	50 Asn –	Asn –	Arg –	Pro –	Ser –	Gly –	Ile –		λ
159 Gly –	160 Val –	Glu –	Thr –	Thr –	Thr –		Pro –	Ser –	Lys –	Gln –		λ
48 Ile –	Tyr –	50 Ala Asp –	Thr Ala –	Ser –	Ser Asn –	Gln Leu –	As, Gly Glu –	Ser Thr –	Gly –	Val		κ
156 Gln Ser –	Asx Gly –	Gly Asn –	Val Ser –	160 Leu Gln –	Glu –	Ser –	Asx Val –	Thr –	Asx Glu –	Trp Gln –		κ
57 Pro –	Asp –	Arg –	60 Phe –	Ser –	Gly –	Ser –	Ser –	Ser –	Gly –	His –	Thr –	λ
169 Ser –	170 Asn –	Asn –	Lys –	Tyr –	Ala –	Ala –	Ser –	Ser –	Tyr –	Leu –	180 Ser –	λ
59 Pro –	Lys, Ala Ser –	Arg –	Phe –	Ser –	Gly –	Ser –	Arg Gly –	Ser Phe –	Gly –	Ser Thr –	70 Asp –	κ
167 Asp –	Ser –	Lys –	170 Asp –	Ser –	Thr –	Tyr –	Ser –	Met Leu –	Ser –	Ser –	Thr –	κ
69 Ala –	70 Ser –	Leu –	Thr –	Ile –	Thr –	Gly –	Ala –	Gln –	Ala –	Glu –	80 Asp –	λ
181 Leu –	Thr –					Pro –	Gln –	Glu –	Trp –	Lys –	Ser –	λ
71 Tyr Phe –	Ser Thr –	Leu Phe –	Asn Thr –	Ile –	His Ser –	Ser, Pro Gly –	Met Leu –	Glu Gln –	80 Ser Pro –	Asx Glu –	Asp –	κ
179 Leu –			180 Thr –	Leu –	Thr Ser –	Lys –	Asx Ala –	Glx Asp –	Tyr –	Glu –	Arg Lys –	κ
81 Glu –	Ala –	Asp –	Tyr –	Tyr –	Cys –	Asn –	Ser –		Arg –	90 Asp –		λ
189 His –	190 Arg	Ser –	Tyr –	Ser –	Cys –	Gln –	Val –	Thr –	His –	Glu –		λ
83 Phe, Thr Ile –	Val Ala –	Asp, Met Thr –	Tyr –	Phe Tyr –	Cys –	Leu Gln –	90 Gln –		Ser Tyr –	Ala, Lys Asp –		κ
189 His –	190 Asx Lys –	Ser Val –	Tyr –	Thr Ala –	Cys –	Glu –	Ala Val –	Thr –	His –	Lys Gln –		κ
91 Ser –	Ser –	Gly –	Lys –	His –	Val –	Leu –	Phe –	Gly –	100 Gly –	Gly –	Thr –	λ
200 Gly –	Ser –	Thr –								Val –	Glu –	λ
93 Ser, Glu Thr –	Ser, Val Leu –	Pro –	Trp Arg –	Thr –			Phe –	Gly –	100 Gly Gln –	Gly –	Thr –	κ
200 Thr Gly –	Ser Leu –	Thr Ser –	Ser –	Pro –						Ile Val –	Val Thr –	κ
103 Lys –	Leu –	Thr –	Val –	Leu –	Gly –	Gln –	110 Pro –	Lys –	Ala –			λ
205 Lys –		Thr –	Val –	Ala –	Pro –	210 Thr –	Glu –	Cys –	213 Ser –	COOH		λ
103 Lys –	Leu –	Glu –	Ile –	Lys –	Arg –	Ala Thr –	Asx Val –	Ala –	112 Ala –			κ
207 Lys –		Ser –	Phe –	210 Asn –	Arg –	Asn Gly –	Glu –	214 Cys –	COOH			κ

involved in antibody synthesis, and Bence Jones proteins are antibody L-chains which are discharged in the urine in plasma-cell myeloma. For L-chains about half the sequence is relatively invariable. This part of the sequence, a little over a hundred amino-acid residues long, runs from the middle of the chain to the carboxyl end and is identical, or almost identical, in L-chains of a given class from a given species. The variable sequence of the L-chain, which is about the same length as the relatively invariable segment, shows differences within the same class of antibody within the same species. This variable sequence is different for antibodies formed against different antigens. The differences are mainly found in three regions of high variability around residues 30–31, 53–56 and 93–96 counting from the N-terminal end of the chain, which may be supposed to be the parts of the L-chain which contribute to the antigen-binding site. For H-chains the variable part of the chain is about the same size as that of light chains, comprising the first 100–120 residues from the N-terminal end.

The variable (V) parts of the L- and H-chains show sequence differences within one species and within one class of immunoglobulins while the remainder of each chain is constant (C). However the constant parts of the chains show differences between classes and between species which are as great as the differences in the V-regions. The changes in amino-acid sequence in both V- and C-regions are mostly consistent with single-base substitutions in the genetic code, and in spite of the variation there remains extensive homology of sequence. For example more than 30 residues are identical in the C-segments of human κ- and λ-chains and more than 50 residues are identical in the C-segments of the κ-chains of mice and men (Table 10.7). There is also sufficient homology between the heavy and light chain sequences and between the variable and constant parts of each type of chain (Table 10.7) to suggest that all the genes specifying antibody polypeptide sequences may have arisen from a common ancestral gene coding for a polypeptide about 110 residues long. By duplication and independent mutation of each segment this gene could have evolved to two genes which at that stage, or later, came to specify the variable and relatively invariable parts of primitive antibody molecules. Further duplication and mutation could then lead to evolution of the two types of chain (H and L) and of the different classes of H- and L-chains (Figs. 10.14 and 10.15).

Progress in sequence determination for the immunoglobulins represents a dramatic breakthrough in immunological science, and the realization that synthesis of a given antibody requires activation of a specific gene, or combination of genes, has had an important impact on thoughts and theories about how antibodies are formed. However, we are still a long way from understanding the immune reaction at a molecular level. When an animal is first exposed to new antigen, either naturally or experimentally, there is a short latent period, then antibody production begins, increases for a week or two, levels off and then slowly declines usually to undetectable levels in a period sometimes as short as a few weeks, sometimes several years. This is the primary response, predominantly involving production of IgM immunoglobulins. A further injection of antigen, after an adequate delay time, leads to more massive and more rapid

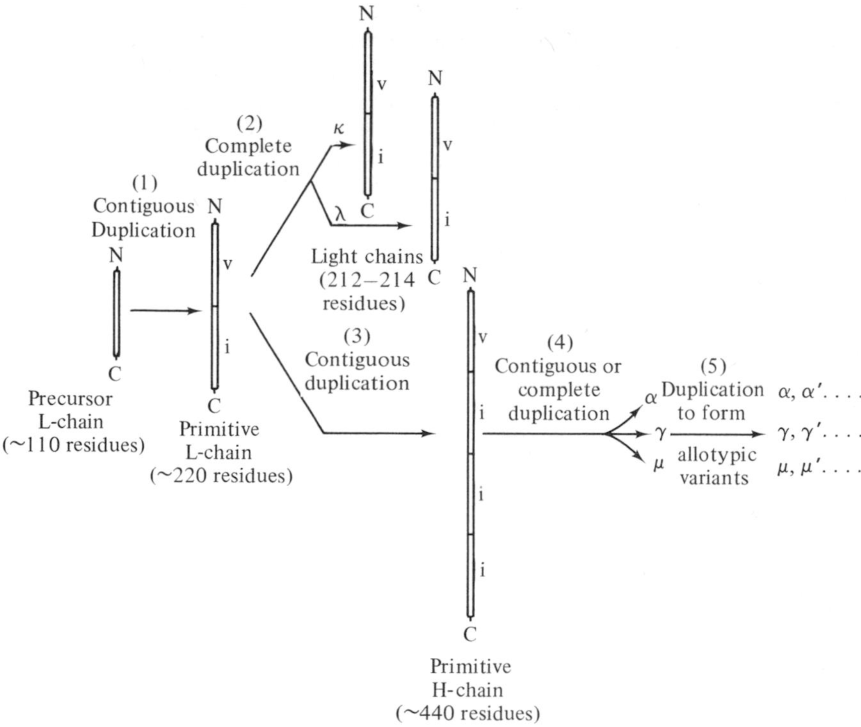

Fig. 10.14 A proposed mechanism for the evolution of the immunoglobulins. (*From* Hill *et al.*, 1967.)

production of antibody (secondary response), the immunoglobulin now being mainly of the IgG type, with multiplication of cells already programmed for specific production of the required antibody. The circulating blood cells which are producing antibody in large amounts may be recognized by their extensive development of rough endoplasmic reticulum (making protein for 'export'). These are the plasma cells. However, since the demonstration by Gowans around 1965, that lymphocytes were involved in the immune response, evidence has grown that these are the potential antibody producing cells. Plasma cells develop from lymphocytes triggered for antibody production.

Two types of lymphocyte can be isolated and their reaction to antigens studied experimentally: T-lymphocytes derived from thymus and B-lymphocytes derived from bone marrow. (These two cell-types are distinguished serologically by a surface group (θ) present on the T-cells but not on the B-cells.) The B-lymphocytes are the precursors of the plasma cells which produce serum antibodies. T-lymphocytes play a helper role in this process, potentiating antibody production by the B-cells, and also play a direct role in cellular immune reactions.

It was suggested by Paul Ehrlich as long ago as 1900, that antibody-forming

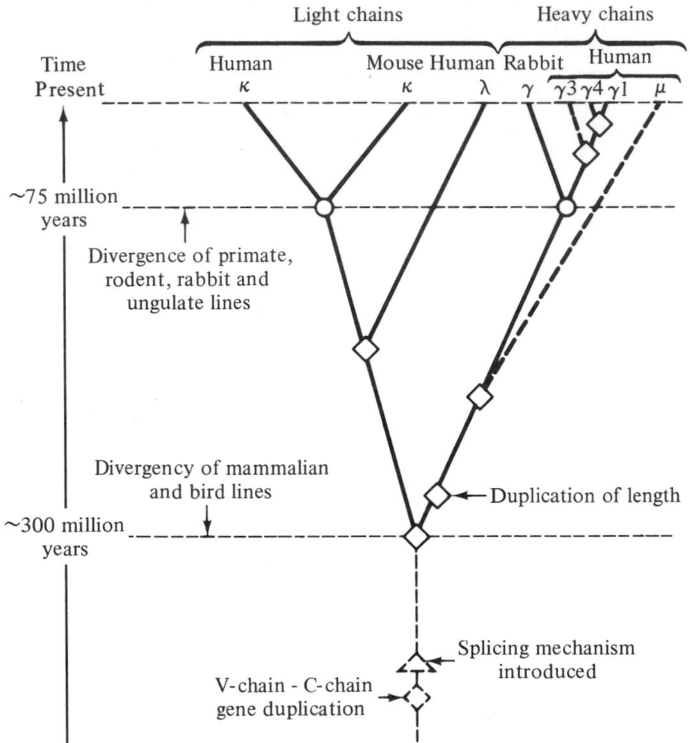

Fig. 10.15 Time scale for the events of Fig. 10.14. (*From* Dayhoff, 1969.)

cells might recognize foreign antigens by means of antibody receptors on the cell surface. It is now possible to show by a variety of techniques, including in favourable cases the direct study of isolated membrane proteins, that both for T- and B-lymphocytes a cell which is capable of responding to a given antigen probably carries the corresponding specific immunoglobulin at its surface. (This is fairly certain for B-cells, less so for T-cells.) Antigen stimulates lymphocyte activity by first forming a complex with immunoglobulin at the lymphocyte surface. This triggers these cells to produce, as well as release, antibody and also to divide, so that clones of cells are formed producing specific antibody. A structural change at the lymphocyte membrane is linked to controls for antibody synthesis and for cell division.

Lymphocytes stimulated to activity by a given antigen produce only antibody specific to this antigen. If we are to account for the known variety of antibody production, we have to consider how a lymphocyte population can be formed, containing 100 000 or so different cell types with different surface antibodies, each sitting ready to respond to the appropriate antigen when challenged. There is actually a further aspect to this question, and that is why the lymphocytes do *not* respond, except in pathological conditions, to the proteins of the animal's own tissues. The success of the immune reaction as a defence against viruses and bacteria depends on the lymphocytes recognizing the surface of these invading organisms as foreign and different from 'self'. If this recognition is defective, or modified, then 'self' may be treated as 'foreign' and antibodies are made against the animal's own tissues (autoimmune diseases) or foreign tissues used in skin or organ grafting may be accepted as 'self' (immunological tolerance). These are areas where increased understanding of the immune response at the molecular and cellular level can contribute directly to the solution of clinical problems.

We have noted earlier two possibilities for the way in which the diversity of antibody genes could have arisen: either through evolution, with a variety of 600 or more genes present in the germ line, or through some special process of somatic mutation or recombination. If the antibody genes are carried in the germ line this would take up only a reasonably small proportion of the mammalian genetic information store (600 or so genes represent only about 0·01 per cent of mammalian DNA). Evolutionary pressures could have allowed variability of genes for the variable parts of the H- and L-chains while keeping the genes for the other parts of the chains (or the other parts of each gene, if H- and L-chains are each coded by a single gene) almost constant within one species. The constant parts of the chains could be coded by a limited number of genes, one for each type of immunoglobulin chain whether γ, α, μ, κ or λ, with a large number of genes coding for the variable (V) parts of the chains. These V-genes could have arisen by multiple duplication of ancestral V-genes, each then undergoing stable, viable, mutation at the usual rare rate, until higher animals now carry 300 or more V-genes for each of the H- and L-chains. The fact that antibody-binding sites are made from two chains not only reduces the number of genes needed to code for a given number of antibodies, but also allows currently unneeded antibody genes to be maintained through evolution. Thus the gene for a given L-chain, L_x, may be maintained because with a given H-chain (H_x) it forms a needed antibody, and L_y because with H_y it forms a needed antibody. At some future time the previously unused antibodies L_xH_y and L_yH_x may be needed and will be available.

If the antibody genes are carried in the germ line the simplest way to envisage formation of a varied lymphocyte population is to suppose that the switching-on of different antibody genes in the lymphocyte precursor cells is a random process leading to a varied population of cells carrying at least a few copies of all the wide range of antibodies that the animal is capable of producing (though only one, or a very limited number are carried on any one cell). A variety of antigen recognition sites are then present, with the appropriate cells ready to respond to the challenge of a given antigen. If the genetic variety arises through a somatic gene-scrambling process, this must be supposed to lead also to a varied cell population carrying a few copies of all the antibodies that the animal is capable of producing. Tolerance to the animal's own tissues, or induced tolerance, involves in some way a paralysis of the immune response of those cells which are capable of making antibody to the tolerated protein. This paralysis may result perhaps from exposure of the potential antibody-producing cells to antigen under conditions which cause the antigen to combine with and inactivate its target sites, so that cells carrying these recognition sites can no longer respond and produce antibody.

If any conclusion can be drawn to end this section, it is that in spite of our new knowledge of the structure of immunoglobulin molecules and of the roles of T- and B-lymphocytes, the central problems of antibody production remain unsolved and challenging. Even if developing lymphocytes contain no novel gene-scrambling mechanism, they at least contain some novel control mechanism for randomly selecting the antibody to be synthesized. The stimulation of lymphocytes to divide involves structural changes in the cell membrane, linked to control of division. Increased understanding of the conditions which lead to immunological tolerance (an active research area because of its importance in tissue grafting) will throw further light on this membrane-antibody-antigen reaction.

Regulatory processes in oocyte development

The ultimate aim, in the study of development, is to understand in molecular detail all the steps that lead from ovum to mature adult, the control, for example, of limb shape and size, or of axon growth and synapse formation in a developing nerve net. The work of Gurdon noted in Chapter 7, and experiments in which plants can be regenerated from single cells provide the evidence that each cell of the mature, or developing, organism carries a full complement of genes (except in special cases, as when the nucleus of the cell is extruded from cells which later develop to red blood cells). Development and differentiation involves switching genes on and off, but of what varied types or how complex these control systems will prove in higher organisms we can hardly guess at present. Already in the development of the amphibian oocyte prior to fertilization we meet a new phenomenon, the rapid *amplification* of the genes coding for ribosomal RNA, that is, multiple replication of the DNA of just this genetic segment. In the oocytes of the toad *Xenopus laevis* 1500–2000 copies of these genes are formed, strung out along long strands of DNA with the genetic sequences for 18 *S* and 28 *S* ribosomal RNA in tandem, separated from the next pair by 'spacer' regions rich in G—C bases. Synthesis of growing lengths of this ribosomal RNA, as RNA polymerase molecules move along the DNA strands, can be beautifully seen in electron micrographs (Fig. 10.16). Regions of ribosomal RNA synthesis (in oocytes and other types of cell) form the *nucleoli* seen in cell nuclei by light microscopy. We do not know whether rapid gene amplification of the type found in oocytes may occur in other developmental situations also. There is great current interest in repetitive sequences that have been found in the DNA of higher organisms. In discussing ribosomal RNA synthesis we must note an expansion of the 'modern definition of a gene' given in Chapter 6, and allow that genes, apart from coding for messenger RNAs, can also code for functional untranslated RNA molecules, such as ribosomal and transfer RNAs.

A little later in the development of the oocyte, active genes can be visualized even in the light microscope for *Xenopus*. These are the so-called 'lamp-brush'

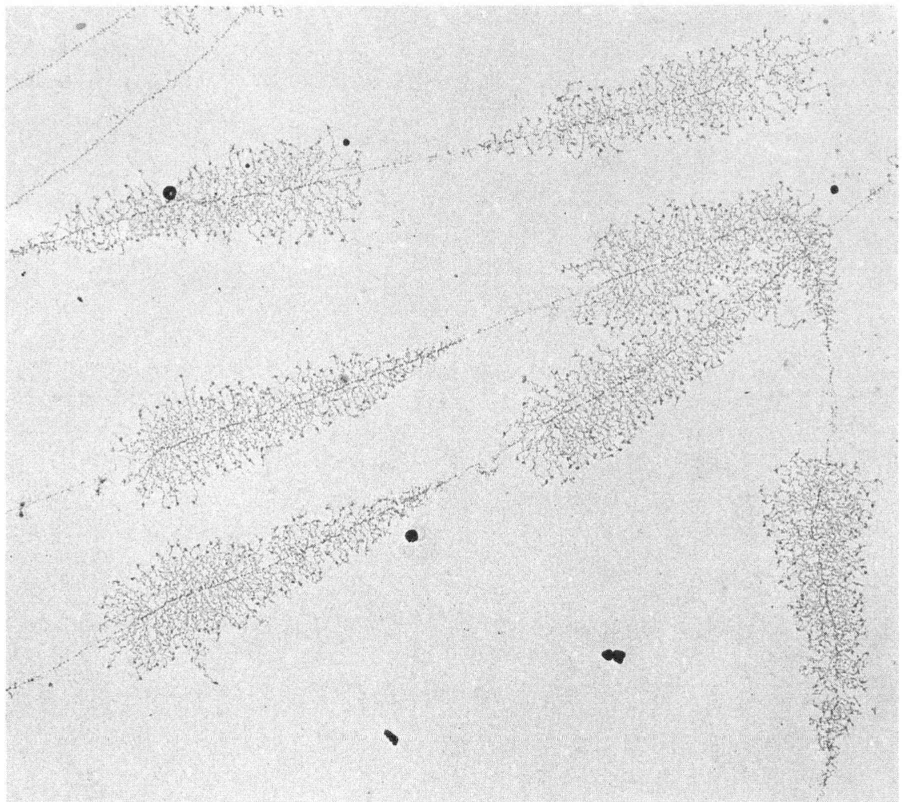

Fig. 10.16. Ribosomal RNA genes isolated from an oocyte of the spotted newt, *Triturus viridescens*. Magnification ×20 000. (*From* Miller and Beatty, 1969.)

chromosomes. In human oocytes a similar type of structure can be seen in electron micrographs (Fig. 10.17). In these active regions, specific staining, and study of structural disintegration brought about by deoxyribonucleases and ribonucleases show that strands of DNA loop out from the main core of the chromosome, probably as a result of uncoiling of the main DNA strand, and that the DNA of these loops is coated with protein and RNA, probably mRNA in course of synthesis. But these observations, of course, still leave unanswered the question of how this activity is triggered.

Interactions between nucleus and cytoplasm in developing oocytes can be studied by transplanting nucleii between oocytes at different development stages. Movement of ^{125}I-labelled cytoplasmic proteins into the transplanted nucleus can be followed by autoradiography. It is becoming clear from this work that interaction between nucleus and cytoplasm is very much a two-way exchange, with mRNA molecules moving from nucleus to cytoplasm to direct protein synthesis, and specific proteins moving from cytoplasm to nucleus, controlling gene expression. The cytoplasm of the ovum, even before fertilization, is not a bag of homogeneously distributed organelles, but is polarized along an axis of symmetry. When the ovum is fertilized, and later divides, the cell nuclei at the 2-cell, 4-cell, 8-cell and later stages of development probably all have the

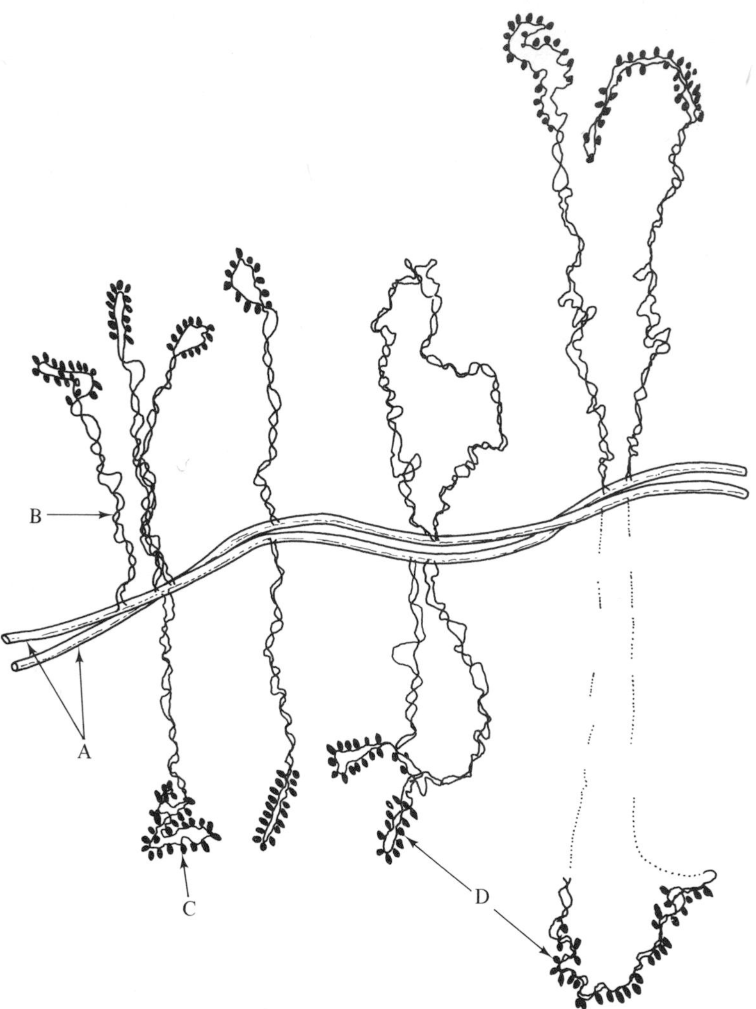

Fig. 10.17 Structure of a chromosome in the human primordial oocyte. A = axial core, B = coiled fibrils forming lateral projections, C = granules at the end of the projections, D = loops formed by reflection of projections back to axis. (*From* Baker and Franchi, 1967.)

same potential, but the cytoplasm in these different cells comes from different regions of the cytoplasm of the original ovum (Fig. 10.18). Differences in the distribution of gene-controlling cytoplasmic proteins, or cytoplasmic mRNAs, in these different cells later become manifest in the development of differentiated

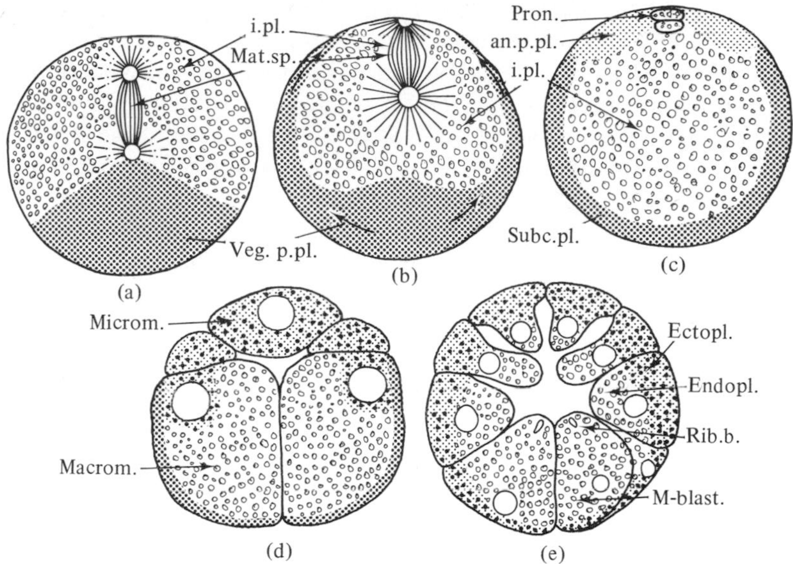

Fig. 10.18 Ooplasmic segregation in *Limnaea stagnalis*. (*a*) Egg immediately after laying, Veg. p. pl. = vegetative pole plasm, i. pl. = inner plasm, mat. sp. = maturation spindle; (*b*) Extension of vegetative pole plasm; (*c*) Formation of animal pole plasm (an. p. pl.) at stage of pronuclei (pron.). The vegetative pole plasm forms a subcortical plasm layer (subc. pl.); (*d*) Eight cell stage (five cells included in plane of section). The plasm of the micromeres (microm.) is mainly formed from animal pole plasm and subcortical plasm. The plasm of the macromeres (macrom.) is mainly formed from inner plasm; (*e*) Bastula stage. The animal pole and subcortical plasms form the ectoplasm (ectopl.) and the inner plasm forms the endoplasm (endopl.) rib. b. = ribonucleic acid bodies, M-blast = mesoblast. (*From* Raven, 1958.)

cell-types in different parts of the embryo. But, of course at these later stages the different types of cell are also interacting with one another in complex ways.

If one begins to think imaginatively about the problems of development it is clear that biology is still a very young, very immature science, in its modern role, that is, of trying to understand, not simply classify, biological phenomena. To take just cell division, with the sequence of events described in Chapter 1, we know so little about how the process is triggered, how the chromosomal material becomes condensed and moves to the equatorial plate, how the fascinating 9-fold structure of the centriole duplicates, or how the chromosomes are drawn to the spindle poles. At the other extreme in development of higher organisms, once the nerve net, the brain, has formed we know that synaptic transmission can be modulated, in the formation of a conditioned reflex, with synaptic activity facilitating transmission. Some molecular change at the synapse resulting from activity makes it transmit more readily – but what this change might be, this fundamental process of human memory, we do not know. Yet this problem is probably not too far beyond the range of present-day experimental techniques involving perhaps proteins released at nerve terminals (along with low-molecular-weight chemical transmitter) which might become incorporated into the postsynaptic membrane, making it more readily depolarizable. The consequences of greater understanding of synaptic facilitation, for understanding of the human mind, and mental disorders such as pathological anxiety or depression, could prove as far reaching as elucidation of the molecular nature of genes.

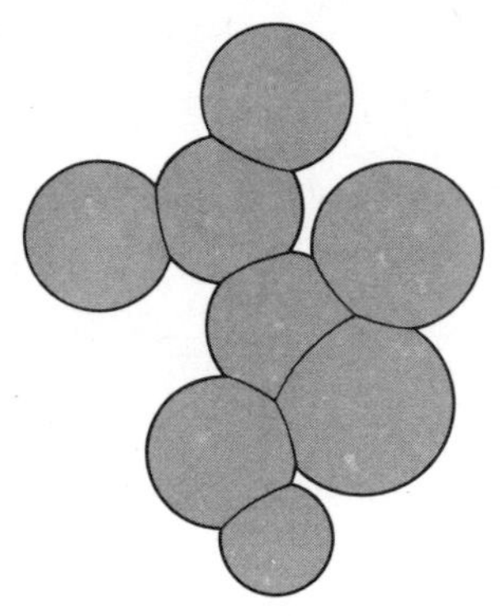

Appendix 1
The X-ray diffraction technique

by David Blow

This appendix begins with a detailed, but non-mathematical description of the properties of a diffraction grating as a foundation for discussion of the diffraction patterns of large biological molecules. A diffraction grating is like a one-dimensional crystal, and many of its properties apply equally well to crystals. But they are much more simply understood without the complication of a three-dimensional lattice.

A diffraction grating can be made by photographing a regularly-spaced series of parallel lines. In Fig. A1.1 we are looking edge-on at such a grating. Each space represents a line where light can pass, and elsewhere the grating is opaque. The grating is being illuminated from below by light of a fixed wavelength, λ.

We usually think that light can travel only in straight lines. The essence of

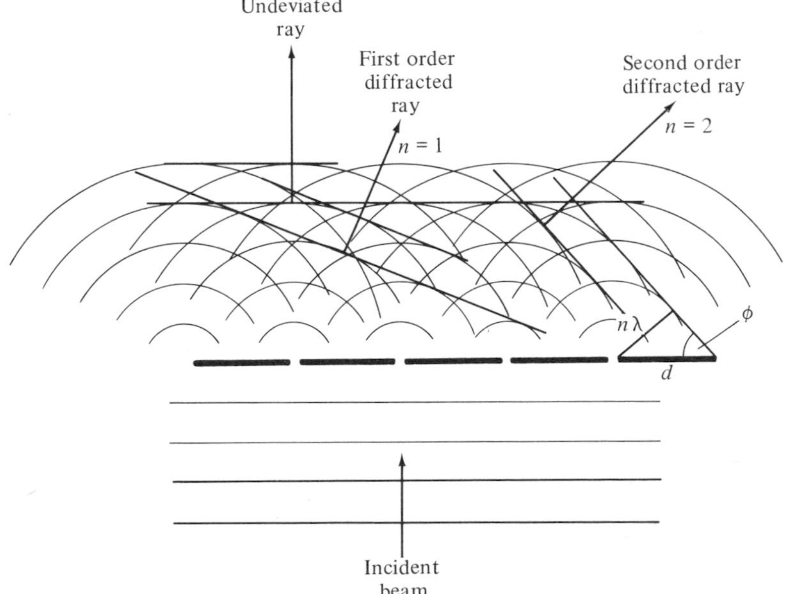

Fig. A1.1 Diffraction from a grating.

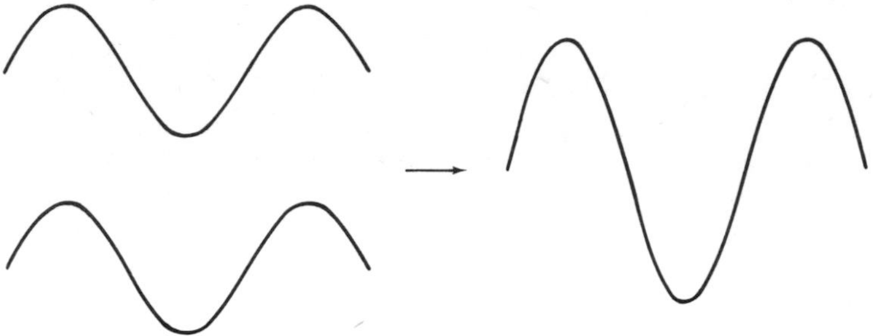

(a) Two waves in phase reinforce each other to give a wave of double the amplitude

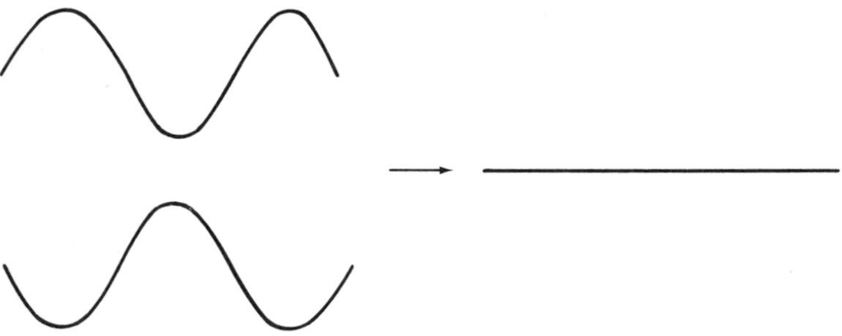

(b) Two equal waves $\lambda/2$ out of phase cancel out

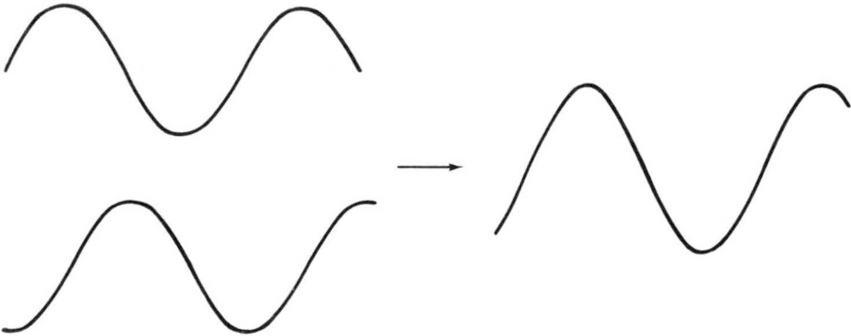

(c) In general two waves out of phase combine to give a wave whose amplitude depends on the phase relation

Fig. A1.2 Interference between waves.

diffraction effects is that when we are dealing with objects not much larger than λ in size, it is important to think of light as a wave motion which can spread out in all directions. If there are no obstacles in the way, waves spreading sideways from different parts of the wavefront cancel each other out, and the wave spreads *as if* the propagation was only in straight lines. We are accustomed to the idea that television signals, which are light waves with a wavelength of a few metres, do not need an exactly straight path from transmitter to receiver.

In Fig. A1.1 each transparent line of the diffraction grating acts as a source of waves. The figure indicates the peaks of these waves at a particular moment. We have assumed that the incident wavefront reaches the whole of the grating at the same moment, so that the waves spread from each of the lines of the grating at the same time, or 'in phase' (Fig. A1.2).

Consider a line far from the grating, and parallel to it. Waves from each of the subsidiary sources arrive at this line at the same time, and generate a wavefront there which is very little different from the incident wavefront, except that it is weaker. This wavefront forms an 'undeviated ray'. There are, however, other directions in which waves from each element of the grating generate a wavefront, because the waves all arrive 'in phase'. These are the 'diffracted beams'. The first of these is in such a direction that light passing through one element of the grating has travelled exactly one wavelength further than light passing through a neighbouring element. The other diffracted beams have path differences of 2λ, 3λ, ... The angle ϕ through which the beam is deviated is given by

$$n\lambda = d \sin \phi \qquad \text{(A1.1)}$$

for the nth order diffracted beam.

To understand how the profile of the rulings of the grating can affect the relative intensities of the different orders of the grating, imagine next a second set of transparent lines added, with the same spacing as the first set, but slightly displaced from them (Fig. A1.3). This set of lines will give rise to another set

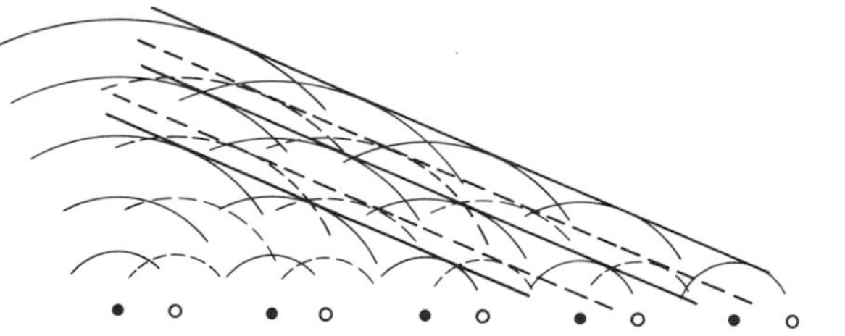

Fig. A1.3 Diffraction from a double-line grating.

of diffracted rays, in the same directions as the original set, because the spacing of the lines is the same. The relative *phases* of the second set of diffracted rays will be different for each order of diffraction: in some cases the two phases will be almost the same, resulting in a reinforcement or strengthening of the diffracted beam; in others, the two rays will be out of phase, and will almost cancel each other out (Fig. A1.2).

To consider a rather special example, imagine the second set of rulings to be placed exactly half-way between the first. The first order diffracted beam will then be *exactly* out of phase with the original set, and will exactly cancel them. (The same will be true of the 3rd, 5th, 7th, . . . order beams.) The second order beams, on the other hand, will be exactly in phase with the original set, and these (together with the 4th, 6th, . . . orders) will be reinforced. What has happened, of course, is that the spacing d of the rulings has been halved. The number of orders of diffraction, n, is also halved (see Eq. (A1.1)).

Any regularly repeating profile of rulings on the grating can be treated in this way as the sum of different components. The *directions* of the diffracted rays depend only on the distance d between the regular repetitions, while the form of the profile will affect the relative *intensities* and *phases* of the diffracted beams.

The overall size of the grating affects the sharpness of the diffracted beams. The direction of the diffracted wavefronts is indicated by making a line tangent to each of the constituent waves from the subsidiary sources. The wavefront moves forward perpendicular to this line as the waves spread out, because along this line the contribution of each element of the grating is 'in phase'. The size of the wavefront is, however, limited in extent by the size of the grating. Consider a slightly different direction from the original wavefront, where at each end we are not more than $\lambda/4$ from the original position (Fig. A1.4). All along this line,

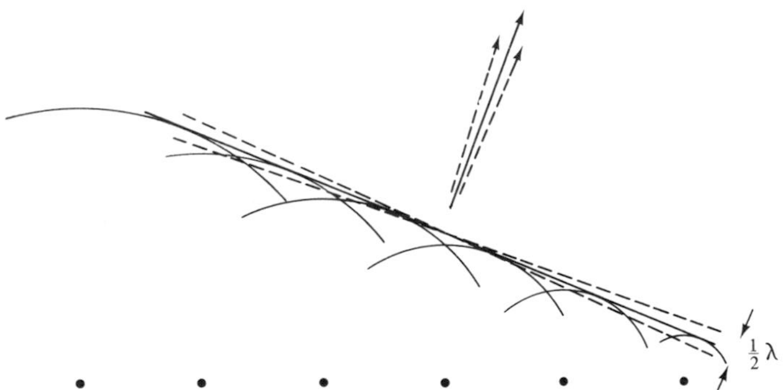

Fig. A1.4 Diffraction from a grating of limited extent.

the constituent waves are nearly in phase, and some energy will spread in this direction also. Far from the grating, each diffracted beam will spread out slightly. As the grating is made smaller, the range of directions for which all constituent waves are nearly in phase will increase, and the tendency for the diffracted beam to spread out will increase. In the same way, the diffraction from specimens composed of very small crystallites consists of diffuse maxima.

To understand the effects of disorder, or thermal motion, in crystals, consider the effect of small displacements of the rulings of the grating from their ideal positions. Clearly this will destroy the 'perfect' phase relationships and weaken the diffracted beams. But they are not all equally affected. The low order diffracted rays, which are only deviated through a small angle, are much less affected in phase by a given displacement, than a high order of diffraction. The amount by which the diffraction dies away at high diffraction angles is thus a measure of the degree of order of the object. In protein crystals the disorder is such that the diffraction pattern usually becomes too weak to observe at a resolution of 0·2 nm or 0·15 nm.

Diffraction and image formation

To see how this discussion relates to structure determination, consider using the scattering of light from a diffraction grating to determine the structure of the rulings of the grating. Suppose we looked at the grating in a microscope. Instead of thinking in the usual way about lenses forming an image of the object, it is instructive to think in terms of diffracted beams. In the optical microscope (Fig. A1.5) the diffracted beams are collected by the objective lens. Each starts out as a parallel beam, and is brought to a focus by the objective lens in its focal plane. Further back, in the image plane, the diffracted beams cross one another. (In the ordinary compound microscope, the image is a virtual image; however, it is simpler for the present purposes to imagine an arrangement where a real image is formed.) The image is generated by the superposition of these diffracted beams, which reinforce or cancel one another in different parts of the image, according to their phase relationships.

If the rulings of the grating are less than a wavelength apart, not even a first-order diffracted beam can exist (when the incident beam is normal to the grating, as in Fig. A1.1). Under these conditions, it is not possible for the image to show the rulings of the grating – they cannot be resolved. In a microscope, the diffracted beam must not only exist, it must go in a direction such that it can be brought back to the image plane by the objective lens. The 'resolving power' is the finest spacing whose first-order diffracted beam can be brought into the image. The resolving power depends not only on the wavelength of illumination, but also on the acceptance angle of the objective lens.

Exactly the same limitations apply to the resolving power of an electron microscope or of an X-ray diffraction study. The energy of the electrons in an electron microscope corresponds to a wavelength of less than one hundredth of a nanometer, but the acceptance angle of available electron lenses is very limited, and the resolving power actually achieved is a few tenths of a nanometer in a well adjusted microscope. The nature of the object usually limits the practical resolution much further than this.

In X-ray diffraction, no lens is employed, but when biological materials are being studied it is often desirable to restrict the observed diffracted rays to a definite limit, and this defines the 'resolution' of the image in exactly the same way. An X-ray image has to be created indirectly, generally by computation, as will be described later.

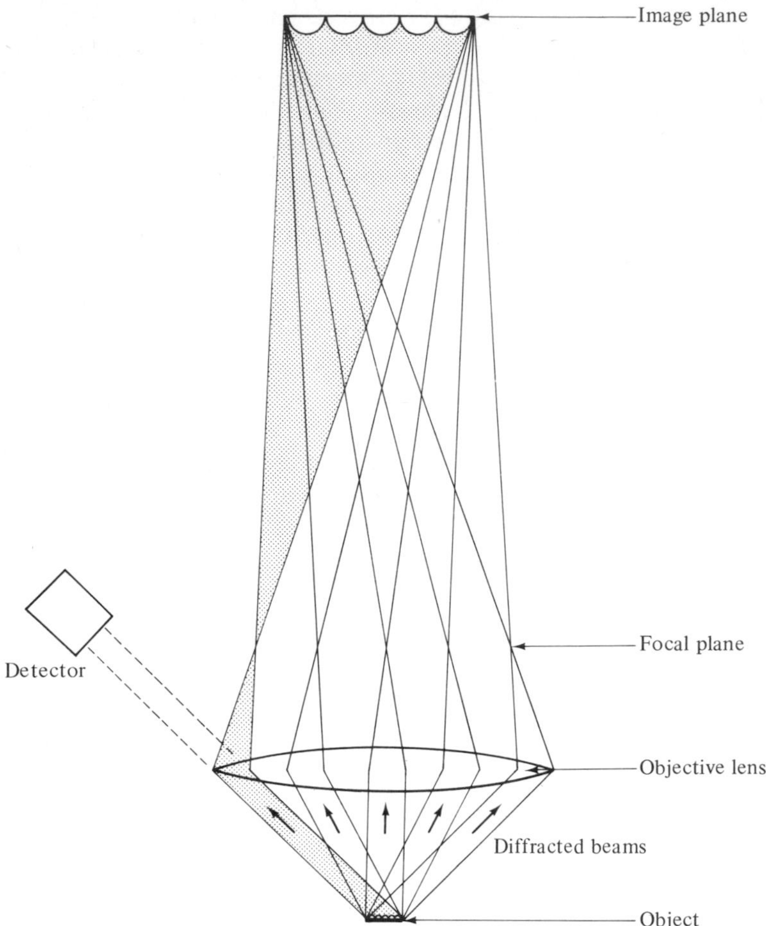

Fig. A1.5 Image formation in a light microscope.

Diffraction by crystals

A diffraction grating, with rulings running only one way, gives rise to a single set of diffracted beams dispersed in one direction. A two-dimensional mesh gives rise to a two-dimensional pattern of diffracted beams which can be seen, for example, when a small monochromatic light source (e.g. a distant sodium street-lamp) is viewed through a finely woven mesh (e.g. a silk handkerchief). When we come to consider a crystal lattice as a three-dimensional diffracting object, a new complication arises. The crystal lattice can be thought of as a stack of two-dimensional lattices. For a particular order of diffraction, a diffracted beam in a particular direction will be generated by each layer of the stack. In considering the resultant diffraction, the relative phase of the contribution from each layer has to be considered. Remembering that there is a very large number of layers, it follows that a significant diffracted beam can only occur if the diffraction from each layer is exactly in phase. The result is that in crystal diffraction, a given diffracted beam can only be generated if the crystal is in the correct orientation, a restriction which has no counterpart in diffraction from one- and two-dimensional lattices.

Von Laue and Ewald have given other formulations of this geometrical restriction, but much the simplest way of expressing the condition has been given by Sir Lawrence Bragg, who introduced the notion of the *reflection* of the incident beam in lattice planes in the crystal. Each order of diffraction can be related to a set of planes passing through lattice points in the crystal. (That there are many types of such planes will be clear to anyone who has travelled past a regularly planted orchard in a train.) Bragg showed that the relation of the incident to the diffracted ray was equivalent to reflection in the corresponding planes. If the *glancing angle*, θ, of the incident ray onto these planes is such that rays from adjacent planes will be in phase, the path difference $2d \sin \theta$ (Fig. A1.6) must be a whole number of wavelengths. Hence

$$n\lambda = 2d \sin \theta \qquad \text{(A1.2)}$$

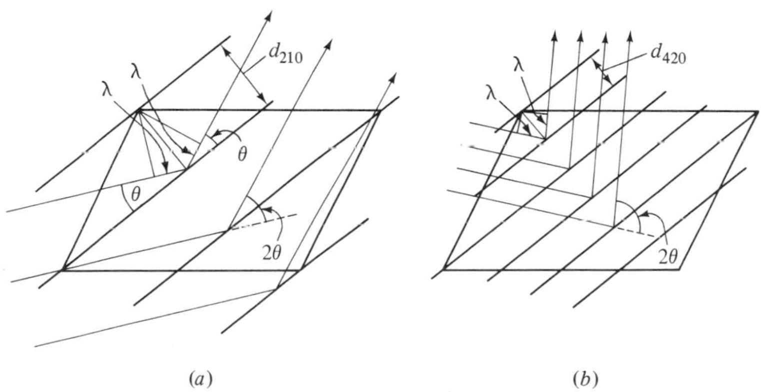

Fig. A1.6 Bragg reflecting planes. *a*. (210), *b*. (420).

In the above equation, d is the spacing between the adjacent planes. By comparison with Eq. (A1.1), it is clear that there is now a series of orders of diffraction ($n = 1, 2, \ldots$) for each type of plane; and as already mentioned, there are many types of plane. This plurality can be somewhat simplified by the way in which the planes are counted.

The repeating block of structure from which the crystal is built up is called the *unit cell*. Its edges define three axial lengths. If one Bragg reflecting plane passes through the corner of the unit cell, the next must cut through the unit cell edges at points which are simple fractions of the axial lengths from the corner (Fig. A1.6). These fractions $1/h$, $1/k$, $1/l$ define three numbers (h, k, l) called the *indices* of these planes. In Fig. A1.6(*a*), the (210) planes are illustrated. (The

0 means that the planes are parallel to the third cell edge, and do not cut through it at all.) The second order of diffraction from these planes can also be called diffraction by the (420) planes, shown in Fig. A1.6(*b*). In this way the factor n in Eq. (A1.2) is not needed and we can write

$$\lambda = 2d_{hkl} \sin \theta \tag{A1.3}$$

This equation is known as Bragg's law. d_{hkl} is the spacing between the set of planes with indices (hkl) which will *reflect* X-rays incident on them at glancing angle θ. If the internal structure of the crystal is such that these planes represent strong planes of electron density, the reflection will be a particularly intense one. Note that the angle through which the diffracted ray is deviated, 2θ, depends only on the wavelength and the spacing of the reflecting planes.

X-ray diffraction and electron microscope techniques

The arrangement of Fig. A1.5 could be used to study the structure of a protein crystal, for example, if the lens were an electron microscope lens, and the incident 'radiation' were a beam of electrons. However, in practice it has proved exceedingly difficult to resolve features less than 2–3 nm apart, from normal types of biological objects. The main difficulty is that the interaction of the electron beam with a single (unstained) protein molecule causes such a slight fraction of the incident energy to be scattered that the image is vanishingly weak.

Suppose we removed the lens, and used a detector to observe the intensity of each order of diffraction. This would give *some* information about the structure of the grating, but not enough to regenerate the image. The lost information is the *phase angle* of the diffracted ray – that is, the time of arrival of the peaks of the diffracted wave, measured relative to the undeviated ray. This information determines exactly how the various waves would recombine to form the image, and it is necessary for the regeneration of the image from the diffracted rays.

If the phase of the diffracted rays were known, the image could be regenerated, for instance by computer calculations. The advantages which follow from such a procedure are:

(1) By looking only at the diffraction maxima, we ignore the differences between the different units of the crystal, and concentrate on features common to every unit.

(2) It is now positively advantageous to use a large specimen, containing very many units. Since we are looking only at common features of the units, the overall result can be considered in terms of only one unit. So by using a large specimen the intensity of the scattered radiation may be greatly enhanced.

(3) By using appropriate methods of computing the image, all difficulties to do with the overlapping of features in a complex structure can be overcome. The computer can be made to work out the structure explicitly in three dimensions, rather than providing the sort of two-dimensional projection of the structure which a picture normally provides.

(4) For similar reasons, problems of contrast in the image are directly overcome. So long as the diffracted beams are strong enough to measure, a structure can be 'seen'.

The price we have to pay for these advantages is in three parts:

(1) The method can only be applied to three-dimensionally ordered systems, such as crystals. Preparation of such a specimen is not technically possible for many types of material.

(2) Instead of simply looking at an image, we have to measure separately all the orders of diffraction, and feed the results to a device, like a computer, capable of working out what the image would be. These problems have accounted for the main operational difficulties of extending the diffraction method to very complex molecules, but with the advent of reliable automatic equipment, these difficulties are becoming less severe.

(3) We have to invent some way of finding out the phase angle of the diffracted wave. This is the 'phase problem' which constitutes the theoretical difficulty of diffraction methods. The way in which this has been done for complex molecules will be outlined in the next section.

Once we decide to use a diffraction method of looking at the specimen, it is generally advantageous to use X-rays rather than electrons. X-rays are much more penetrating than electrons, and interact much more weakly with the matter they pass through. So we need a much bigger crystal, perhaps half a millimetre across. The penetrating nature of X-rays makes it easy to put the specimen in its natural surroundings (water). Methods for keeping a specimen wet in the high vacuum of an electron microscope are being developed but are not yet widely available.

Study of wet specimens is particularly important in the case of proteins, whose normal properties are preserved only when the outside of each molecule can be immersed in its natural aqueous environment.

Determining phases by isomorphous replacement

The isomorphous replacement method depends on the preparation of a pair of crystals, which differ only in the introduction of a strong X-ray scattering centre (that is, a heavy atom) at a unique site in the crystal cell – or, at worst, at a small number of sites. It is essential that the structure of the remainder of the crystal should be disturbed as little as possible, so that the scattering by the heavy atom-substituted crystal can be regarded as (scattering by the unsubstituted crystal) + (scattering by heavy atom) with reasonable accuracy. This is the implication of the word 'isomorphous'.

The first step in the application of the method is to locate the heavy atom sites. This is done by recording the difference in the amplitude of scattering of each reflexion between the isomorphous pair (amplitude $= \sqrt{\text{intensity}}$). The problem is then tackled as though this difference of amplitude was the same as the amplitude of scattering by the heavy atoms alone. An approximation enters here, as this will only be the case if the scattering of the heavy atoms is exactly 'in' or 'out' of phase with the remaining scattering, even given perfect isomorphism.

Once this relatively simple problem of finding the heavy atom arrangement is solved, it is possible to calculate the amplitude and phase of the heavy atom contribution to each reflexion. This now leads to information about the phase of the scattering from the unsubstituted crystal. If, for example, the presence of the heavy atom increases the scattering by exactly the amount of the calculated heavy atom contribution, this implies that the phase of scattering from the unsubstituted crystal is the same as that of the heavy atom scattering so that it is reinforced. On the other hand, if the scattering is reduced by this amount, it implies that the heavy atom scattering is just 'out of phase' with the crystal scattering. More frequently, intermediate cases will occur, but in these cases it is not possible to tell whether the phase of the crystal scattering 'leads' or 'lags' with respect to the heavy atom scattering.

These ambiguities can be solved, in principle, by the use of a second isomorphous replacement. In practice, due to the effects of non-isomorphism, and also because the scattering of the heavy atom is likely to be quite small by comparison with the protein scattering, at least three different isomorphous substitutions are generally needed to make a useful estimate of the phase.

Diffraction by fibres

A fibre diffraction specimen is one in which the structure is ordered in one direction, the *fibre axis*, but not in directions perpendicular to this. Several different types of order can exist. At one extreme, the specimen may be built from crystallites, oriented along the fibre axis, but rotationally disordered. In this case, the diffraction pattern is the same as would be observed from a crystal which is rotated about the fibre axis. The crystal spacing along the fibre axis causes spacings to appear on a diffraction photograph in the direction corresponding to the fibre axis, called the *meridian*. The rest of the diffraction pattern is localized on lines known as *layer lines*, perpendicular to the fibre axis. The central layer line of the pattern is known as the *equator*. A fibre specimen made up of crystallites causes individual reflections to appear on the layer lines (Fig. 2.17). The larger the crystallites, the sharper the reflections will be (Fig. A1.4). In most fibre specimens, the fibre axes are not exactly aligned, so that these sharp reflexions form arcs rather than single points.

The other extreme type of order is when each component molecule has a regular structure along the fibre axis, but in perpendicular directions the arrangement of molecules is completely random. Under these conditions, the diffraction pattern no longer consists of individual reflexions, but spreads into a continuous distribution across each layer line.

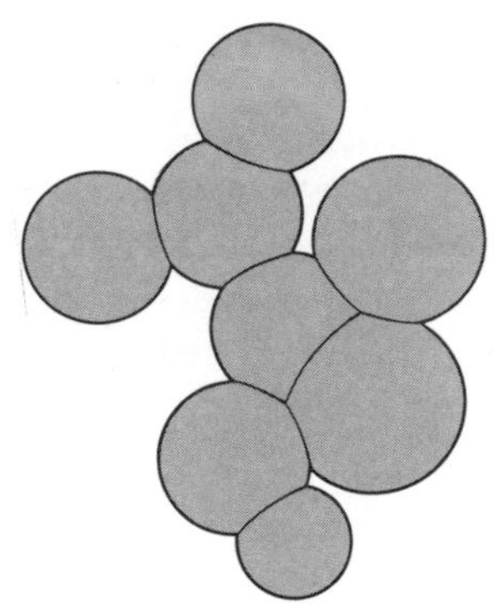

Appendix 2
Hydrogen bonds. Atomic and submolecular structure

Hydrogen bonds play an important role in the structure of proteins and nucleic acids. In this appendix we shall discuss the nature of these bonds, in so far as this is possible in pictorial and non-mathematical terms. We shall try at the same time to point out, in general, the limitations of the pictorial approach used throughout the main text of this book, in which molecular structures, and interactions between molecular groups, are illustrated by drawings of molecular models.

For the discussion of many aspects of molecular behaviour this pictorial description is completely adequate. In certain cases, however, it becomes less adequate, or sometimes completely inadequate, and in the main text we have simply avoided pursuing certain topics in this amount of detail. The critical reader will be interested to know where our discussion is limited in this way, and to be given some impression of the methods used, and the problems that arise, in more detailed treatment. Our aim in this appendix is to indicate how aspects of the behaviour of large molecules discussed in the main text arise from the fundamental properties of atoms and molecules and from the force fields of electrons and atomic nuclei.

Diagrams of atomic structure, or our mental pictures of atoms based for the most part on popular illustrations, represent only certain facets of atomic detail. Electrons and atomic nuclei do not behave like the large objects on which our visual images and language are based. No single picture of an electron or of atomic nuclei covers all facets of their behaviour, and the picture which is most appropriate depends on the type of experiment or atomic event which we are interested to discuss.

In most biological or biochemical events we are not concerned with nuclear interactions and we can picture the atomic nuclei as tiny dots at the centre of each atom. A further simplification arises from the fact that the convention of regarding atomic nuclei as 'positively charged' and electrons as 'negatively charged' provides a simple and accurate way of describing the force of attraction between atomic nuclei and electrons, and the force of mutual repulsion between electrons, provided these sub-atomic entities do not approach one another too closely. The interactions between molecules, and between the parts of a large molecule, can be to a large extent accurately described in electrostatic terms.

At the short ranges prevailing *within* small molecules, or within the molecular groups of a large molecule, interactions between electrons and atomic nuclei become more complex. The details of electron distributions within atoms and molecules can be described adequately only by a branch of mathematics known as wave mechanics. Electrons within atoms and molecules can no longer be adequately considered as particles. For our purpose, electrons within molecules should be visualized, if at all, only as a whirling cloud of 'negative charge' surrounding atomic nuclei and filling the volume of the atom.

Nevertheless the outer boundary of the electron cloud which defines the volume of an atom or molecule can be fairly precisely visualized. When two atoms approach one another, strong repulsive forces come into play as their electron clouds interpenetrate. If the electrons can redistribute themselves in a satisfactory way, the force of repulsion may be overcome by a force of attraction as the electron clouds interpenetrate to form a *covalent* chemical bond. If no chemical bond is formed, the repulsive force prevents more than slight interpenetration. Measurement of the distances between small molecules in crystals (extensively studied by X-ray diffraction since about 1920) allowed the outer boundaries of their constituent atoms, i.e. the limit of the electron cloud, to be mapped in a precise way. It is this outer boundary which is illustrated by the molecular drawings of this book. The boundary is not in reality quite so sharp as these drawings suggest, but the molecular models from which these drawings are made do give a very good representation of the *steric* (i.e. spatial, or space filling) properties of molecular groups.

We consider now the distribution of charge (electrons) within a polypeptide chain, illustrated by the + and − signs of Fig. 2.1. Although, as may be appreciated from what has been said already, the only really adequate way to discuss this question would be mathematical, a description of sorts may be given in terms of the concept of *resonance* within a molecule. A simple example with which to illustrate this idea is provided by the ionized carboxyl group ($—COO^-$). This group is in reality symmetrical, with the single unit of negative charge distributed equally between the two oxygens. However this situation cannot be represented by conventional chemical formulae, and if we wish to use such formulae, we have to suppose the group to be resonating between two forms:

$$—\mathrm{C}{\Large\langle}^{\mathrm{O}\ (=)}_{\mathrm{O}^-} \rightleftharpoons —\mathrm{C}{\Large\langle}^{\mathrm{O}^-}_{\mathrm{O}\ (=)}$$

We imagine the group spending half of its time in each form.

The peptide backbone group may be written in a similar way in two forms:

$$\begin{matrix}\mathrm{O} \\ \| \\ \mathrm{C} \\ /\ \ \backslash\ \ \ / \\ \mathrm{N} \\ | \\ \mathrm{H}\end{matrix} \rightleftharpoons \begin{matrix}\mathrm{O}^- \\ | \\ \mathrm{C} \\ /\ \ \backslash\!\!\backslash\ \ / \\ \overset{+}{\mathrm{N}} \\ | \\ \mathrm{H}\end{matrix}$$

In their detailed study of the distances between atoms in crystals of small peptides (see Chapter 2) Pauling and Corey found that the C–N distance in this group was intermediate between the distances characteristic of single and double bonds. They found that this partial double-bond character of the C–N bond prevented free rotation about this bond, and caused these C, O, N and H atoms to all lie in one plane. They estimated from the C–N bond length that the group, on this simple resonating picture, must be spending just over one-third of its time in the charged form. In other words, abandoning now the oversimple resonance picture, the electron distribution within the peptide backbone group is such that the electron cloud is drawn over somewhat towards the oxygen atom, which carries a net negative charge of rather less than half a unit. This charge is drawn from the nitrogen and hydrogen atoms, which are thus left with a similar net positive charge (atomic unit of charge = charge on one electron).

The formation of hydrogen bonds is always a result of the drawing away of electrons from a hydrogen atom, which is left with net positive charge and is attracted to the negatively charged region of an oxygen or nitrogen atom of a neighbouring molecule. Hydrogen bonding between water molecules is illustrated in Fig. A2.1. The hydrogen atoms of a water molecule carry a small net

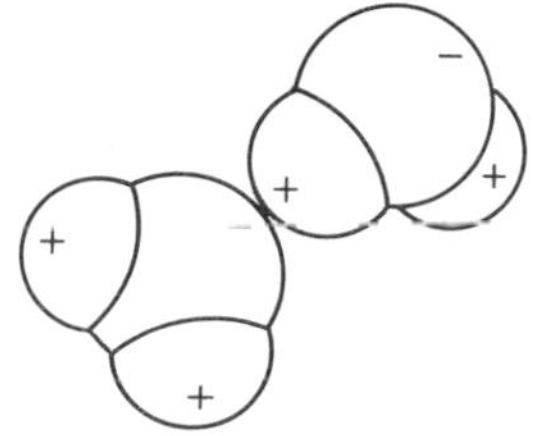

Fig. A2.1 Hydrogen bonding between water molecules

positive charge, as a result of the electronegativity (i.e. electron drawing power) of the oxygen atom. The hydrogen atoms form hydrogen bonds to the oxygen atoms of neighbouring molecules. (It is this force of attraction between water molecules which leads to the high latent heat and high surface tension of water, since a relatively large amount of energy is needed to separate a water molecule from its neighbours.) Other types of hydrogen bonds are listed in Table A2.

By contrast with —OH and —NH groups, —CH groups do not form hydrogen bonds. The electronegativity values of carbon and hydrogen atoms are rather similar, and the electrons are not drawn away from the hydrogen in the way that they are in —OH and —NH groups. (The atomic nuclei of oxygen and nitrogen which carry positive charges of +8 and +7 respectively, have greater electron drawing power than carbon nuclei which carry a positive charge of +6.)

This simple difference between carbon atoms on the one hand, and oxygen and nitrogen atoms on the other, has profound consequences. Since —CH groups cannot form hydrogen bonds, hydrocarbon chains penetrate into water only with difficulty. To penetrate they have to break water–water bonds and modify the structure of the surrounding water in ways discussed in Appendix 4, and no comparable hydrogen bonds can form between the water molecules and the hydrocarbons. Hydrocarbons are thus *hydrophobic* or water-fearing, and fats and oils are very insoluble in water. Molecules such as methyl and ethyl alcohol, on the other hand, are freely miscible with water, since they carry a hydrogen bonding, or *hydrophilic* (water-loving) —OH group which can form hydrogen bonds to water molecules. Hydrogen bonding groups are usually

TABLE A2 Different types of hydrogen bond.

O—H...O=	N—H...O=
O—H...O	N—H...O
O—H...N—	N—H...N—

referred to as *polar* groups (because of their electric dipole moment). Hydrophobic effects are important in determining details of protein and nucleic acid structure (see later sections of Chapter 2 and Chapter 7) and these points are relevant also to the discussion of membrane structure and permeability (Chapter 4).

The drawing of Fig. 2.1, with its + and − signs, thus gives a valid representation of the shape of the molecule, and of that part of the electron distribution within the molecule which is important in hydrogen bonding, although one minor point that we have so far ignored should be mentioned here. Most (90 per cent) of the force of attraction leading to the formation of a hydrogen bond is electrostatic, but there is also thought to be a small covalent component, due to partial interpenetration of the electron clouds round the bonded hydrogen and oxygen atoms. In general, the molecular drawings of Chapters 2, 4 and 7 (similar to those of Figs. 2.1 and 2.7) are adequate to show the physical structure of proteins, lipid membranes, nucleic acids and nucleoproteins, since these structures are determined mainly by hydrogen bonds, steric effects, and attractions or repulsions between ionic groups (e.g. $—COO^-$ or $—NH_3^+$).

We conclude this appendix with a note on the way acidic and basic groups are represented in the drawings of molecular models throughout the book, including a brief definition of 'pH' and 'p*K*'.

In pure water there is a tendency for protons (the nuclei of hydrogen atoms) to move from one water molecule to another:

$$H_2O + H_2O \rightleftharpoons H_3^+O + OH^-$$

The concentration of H_3^+O ions in pure water at 20° C is approximately 10^{-7} g. ion per litre (pH ~ 7). If molecules with acidic groups (e.g. —COOH) are dissolved in the water these groups tend to transfer protons to water molecules:

$$—COOH + H_2O \rightleftharpoons —COO^- + H_3^+O$$

and hence to increase the H_3^+O ion concentration. An H_3^+O ion concentration of 10^{-6} g. ion per litre corresponds to pH 6, 10^{-5} g. ion per litre to pH 5, etc. Acid solutions thus have pH values lower than pH 7.

If molecules with basic groups (e.g. —NH_2) are dissolved in water they tend to mop up protons:

$$—NH_2 + H_3^+O \rightleftharpoons —N^+H_3 + H_2O$$

Bases thus lower the concentration of H_3^+O ions. An H_3^+O ion concentration of 10^{-8} g. ion per litre corresponds to pH 8, etc. Alkaline solutions have pH values higher than pH 7.

At pH ~ 7 —COOH and —NH_2 groups are almost all in the —COO^- and NH_3^+ form, and in the molecular drawings throughout the book all groups have been shown as they exist in aqueous solution at pH ~ 7. COOH groups have a pK ~ 4. This means that as a protein solution is made more acid, from pH ~ 7, an increasing proportion of the —COO^- groups will be found in the —COOH form as the pH approaches pH ~ 4. At a pH equal to the pK value half these groups will be ionized. As the pH is lowered below the pK value, an increasing majority of the groups will be in the unionized form.

NH_2 groups have a pK ~ 9; as the pH of a protein solution is made more alkaline from pH ~ 7 an increasing proportion of these groups will go over to the unionized form in the pH range 8–10. The net charge on a protein molecule thus depends on pH; it is positive in sufficiently acid solutions and negative in sufficiently alkaline solutions. The pH at which a given protein molecule carries no net charge defines the isoelectric point for this protein.

Since the pH of cell cytoplasm and of extracellular media almost always remain within the pH range 5–8, the drawings of molecular models with acidic and basic groups in ionized form give the correct representation of the state of the molecules under normal physiological conditions.

The term 'buffered solution' is occasionally used in the text. A buffered solution is a solution containing a mixture of a weak acid or base and its salt designed to keep the pH relatively constant on addition of limited amounts of acids or bases. (Weak acids and bases are those with pK values nearer to 7.)

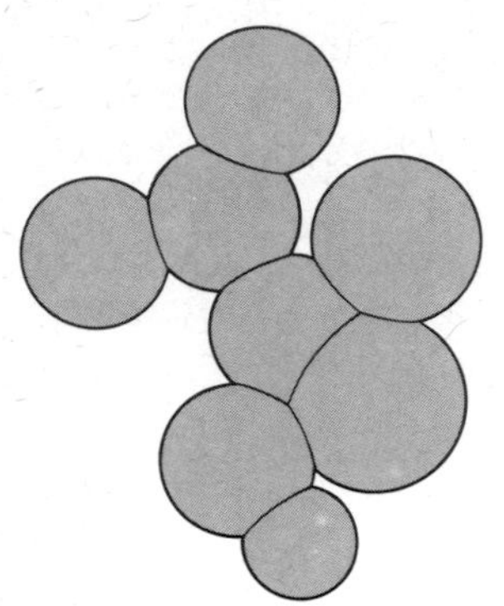

Appendix 3
Fundamentals of thermodynamics

The foundations of thermodynamics were laid in the nineteenth century in the study of engines converting heat to mechanical energy. Heat engines were of more general interest in those days, the great days of steam, than they are today, but it happens that the ideas developed from their study have a central relevance to modern molecular biology. For thermodynamics explores the relations between heat and other forms of energy. The molecules in living cells are constantly undergoing chemical transformation, and also aggregating, splitting, folding and unfolding, all as a result of the energy of random thermal movement. Thermal energy interacts with the chemical energy of molecular bonds, the energy of light in photosynthesis, the energy of mechanical work in muscle, and osmotic work in moving ions across membranes. Important understanding of these processes comes from an understanding of the basic principles of thermodynamics.

Although heat is a form of energy (kinetic and vibrational energy of atoms and molecules) there is a difference between heat and the other forms of energy. Energy in other forms can be readily and completely converted to heat, but heat energy cannot be so readily or completely converted back again. This is because the molecular movement of heat energy is a random movement. For a stone flying through the air with velocity x each molecule has a velocity x in the same direction (in addition to its random thermal motion). When the stone hits the ground the kinetic energy of flight is converted to random molecular movement, warming the stone and the ground where it lands. When mechanical energy is turned into heat energy, the total energy of the system is not changed, but there is an increase in the *disorder* of the system.

As an illustration of the basic principles of the simplest sort of heat engine consider an ideal gas confined in a cylinder fitted with a movable piston with which we can compress the gas (Fig. A3.1). Suppose initially one mole of gas is contained in the cylinder occupying a volume V_1 (left-hand drawing of Fig. A3.1), and suppose the cylinder is made of metal, so that heat can readily flow to or from the gas, which therefore remains throughout the experiment at the environmental temperature T. We now compress the gas, supposing the piston rod to slide through a frictionless seal, so that a vacuum is formed behind the

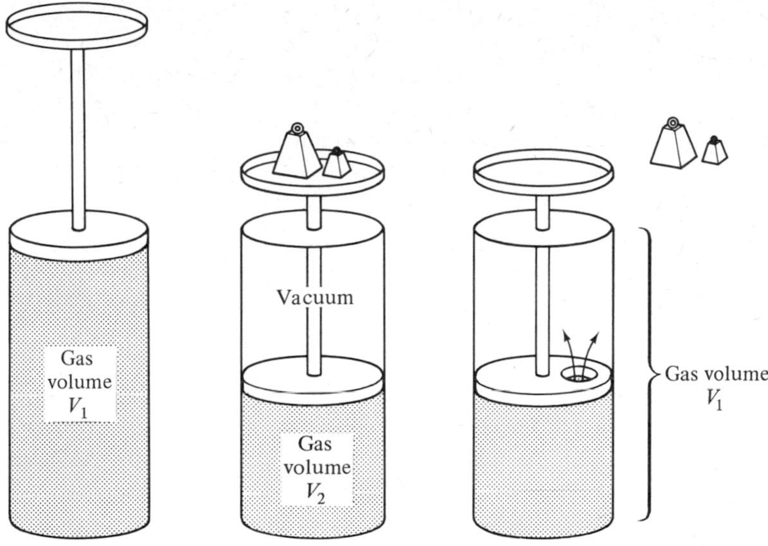

Fig. A3.1 Compression of a gas in a cylinder (see text).

piston. If the piston area is A the work done in moving it in a small distance dl is:

$$\text{Force} \times \text{distance} = pA\,\mathrm{d}l = p\,\mathrm{d}V$$

where dV is the corresponding volume change, and p the gas pressure. Thus the work done in compressing the gas from volume V_1 to volume V_2 is:

$$\int_{V_2}^{V_1} p\,\mathrm{d}V$$

Since $pV = RT$, the work done is

$$RT\int_{V_2}^{V_1} \frac{1}{V}\,\mathrm{d}V = RT\log_e \frac{V_1}{V_2}$$

where R = the gas constant (in ergs/degree/mole) and T = absolute temperature. (All logarithms in this Appendix are to base e.)

During the compression the gas remains at temperature T. However, the moving piston transfers energy to gas molecules. These gain energy temporarily which they subsequently lose by collision with the cylinder walls. Thus the energy $RT\log(V_1/V_2)$ is converted into a gain in thermal energy of the environment (though to keep the argument simple and T constant we suppose the heat capacity of the environment is so large that this heat gain produces only a negligibly small increase in T). If the piston is frictionless and the force applied to the piston rod is only just sufficient to move it in slowly against the gas pressure, this experiment is *reversible*. The gas can be allowed to expand slowly again by gradual reduction of the force on the piston rod and the energy $RT \log (V_1/V_2)$ previously converted into heat energy can be fully converted back again into mechanical energy. For example, the gas might have been compressed by adding weights gradually to a weight pan at the top of the piston rod (Fig. A3.1) and during the expansion each weight would be brought back to its initial height (same gravitational potential energy).

Consider now the alternative experiment shown in the right-hand drawing of Fig. A3.1. After the gas has been compressed to volume V_2 a small hole is made in the piston so that the gas expands into the vacuum to fill the volume V_1. For an ideal gas, no heat will be withdrawn from the environment in this process. The gas molecules will remain at the same temperature and simply move by random thermal motion into the larger volume now available to them. This is an *irreversible* process. To raise the weights and the piston to their original position will require an amount of work $RT \log (V_1/V_2)$. This energy has to be taken from somewhere, and although the gas, the piston and the weights can be brought back to their original state, the total effect of the experiment has been the conversion of an amount of mechanical energy $RT \log (V_1/V_2)$, from somewhere, into a gain in thermal energy of the environment. This is a change which, as we shall now see from fuller discussion, cannot be completely reversed.

Heat energy can be converted into mechanical energy, during expansion of a gas, but it is not possible to operate our piston in a continuous repeating cycle bringing every component back, after each cycle, to its original position and continuously convert heat energy to mechanical energy. The best that can be achieved, in a repeating cycle, is expansion at temperature T_1 taking an amount of heat Q_1 from a heat store at temperature T_1 and compression at a lower temperature T_2 giving up heat Q_2 to a store at T_2, such that $Q_1/T_1 = Q_2/T_2$. A net amount of mechanical work $= Q_1 - Q_2$ can then be done by the engine per cycle associated with this transfer of heat from the store at higher temperature to the one at lower temperature, but if a certain amount of mechanical energy has been dissipated, with no other result than slight warming of the environment, the situation is irreversible. A heat engine could withdraw the appropriate amount of heat from the environment but could only partly convert the heat back to mechanical work, since to bring the heat engine back to its initial condition a certain amount of this energy has to be given up as heat to a heat store at lower temperature.

In order to deal with the irreversibility of natural events and the peculiar characteristics of heat energy, it was found necessary in the development of thermodynamics to introduce the concept of *entropy*. The change in entropy of a system when it goes by a series of *reversible* steps from state A to state B is defined as $\int_A^B \mathrm{d}Q/T$ where $\mathrm{d}Q$ is the amount of heat gained by the body at temperature T, and the integral adds together a series of terms, for all the little bits of heat that may be taken up by the system at different temperatures as it goes from A to B. To consider what this concept means and why it is useful let us return to the reversible gas compression experiment. When the gas is compressed from V_1 to V_2 the environment gains an amount of heat $RT \log (V_1/V_2)$ and hence gains an amount of entropy $R \log (V_1/V_2)$. The gas gives up the same amount of heat and hence loses this amount of entropy. In general for reversible reactions the total entropy remains constant. The gas gains an amount of entropy $R \log (V_1/V_2)$ when it expands *reversibly* from V_2 to V_1 and the environment loses this amount of entropy. The gas gains the same amount of entropy when it expands *irreversibly* into a vacuum (right-hand drawing of Fig. A3.1) for, in an irreversible reaction, the entropy change is the entropy difference between the initial and final states, as measured when the change is made reversibly. In an irreversible reaction the total entropy does *not* remain constant. The gas gains entropy while the environmental entropy remains unchanged. If the net result of an irreversible experiment is the conversion of an amount of mechanical work Q into a gain in heat energy of the environment at temperature T there has been a net increase in entropy of amount Q/T. Entropy gives us a measure of the irreversibility of reactions.

Thermodynamics is a science, like genetics, which can be pursued, often with powerful results, in a purely formal way to a certain point, without regard to the nature of atoms and molecules. However consideration of the atomic nature of molecules (and genes) can often complement the formal viewpoint and in thermodynamics allows much greater understanding of the nature of entropy. The study of the movement of assemblies of molecules is called statistical mechanics, and to develop a statistical theory of gases the only initial assumption that is necessary is to suppose the molecules move in a random way. Consider one mole of a gas confined in a volume V, and suppose this volume divided up into a large number of little cubic boxes of volume v. (The number of molecules in a mole is $\simeq 6 \times 10^{23}$ so that if we divide the volume into, say 10^{12} boxes there are still a large number of molecules in each box, on average 6×10^{11}.) Our initial assumption implies simply that there is an equal probability of any given molecule being in any box.

Figure A3.2 illustrates a simple example of the results that follow from an assumption of this kind. If four balls are thrown at random into two boxes the relative probabilities of obtaining the different possible distributions are:

all 4 in left-hand box ... 1
all 4 in right-hand box ... 1
three in left-hand box, one in right ... 4
three in right-hand box, one in left ... 4
two in each box ... 6

In general, if we have N balls and i boxes, the relative probability (W) of obtaining a given conformation with N_1 balls in box 1, N_2 balls in box 2, etc., is given by

$$W = \frac{N!}{N_1!N_2!\ldots N_i!}$$

$[N! = N(N-1)(N-2)\ldots 1 \text{ and } 0! = 1]$.

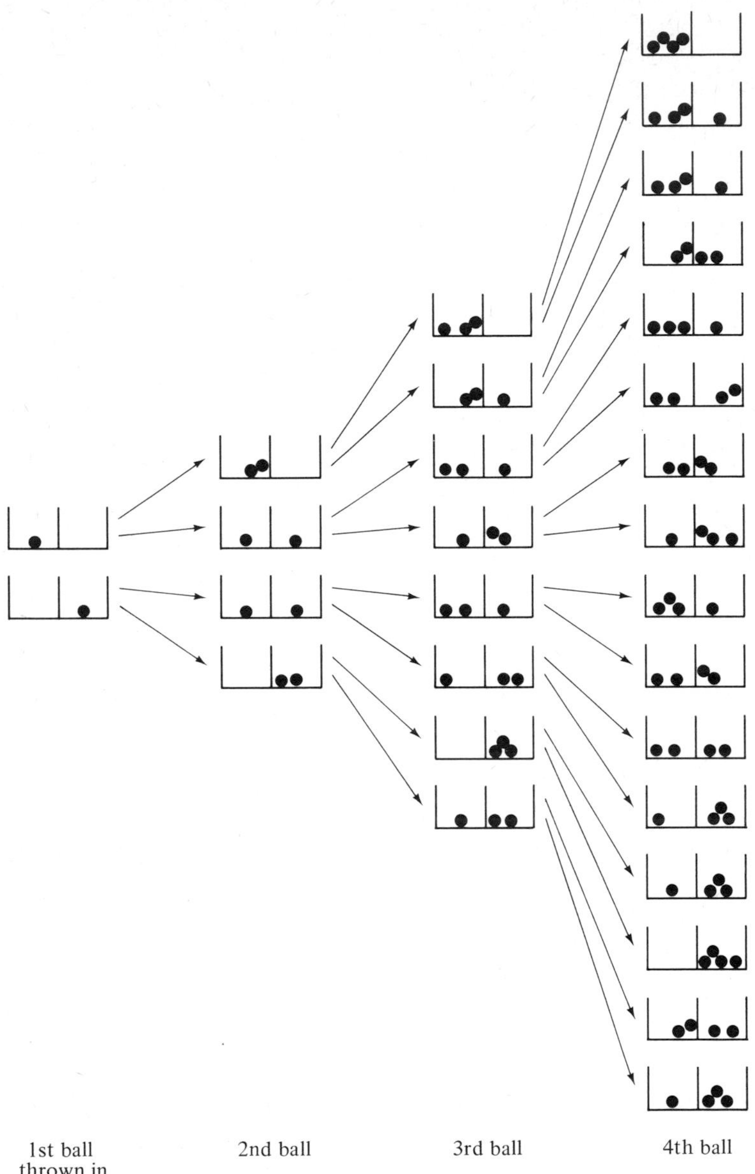

Fig. A3.2 The possible ways in which four balls can be thrown into two boxes.

Thus in the example given above:

$$W_{4:0} = \frac{4!}{4!0!} = 1$$

$$W_{3:1} = \frac{4!}{3!1!} = 4$$

$$W_{2:2} = \frac{4!}{2!2!} = 6$$

Viewing a gas as an assembly of molecules randomly thrown into the little boxes of volume v we will now show that the entropy (S) of the gas in any given conformation with N_1 molecules in box 1, N_2 in box 2 and so on, is related to the probability function W by the equation:

$$S = k \log_e W$$

where k is a constant, called Boltzmann's constant. In the example of Fig. A3.2, the most probable distribution is the even distribution 2:2, and for large numbers this effect is enhanced, so that the most probable distribution of gas molecules (the one that is actually found) is one of uniform density and approximately the same number of molecules in each box. We can thus take

$$N_1 = N_2 = N_3 = \cdots = N_i$$

Further,

$$N_1 + N_2 + \cdots + N_i = N$$

and therefore

$$N_1 = N_2 = N_3 = \cdots = N_i = \frac{N}{i}$$

$$W = \frac{N!}{N_1!N_2!\ldots N_i!} = \frac{N!}{[(N/i)!]^i}$$

(for one mole of gas N is now Avogadro's Number).

With very large numbers for N and for N/i we can use with good accuracy an approximation known as Stirling's formula according to which

$$\log_e N! = N \log_e N - N$$

Thus
$$\log W = N \log_e N - N - i\left[\frac{N}{i}\log_e \frac{N}{i} - \frac{N}{i}\right]$$

$$= N \log_e N - N \log_e \frac{N}{i}$$

$$= N \log_e i.$$

But $i = \dfrac{V}{v}$ so $\log W = N \log \dfrac{V}{v}\cdot$

When a gas expands from volume V_2 to volume V_1, if we take

$$S = k \log_e W$$

the entropy change is given by

$$\Delta S = kN\left[\log\frac{V_1}{v} - \log\frac{V_2}{v}\right]$$

$$= kN \log\frac{V_1}{V_2}$$

This is the entropy change calculated above. Boltzmann's constant, k, is equal to R/N where R=the gas constant and N=Avogadro's number ($\simeq 6 \times 10^{23}$) the number of molecules in one mole. At the instant the hole is made, in the experiment illustrated in the right-hand drawing of Fig. A3.1 (and the gas is all in V_2, though it has V_1 available to it) the gas is in a state of low probability for the new condition (V_1 available). It moves to a state of higher probability with an entropy change related to the probability change.

So far we have included in W a probability function which describes the way randomly moving molecules tend to spread out evenly in the space available to them. To complete the story we now have to include a further probability function which describes the way the total *energy* of an assembly of molecules tends to spread evenly through the assembly, as when gases at T_1 and a lower temperature T_2 are brought into thermal contact (Fig. A3.3) and come to an intermediate temperature. Like the expansion of a gas into a vacuum this heat spreading is also an irreversible process, associated with increase in entropy, for if instead of bringing the gases into thermal contact we had used a heat

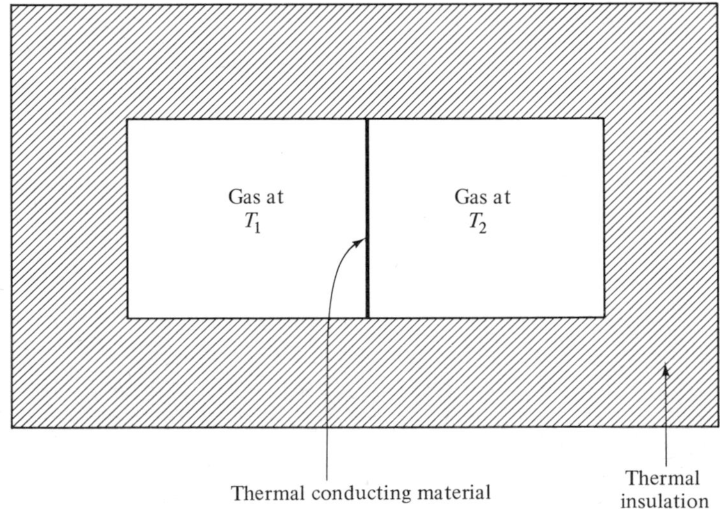

Fig. A3.3 Heat tends to spread evenly through an assembly of molecules (see text).

engine to withdraw heat from the gas at T_1 and transfer the heat to T_2, we could have got some mechanical work out of the system, which we cannot do once they have come to thermal equilibrium.

Thus
$$S = k \log W_s W_E$$

where W_s is the probability function for the spatial distribution and W_E is the probability function for energy distribution. This amendment does not affect the value derived above for ΔS when an ideal gas expands from V_2 to V_1 since the energy distribution and W_E remain unchanged so that

$$\Delta S = kN \log \frac{V_1}{v} + k \log W_E - \left[kN \log \frac{V_2}{v} + k \log W_E \right]$$

$$= kN \log \frac{V_1}{V_2} \text{ as derived above}$$

However, in other situations it is necessary to work out the probability function W_E. For a gas this is derived as follows: each molecule has at any instant a given velocity w in a given direction. This velocity can be resolved into three components w_x, w_y, w_z along the directions of perpendicular axes x, y, z. We can then take a point $x=w_x$, $y=w_y$, $z=w_z$ as representing the position of the molecule in *velocity space*. Now all we need to do is to divide velocity space as we divided real space into little boxes and make the initial assumption that it is equally probable for a molecule to be in any of the little boxes of velocity space, subject to the condition that the total energy of the assembly of molecules remains constant (for a gas at a given temperature). The most probable distribution is found to be the distribution in which the molecules are spread out as evenly as possible in velocity space (just as they spread out as evenly as possible in real space) subject to the total energy condition.

The mechanism underlying thermodynamic discussion is a very simple one – the constant, tireless, frictionless, perpetual movement of molecules. The observed fact that entropy is always increasing in the universe is a simple consequence of the tendency for all systems, as soon as any constraint is removed or altered, to move from a less probable to a more probable arrangement – more probable, that is, under conditions of thermal molecular movement. To impose constraints, and bring a system into a more ordered state, work must be done on the system. The entropy of a system in a given state may be written, in general, $S=k \log W$, *where* W *is now a total probability function for the system and a measure of the probability of the molecules of the system adopting the spatial and energy distribution found in the given state.*

We have so far been considering the simple system of an ideal gas, but it may be noted at this point that an aqueous solution can be treated in exactly the same way. The movement of solute molecules in dilute solutions (for which solute–solute interactions can be neglected) is formally equivalent to the movement of molecules of an ideal gas. Also, if a number of different types of gas molecules or solute molecules are present (again, for ideal gases, or low solute concentrations) each molecular species present can be treated independently. Thus the entropy increase on mixing two gases (Fig. A3.4) is obtained by adding the entropy increase of gas 1, as it expands to fill the whole volume, to the entropy increase of expansion of gas 2. The tendency of solute molecules to

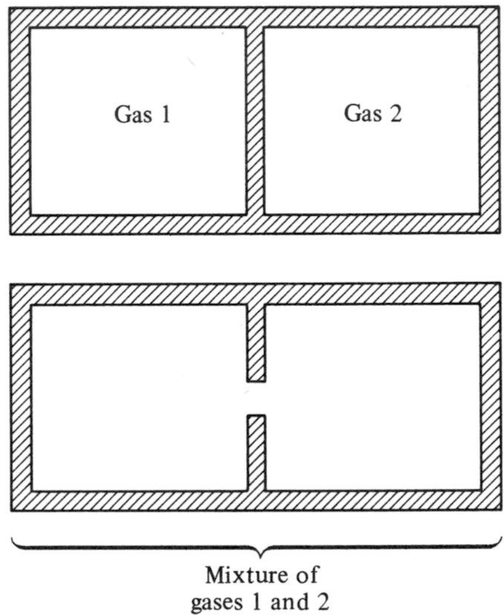

Fig. A3.4 Mixing of two gases (see text).

mix and spread by diffusion over the space available to them is an important feature of the milieu in which living cells must operate. In order to function, cells must do work to maintain or build up their internal concentration of certain solutes, and to maintain ionic and solute concentration differences across membranes. The work necessary can be calculated from the entropy change associated with the diffusion process – a process which is irreversible in the sense defined above. We return to this topic in Appendix 4, but for the moment we must first introduce the concept of free energy and discuss a simple example of interaction between two solutes in solution.

When any system such as a mechanical system, or a chemical solution, undergoes irreversible change with increase in entropy, there is a reduction in the amount of work that the system can perform, even if there is no change in the total energy of the system. This leads to the concept of 'free energy' as a measure of the *amount of energy available to do work*. If a system undergoes a change in conditions at a constant temperature T, the free energy change (ΔG) is related to the total energy change (ΔH) and the entropy change (ΔS) by the equation

$$\Delta G = \Delta H - T\Delta S$$

(ΔH includes both external work and work done against forces of molecular attraction, as discussed below.) It will be noted that a *change* in conditions (and energy and entropy changes) have been considered here, for energy and entropy are easier to treat in a relative rather than an absolute way. One advantage of considering only energy changes is that irrelevant factors are left out of account. For example, the total energy of a molecule includes the energy of atomic nuclei. In a normal chemical reaction the nuclei do not alter, and in calculating the energy change only atomic interactions need be considered. Also, if only a change of entropy is considered the constant part of the entropy probability function can be left out of account, as in the treatment given above of expansion of a gas at constant temperature.

For a mechanical system, say of pulleys and strings, or a ball on sloping ground, the equilibrium position is the position of minimum gravitational potential energy. The system comes to a position in which its capacity for doing work is a minimum. For a thermodynamic system in thermal equilibrium with an environment at constant temperature *the equilibrium condition is the condition of minimum free energy* (as defined above). If the system starts initially in a non-equilibrium condition the molecules move around in a random way so far as they have freedom to do so, and the randomly moving molecules increase their state of disorder until the condition of minimum free energy is reached. Thermal movement makes the thermodynamic system analogous to a mechanical system of strings or balls which is constantly being shaken – mildly at low temperature and more fiercely at higher temperatures. If conditions allow a number of energy states of low free energy, and the energy barrier for movement from one state to another is smaller than, or comparable with, the energy of thermal motion then the system will move around among the available low energy states.

We can now discuss some of these points in more detail, and show that the condition of minimum free energy is in fact the equilibrium condition for solute molecules interacting together in aqueous solution. To keep the treatment as simple as possible we consider only dilute solutions. Solute molecules moving around in dilute aqueous solution behave like the molecules of an ideal gas. The experiments previously described can now be performed with the apparatus shown in Fig. A3.5. If the piston is moved slowly enough for the water pressure difference across the piston to be always negligible then the pressure exerted on the piston is the pressure due to the solute molecules:

$$RTc$$

where c is solute concentration below the piston, in moles/cm^3.

Using the same sort of argument as is given above for expansion of a gas it can be readily shown that the entropy per mole (S_c) of a solution containing solute at concentration c is

$$S_c = S_0 - RT \log c$$

where S_0 is the entropy of one mole of solute in one litre of solution. (This argument is given more fully in Appendix 4.)

We now consider a solution containing two types of solute molecule A and

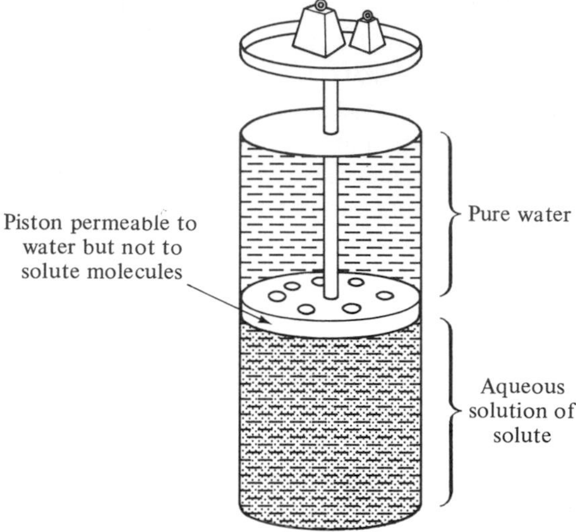

Fig. A3.5 The experiment of Fig. A3.1 repeated for an aqueous solution.

B and suppose these molecules have a tendency to come together to form a complex AB

$$A + B \rightleftharpoons AB$$

AB may be a complex formed by physical forces, such as hydrophobic effects, as when protein sub-units of two different kinds aggregate to form a dimer. Alternatively A and B may be small molecules joined by a chemical bond in a chemical reaction. However, we will consider here a simple physical association, for then no enzyme is needed to bring the components to equilibrium at normal environmental temperature.

As in the case of gas molecules in a box, thermal motion will ensure that the system

$$A + B \rightleftharpoons AB$$

will come to an equilibrium in which the molecules are spread as evenly as possible over the space conformations and energy states available, so that the total entropy of solution and environment is a maximum.

We will suppose A and B are in a solution in thermal equilibrium with an environment at constant temperature T and that the solution can expand or contract freely to remain at constant (atmospheric) pressure. Suppose initially the concentrations of A, B and AB are not at equilibrium. Then if some molecules of A and B associate there will be a change in the total entropy of solution and environment. If this entropy change is positive the association will be spontaneous. If it is negative then the reaction will go the other way, and there will be spontaneous dissociation of AB. This association or dissociation will go on spontaneously until the concentrations of A, B and AB adjust to values which make the total entropy a maximum. This will be a state of dynamic equilibrium with association and dissociation going on all the time, but at equal rates, so that there is no net change in concentration levels. At maximum entropy the total entropy *change*, of solution and environment, must be *zero* for a small dissociation of AB. In mathematical terms, the total entropy of solution and environment increases to a maximum at which $dS/dc_{AB}=0$, where dS is the change in total entropy arising from a small change dc_{AB} in the concentration of AB.

We write c_A, c_B, c_{AB} for the concentrations (in moles/litre) of A, B and AB at equilibrium. Then

$$Sc_A = S_A - R \log c_A$$

$$Sc_B = S_B - R \log c_B$$

$$Sc_{AB} = S_{AB} - R \log c_{AB}$$

where Sc_A, Sc_B, Sc_{AB} are the entropies/mole of A, B and AB at the equilibrium concentrations, and S_A, S_B, S_{AB} are the entropies/mole of A, B and AB when each are at concentrations of 1 mole/litre. (It may be noted that we have again avoided defining absolute entropy values, by considering only the entropy changes $Sc_A - S_A$, etc., resulting from concentration changes.)

Suppose now that a fraction of a mole (f) of AB dissociates, with f so very small (say $f=10^{-6}$) that this has only a very small effect on the concentrations of A, B and AB present. The entropy change of the solution is

$$fSc_A + fSc_B - fSc_{AB}$$

The entropy change of the environment is

$$-f\frac{\Delta E}{T} - f\frac{p\,\Delta V}{T}$$

where ΔE is the energy of association per mole of AB formed, $f\Delta E$ is the heat taken up from the environment when a mole fraction f dissociates, and a small amount of energy, which we have written as $fp\,\Delta V$ must be included for the volume expansion of a solution held at constant pressure, this energy also being taken up as heat energy from the environment.

At equilibrium

$$fSc_A + fSc_B - fSc_{AB} - f\frac{\Delta E}{T} - f\frac{p\,\Delta V}{T} = 0$$

$$(S_A + S_B - S_{AB}) - R \log\left(\frac{c_A c_B}{c_{AB}}\right) - \frac{\Delta E}{T} - \frac{p\,\Delta V}{T} = 0$$

$$-RT \log\left(\frac{c_A c_B}{c_{AB}}\right) = \Delta E + p\,\Delta V - T\,\Delta S$$

where $\Delta S = S_A + S_B - S_{AB}$.

We may now write

$$\Delta H = \Delta E + p\,\Delta V$$

so that ΔH (the change in *enthalpy* of the reaction) includes the two straightforward energy changes in the reaction, viz. the work done against the forces of molecular attraction in dissociating one mole of A and B (ΔE) and the work done ($p\,\Delta V$) against the external pressure as a result of the associated volume change (ΔV). For reactions in aqueous solution ΔV is small and ΔH is approximately equal to ΔE.

We now have

$$-RT \log\left(\frac{c_A c_B}{c_{AB}}\right) = \Delta H - T\,\Delta S$$

The *standard free energy* change of the reaction is defined as above

$$\Delta G^0 = \Delta H - T\,\Delta S$$

Hence at equilibrium

$$-RT \log\left(\frac{c_A c_B}{c_{AB}}\right) = \Delta G^0$$

(In full discussion of thermodynamic processes ΔG, the free energy change for a reaction taking place at constant temperature and pressure, has to be given a name to distinguish it from free energy changes for reactions taking place at constant temperature and volume. ΔG is more precisely the Gibbs free energy change. In the discussion in the text of this book and in the Appendices we are concerned only with reactions taking place at constant temperature and pressure, and where 'free energy change' is mentioned this refers in every case to the Gibbs free energy change.)

From the equation derived above we see that the equilibrium concentrations should be related by the equation

$$\frac{c_A c_B}{c_{AB}} = e^{\frac{-\Delta G^0}{RT}}$$

This is the Law of Mass Action found experimentally for all physical and chemical reactions at equilibrium, which allows ΔG^0 to be determined from the equilibrium concentrations. Consideration of the equations given above shows that at equilibrium, when the entropy of solution and environment is a *maximum* the free energy of the solution is at a *minimum* value.

The astute reader (or the average-mathematical biologist who mulls over these equations, trying to really understand how living cells create order out of the chaos of our universe) will note that the equations derived in this last section are not dimensionally correct, so that the argument given, although illustrating the main principles involved, has been oversimplified at some point. (In textbook of physical chemistry this difficulty is usually overcome by considering a reaction $A + B \rightleftharpoons C + D$.) If A and B represent protein sub-units aggregating to form a dimer, we have, in fact, neglected the water bound to A and B, which is released at the area of contact when the dimer is formed. (The role of water in determining protein structure and aggregation is discussed further in the next Appendix.) By measuring equilibrium concentrations and hence ΔG^0, at two different temperatures, it is possible to determine separately the two component parts of the free energy change, ΔH and ΔS, for any reaction under study. For aggregation of protein sub-units it is often found that ΔS is the larger component, the aggregated form being the form of *greater* freedom. In these cases the increased freedom of water molecules, when sub-units come together, is the main factor favouring aggregation.

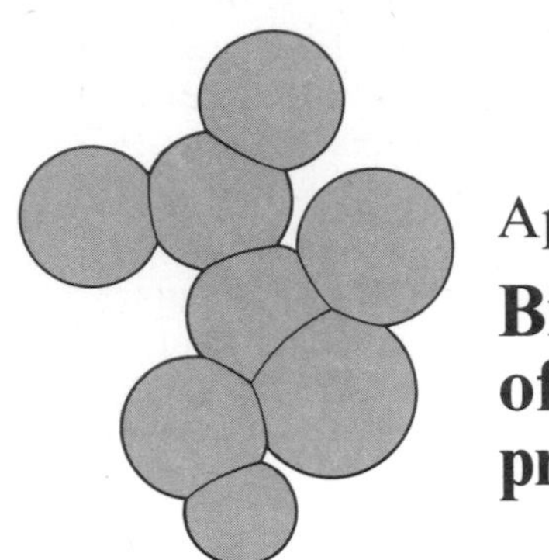

Appendix 4

Biological applications of thermodynamic principles

The stability of the α-helix

This appendix is written as a continuation of Appendices 2 and 3. It includes some comments on the strength of hydrogen bonds, as well as a more detailed consideration of their role in determining macromolecular structure, illustrated by a discussion of the coiling of a polypeptide chain into an α-helix. Hydrogen bonds are weak bonds (1–7 kcal/mole) compared with covalent bonds (15–150 kcal/mole). Their importance in biomolecular structure lies precisely in the fact they are not much stronger than the disruptive forces of thermal agitation at 20–37°C (~1 kcal/mole). The additive effect of a number of hydrogen bonds in a macromolecule may lead to a structure which is stable at 37°C, under appropriate conditions. But quite small changes in conditions, such as a change in the concentration of Mg^{++} ions, may make the structure unstable. Biological molecules, as noted in Chapter 7, are not rigid and unchangeable, but are capable of subtle change, even within the normal physiological range of conditions.

In general, the more the electrons are drawn away from a hydrogen atom in an —NH or —OH group, the stronger will be the hydrogen bond which it forms. The strength of the hydrogen bonds between water molecules has been estimated to be ~5 kcal/mole, and the hydrogen bonds formed by polypeptide backbone groups must be rather stronger than this (as a result of the resonance effects discussed in Appendix 2).

In considering the effect of intrachain hydrogen bonds in stabilizing the α-helix it must be remembered that when the chain coils into a helix, in an aqueous environment, intrachain hydrogen bonds are formed, but hydrogen bonds are *broken* between backbone —NH and =O groups and surrounding water molecules (Fig. A4.1). In general, for polypeptide chains in different solutions, the coiled helical form will be the energetically more favourable one (and hence the one adopted by the molecule) only if the intrachain bond attraction is stronger than the force of attraction of the peptide groups for the solvent molecules.

It is found, for example, in the study of synthetic polypeptides, that α-helices are very stable when dissolved in chloroform, or some other solvent which cannot form hydrogen bonds. In aqueous solution the strength of intrachain hydrogen bonds is such that they confer a marginal stability on the α-helix (as discussed in Chapter 2). If urea is added to aqueous solutions of polypeptides or proteins it is found that the α-helices tend to come unfolded as the urea concentration is increased. Urea molecules form rather stronger hydrogen bonds than water molecules, as a result of the electron distribution (resonance) within the urea molecule, which is comparable with that of polypeptide backbone groups. Unfolding in urea solutions the polypeptide backbone groups can form strong hydrogen bonds with urea molecules, and the helix is no longer the energetically favourable form. Proteins are thus denatured in urea solutions of sufficient concentration.

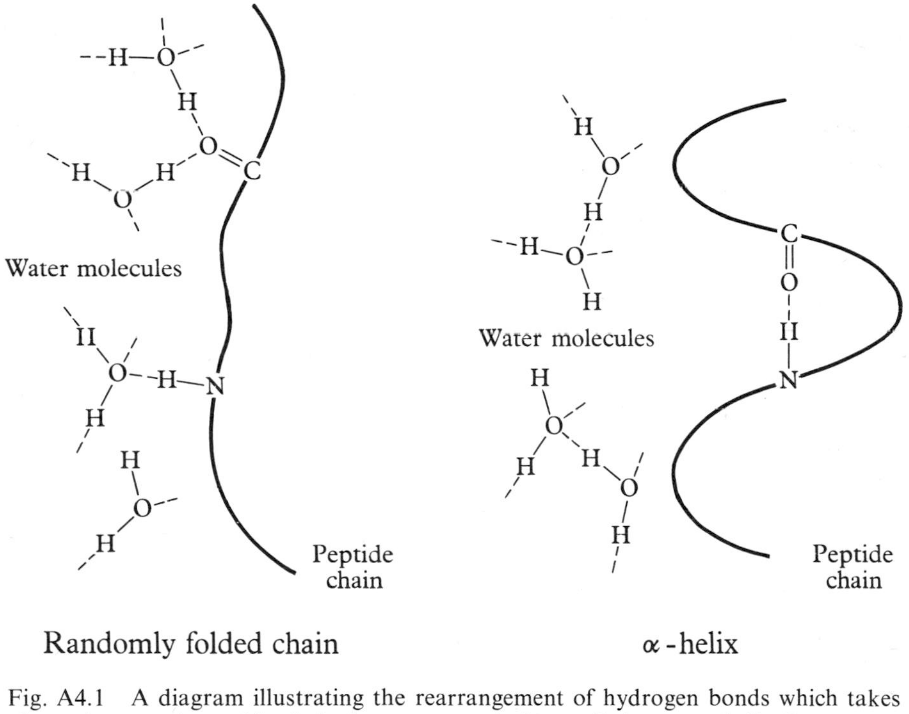

Water molecules

Peptide chain

Randomly folded chain

Water molecules

Peptide chain

α-helix

Fig. A4.1 A diagram illustrating the rearrangement of hydrogen bonds which takes place when a polypeptide chain, in aqueous solution, coils into an α-helix. In aqueous solution the backbone groups of the uncoiled chain form hydrogen bonds with water molecules

This discussion of the coiling and uncoiling of the α-helix has led us to a point of fundamental importance in molecular biology: the form adopted by a molecular chain, or by the molecular aggregate which makes up a chromosome, or a cell membrane, is the one which is energetically most favourable – the form in which the *free* energy is a minimum (compare the way in which a mechanical system moves to a state of minimum potential energy). This form will be stable if the energy required for the molecule to move to an alternative form is greater than the energy it can temporarily acquire in random thermal motion. If, on the other hand, the differences between the free energy levels for several alternative forms are smaller than, or comparable with, the energy of thermal motion, then the chain, or molecular aggregate, will spend part of its time in each of these alternative forms, and the greater proportion of its time in those of lower potential energy.

Free energy includes both the inherent energy, or energy of attraction, in the molecular bonds, and also an *entropy* term as discussed in Appendix 3 which is a measure of the freedom of movement of the molecules. Suppose, for example, that a molecular chain were able to fold up in two different ways, and that in both forms the attractive forces or inherent energy of the bonds were the same. If one form allowed the molecular chain more freedom to flop about, then it is this floppy form which would be adopted. This is a result of the fundamental property of matter: entropy, in a closed system, always tends to increase. There is a trend in random molecular movement towards greater disorder. It is perfectly possible in fact for a molecule, or molecular aggregate, to take up a form in which the forces of attraction are less well satisfied, but which allows greater freedom of movement. A comparable balancing of competing factors sometimes leads a scientist to accept a post of relatively low salary for the sake of greater freedom in research activity.

The concept of free energy is introduced to take account of the relative contributions of inherent energy and entropy, and is so defined that the condition of minimum free energy corresponds to the equilibrium form for the system. The change in free energy associated with any given rearrangement (ΔG), as discussed in Appendix 3, is defined by the equation:

$$\Delta G = \Delta H - T\Delta S$$

where ΔH = change in inherent energy in this rearrangement
ΔS = change in entropy in the rearrangement
T = absolute temperature.

The form adopted by a molecular chain, the conformation of minimum free energy, can be determined in principle by the statistical methods introduced in Appendix 3, and the entropy can be calculated as a probability function for different chain conformations. If the chain adopts one particular conformation this is analogous to all balls being in one box in Fig. A3.2, while random folding is analogous to a more even distribution of the balls. If energy considerations are equal the random folding is more probable.

In a full discussion of the factors which determine whether or not a polypeptide chain will coil up into an α-helix, or whether the polypeptide of a globular protein will coil and fold into the unique tertiary structure characteristic of that protein, the loss of freedom of movement on coiling must be taken into account, and also the increased freedom of water molecules released by the formation of intrachain hydrogen bonds and by hydrophobic interaction between side-chain groups. The electrostatic attractions of intrachain hydrogen bonding represented in Fig. 2.7 are thus only one factor contributing to the stability of the α-helix – the increased freedom of the released water molecules is perhaps of equal, or greater, importance. The freedom of movement of water molecules is an important factor making for the insolubility of short hydrocarbon chains in aqueous solution. Long hydrocarbon chains are insoluble because their insertion into the water structure requires the breaking of hydrogen bonds; short hydrocarbon chains are insoluble for the more subtle reason that they restrict the freedom of movement of surrounding water molecules.

Although, as we have noted, the inherent energy of polypeptide intrachain hydrogen bonds is greater than 5 kcal/mole, the change in free energy per peptide group when the chain coils into an α-helix in an aqueous environment has been estimated to be $\sim 1{\cdot}5$ kcal/mole. If this figure is compared with the figure we have given for the average disruptive force of thermal agitation (~ 1 kcal/mole), it will be appreciated that intrachain hydrogen bonding confers only a marginal stability on the α-helix. As discussed in the later sections of Chapter 2, interactions between side chains become important in determining the overall conformation of a protein molecule.

Osmotic pressure. Ionic equilibrium across a membrane

If an aqueous solution is separated from pure water by a semipermeable membrane – permeable to water, but not to the solute – and the total pressure is initially the same on each side of the membrane, there will be a movement of water across the membrane making the solution more dilute. The total pressure on the solution side is contributed partly by water and partly by solute, but the total pressure on the pure water side is contributed by water alone. The water pressure is therefore higher initially on the pure water side, more water molecules per second bombard the membrane, and more cross the membrane, from this side, so that there is net flow.

The system can come to equilibrium only if the total pressure builds up on the solution side to make the pressure of water the same on the two sides of the membrane. When this equilibrium is achieved, the difference in total pressure on the two sides of the membrane, termed the *osmotic pressure* of the solution, is the pressure exerted by the solute molecules. At concentrations sufficiently low for interaction between solute molecules to be neglected, this is the same as the pressure exerted by a perfect gas at the same temperature and concentration.

Osmotic pressure (in dynes/cm^2) $= RTc$

where c = solute concentration in moles/cm^3

R = the gas constant (in ergs/degree/mole)

T = absolute temperature.

For a solution containing a number of different solutes, or dissociated salts, the total osmotic pressure is the sum of the contributions due to each solute or ion type (e.g. K^+), just as the total pressure exerted by a mixture of gases is the sum of their partial pressures. The term *isotonic solution* is used for a solution (e.g. 0·15 molar sodium chloride for mammalian cells) exerting the same osmotic pressure as cell cytoplasm. The isotonic solute concentration, for a given cell type, is determined experimentally by observation of the behaviour of the cells in solutions of varying solute concentration. Cells placed in hypertonic solutions lose water and shrink; bacteria and plant cells in hypertonic solutions shrink away from their rigid surrounding walls. In hypotonic solutions cells without rigid surrounding walls take up water, become spherical, then burst.

To derive an equation relating electrical potential and ionic concentration differences across a membrane, or across an interface between two phases, we first calculate the free energy of a solution (Eq. (A4.1)). This is a reworking in slightly more detail of some of the ideas of Appendix 3. The non-mathematical reader may skip lightly through this argument, perhaps following the principles behind it. Accepting Eq. (A4.1) as established he can still see how the equilibrium condition Eq. (A4.2) follows from the necessity for energy balance across the membrane.

To calculate the free energy inherently present in a simple solution we have to remember that any solution can exert an osmotic pressure, and that this pressure can, in principle, be used to perform useful work. The sort of apparatus we should need to perform work in this way would be a cylindrical vessel, with a movable piston made of a semipermeable material, permeable to the solvent and impermeable to the solute (Fig. A4.2). If the solution is contained in the

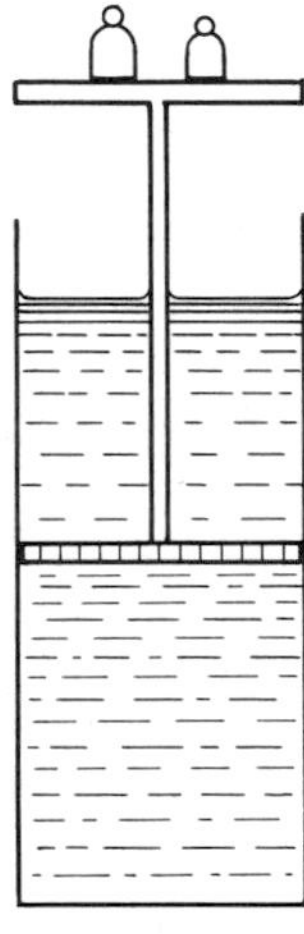

Fig. A4.2 Experiment to show that a solution can be made to do physical work. The solution is contained in a cylindrical vessel, below a piston made of semi-permeable material: permeable to the solvent, but not to the solute. The part of the cylinder above the semi-permeable membrane is filled with pure solvent. Osmotic pressure causes the piston to rise, and lift the weights on a pan (connected to the piston by a rod)

lower part of the cylinder of Fig. A4.2, and pure solvent is present in the vessel above the semipermeable piston, then osmotic pressure will cause the piston to rise, and the weights on the pan attached to the piston will acquire increased potential energy. The solution below the piston will become more dilute. The increased potential energy of the weights is a measure of the work done, and the maximum amount of work that can be done defines the free energy change of the solution, associated with change in concentration.

If the osmotic pressure is P when the piston is a height x above the base of the cylinder, and if the area of the piston is A, the maximum work it can do in rising a small distance $\mathrm{d}x$ is:

$$PA\,\mathrm{d}x$$

If it is initially at a height x_0, and at the end of the experiment at height x_1, then the maximum work it can do is:

$$\int_{x_0}^{x_1} PA\,\mathrm{d}x = RT\int_{x_0}^{x_1} Ac\,\mathrm{d}x$$

(provided the concentration is low enough for the osmotic pressure equation given above to be valid).

Concentration is related to x by the equation:

$$cx = c_0 x_0$$

where c_0 is the initial concentration. If the total amount of solute present is one mole then:

$$c_0 x_0 A = 1, \text{ therefore } cxA = 1 \text{ and } Ac = \frac{1}{x}$$

The maximum work done is:

$$RT\int_{x_0}^{x_1} \frac{1}{x}\,\mathrm{d}x = RT\log_e\frac{x_1}{x_0} = RT\log_e\frac{c_0}{c_1}$$

$$= RT(\log_e c_0 - \log_e c_1)$$

where c_1 is the solute concentration at the end of the experiment. Thus free energy change per mole

$$= RT\,(\text{change in } \log_e c) \qquad \text{Eq. (A4.1)}$$

If there is equilibrium across a membrane then each type of ion (e.g. K^+) to which the membrane is at all permeable must be in equilibrium across the membrane. This means that the free energy difference between the two sides of the membrane, for each ion type to which the membrane is permeable, must be equal to the work done in taking one ion-mole of this ion type across the membrane. If the electrical potential difference across the membrane is E volts the work done is:

$$nEF \text{ joules}$$

where F=Faraday's constant (the charge in coulombs on one ion-mole of a univalent ion) and where n is the ion valency (for example $n=2$ for Ca^{++} and Mg^{++}, $n=1$ for Na^+ and Cl^-). If we specify that the ion-mole is taken from the low voltage to the high voltage side of the membrane, the work done is positive for positive ions, and negative for negative ions.

For each ion to which the membrane is permeable:

$$nEF = RT \log_e \left(\frac{\text{concentration outside}}{\text{concentration inside}} \right)$$

$$E = \frac{RT}{nF} \log_e \left(\frac{\text{concentration outside}}{\text{concentration inside}} \right) \qquad \text{Eq. (A4.2)}$$

For univalent positive ions, at 20°C, substituting numerical values for R (in joules/degree/mole) and F, and adopting the normal sign convention, in which the membrane potential is taken to be negative if the interior of the cell is at a lower potential than the exterior:

$$E = 58 \log_{10} \left(\frac{\text{concentration outside}}{\text{concentration inside}} \right) \text{mV}$$

For univalent negative ions at 20°C

$$E = 58 \log_{10} \left(\frac{\text{concentration inside}}{\text{concentration outside}} \right) \text{mV}$$

Sources of Figures

In addition to providing further information, experimental details, etc., for each figure reproduced from the literature, the papers listed in this section form the main list recommended for further reading, supplemented by papers listed in the Bibliography.

Cover illustration adapted from KENDREWS, J. C. (1961), *Scientific American* (December).
Fig. 1.2 RUSTAD, R.C. (1961) *Scientific American* (April).
Figs 1.7, 1.9, 1.10 WILLIAMS, C.A. (1960) *Scientific American* (March).
Fig. 1.8 SMITHIES, O. (1959) *Ad. Protein Chem.*, **14**, 65.
Fig. 2.2 MOROWITZ, H.J. (1958) in *Microsomal Particles and Protein Synthesis*, ed. R.B. Roberts. Oxford, Pergamon.
Fig. 2.5 SPACKMAN, D.H., STEIN, W.H. and MOORE, S. (1960) *J. biol. Chem.*, **235**, 648.
Fig. 2.9 DICKERSON, R.E. and GEIS, I. (1969) *The Structure and Action of Proteins*. W. A. Benjamin Co., Menlo Park. Copyright 1969 by Dickerson and Geis.
Fig. 2.10 DOTY, P. (1959) in *Biophysical Science*, ed. J.L. Oncley. New York, Wiley.
Fig. 2.11 KENDREW, J.C. (1961) *Scientific American* (December).
Fig. 2.12 DICKERSON, R.E. (1963) in *The Proteins*, ed. H. Neurath, 2nd edition, Vol. 2.
Fig. 2.13 JOLLES, P., CHARLEMAGNE, D., PETIT, J.-F., MAIRE, A.-C. and JOLLES, J. (1965) *Bull. Soc. Chim. Biol.*, **47**, 2241.
Fig. 2.14 DAYHOFF, Margaret O. and ECK, R.V. (1968) *Atlas of Protein Sequence and Structure 1967–68*, National Biomedical Research Foundation, Washington, D.C. The drawing was made by Irving Geis based on his perspective painting of the molecule which appeared in *Scientific American*, November 1966. The painting was made of an actual three-dimensional model assembled at the Royal Institution, London, by D.C. Phillips and his colleagues, based on their X-ray crystallography results.
Fig. 2.15 PERUTZ, M.F. *et al.* (1960) *Nature, Lond.* **185**, 416.
Fig. 2.16 HUXLEY, H.E. (1969) Science, **164**, 1356; FINEAN, J.B. (1962) *Circulation*, **26**, 1151; WILKINS, M. H. F. (1956) *Cold Spring Harbor Symposia quant. Biol.*, **21**, 75
Fig. 2.17 LANGRIDGE, R. *et al.* (1957) *J. biophys. biochem. Cytol.* **3**, 767.
Figs. 2.18, 3.1 FRANKLIN, R.E. and HOLMES, K.C. (1958) *Acta cryst.* **11**, 213.
Fig. 2.22 RAMACHANDRAN, G.N., RAMAKRISHNAN, C. and SASISEKHARAN, V. (1963) *J. mol. Biol.*, **7**, 95.
Fig. 2.23 DAVIES, D.R. (1965) *Prog. Biophys.*, **15**, 189.
Fig. 3.2 KLUG, A. and CASPAR, D.L.D. (1960) *Advances in Virus Research*, **7**, 225.
Fig. 3.3 NIXON, H.L. and WOODS, R.D. (1960) *Virology*, **10**, 157.
Fig. 3.4 NIXON, H.L. and GIBBS, A.J. (1960) *J. mol. Biol.*, **2**, 197.
Fig. 3.7 WILDY, P., RUSSELL, W.C. and HORNE, R.W. (1960) *Virology*, **12**, 204.
Fig. 3.8 HORNE, R.W., BRENNER, S., WATERSON, A.P. and WILDY, P. (1959) *J. mol. Biol.*, **1**, 84.

Fig. 3.9 HORNE, R.W. and WATERSON, A.P. (1960) *J. mol. Biol.*, **2**, 75.
Fig. 3.10 BRENNER, S. *et al.* (1959) *J. mol. Biol.*, **1**, 281.
Fig. 3.11 SCHMITT, F.O. (1959) in *Biophysical Science*, ed. J.L. Oncley. New York, Wiley.
Fig. 3.12 GRANT, R.A., HORNE, R.W. and COX, R.W. (1965) *Nature, Lond.*, **207**, 822.
Fig. 3.13 FITTON-JACKSON, S. (1957) *Proc. roy. Soc.* **B146**, 270.
Fig. 3.14 LAKI, K., GLADNER, J.A. and FOLK, J.E. (1960) *Nature, Lond.*, **187**, 758.
Fig. 3.16 PHILLIPS, D.C. (1966) *Scientific American* (November). Copyright © 1966 by Scientific American, Inc. All rights reserved.
Fig. 3.17 MONOD, J., CHANGEUX, J.P. and JACOB, F. (1963) *J. mol. Biol.*, **6**, 306.
Figs. 3.19, 3.20 LYNEN, F. (1964) in *New Perspectives in Biology*, ed. M. Sela. Amsterdam, Elsevier, **4**, 140.
Fig. 3.22 ISHIKAWA, E., OLIVER, R.M. and REED, L.J. (1966) *Proc. nat. Acad. Sci. Wash.*, **56**, 534.
Figs. 3.24, 3.25 HUXLEY, H.E. (1957) *J. biophys. biochem. Cytol.*, **3**, 631.
Fig. 3.26 HUXLEY, H.E. and HANSON, J. (1960) in *The Structure and Function of Muscle*. Vol. 1, ed. G.H. Bourne. New York, Academic Press.
Fig. 3.27 MUIR, A.R. (1967) *Veterinary Record*, **80**, 456.
Fig. 3.28 HANSON, Jean and LOWY, J. (1963) *J. mol. Biol.*, **6**, 46.
Fig. 3.29 MUIR, A.R. (1971) in *Companion to Medical Studies*, eds. R. Passmore and J.S. Robson, Revised printing, Vol. 1. Oxford, Blackwell.
Fig. 3.30 HUXLEY, H.E. (1969) *Science*, **164**, 1356.
Fig. 3.31 EBASHI, S., ENDO, M. and OHTSUKI, I. (1969) *Quart. Revs. Biophys.*, **2**, 351.
Fig. 3.32 KNAPPEIS, G.G. and CARLSEN, F. (1962). *J. Cell Biol.*, **13**, 323.
Fig. 3.33 HUXLEY, H.E. (1967) *J. mol. Biol.*, **30**, 383.
Fig. 4.2 FARQUAR, Marilyn G. and PALADE, G.E. (1963) *J. Cell Biol.*, **17**, 375.
Fig. 4.3 ITO, S. (1965) *J. Cell Biol.*, **27**, 475.
Fig. 4.5 FAWCETT, D.W. (1969) *The Cell.* Philadelphia, Saunders.
Fig. 4.9 JACOBS, M.H., GLASSMAN, H.N. and PAPPART, A.K. (1935) *J. cell. comp. Physiol.*, **7**, 197; and COE, E. L. and COE, MARY H. (1965) *J. theoret. Biol.*, **8**, 327.
Fig. 4.13 LUZZATI, V., MUSTACCHI, H. and SKOULIOS, A. (1958) *Farad. Soc. Disc. No.* **25**, 43.
Fig. 4.15 VANDENHEUVEL, F.A. (1965) *Ann. New York Acad. Sci.*, **122**, 57.
Fig. 4.16 STOECKENIUS, W. (1962) in *The Interpretation of Ultrastructure*, ed. R.J.C. Harris. New York, Academic Press.
Fig. 4.18 SINGER, S. J. (1972) *Ann. N.Y. Acad. Sci.* **195**, 16.
Fig. 5.1 ANDERSSON-CEDERGREN, Ebba (1959) *J. Ultrastructure Research, Suppl. 1*; and STOECKENIUS, W. (1963) *J. Cell Biol.*, **17**, 443.
Fig. 5.2 LEHNINGER, A.L. (1959) in *Biophysical Science*, ed. J.L. Oncley. New York, Wiley.
Fig. 5.3 LEHNINGER, A.L. (1953–1954) *Harvey Lectures*, **49**, 176.
Fig. 5.4 LUNDEGARDH, H. (1945) *Arkiv. Bot.* **32*A*** (No. 12) 1; and MITCHELL, P. (1967) *Ad. Enzymol.*, **29**, 33.
Fig. 5.5 JAGENDORF, A.T. and URIBE, E. (1967) *Brookhaven Symposium*, **19**, 215.
Fig. 5.7 KNAFF, D.B. and ARNON, D.I. (1969) *Proc. nat. Acad. Sci. Wash.*, **64**, 715.
Figs. 5.11, 5.12 REVEL, J.P. (1962) *J. Cell Biol.*, **12**, 571.
Fig. 5.16 DE ROBERTIS, E. (1960) *J. gen. Physiol.*, **43**, *Suppl. 2*, 1.
Fig. 6.11 BENZER, S. (1962) *Scientific American* (January).
Fig. 7.3 FEUGHELMAN, M. *et al.* (1955) *Nature, Lond.*, **175**, 834.
Fig. 7.5 CAIRNS, J. (1966) *Scientific American* (January).
Fig. 7.7 FELSENFELD, G. and RICH, A. (1957) *Biochim. biophys. Acta*, **26**, 457.
Fig. 7.8 DOTY, P., BOEDTKER, H., FRESCO, J.R., HASELKORN, R. and LITT, M. (1959) *Proc. nat. Acad. Sci. Wash.*, **45**, 482.
Fig. 7.9 FRESCO, J.R., ALBERTS, B.M. and DOTY, P. (1960) *Nature, Lond.*, **188**, 98.
Fig. 7.10 WILKINS, M.H.F. (1956) *Cold Spring Harbor Symposia quant. Biol.*, **21**, 75.

Fig. 7.11 WILHELM, J.A., SPELSBERG, T.C. and HNILICA, L.S. (1971) *Subcellular Biochem.*, **1**, 39.
Fig. 7.12 ROBBINS, E. and GONATAS, N.K. (1964) *J. Cell Biol.*, **21**, 429.
Fig. 7.13 HUXLEY, H.E. and ZUBAY, G. (1960) *J. mol. Biol.*, **2**, 10.
Figs. 8.1, 8.5, 9.15, 9.16, 9.17, 10.15 DAYHOFF, Margaret O. (1969) *Atlas of Protein Sequence and Structure* Vol. 4, National Biomedical Research Foundation, Silver Spring, Maryland.
Fig. 8.3 DINTZIS, H. (1961) *Proc. nat. Acad. Sci. Wash.*, **47**, 247.
Fig. 8.4 STENT, G.S. (1966) *Proc. roy. Soc. Lond.*, **B164**, 181
Fig. 8.6 CRICK, F.H.C. (1966) *J. mol. Biol.*, **19**, 548.
Fig. 8.7 RICH, A. (1962) in *Horizons in Biochemistry*, ed. M. Kasha and B. Pullman. New York, Academic Press, and WATSON, J.D. (1963) *Science*, **140**, 17.
Fig. 9.2 BESSIS, M. (1956) *Cytology of the Blood and Blood-forming Organs*. New York, Grune and Stratton.
Fig. 9.3 PAULING, L., ITANO, H.A., SINGER, S.J. and WELLS, I.C. (1949) *Science*, **110**, 543.
Fig. 9.4 HARRIS, H. (1959) *Human Biochemical Genetics*. Cambridge University Press.
Fig. 9.5 PERUTZ, M.F. and MITCHISON, J.M. (1960) *Nature, Lond.*, **166**, 677.
Figs. 9.6, 9.7 INGRAM, V.M. (1961) *Haemoglobin and its Abnormalities*. Springfield, U.S.A., Charles C. Thomas.
Fig. 9.8 LEHMANN, H. (1959) *Brit. med. Bull.*, **15**, 40.
Fig. 9.9 HUISMAN, T.H.J. (1959) in *Abnormal Haemoglobin*, ed. J.H.P. Jonxis and J.F. Delafresnaye, Oxford, Blackwell.
Fig. 9.10 WALKER, J. and TURNBULL, E.P.N. (1955) *Arch. Dis. Childh.*, **30**, 111.
Figs. 9.11, 9.12 PERUTZ, M.F. (1970) *Nature, Lond.*, **228**, 726.
Fig. 9.13 JONXIS, J.H.P. (1959) in *Abnormal Haemoglobins*, ed. J.H.P. Jonxis and J.F. Delafresnaye, Oxford, Blackwell.
Fig. 9.18 DICKERSON, R.E. *et al.* (1971) *J. biol. Chem.*, **246**, 1511.
Fig. 10.4 HAYASHI, Y. and HAYASHI, M. (1970) *Cold Spring Harbor Symposia quant. Biol.*, **35**, 171.
Fig. 10.5 EDGAR, R.S. and EPSTEIN, R.H. (1965) *Scientific American* (February). Copyright © 1965 by Scientific American, Inc. All rights reserved.
Fig. 10.6 EDGAR, R.S. (1967–1968) *Harvey Lectures*, **63**, 263.
Fig. 10.7 SIMON, L.D. and ANDERSON, T.F. (1967) *Virology*, **32**, 279.
Fig. 10.8 REICHART, L. and KAISER, A.D. (1971) *Proc. nat. Acad. Sci. Wash.*, **68**, 2185.
Fig. 10.12 SARMA, V.R. *et al.* (1971) *J. biol. Chem.*, **246**, 3753.
Fig. 10.13 KABAT, E.A. (1968) *Structural Concepts in Immunology and Immunochemistry*. New York, Holt Rinehart and Winston.
Fig. 10.14 HILL, R.L. *et al.* (1967) in *Third Nobel Symposium*, ed. J. Killander. New York, Interscience.
Fig. 10.16 MILLER, O.R. Jr. and BEATTY, Barbara R. (1969) *Science*, **164**, 955. Copyright 1969 by the American Association for the Advancement of Science.
Fig. 10.17 BAKER, T.G. and FRANCHI, L.L. (1967) *Chromosoma*, **22**, 358.
Fig. 10.18 RAVEN, C.P. (1958) *Morphogenesis: The Analysis of Molluscan Development*. London, Pergamon.

Bibliography

CHAPTER 1

Textbooks of cytology and histology:

DE ROBERTIS, E. D. P., NOWINSKI, W. W. and SAEZ, F. A. (1970) *Cell Biology*, 5th edn. Philadelphia, Saunders.

BRACHET, J. and MIRSKY, A. E., eds. (1959) *The Cell*. 5 vols. New York, Academic Press.

PICKEN, L. E. R. (1960) *The Organisation of Cells and other Organisms*. Oxford, Clarendon Press.

BLOOM, W. and FAWCETT, D. W. (1968) *A Textbook of Histology*, 9th edn. Philadelphia, Saunders.

The historical development of cytology and micro-anatomy:

HUGHES, A. F. W. (1959) *A History of Cytology*. London, Abelard Schuman.

SINGER, C. and UNDERWOOD, E. A. (1962) *A Short History of Medicine*. 2nd edn. Oxford, Clarendon Press.

Cell movement:

Cell Movement and Cell Contact. *Exp. Cell Res., Suppl. 8*, 1961.

Contact inhibition of movement:

ABERCROMBIE, M. and AMBROSE, E. J. (1958) *Exp. Cell Res.* **15**, 332.

Aggregation of dissociated cells:

MOSCONA, A. and MOSCONA, H. (1952) *J. Anat.* **86**, 287.

MOSCONA, A. (1961) *Exp. Cell Res.* **22**, 455.

Pinocytosis:

BRANDT, P. W. (1958) *Exp. Cell Res.* **15**, 300.

The nucleus and mitosis:

MAZIA, D. (1959) in *The Cell*, eds. J. Brachet and A. E. Mirsky. Vol. 3. New York, Academic Press.

HUGHES, A. F. W. (1952) *The Mitotic Cycle*. London, Butterworth.

Meiosis:

WHITE, M. J. D. (1961) *The Chromosomes*, 5th edn. London, Methuen.

Electrophoresis:

TISELIUS, A. (1937) *Biochem. J.* **31**, 313.

BIER, M. ed. (1959) *Electrophoresis*. New York, Academic Press.

WOLSTENHOLME, G. E. W. and MILLAR, E. C. P., eds. (1956) *Paper electrophoresis.* London, Churchill. *CIBA Foundation Symposium.*

Polyacrylamide gel electrophoresis:
REISFELD, R. A., LEWIS, U. J. and WILLIAMS, D. E. (1962) *Nature, Lond.* **195**, 281.
KALTSCHMIDT, E. and WITTMANN, H. G. (1970) *Analytical Biochem.* **36**, 401.

Oudin-Ouchterlony gel-diffusion and immuno-electrophoresis:
CROWLE, A. J. (1961) *Immunodiffusion.* New York, Academic Press.

Enzymes:
DIXON, M. and WEBB, E. C. (1964) *Enzymes*, 2nd edn. London, Longmans.
BOYER, P. D., LARDY, H. and MYRBÄCK, K., eds. (1959) *The Enzymes*, 8 vols. 2nd edn. New York, Academic Press.

The role of enzymes in metabolism:
BALDWIN, E. (1957) *Dynamic Aspects of Biochemistry*, 3rd edn. Cambridge University Press.

Differential centrifugation:
CLAUDE, A. (1954) *Proc. roy. Soc.* B**142**, 177.
ALLFREY, V. (1959) in *The Cell*, eds. J. Brachet and A. E. Mirsky. Vol. 1. New York, Academic Press.

Centrifuged amoebae:
HOLTER, H. (1954) *Proc. roy. Soc.* B**142**, 140.

CHAPTER 2

Physical and chemical properties of proteins:
NEURATH, H. ed. (1965) *The Proteins*, 2nd edn. (4 vols.) New York, Academic Press.

Determination of the primary structure of insulin:
SANGER, F. (1956) *Currents in Biochemical Research*, ed. D. E. Green. New York, Interscience.

Early studies of fibrous proteins:
ASTBURY, W. T. and WOODS, H. J. (1934) *Phil. Trans.* A**232**, 333.

Secondary structures proposed for peptide chains prior to 1950:
BRAGG, W. L., KENDREW, J. C. and PERUTZ, M. F. (1950) *Proc. roy. Soc.* A**203**, 321.

The α-helix:
PAULING, L., COREY, R. B. and BRANSON, H. R. (1951) *Proc. nat. Acad. Sci. Wash.* **37**, 205.

Studies of the optical rotation of peptide solutions:
DOTY, P. (1959) in *Biophysical Science*, ed. J. L. Oncley. New York, Wiley.

X-ray diffraction studies of myoglobin and haemoglobin:
KENDREW, J. C. *et al.* (1958) *Nature, Lond.* **181**, 662.
PERUTZ, M. F. *et al.* (1960) *Nature, Lond.* **185**, 416.
KENDREW, J. C. *et al.* (1960) *Nature, Lond.* **185**, 422.
PERUTZ, M. F. (1962) *Nature, Lond.* **194**, 914.
PERUTZ, M. F. (1962) *Proteins and Nucleic Acids.* Amsterdam, Elsevier.

Review of X-ray studies of crystalline proteins:
NORTH, A. C. T. and PHILLIPS, D. C. (1969) *Prog. Biophys.* **19**, 1.

The unfolding of protein molecules at interfaces:
CHEESMAN, D. F. and DAVIES, J. T. (1954) *Ad. Protein Chem.* **9**, 439.

Evidence that α-helical structure may be preserved at interfaces:
MALCOLM, B. R. (1962) *Nature, Lond.* **195**, 901.

The forces determining tertiary and quaternary structure:
TANFORD, C. (1958) in *Symposium on Protein Structure*, ed. A. Neuberger. London, Methuen.
KAUZMANN, W. (1959) *Ad. Protein Chem.* **14**, 1.
LINDERSTRØM-LANG, K. and SCHELLMANN, J. A. (1959) in *The Enzymes*, eds. P. D. Boyer, H. Lardy and K. Myrbäck. 2nd edn. Vol. 1. New York, Academic Press.

Side-chain interactions in myoglobin:
KENDREW, J. C. (1962) *Brookhaven Symposia* **15**, 216.

Conformations of polypeptide chains in proteins:
DAVIES, D. R. (1965) *Prog. Biophys.* **15**, 189.

CHAPTER 3

General reference:
Principles of Biomolecular Organization (1966), eds. G. E. W. Wolstenholme and Maeve O'Connor. London, Churchill. *CIBA Foundation Symposium.*

TMV:
GIERER, A. (1960) *Prog. Biophys.* **10**, 299.
FRAENKEL-CONRAT, H. (1962) *Design and Function at the Threshold of Life.* New York, Academic Press.

Early work on TYMV, and discussion of the physico-chemical properties of TMV solutions:
MARKHAM, R. (1953) *Prog. Biophys.* **3**, 61.

General reviews of the structure of small viruses:
KLUG, A. and CASPAR, D. L. D. (1961) *Ad. Virus Research* **7**, 225.
CASPAR, D. L. D. and KLUG, A. (1962) *Cold Spring Harbor Symposia quant. Biol.* **27**, 1.

Speculation about virus structure:
CRICK, F. H. C. and WATSON, J. D. (1956) *Nature, Lond.* **177**, 473.

A review covering electron microscope studies of viruses by the 'negative-contrast' technique:
HORNE, R. W. and WILDY, P. (1963) *Ad. Virus Research* **10**, 101.

α-helical content of TMV protein:
SIMMONS, N. S. and BLOUT, E. R. (1960) *Biophys. J.* **1**, 55.

The effect of the TMV RNA chain on the stability of the aggregated virus protein:
HART, R. G. and SMITH, J. D. (1956) *Nature, Lond.* **178**, 739.

Theoretical discussion of the formation of helices:
CRANE, H. R. (1950) *Scientific Monthly N.Y.* **70**, 376 (particularly pp. 382–3).

Preparative techniques used in electron microscopy:
HAGGIS, G. H. (1966) *The Electron Microscope in Molecular Biology.* London, Longman.

ϕX174 structure:
EDGELL, M. H., HUTCHISON, C. A. and SINSHEIMER, R. L. (1969) *J. mol. Biol.* **42**, 547.
BURGESS, ANN B. (1969) *Proc. nat. Acad. Sci. Wash.* **64**, 613.

Silk proteins:
LUCAS, F., SHAW, J. T. B. and SMITH, S. G. (1958) *Ad. Protein Chem.* **13**, 108.

Keratins:
CREWTHER, W. G., FRASER, R. D. B., LENNOX, F. G. and LINDLEY, H. (1965) *Ad. Protein Chem.* **20**, 191.

Collagen:
RICH, A. and CRICK, F. H. C. (1961) *J. mol. Biol.* **3**, 483.
TRAUB, W. and PIEZ, K. A. (1971) *Ad. Protein Chem.* **25**, 243.

Fibrinogen-fibrin conversion:
LAKI, K. and GLADNER, J. A. (1964) *Physiol. Revs.* **44**, 127.

The structure and activation of chymotrypsin:
MATTHEWS, B. W., SIGLER, P. B., HENDERSON, R. and BLOW, D. M. (1967) *Nature, Lond.* **214**, 652.

Reviews of muscle structure and function:
HUXLEY, H. E. (1969) *Science* **164**, 1356.
BOURNE, G. H., ed. (1960) *Structure and Function of Muscle*. 3 Vols. New York, Academic Press.

Early evidence for the 'sliding filament' model of muscular contraction:
HUXLEY, H. E., HANSON, J., HUXLEY, A. F. and NIEDERGERKE, R. (1954) *Nature, Lond.* **173**, pp. 971–6.

The role of calcium in muscle:
EBASHI, S. and ENDO, M. (1968) *Prog. Biophys.* **18**, 123.

Structure of tropomyosin:
SODEK, J. *et al.* (1972) *Proc. nat. Acad. Sci. Wash.* **69**, 3800.

Suggested mechanism by which movement of cross-bridges can generate tension:
HUXLEY, A. F. and SIMMONS, R. M. (1971) *Nature, Lond.* **233**, 533.

Smooth muscle:
PANNER, B. J. and HONIG, C. R. (1967) *J. Cell Biol.* **35**, 303.

CHAPTER 4

General references:
FAWCETT, D. W. (1966) *The Cell*. Philadelphia, Saunders.
AFZELIUS, B. (1966) *Anatomy of the Cell*, tr. Birgit Satir. Chicago, University Press.

Physiological evidence for a 30 nm aqueous channel between axons and Schwann cells:
FRANKENHAEUSER, B. and HODGKIN, A. L. (1956) *J. Physiol.* **131**, 341.

Bacterial membranes:
GLAUERT, A. M. (1962) *Brit. med. Bull.* **18**, 245.

Desmosomes and terminal bars:
FAWCETT, D. W. (1958) in *Frontiers in Cytology*, ed. S. L. Palay. New Haven, Yale University Press.
FARQUHAR, M. G. and PALADE, G. E. (1963) *J. Cell Biol.* **17**, 375.

FAWCETT, D. W. (1961) *Exp. Cell Res., Suppl. 8*, 174.
This supplement also covers other intercellular relationships.

Ion-conducting contacts between liver cells:
LOEWENSTEIN, W. R. and KANNO, Y. (1967) *J. Cell Biol.* **33**, 225.

Phagocytosis and pinocytosis:
HOLTER, H. (1959) *Internat. Rev. Cytol.* **8**, 481.

Nerve myelin:
GEREN, B. B. (1954) *Exp. Cell Res.* **7**, 558.
ROBERTSON, J. D. (1960) *Prog. Biophys.* **10**, 344.

Permeability of membranes:
DAVSON, H. and DANIELLI, J. F. (1952) *The Permeability of Natural Membranes,* 2nd edn. Cambridge University Press.
DAVSON, H. (1959) *Textbook of General Physiology*, 2nd edn. London, Churchill.

Active transport and facilitated diffusion:
KLEINZELLER, A. and KOTYK, A. eds. (1961) *Membrane Transport and Metabolism.* New York, Academic Press.
JUDAH, J. D. and AHMED, K. (1964) *Biol. Revs.* **39**, 160.
STEIN, W. D. (1964) *Recent Prog. Surface Science*, **1**, 300.

Nerve and muscle ion movements:
HUXLEY, A. F. (1954) in *Ion Transport across Membranes*, ed. H. T. Clarke. New York, Academic Press.
HODGKIN, A. L. (1958) *Proc. roy. Soc.* B**148**, 1.

Red cell and red cell ghost ion movements:
WHITTAM, R. (1964) *Transport and Diffusion in Red Blood Cells*, London, Arnold.

Acetylcholine at the neuromuscular junction:
KATZ, B. (1962) *Proc. roy. Soc.* B**155**, 455.

Properties of polar lipids:
DEUEL, H. T. jr. (1951) *The Lipids*, 3 vols. New York, Interscience.
Macromolecules and Liquid Crystals. *Farad. Soc. Disc. No. 25*, 1958.
LUZZATI, V. and HUSSON, F. (1962) *J. Cell Biol.* **12**, 207.
HAYDON, D. A. and TAYLOR, J. (1963) *J. theoret. Biol.* **4**, 281.

Early membrane models:
GORTNER, E. and GRENDEL, R. (1925) *J. exp. Med.* **41**, 439.
DANIELLI, J. F. and DAVSON, H. (1935) *J. cell. comp. Physiol.* **5**, 495.

Area of lipid bilayer in a red cell:
BAR, R. S., DEAMER, D. W. and CORNWELL, D. G. (1966) *Science* **153**, 1010.

High frequency membrane capacity measurements:
SCHWANN, H. P. (1957) *Ad. Biol. and Med. Phys.* **5**, 147.

Electron microscopic studies of membrane structure:
ROBERTSON, J. D. (1959) *Biochem. Soc. Symposia No. 16*, 3.

In vivo dimensions of myelin, and changes during fixation:
FINEAN, J. B. (1958) *Exp. Cell Res., Suppl. 5*, 18.

Molecular structure of myelin sheath deduced from X-ray diffraction studies:
ENGSTRON, A. and FINEAN, J. B. (1958) *Biological Ultrastructure.* New York, Academic Press.

Electron microscope studies of lipid-water systems:
STOECKENIUS, W. (1962) *J. Cell Biol.* **12**, 221.

Interaction between proteins and lipids:
Lipoproteins. *Farad. Soc. Disc. No. 6*, 1949.
Membrane Phenomena. *Farad. Soc. Disc. No. 21*, 1956.

DAWSON, R. M. C. and QUINN, P. J. (1971) in *Membrane-bound Enzymes*, eds. G. Porcellati and F. di Jeso. New York, Plenum.

Reviews covering membrane structure:
KORN, E. D. (1966) *Science* **153**, 1491.
DEWEY, M. M. and BARR, L. (1970) *Current Topics in Membranes and Transport* **1**, 1.

Freeze-etch study of particles within membranes:
BRANTON, D. (1971) in *New Developments in Electron Microscopy*, eds. H. E. Huxley and A. Klug. London, Royal Society.

Effect of valinomycin on membrane permeability:
TOSTESON, D. C. *et al.* (1967) *J. gen. Physiol.* **50**, 2513.

Protein spanning the red cell membrane:
MARCHESI, V. T., SEGREST, J. P. and KAHANE, I. (1972) in *Membrane Research* Ed. C. F. Fox. New York, Academic Press.

A stimulating article on some current ideas about membrane structure:
BRETSCHER, M. S. (1973) *Science* **181**, 622.

CHAPTER 5

Review of mitochondrial structure with three-dimensional reconstructions:
ANDERSSON-CEDERGREN, E. (1959) *J. Ultrastructure Res., Suppl. 1*, 97.

Mitochondrial electron transport:
RACKER, E. (1970) *Essays in Biochemistry* **6**, 1.

PRESSMAN, B. C. (1970) and other articles in *Membranes of Mitochondria and Chloroplasts*, ed. E. Racker, New York, Van Nostrand Reinhold.

Chloroplast ion and electron transport:
DILLEY, R. A. (1971) *Current Topics in Bioenergetics* **4**, 237; also reviews and papers in *Brookhaven Symposium* **19**, 1969.

Freeze-etch study of chloroplasts:
ARNTZEN, C. J., DILLEY, R. A. and CRANE, F. L. (1969) *J. Cell Biol.* **43**, 16.

Lysozomes:
DE DUVE, C. and WATTIAUX, R. (1966) *Ann. Revs. Physiol.* **28**, 435.

Combined electron microscopic and biochemical study of the exocrine pancreas of the guinea-pig is reviewed by:

PALADE, G. E. (1961) in *Electron Microscopy in Anatomy*, eds. J. D. Boyd, F. R. Johnson and J. D. Lever. London, Edward Arnold.

Vesicle movement and discharge of wall material in plant cells:
NORTHCOTE, D. H. (1969) *Essays in Biochemistry* **5**, 89.

Electron microscopic appearance of spindle fibres:
ROTH, L. E. and DANIELS, E. W. (1962) *J. Cell Biol.* **12**, 57.

The endoplasmic reticulum in a striated muscle cell:
The Sarcoplasmic Reticulum, *J. biophys. biochem. Cytol.* **10**, *Suppl.*, 1961.
This supplement includes a translation of the original paper of Veratti (1902).

Infolding of the plasma membrane to form the T-system of muscle:
FRANZINI-ARMSTRONG, C. and PORTER, K. R. (1964) *J. Cell Biol.* **22**, 675.
HUXLEY, H. E. (1964) *Nature*, **202**, 1067.

Speed of muscle response to stimulation:
HILL, A. V. (1949) *Proc. roy. Soc.* B**136**, 399.

Local stimulation of frog and lizard cells:
HUXLEY, A. F. (1959) *Ann. N.Y. Acad. Sci.* **81**, 446.

The pumping of Ca^{++} into the vesicles of fragmented sarcoplasmic reticulum:
HASSELBACH, W. and MAKINOSE, M. (1962) *Biochem. biophys. Res. Comm.* **7**, 132.

The role of the reticulum in relaxation:
EBASHI, S. *et al.* (1960) *J. Biochem.* **47**, 54.
PARKER, C. J. and GERGELY, J. (1961) *J. biol. Chem.* **236**, 411.

Ultrastructure of photoreceptors:
DE ROBERTIS, E. (1960) *J. gen. Physiol.* **43**, *Suppl. 2*, 1.

Response of visual receptors to light:
WALD, G. (1968) *Nature, Lond.* **219**, 800.

CHAPTER 6

Textbook of genetics at the cytological level:
SAGER, R. and RYAN, F. J. (1961) *Cell Heredity*. New York, Wiley.

Early ideas:
DARWIN, C. (1875) *The Variation of Animals and Plants under Domestication*. London.

Mendel's discoveries:
MENDEL, G. (1865) *Verh. naturf. Ver. in Brunn, Abhandlungen* **4**. An English Translation is printed as an appendix to SINNOTT, E. W., DUNN, L. C. and DOBZHANSKY, T. (1950) *Principles of Genetics*. New York, McGraw-Hill.

Early proposal of a genetic role for the chromosomes:
SUTTON, W. S. (1903) *Biological Bulletin* **4**, 231.

First demonstration of the linear arrangement of the genes:
STURTEVANT, A. H. (1913) *J. exp. Zool.* **14**, 43.

The essentials of the modern concept of the gene were clearly set forth as early as 1952 in:
PONTECORVO, G. (1952) *Ad. Enzymol.* **13**, 121.

The parasexual cycle in fungi is described in:
PONTECORVO, G. (1956) *Ann. Revs. Microbiol.* **10**, 393.

A more general review:
PONTECORVO, G. (1958) *Trends in Genetic Analysis*. New York, Columbia University Press.

An outline account of the genetics of *Aspergillus nidulans*, with references to original papers, will be found in:
FINCHAM, J. R. S. and DAY, P. R. (1963) *Fungal Genetics*. Oxford, Blackwell.

Analysis of genetic fine structure in bacteriophage:
BENZER, S. (1955) *Proc. nat. Acad. Sci. Wash.* **41**, 344.

A textbook of bacterial and phage genetics:
HAYES, W. (1968) *The Genetics of Bacteria and their Viruses*, 2nd edn. New York, Wiley.

Insertion and excision of prophage:
SINGER, E. R. (1968) *Ann. Revs. Microbiol.* **22**, 451.

CHAPTER 7

Discovery of bacterial transformation:
GRIFFITH, F. J. (1928) *J, Hyg.* **27**, 113.

Chemical characterization of the transforming principle:
AVERY, O. T., MACLEOD, C. M. and MCCARTY, M. (1944) *J. exp. Med.* **79**, 137.

The quotation from Avery's letter is taken from:
Phage and the Origins of Molecular Biology (1966) eds. J. Cairns, G. S. Stent and J. D. Watson, Pub: Cold Spring Harbor Laboratory.

Role of nucleic acid in phage infection:
HERSHEY, A. D. and CHASE, M. (1952) *J, gen. Physiol.* **36**, 39.

Chemistry of nucleotides and nucleic acids:
DAVIDSON, J. N. (1972) *The Biochemistry of the Nucleic Acids.* 7th edn. London, Methuen.

Secondary structure of DNA:
ASTBURY, W.T. and BELL, F. O. (1938) *Cold Spring Harbor Symposia quant. Biol.* **6**, 109.
ASTBURY, W. T. (1947) *Soc. Exp. Biol. Symposia* **1**, 66.
FURBERG, S. (1952) *Acta Chem. Scand.* **6**, 634.
PAULING, L. and COREY, R. B. (1953) *Proc. nat. Acad. Sci. Wash.* **39**, 84.
WATSON, J. D. and CRICK, F. H. C. (1953) *Nature, Lond.* **171**, 737.
WILKINS, M. H. F., STOKES, A. R. and WILSON, H. R. (1953) *Nature, Lond.* **171**, 738.
FRANKLIN, R. E. and GOSLING, R. G. (1953) *Nature, Lond.* **171**, 740.
FEUGHELMAN, M. *et al.* (1955) *Nature, Lond.* **175**, 834.

Base-ratio studies:
CHARGAFF, E. (1951) *Fed. Proc.* **10**, 654.

Early studies of *in vitro* synthesis of DNA:
KORNBERG, A. (1959) in *Biophysical Science,* ed. J. L. Oncley. New York, Wiley.

Newly-synthesised DNA of low molecular weight:
OKAZAKI, R. *et al.* (1968) *Proc. nat. Acad. Sci. Wash.* **59**, 598.

Evidence for semi-conservative replication of DNA in *E. coli*:
MESELSON, M. and STAHL, F. W. (1958) *Proc. nat. Acad. Sci. Wash.* **44**, 671.

Control of DNA duplication in *E. coli*:
JACOB, F., BRENNER, S. and CUZIN, F. (1963) *Cold Spring Harbor Symposia quant. Biol.* **28**, 329.

Polynucleotide phosphorylase:
GRUNBERG-MANAGO, M. and OCHOA, S. (1955) *J. Am. chem. Soc.* **77**, 3165.

Interactions between polynucleotides:
DOTY, P. (1962) *Biochem. Soc. Symposia No. 21*, 8.

Studies of base-pairing with molecular models:
DONOHUE, J. (1956) *Proc. nat. Acad. Sci. Wash.* **42**, 60.
DONOHUE, J. and TRUEBLOOD, K. N. (1960) *J. mol. Biol.* **2**, 363.

Proposal for secondary structure of a single-stranded chain of random base sequence:
FRESCO, J. R., ALBERTS, B. M. and DOTY, P. (1960) *Nature, Lond.* **183**, 98.

Estimate for the relative contribution of hydrophobic effects and hydrogen-bonding to stability of double-helices:
ZUBAY, G. and WILKINS, M. H. F. (1960) *J. mol. Biol.* **2**, 105.

Repair of DNA after ultraviolet irradiation:
SEKIGUCHI, M. *et al.* (1970) *J. mol. Biol.* **47**, 231.

Histones:
WILHELM, J. A., SPELSBERG, T. C. and HNILICA, L. S. (1971) *Sub-cellular Biochem.* **1**, 39.
PHILLIPS, D. M. P. ed. (1971) *Histones and Nucleohistones*. London, Plenum.

X-ray diffraction studies of nucleoproteins, sperm heads, etc.:
FEUGHELMAN, M. *et al.* (1955) *Nature, Lond.* **175**, 834.
WILKINS, M. H. F. (1956) *Cold Spring Harbor Symposia quant. Biol.* **21**, 75.

Evidence that the genes preserve their structural integrity throughout cell-differentiation:
GURDON, J. B. (1962) *J. Embryol. exp. Morph.* **10**, 622.

Ribosome structure:
SPIRIN, A. S. (1969) *Prog. Biophys.* **19**, 133.
KAJI, H. (1970) *Internat. Rev. Cytol.* **29**, 169.

Ribosomal proteins of *E. coli*:
KING, H. W. S., GOULD, H. J. and SHEARMAN, J. J. (1971) *J. mol. Biol.* **61**, 143.

Attachment of ribosomes to membranes:
SABATINI, D. D., TASHIRO, Y. and PALADE, G. E. (1966) *J. mol. Biol.* **19**, 503.

CHAPTER 8

Reviews tracing the development of modern ideas on protein synthesis:
LOFTFIELD, R. B. (1957) *Prog. Biophys.* **8**, 348.
LIPMANN, F. (1969) *Science* **164**, 1024.
BOSCH, L., ed. (1972) *The Mechanism of Protein Synthesis and its Regulation,* New York. American Elsevier.

Three-dimensional structure of yeast phenylalanine tRNA:
KIM, S. H. *et al* (1973) *Science* **179**, 285. (This is the first paper giving detailed results of X-ray diffraction analysis of tRNA structure, showing, as predicted in the text on p. 259, how the loops of the molecule are arranged in three dimensions.)

Demonstration of existence of messenger-RNA:
HALL, B. D. and SPIEGELMAN, S. (1961) *Proc. nat. Acad. Sci. Wash.* **47**, 137.
BRENNER, S., JACOB, F. and MESELSON, M. (1961) *Nature, Lond.* **190**, 576.

The nature of the code:
CRICK, F. H. C., BARNETT, L., BRENNER, S. and WATTS-TOBIN, R. J. (1961) *Nature, Lond.* **192**, 1227.
CRICK, F. H. C. (1963) The recent excitement in the coding problem. *Prog. Nucleic Acid Research*, **1**, 164.

Breaking the code:
NIRENBERG, M. W. and MATTHAEI, J. H. (1961) *Proc. nat. Acad. Sci. Wash.* **47**, 1588.

Coding triplets:
JONES, O. W. and NIRENBERG, M. W. (1962) *Proc. nat. Acad. Sci. Wash.* **48**, 2115.
WAHBA, A. J. *et al.* (1963) *Proc. nat. Acad. Sci. Wash.* **49**, 116.

Chemically evoked mutants of TMV:
TSUGITA, A. and FRAENKEL-CONRAT, H. (1962) *J. mol. Biol.* **4**, 73.

The mechanism of recognition of amino acids by transfer-RNA:
CHAPEVILLE, F. *et al.* (1962) *Proc. nat. Acad. Sci. Wash.* **48**, 1088.
WEISBLUM, B., BENZER, S. and HOLLEY, R. W. (1962) *Proc. nat. Acad. Sci. Wash.* **48**, 1449.

Role of T_sT_u and G factors and GTP in chain elongation:
GUPTA, S. L. *et al.* (1971) *Biochemistry* **10**, 4410.

Possible role of 5 *S* RNA in chain elongation:
RAACKE, ILSE D. (1971) *Proc. nat. Acad. Sci. Wash.* **68**, 2357.

Role of F_3 in chain initiation:
SABOL, S., LEE-HUANG, SYLVIA and OCHOA, S. (1971) *Nature, New Biology* **234**, 233 and 236 (2 papers).

Refolding of ribonuclease:
ANFINSEN, C. B., HABER, E., SELA, M. and WHITE, F. H. (1961) *Proc. nat. Acad. Sci. Wash.* **47**, 1309.

Electron microscope studies of mitochondrial DNA:
NASS, MARGIT M. K. (1969) *J. mol. Biol.* **42**, 521 and 529 (2 papers).

Biogenesis of mitochondria and chloroplasts:
GETZ, G. S. (1972) in *Membrane Molecular Biology*, eds. C. F. Fox and A. D. Keith. Stamford (Connecticut), Sinauer.

Genetics of mitochondria and chloroplasts:
SAGER, Ruth (1972) *Cytoplasmic Genes and Organelles*. New York, Academic Press.

CHAPTER 9

The inheritance of sickle-cell anaemia:
NEEL, J. V. (1949) *Science* **110**, 64.

'Finger-printing' of HbA and HbS:
INGRAM, V. M. (1958) *Biochim. biophys. Acta* **28**, 539.

Identification of altered amino acid in HbS:
HUNT, J. A. and INGRAM, V. M. (1959) in *Biochemistry of Human Genetics*. London, Churchill. *CIBA Foundation Symposium*.

Malaria and the sickle-cell trait:
ALLISON, A. C. (1954) *Brit. med. J.* **1**, 290.

Four adult human haemoglobins in one person:
BAGLIONI, C. and INGRAM, V. M. (1961) *Nature, Lond.* **189**, 465.

Studies of the origin of life:
FLORKIN, M. Ed. (1960) *Aspects of the Origin of Life.* Oxford, Pergamon.
BUVET, R. and PONNAMPERUMA, C. eds. (1971) *Chemical Evolution and the Origin of Life.* Amsterdam, North Holland.
CAIRNS-SMITH, A. G. (1971) *The Life Puzzle.* Edinburgh, Oliver and Boyd.

Experiments with gas discharges:
MILLER, S. L. (1957) *Biochim. biophys. Acta* **23**, 480.
WILSON, A. T. (1960) *Nature, Lond.* **188**, 1007.

The formation of primitive cells from surface films:
GOLDACRE, R. J. (1958) in *Surface Phenomena in Chemistry and Biology*, eds J. F. Danielli *et al.* Oxford, Pergamon.

Evolution of proteins:
MARGOLIASH, E. (1970) *Harvey Lectures* **66**, 177.
DICKERSON, R. E. and GEIS, I. (1969) *The Structure and Action of Proteins.* New York, Harper and Row.

CHAPTER 10

The *lac* operon:
JACOB, F. and MONOD, J. (1961) *J. mol. Biol.* **3**, 318.
MONOD, J., CHANGEUX, J. P. and JACOB, F. (1963) *J. mol. Biol.* **6**, 306.
BECKWITH, J. R. and ZIPSER, D. (1970) *The Lactose Operon.* Pub: Cold Spring Harbor Laboratory.
BRETSCHER, M. S. (1968) *Nature, Lond.* **217**, 509.
SHAPIRO, J. *et al.* (1969) *Nature, Lond.* **224**, 768.

Isolation of *lac* repressor:
GILBERT, W. and MÜLLER-HILL, B. (1966) *Proc. nat. Acad. Sci. Wash.* **56**, 1891.

Transcription of R17 and Qβ RNAs:
A number of papers in *Cold Spring Harbor Symposia quant. Biol.* **35**, (1970).

Some aspects of Qβ replication:
KOLAKOFSKY, D. and WEISSMANN, C. (1971) *Nature, New Biology* **231**, 42.

Sequence studies of R17 and Qβ RNAs:
STEITZ, JOAN A. *et al.* (1969) *Nature, Lond.* **224**, 957, 964 and 1083 (3 papers).

ϕX174:
SINSHEIMER, R. L. (1970) *Harvey Lectures* **64**, 69.

DNA viruses:
STONE, A. B. (1970) *Prog. Biophys.* **20**, 135.

Infection by λ phage:
HERSHEY, A. D., ed. (1971) *The Bacteriophage Lambda.* Pub: Cold Spring Harbor Laboratory.
SZYBALSKI, W. *et al.* (1970) *Cold Spring Harbor Symposia quant. Biol.* **35**, 341.
GROS, F. *et al.* (1970) *Proc roy. Soc. Lond.* **B176**, 251.

Isolation of λ suppressor:
PTASHNE, M. (1967) *Proc. nat. Acad. Sci. Wash.* **57**, 306.

The mechanism of T-even phage infection:
LWOFF, A. (1961) *Proc. roy. Soc.* **B154**, 1.

Assembly of T_4 tail proteins:
KING, J. (1968) *J. mol. Biol.* **32**, 231.

Regulation of RNA synthesis:
BAUTZ, E. K. F. (1972) *Prog. Nucleic Acid Research.* **12**, 129.

Control processes in synthesis and assembly of protein subunits:
WILLIAMSON, A. R. (1969) *Essays in Biochemistry* **5**, 139.

Early work on tumour viruses:
SHOPE, R. E. (1966) *Perspectives in Biology and Medicine* **9**, 258.

Transformation by polyoma and SV40:
SAMBROOK, J. (1972) *Ad. Cancer Res.* **16**, 141.

Role of reverse transcriptase in transformation:
TEMIN, H. M. (1972) *Proc. nat. Acad. Sci. Wash.* **69**, 1016.

Virus-induced tumour antigens:
KLEIN, G. (1971) in *The Strategy of the Viral Genome,* eds. G. E. Wolstenholme and Maeve O'Connor. Edinburgh, Churchill and Livingstone.

Membrane changes in transformation:
MEYER, G. (1971) *Ad. Cancer Res.* **14**, 71.

Role of cyclic AMP in normal and transformed fibroblasts:
JOHNSON, G. S. and PASTAN, I. (1972) *J. nat. Cancer Inst.* **48**, 1377.

The viral oncogene hypothesis:
TODARO, G. J. and HUEBNER, R. J. (1972) *Proc. nat. Acad. Sci. Wash.* **69**, 1009.

Viruses and human cancer:
ALLEN, D. W. and COLE, P. (1972) *New England J. Med.* **286**, 70.

The study of one human lymphoma and its possible viral carcinogen:
GLEMSER, B. (1970) *Mr. Burkitt and Africa.* New York, World Publishers.

Introduction to molecular immunochemistry:
KABAT, E. A. (1968) *Structural concepts in Immunology and Immunochemistry.* New York, Holt Rinehart and Winston.

History of development of ideas about antibody formation:
HAUROWITZ, F. (1968) *Cold Spring Harbor Symposia quant. Biol.* **32**, 559.

Symposium covering aspects of modern work:
FRANEK, F. and SHUGAR, D., eds. (1969) *Gamma Globulins.* New York, Academic Press.

Review of biosynthesis of immunoglobulins:
BEVAN, M. J. *et al.* (1972) *Prog. Biophys.* **25**, 131.

Unfolding of immunoglobulin in urea and refolding with recovery of antigen-binding properties:
HABER, E. (1964) *Proc. nat. Acad. Sci. Wash.* **52**, 1099.

Early evidence for role of lymphocytes in immune reaction:
GOWANS, J. L. and MCGREGOR, D. D. (1965) *Progress in Allergy*, **9**, 1.

Cell-surface immunoglobulins isolated from T-lymphocytes:
CONE, R. E., SPRENT, J. and MARCHALONIS, J. J. (1972) *Proc. nat. Acad. Sci. Wash.* **69**, 2556.

Nucleocytoplasmic interaction in oocyte development:
BRACHET, J. and MALPOIX, P. (1971) *Advances in Morphogenesis* **9**, 263.

Regeneration of plants from single cells:
STEWARD, F. C., KENT, A. E. and MAPES, M. O. (1966) *Current Topics in Developmental Biology* **1**, 113.

Repetitive sequences in *Xenopus* DNA:
DAVIDSON, E. H. *et al.* (1973) *J. mol. Biol.* **77**, 1.

Nuclear transplantation in developing oocytes:
GURDON, J. B. and WOODLAND, H. R. (1970) *Current Topics in Developmental Biology* **5**, 39.

APPENDICES

Introductory accounts of the X-ray diffraction technique will be found in:
BRAND, J. C. D. and SPEAKMAN, J. C. (1960) *Molecular Structure*. London, Arnold.
VAN HOLDE, K. E. (1971) *Physical Biochemistry*. New Jersey, Prentice Hall.

A relatively non-technical discussion of the methods used in the analysis of the diffraction patterns of globular protein crystals:

KENDREW, J. C. and PERUTZ, M. F. (1949) in *Haemoglobin*, eds. F. J. W. Roughton and J. C. Kendrew. London, Butterworth.

Molecular structure:
PAULING, L. (1960) *The Nature of the Chemical Bond*, 3rd edn. New York, Cornell University Press.

Fuller discussion of the physical chemistry of acids and bases will be found in:
GLASSTONE, S. and LEWIS, D. (1960) *Elements of Physical Chemistry*. London, Macmillan.

An introduction to thermodynamics, with some special reference to physiological problems, will be found in the opening chapter of:
BAYLISS, L. E. (1959) *Principles of General Physiology*, 5th edn. Vol. 1. London, Longmans.

The stability of the α-helix:
SCHELLMANN, J. A. (1955) *Comptes Rendues Lab. Carlsberg* **29**, 223.
HARRINGTON, W. F. and SCHELLMANN, J. A. (1956) *Comptes Rendues Lab. Carlsberg* **30**, 21.

Osmotic effects:
DAVSON, H. (1959) *Textbook of General Physiology*, 2nd edn. London, Churchill.

Membrane potentials:
HÖBER, R., ed. (1947) *Physical Chemistry of Cells and Tissues*. London, Churchill.

Index